Lecture Notes in Computer Science   9098

Commenced Publication in 1973
Founding and Former Series Editors:
Gerhard Goos, Juris Hartmanis, and Jan van Leeuwen

More information about this series at http://www.springer.com/series/7409

Jian Li · Yizhou Sun (Eds.)

Web-Age
Information Management

16th International Conference, WAIM 2015
Qingdao, China, June 8–10, 2015
Proceedings

 Springer

Editors
Jian Li
Tsinghua University
Beijing
China

Yizhou Sun
Northeastern University
Boston
USA

ISSN 0302-9743 ISSN 1611-3349 (electronic)
Lecture Notes in Computer Science
ISBN 978-3-319-21041-4 ISBN 978-3-319-21042-1 (eBook)
DOI 10.1007/978-3-319-21042-1

Library of Congress Control Number: 2015942482

LNCS Sublibrary: SL3 – Information Systems and Applications, incl. Internet/Web, and HCI

Springer Cham Heidelberg New York Dordrecht London
© Springer International Publishing Switzerland 2015

Printed on acid-free paper

Springer International Publishing AG Switzerland is part of Springer Science+Business Media
(www.springer.com)

Preface

We are delighted to welcome you to the proceedings of 16th International Conference on Web-Age Information Management (WAIM 2015), which was held in the beautiful coastal city of Qingdao, China. WAIM is a premier international conference for researchers, practitioners, developers, and users to share and exchange ideas, results, experience, techniques, and tools in connection with all aspects of data management.

As the 16th installment of this increasingly popular series, WAIM 2015 attracted submissions from researchers around the globe—Australia, Germany, The Netherlands, Japan, USA, and China. Out of the 164 submissions to the research track and 15 to the demonstration track, the conference accepted 33 full research papers, 31 short research papers, and 6 demonstrations. Our sincere thanks go out to all authors for their submissions, and to all Program Committee members, who worked hard in reviewing submissions and providing suggestions for improvements.

The technical program of WAIM 2015 also included three keynote talks by Profs. Donald Kossman (ETH Zurich and Microsoft Research), Michael Benedikt (Oxford University), and Wang-Chiew Tan (University of California at Santa Cruz); three talks in the Distinguished Young Lecturer Series by Profs. Lei Zou (Peking University), Floris Geerts (University of Antwerp), and Feida Zhu (Singapore Management University); and a panel moderated by Prof. Feifei Li (University of Utah). We are immensely grateful to these distinguished guests for their invaluable contributions to the conference program.

Our deepest thanks go to the members of the Organizing Committee for their tremendous efforts in making WAIM 2015 a success. In particular, we thank the workshop co-chairs, Xiaokui Xiao (Nanyang Technological University) and Zhenjie Zhang (Advanced Digital Sciences Center, Singapore); demonstration co-chairs, Nan Zhang (George Washington University) and Wook-shin Han (POSTECH, South Korea); Distinguished Young Lecturer Series co-chairs, Yufei Tao (Chinese University of Hong Kong), Lizhu Zhou (Tsinghua University), and Wenfei Fan (University of Edinburgh); industry chair, Wei Fan (Huawei); Best Paper Award co-chairs, Xiaofang Zhou (University of Queensland), Kyuseok Shim (Seoul National University), and Aoying Zhou (East China Normal University); publicity co-chairs, Zhenhui Jessie Li (Penn State University), Yang Liu (Shandong University), and Yongluan Zhou (University of Southern Denmark); local organization co-chairs, Lizhen Cui (Shandong University), Zhenbo Guo (Qingdao University), and Lei Liu (Shandong University); registration chair and finance co-chair, Zhaohui Peng (Shandong University); finance co-chair, Howard Leung (City University of Hong Kong); Web chair, Xingcan Cui (Shandong University); and WAIM/SAP Summer School co-chairs, Xiaofeng Meng (Renmin University), Lei Zou (Peking University), and Guoliang Li (Tsinghua University). We also thank our tireless liaisons: Weiyi Meng

(Binghamton University) to the WAIM Steering Committee; Xiaofeng Meng (Renmin University) to the China Computer Federation Technical Committee on Database; Xuemin Lin (University of New South Wales) and Yanchun Zhang (Victoria University) to our sister conferences APWeb and WISE.

We thank the many WAIM 2015 supporters, without whose contributions the conference would not have been possible. Shandong University and Qingdao University were wonderful hosting institutions. SAP and Inspur were our generous sponsors.

We hope you will enjoy the proceedings of WAIM 2015!

June 2015

Fengjing Shao
Jun Yang
Xin Luna Dong
Xiaohui Yu
Jian Li
Yizhou Sun

Organization

Organizing Committee

General Co-chairs

Jun Yang — Duke University, USA
Fengjing Shao — Qingdao University, China

Program Co-chairs

Xin Luna Dong — Google, USA
Xiaohui Yu — Shandong University, China

Workshop Co-chairs

Xiaokui Xiao — Nanyang Technological University, Singapore
Zhenjie Zhang — Advanced Digital Sciences Center, Singapore

Demo Co-chairs

Nan Zhang — George Washington University, USA
Wook-shin Han — POSTECH, South Korea

Panel Chair

Feifei Li — University of Utah, USA

DYL Series Co-chairs (Distinguished Young Lecturer)

Yufei Tao — Chinese University of Hong Kong, SAR China
Lizhu Zhou — Tsinghua University, China
Wenfei Fan — University of Edinburgh, UK

Industry Chairs

Wei Fan — Huawei

Best Paper Award Co-chairs

Xiaofang Zhou — University of Queensland, Australia
Kyuseok Shim — Seoul National University, Korea
Aoying Zhou — East China Normal University, China

Proceedings Co-chairs

Jian Li — Tsinghua University, China
Yizhou Sun — Northeastern University, USA

Publicity Co-chairs

Zhenhui Jessie Li	Penn State University, USA
Yang Liu	Shandong University, China
Yongluan Zhou	University of Southern Denmark

Local Organization Co-chairs

Lizhen Cui	Shandong University, China
Zhenbo Guo	Qingdao University, China
Lei Liu	Shandong University, China

Registration Chair

| Zhaohui Peng | Shandong University, China |

Finance Co-chairs

| Zhaohui Peng | Shandong University, China |
| Howard Leung | City University of Hong Kong, SAR China |

Web Chair

| Xingcan Cui | Shandong University, China |

Steering Committee Liaison

| Weiyi Meng | Binghamton University, USA |

CCF DBS Liaison

| Xiaofeng Meng | Renmin University, China |

Summer School Chair

Xiaofeng Meng	Renmin University, China
Lei Zou	Peking University, China
Guoliang Li	Tsinghua University, China

APWeb Liaison

| Xuemin Lin | University of New South Wales, Australia |

WISE Liaison

| Yanchun Zhang | Victoria University, Australia |

Program Committee

Wolf-Tilo Balke	TU-Braunschweig, Germany
Zhifeng Bao	University of Tasmania, Australia
Sourav Bhowmick	National Taiwan University

Gang Chen	Zhejiang University, China
Enhong Chen	University of Science and Technology of China
Shimin Chen	Institute of Computing Technology, Chinese Academy of Sciences, China
Yueguo Chen	Renmin University of China
Hong Chen	Renmin University of China
Jian Chen	South China University of Technology, China
Jinchuan Chen	Renmin University of China
Ling Chen	University of Technology, Sydney, Australia
Reynold Cheng	The University of Hong Kong, SAR China
James Cheng	The Chinese University of Hong Kong, SAR China
David Cheung	The University of Hong Kong, SAR China
Fei Cheung	McMaster University, Canada
Gao Cong	Nanyang Technological University, Singapore
Bin Cui	Peking University, China
Ting Deng	Beihang University, China
Dejing Dou	University of Oregon, USA
Xiaoyong Du	Renmin University of China
Ju Fan	National University of Singapore
Yaokai Feng	Kyushu University, Japan
Jun Gao	Peking University, China
Yunjun Gao	Zhejiang University, China
Hong Gao	Harbin Institute of Technology, China
Zhiguo Gong	University of Macau, China
Giovanna Guerrini	Università di Genova, Italy
Jingfeng Guo	Yanshan University, China
Takahiro Hara	Osaka University, Japan
Qinmin Hu	East China Normal University, China
Luke Huan	University of Kansas, USA
Jianbin Huang	Xidian University, China
Xuanjing Huang	Fudan University, China
Seung-won Hwang	Pohang University of Science and Technology POSTECH, South Korea
Yan Jia	National University of Defense Technology, China
Lili Jiang	Max Planck Institute for Informatics, Germany
Peiquan Jin	University of Science and Technology of China
Wookey Lee	Inha University, Korea
Carson Leung	University of Manitoba, Canada
Chengkai Li	University of Texas at Arlington, USA
Cuiping Li	Renmin University of China
Guohui Li	Huazhong University of Science and Technology, China
Jian Li	Tsinghua University, China
Chuan Li	Sichuan University, China
Guoliang Li	Tsinghua University, China
Zhanhuai Li	Northwestern Polytechnical University, China
Zhoujun Li	Beihang University, China

Qingzhong Li	Shandong University, China
Tao Li	Florida International University, USA
Xiang Lian	University of Texas - Pan American, USA
Guoqiong Liao	Jiangxi University of Finance and Economics, China
Hongyan Liu	Tsinghua University, China
Qi Liu	University of Science and Technology of China
Yang Liu	Shandong University, China
Jiaheng Lu	Renmin University of China
Jizhou Luo	Harbin Institute of Technology, China
Shuai Ma	Beihang University, China
Xiaofeng Meng	Renmin University of China
Yang-Sae Moon	Kangwon National University, Korea
Shinsuke Nakajima	Kyoto Sangyo University, Japan
Weiwei Ni	Southeast University, China
Baoning Niu	Taiyuan University of Technology, China
Hiroaki Ohshima	Kyoto University, Japan
Zhiyong Peng	Wuhan University, China
Ken Pu	University of Ontario Institute of Technology, Canada
Jianbin Qin	The University of New South Wales, Australia
Jie Shao	University of Electronic Science and Technology of China
Kyuseok Shim	Seoul National University, South Korea
Lidan Shou	Zhejiang University, China
Hailong Sun	Beihang University, China
Weiwei Sun	Fudan University, China
Chih-Hua Tai	National Taipei University
Alex Thomo	University of Victoria, Canada
Taketoshi Ushiama	Kyushu University, Japan
Jianmin Wang	Tsinghua University, China
Jiannan Wang	University of California - Berkeley, USA
Peng Wang	Fudan University, China
Yijie Wang	National University of Defense Technology, China
Hongzhi Wang	Harbin Institute of Technology, China
Wei Wang	University of New South Wales, Australia
Wei Wang	Fudan University, China
Qiang Wei	Tsinghua University, China
Shengli Wu	Jiangsu University, China
Junjie Wu	Beihang University, China
Yinghui Wu	University of California, Santa Barbara, USA
Yanghua Xiao	Fudan University, China
Jianliang Xu	Hong Kong Baptist University, SAR China
Lianghuai Yang	Zhejiang University of Technology, China
Xiaochun Yang	Northeast University, China
Bin Yao	Shanghai Jiaotong University, China
Zheng Ye	York University, Canada
Ke Yi	Hong Kong University of Science and Technology, SAR China

Jian Yin	Zhongshan University, China
Jeffrey Yu	Chinese University of Hong Kong, SAR China
Ming Zhang	Peking University, China
Xiangliang Zhang	King Abdullah University of Science and Technology, KAUST, Saudi Arabia
Dongxiang Zhang	National University of Singapore
Richong Zhang	Beihang University, China
Yong Zhang	Tsinghua University, China
Ying Zhao	Tsinghua University, China
Wenjun Zhou	University of Tennessee, Knoxville, USA
Xuan Zhou	Renmin University of China
Junfeng Zhou	Yanshan University, China
Feida Zhu	Singapore Management University, Singapore
Xingquan Zhu	Florida Atlantic University, USA
Yi Zhuang	Zhejiang Gongshang University, China
Fuzhen Zhuang	ICT, Chinese Academy of Sciences, China
Zhaonian Zou	Harbin Institute of Technology, China
Lei Zou	Peking University, China
Quan Zou	Xiamen University, China

Contents

Graph and Social Network

Information and Knowledge

Recommender Systems

Big Data

Short Papers

Demo Papers

Graph and Social Network

An Influence Field Perspective on Predicting User's Retweeting Behavior

Yi Shen[1,2], Jianjun Yu[2], Kejun Dong[2], Juan Zhao[2], and Kai Nan[2(✉)]

[1] University of Chinese Academy of Sciences, Beijing, China
shenyi@cnic.ac.cn
[2] Computer Network Information Center,
Chinese Academy of Sciences, Beijing, China
{yujj,kevin,zhaojuan,nankai}@cnic.ac.cn

Abstract. User's retweeting behavior, which is the key mechanism for information diffusion in the micro-blogging systems, has been widely employed as an important profile for personalized recommendation and many other tasks. Retweeting prediction is of great significance. In this paper, we believe that user's retweeting behavior is synthetically caused by the influence from other users and the post. By analogy with the concept of electric field in physics, we propose a new conception named "influence field" which is able to incorporate different types of potential influence. Based on this conception, we provide a novel approach to predict user's retweeting behavior. The experimental results demonstrate the effectiveness of our approach.

Keywords: Retweet · Field theory · Influence · Electric field strength

1 Introduction

Due to its dual role of social network and news media [1], the micro-blogging systems have become an important platform for people to acquire information. In a typical micro-blogging system (e.g., Twitter), users are allowed to publish short messages with a limitation of 140 characters (i.e., posts or tweets). Unlike other social-networking services, micro-blogging introduces a new relationship named "follow", which enables users to access a filtered timeline of posts from anyone else they care about without any permission. The retweet mechanism, which provides a convenient function for users to share information with their followers, has become a hot spot in the field of social network analysis.

There are many previous studies focus on how to employ diverse features to address the retweeting prediction problem. Intuitively, people tend to retweet the posts with the content they are interested in. Therefore, the content-based

Y. Shen—This research was supported by Special Items of Information, Chinese Academy of Sciences under Grant XXH12503; and by Around Five Top Priorities of "One-Three-Five" Strategic Planning, CNIC under Grant CNIC_PY-14XX; and by NSFC Grant No.61202408.

© Springer International Publishing Switzerland 2015
J. Li and Y. Sun (Eds.): WAIM 2015, LNCS 9098, pp. 3–16, 2015.
DOI: 10.1007/978-3-319-21042-1_1

features are widely used in the existing works [2–4]. Actually, the reality is much more complicated because besides users' intrinsic interests, users' behaviors on social networks may be also caused by the influence of other users [5,6]. For example, a user retweets a post about "World Cup 2014", the reason behind this retweeting might be: (1) he is a soccer fan and he is interested in "World Cup 2014" (2) he is attracted by a post about "World Cup 2014" because it has been retweeted by his intimate friends. As a result, social-based features are also important for addressing the retweeting prediction task [7,8]. However, the social-based features applied in the previous studies are mainly author-oriented, in other words, they mainly focus on the relationship between the current user and the author of the post, while the influence from other influential users (especially the close friends of the current user) is neglected.

In this paper, we interpret retweeting behavior as an outcome of the influence. In fact, everyone has a certain amount of potential influence [9], and the influence of user A to user B is negatively correlated with the "distance" between them. For instance, the influence of A to one of his closest friends is likely to be greater than the influence to a long-forgotten acquaintance. Inspired by this, we make an analogy between the users in the micro-blogging environment and the electric charges in an electric field and assume that every user has an "influence field" around himself, then we employ the field theory in physics to interpret users' retweeting behavior.

The major contributions of our paper are as follows:

(1) We propose the conception of "influence field" and borrow the field theory in traditional physics to model the influence between the users in the micro-blogging.

(2) By defining different types of influence, we apply the "influence field" to address the retweeting prediction task. Our "influence field" model not only considers the influence of all the potential influential-users, but also takes account of the inherent relationship among different features. We evaluate our approach on a large dataset from Sina Weibo, which is the most popular micro-blogging in China. The experimental results demonstrate that our "influence field" model is indeed effective.

(3) Although our work has been done in the context of Sina Weibo, we expect that the "influence field" conception would hold for many other social-network platforms(e.g., Twitter and Facebook).

2   Related Work

A better understanding of the user retweeting behavior is of great significance. Since retweeting plays a crucial role in the information propagation in the micro-blogging systems, researchers have completed a lot of interesting work through analyzing users' retweeting behavior, such as rumor identification [10] and break news detection [11]. Furthermore, retweet is also employed as an important profile to build the user interest model in many personalized applications such as recommender system [12].

A bulk of studies try to understand why people retweet. For example, Boyd et al. [13] pointed out that there are diverse motivations for users to retweet, such as for spreading information to other users, saving the valuable posts for future personal access, commenting on someone's post in the form of incidental retweeting, etc. Macskassy and Michelson [14] studied a set of retweets from Twitter, they claimed that the content based propagation models could explain the majority of retweeting behaviors they saw in their data.

Plenty of previous works focus on extracting various kinds of features to predict the users' retweeting behavior as well as analyze the effects caused by different features. Some of these works [15–17] predicted retweeting from a global perspective which aimed to forecast whether a post will be popular in the future. There are also some studies which conducted the prediction to the individual level and manage to answer the question whether a post will be retweeted by a specific user. For example, Luo et al. [3] proposed a learning to rank based framework to discover the users who are most likely to retweet a specific post. Peng et al [2] applied conditional random fields(CRF) to model and predict the users' retweet patterns. The authors of [18] address this problem by means of executing constrained optimization on a factor graph model. There are also many works [4,7,8] in which the authors considered the individual level prediction as a classification task and then built effective classifiers to address this problem.

The difference between our work and most of the previous studies is mainly reflected in two aspects. One is that we emphasize the impact of all the influential users rather than only the author of the post, the other is that we integrate diverse features as well as their correlation to model user's retweeting behavior.

3 Methodology

In this section, we first describe the conception of "influence field" in detail, and then present the calculation of the elements in the "influence field" model. Finally, we introduce the classifier based on "influence field" to address the retweet prediction task.

3.1 The Conception of Influence Field

The intuitive idea behind our approach is that a user's behavior is usually influenced by others. Take retweeting as an example, we suppose that user A has a certain probability $P_{u_A}^{po}$ to retweet a post po about "data mining". If A notices that another user B has retweeted po, then $P_{u_A}^{po}$ may increase due to the influence of B. If B is a famous expert on "data mining", $P_{u_A}^{po}$ will be even larger. Moreover, as people are easy to be affected by their close friends in many cases [19], for this reason, if B happens to be a good friend of A, $P_{u_A}^{po}$ may be significantly increased. To sum up, user's (e.g., B) impact on another user (e.g., A) is positively correlated to B's influence and negatively correlated to the "distance" between them. Thus, this effect can be written as follows:

$$E = K\frac{I_u}{R^\rho} \tag{1}$$

Where E is used to measure the impact caused by u. K is a constant. I_u is the global influence of the influential user. R represents the distance between the two users. ρ is a coefficient to tune the effect of the distance. It is worth noting that this formula is very similar to the definition of "electric field strength" in physics. As depicted in Figure 1(a), an electric charge with power Q will generate an electric field around itself, and the field strength at a point r far away is calculated as: $E' = k\frac{Q}{r^2}$. Inspired by the electric field theory, we assume that everyone on micro-blogging has his/her own "influence field", which makes himself/herself as the center.

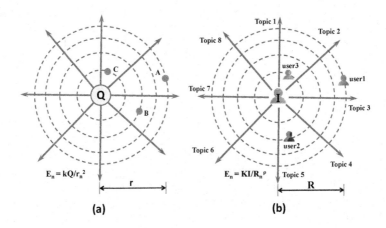

Fig. 1. Examples of the electric field (a) and the influence field (b)

As portrayed in Figure 1(b), the user at the center will affect the behaviors of user1-user3 through the influence field of himself and the strength will attenuate as the distance increases. The field strength of the "influence field" can be defined according to Equation 1.

The electric field strength in physics is a vector with its direction. In order to be consistent with this concept, we define "topic" as the direction of "influence field". This design is to match the fact that each user may has uneven influence on different topics. For example, David Beckham is a famous soccer player, and his influence on the topic "sports" is far greater than on "cloud computing".

We try to employ "influence field" discussed above to interpret and predict users' retweeting behavior in the micro-blogging systems. We assume that user u will be affected by a force (f_u) when he/she is reading post po, and u has a threshold θ_u^{po} for retweeting it. Besides the influential users, the influence from po should also be considered. Finally we utilize an inequality to predict whether u will retweet po: $f_u \geq \theta_u^{po}$

Based on Equation 1, we define f_u as follows:

$$f_u = \sum_{u' \in U} \frac{K I_{u'} I_{po}}{R(u', u)^\rho} \qquad (2)$$

Where U is the collection of the users who are able to influence u, I_{po} is the influence of po ,which is used to measure the importance of po. Generally speaking, the posts with high influence are more likely to be retweeted. We can see that Equation 2 shares the similar structure with the famous Coulomb's law($F = kQq/r^2$) in physics.

3.2   The Calculation of the Elements

In this subsection, we mainly introduce how to calculate the elements of the "influence field". First, we will talk about how to identify those users (U) who will influence the current user u when he/she is reading a post. Next, we will elaborate the calculation of I_u, $R(u', u)$, I_{po} and θ_u^{po} respectively.

Identify the Influential Users (U). As depicted in Figure 2, three types of users as following will influence the current user u when po is exposed to him/her.

The author of po. Consider the simplest case shown in Figure 2(a), both user B and user C follow user A. Once A has published a post po, B and C will be influenced by user A to some extent since po will appear in their separate timelines.

The followees of u . There are also many other cases that u does not follow the author of po directly. As portrayed in Figure 2(b), user A is the author of po, B only follows user D and receives po through the "retweeting-path" from A to D. In this case, we consider that both A and D will influence B because B will perceive their retweeting behaviors. In Figure 2(c), two followees of B have retweeted po and we believe that both of them(i.e., E and F) will influence B.

The mentioned users in po . Besides the author and the followees, there is another type of users on the "retweeting-path". Take user G in Figure 2(c) as an instance, although he is neither the author nor the followee of B, he is also able to influence B because his nickname will be mentioned in po with a prefix of "@" symbol.

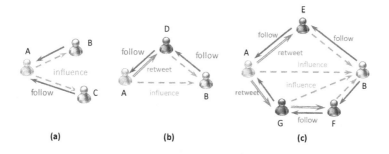

Fig. 2. Different types of influential users

To sum up, the influential users U includes all those users appear on the "retweeting-paths" from the author of po to u. The duplicate users are removed.

Calculate the Influence of User(I_u). Since we have defined the directions of "influence field" as the topics, it is necessary for us to find some way to measure user's influence on each topic. TwitterRank [20] provided an approach based on the PageRank algorithm and the topic model which is able to calculate the influence on a single topic for each user on Twitter. It measures the influence taking both the link structure and the topical similarity between users into account. Since TwitterRank has proved its effectiveness on a large dataset of a typical micro-blogging platform (i.e., Twitter), we decide to adopt it to calculate the users' topic-level influence in this paper.

The core idea of TwitterRank is to make an analogy between the influence of a user and the authority of a web page, then extend the PageRank algorithm with topical information to calculate the user's topic-specific influence.In detail, the TwitterRank algorithm consists of the following steps:

1. First of all, The Latent Dirichlet Allocation (LDA) [21] model is applied to distill the topics that users are interested in based on the posts they have published.
2. Secondly, a directed graph of users is formed according to the following relationships among them.
3. Each topic corresponds to a probability transition matrix P_t, and the specific transition probability of the random walk from u_i to his followee u_j is defined as:

$$P_t(i,j) = \frac{|\tau_j|}{\sum_{a:a\in i's followee} |\tau_a|} * sim_t(i,j) \tag{3}$$

Where $|\tau_j|$ denotes the number of posts published by u_j and the denominator part represents the total number of posts published by all u_i's followees. $sim_t(i,j)$ is the similarity of u_i and u_j in topic t, which can be calculated as:

$$sim_t(i,j) = 1 - |P_{u_i}^t - P_{u_j}^t| \tag{4}$$

Here, $P_{u_i}^t$ and $P_{u_j}^t$ are both topic distributions generated by LDA.
4. Finally, based on the transition probability matrix, the topic-specific influence of the users, which is denoted as $\overrightarrow{I_t}$, can be iteratively calculated by:

$$\overrightarrow{I_t} = \gamma P_t \times \overrightarrow{I_t} + (1-\gamma)E_t \tag{5}$$

E_t is the teleportation vector which is introduced to tackle the case in which the users may follow one another in a loop. The authors normalized the t-th column of the user-topic matrix obtained by LDA so as to represent E_t. γ is a parameter between 0 and 1 to tune the probability of teleportation.

As presented in [20], the user's global influence can be measured as an aggregation of his/her influence in different topics, which can be calculated as:

$$I_u = \sum_t r_t \cdot I_u^t \tag{6}$$

I_u^t is the corresponding element of the user u in vector $\overrightarrow{I_t}$. r_t is the weight assigned to topic t, which is the probability of user u is interested in t, i.e., P_u^t generated by LDA.

Measure the Distance between Users($R(u_1, u_2)$). The distance between two users is largely determined by their relationship. In the previous research, the relationship is measured by "tie strength" [22]. The relationships between users can be divided into two categories: the weak ties and the strong ties [23–25]. The weak ties include our loose acquaintance, new friends and our 2-degree friends (i.e., friends of our friends). Strong ties refer those people we are most concerned, such as our family and our trusted friends. In fact, most of our communication on social networks is with our strong ties [26]. Not surprisingly, people are disproportionately influenced by the strong ties, and the strongest influence is between mutual best friends [27]. As a result, we believe that the users' retweeting behavior is also influenced by the strong ties. Intuitively, the more frequent the interaction between two users, the stronger their tie strength. In addition, a large number of common friends may also mean strong ties [28]. In micro-blogging, we define friends as the users who have followed with each other.

$$R(u_1, u_2) = \frac{\lambda r(u_1, u_2)}{lg[(N_0 + 1)\sqrt{(N_1 + 1)(N_2 + 1)}]} \tag{7}$$

Finally, we model the distance between two users according to Equation 7, where λ is a constant coefficient, $r(u_1, u_2)$ is the "router distance" between u_1 and u_2. For example, if A follows B, then $r(A, B) = 1$, and more, if B follows C and A doesn't follow C, then $r(A, C) = 2$, etc. N_0 is the number of the common friends of u_1 and u_2. N_1 is the frequency of interaction from u_1 to u_2 and N_2 denotes the frequency of interaction from u_2 to u_1. The interaction here refers the behaviors include retweeting, mention and comment on the micro-blogging system. The closer N_1 and N_2, the stronger the tie strength, then the nearer the distance. The square root and the logarithmic function are used to smooth the final result.

Calculate the Influence of Post (I_{po}). Intuitively, a post with rich information usually has a strong influence. We model the influence of a post also at the topic granularity. We employ "topic entropy" to measure the amount of information contained in po on each topic and the higher the value of "topic entropy", the stronger the influence. We consider each post as a document, based on the "bag of words" assumption, LDA is applied to represent each post as a probability distribution over a certain number of topics, while each topic is expressed as a probability distribution over the words. Finally, the topic-level influence of post po on topic t is calculated as the following equation:

$$H_{po}^t = -\sum_{i=1}^{K} P(w_i|t) \log_2 P(w_i|t) \tag{8}$$

Where w_i denotes the words in po. K is the total number of words in po.

For each retweet, we calculate the intervals between the retweeting timestamp and the generation timestamp of corresponding po in our dataset. According to our statistics, most retweeting behaviors happened during a short period after the original posts have been published. About 50% of the intervals are less than 1 hour and over 90% of them are less than 1 day(1440 minutes), which means the probability of user to retweet po will gradually diminish over time. For this reason, we apply an exponential time factor to discount the influence of those old posts as Equation 9, where Δt_0 is the interval between the published timestamp of po and the current timestamp, μ is a decay coefficient.

$$I_{po} = \sum_{t \in T} I_{po}^t = \sum_{t \in T} H_{po}^t \times e^{-\mu \Delta t_0} \tag{9}$$

The Threshold for User to Retweet a Post (θ_u^{po}). Different users have their respective "accepted thresholds" even for the same post, additionally, single user has different "accepted thresholds" for different posts [19]. Therefore, the threshold θ_u^{po} should be determined by both the current user and the post.

In general, for a specific user u, the threshold of the possibility to retweet post po is positively correlated to their divergence. We still employ LDA to generate latent topic distributions of u and po respectively, then adopt Kullback Leibler divergence as Equation 10 to measure the distance between them.

Finally, the threshold θ_u^{po} is defined as Equation 11, where σ is a constant, $freq(u)$ stands for the percentage of retweeted posts in the latest 100 posts of u, which is used to describe how much a user prefer to retweet posts recently.

$$D_{KL}(u||po) = \sum_{t \in T} P(t|u) \cdot \log \frac{P(t|u)}{P(t|po)} \tag{10}$$

$$\theta_u^{po} = \frac{\sigma D_{KL}(u||po)}{freq(u)} \tag{11}$$

In summary, there are four elements in the "influence field" model. I_u and I_{po} represent the global influence of the influential users and the post respectively. The distance R is applied to model the relationship between users, which projects I_u to an individual level. Analogously, the θ_u^{po} stands for the correlation between the post and the current user, which captures the effect of I_{po} at an individual level.

3.3   Predict the Retweeting Behavior

The retweeting behavior prediction can be considered as a classification task. For a given triplet (u, po, t_0), our goal is to correctly categorize it. The outcome is denoted as Y_{u,po,t_0}. $Y_{u,po,t_0} = 1$ means that user u will retweet the post po before timestamp t_0, and $Y_{u,po,t_0} = 0$ otherwise.

Since we have defined the calculation of all the elements necessary in the model, the decision inequality $f_u \geq \theta_u^{po}$ could be written as:

$$Q(u, po, t_0) = \frac{f_u}{\theta_u^{po}} - 1 = \sum_{u' \in U} \sum_{t \in T} \frac{K r_t \cdot I_u^t I_{po}^t}{\theta_u^{po} R(u', u)^\rho} - 1 \geq 0 \qquad (12)$$

We merge the constant coefficients, and $Q(u, po, t_0)$ could be written as:

$$Q(u, po, t_0) = \phi Q'(u, po, t_0) - 1 \qquad (13)$$

Where $\phi = \frac{K\lambda}{\sigma}$.

We employ the logistic regression model as our classifier for learning the value of ϕ. We make use of $Q(u, po, t_0)$ as the decision boundary, then the logistic function can be written as:

$$P(Y_{u,po,t_0} = 1 | Q(u, po, t_0)) = \frac{1}{1 + e^{-(\omega_1 Q(u,po,t_0) + \omega_0)}} \qquad (14)$$

Where ω_1 is the weight coefficient and ω_0 is the bias, both of which can be learned by minimizing the cost function of the logistic regression model. For convenience, we modify Equation 13 as:

$$P(Y_{u,po,t_0} = 1 | Q'(u, po, t_0)) = \frac{1}{1 + e^{-(\omega_1' Q'(u,po,t_0) + \omega_0')}} \qquad (15)$$

Where $\omega_1' = \phi\omega_1$ and $\omega_0' = \omega_0 - 1$.

4   Experimental Evaluation

4.1   Data Preparation and Experimental Setting

We crawled the latest 200 posts of 61,736 users as well as their follow-relationship from Sina Weibo. The dates of these posts range from 2012.3.2 to 2013.8.28. After removing the inactive users and the fake accounts, 37,931 users are left. Here, the inactive users refer to those users whose total posts number is less than 20 or the followee number is less than 10. We employ the approach introduced in [29] to detect those fake accounts.

We select 300 popular posts which are published on 2013.05.01, 2013.05.11 and 2013.05.21 (100 posts each day). By analyzing these posts, we obtain 15,831 retweets and 650,212 non-retweets and merge them as the dataset for the experiment. For a triple (u, po, t_0), we consider it as a non-retweet if it meets four conditions:

(1) po is exposed to u.
(2) u publishes or retweets another post at t_0.
(3) po is published with 5 hours before t_0.
(4) po is not retweeted by u before t_0.

We first treat every post as a document and apply LDA to calculate the probability of generating a post from each topic. Users' topical distributions are also estimated by the same LDA model, then the TwitterRank algorithm is executed on the global dataset to generate users' topical influence. During these processes, the teleportation parameter(γ) in TwitterRank is set to 0.85 as suggested in [20]. There are three parameters in LDA, i.e., the topic number T and two Dirichlet hyper-parameters α, β, they are set as $T = 50$, $\alpha = 50/T$, $\beta = 0.1$ respectively. When calculating the field strength of influence field, ρ is initially set to 2 via making an analogy with the electric field and the decay coefficient μ is heuristically set to 0.6.

For each triple (u, po, t_0), we first collect corresponding influential user set which is denoted as U for each u. Then we calculate $Q'(u, po, t_0)$ as elaborated in Section 3. Finally, 10-fold validation for the logistic regression classifier is executed on the dataset. u will be predicted to retweet po before t_0 if the probability returned by logistic regression is larger than 0.5.

4.2   Performance of the Proposed Approach

In order to verify the effectiveness of our model based on "influence field", we utilize the logistic function as the classifier, employ the force function $Q'(u, po, t_0)$ as the feature, then compare with several baseline approaches as follows:

(1)Only consider the influence of the author in the force function $(Q'_A(u, po, t_0))$.

(2)Consider the influence of the author and the mentioned users in the force function $(Q'_{A+M}(u, po, t_0))$.

(3)Consider the influence of the author and u's followees as the influential users in the force function $(Q'_{A+F}(u, po, t_0))$.

(4)Use the features listed in [7] and [2] as the basic features.

(5)Combine $Q'(u, po, t_0)$ and the basic features as the advanced features.

The evaluation results are listed in Table 1. We can see that $Q'_{A+F}(u, po, t_0)$ obviously outperforms the result of the case in which only consider the influence of the author, which means beside the author, the followees who have retweet po indeed play an important role to influence u. This may be caused by two reasons: for one reason, users tend to trust their close friends as well as the posts retweeted by these friends, and most of these intimate friends are usually in their respective followee-lists; for the other reason, we tend to subconsciously imitate other people's behavior, especially those people we perceive to be like us [19]. Since people usually follow many users who share similar preferences with themselves due to the homophily [30], as a result, these followees may influence the current user to retweet a post which matches all their tastes. For example, a

PhD student on data mining may retweet a post about a new algorithm because he noticed that many famous data scientists he has followed have retweeted this post. However, those "mentioned users" can hardly improve the performance. This is probably because u is not familiar with these users. Moreover, only their "nickname" appended before the original post is visible to u, therefore, it may be difficult for them to catch u's attention.

We also notice that the approach using $Q'(u, po, t_0)$ with all the influential users does not significantly outperforms the approach with basic features. The reason may be that the basic features contain some other factors associated with users' retweeting behavior. For example, the features "how many times po has been retweeted" and "whether po contains a hashtag" may reflect the impact of some breaking news. However, we have not considered them yet. After incorporating $Q'(u, po, t_0)$ and the basic features, we are able to obtain an improvement of performance with about 3.7% in terms of F1 value. It means that our force function $Q'(u, po, t_0)$ is significant for retweeting prediction.

Table 1. The results of approaches with different features

Features	Precision(%)	Recall(%)	F1(%)
$Q'_A(u, po, t_0)$	59.74	62.32	61.01
$Q'_{A+M}(u, po, t_0)$	60.03	62.19	61.09
$Q'_{A+F}(u, po, t_0)$	62.56	65.67	64.08
$Q'(u, po, t_0)$	62.78	65.63	64.17
basic features (BF)	61.41	67.65	64.38
$Q'(u, po, t_0)$ + BF	**65.15**	**71.27**	**68.07**

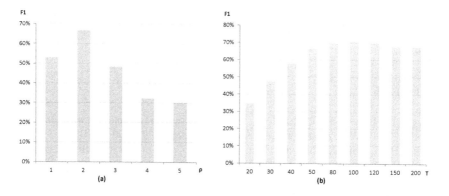

Fig. 3. The impact of distance coefficient (a) and topic number (b)

4.3   Impact of Model Parameters

We investigate the impact of distance tuning coefficient ρ and the topic number T on the model performance in this subsection. The results of different ρ are portrayed in Figure 3(a). The vertical axis is the F1 value. It is obvious that result outperforms others when ρ is set to 2. We note that it is exactly the same coefficient in Coulomb's law.

The number of topics would also influence the result. As depicted in Figure 3(b), the performance of our model tends to be stable when $T \geqslant 50$. However, we should note that the complexity of the model will increase as we add more topics, as a result, we decide to set $T = 50$.

5   Conclusion

In summary, inspired by the field theory in physics, we present a novel approach based on "influence field" with various features to predict the retweeting behavior in the micro-blogging system. We interpret retweeting behavior as an outcome of the influence from the post and the influential users. During our approach, many features are integrated together to generate the force which effects on the user and the inherent correlations among the features are also considered. The experimental results show that our strategy is indeed effective for the retweet prediction task.

There are also some limitations of our method. Firstly, because we are not able to get the timestamp of user's following behavior, as a result, the global topology structure of users' following relationship we used in the TwitterRank algorithm may deviate from the exact state at the timestamp of users' retweeting behavior in our experiment. Secondly, during our experiments, we simply assume that all the posts exposed to the current user have been read. In fact, we do not know whether the user did not want to retweet a post or he did not even see this post. These may hurt the overall performance to some extent. We will attempt to address these problems in the future.

References

1. Kwak, H., Lee, C., Park, H., Moon, S.: What is twitter, a social network or a news media? In: WWW 2010 Conference Proceedings, pp. 591–600 (2010)
2. Peng, H.K., Zhu, J., Piao, D., Yan, R., Zhang, Y.: Retweet modeling using conditional random fields. In: ICDMW 2011 Conference Proceedings, pp. 336–343 (2011)
3. Luo, Z., Osborne, M., Tang, J., Wang, T.: Who will retweet me?: finding retweeters in twitter. In: SIGIR 2013 Conference Proceedings, pp. 869–872 (2013)
4. Uysal, I., Croft, W.B.: User oriented tweet ranking: a filtering approach to microblogs. In: CIKM 2011 Conference Proceedings, pp. 2261–2264 (2011)
5. Xu, Z., Zhang, Y., Wu, Y., Yang, Q.: Modeling user posting behavior on social media. In: SIGIR 2012 Conference Proceedings, pp. 545–554 (2012)

6. Zhang, J., Liu, B., Tang, J., Chen, T., Li, J.: Social influence locality for modeling retweeting behaviors. In: IJCAI 2013 Conference Proceedings, pp. 2761–2767 (2013)
7. Xu, Z., Yang, Q.: Analyzing user retweet behavior on twitter. In: ASONAM 2012 Conference Proceedings, pp. 46–50 (2012)
8. Petrovic, S., Osborne, M., Lavrenko, V.: Rt to win! predicting message propagation in twitter. In: ICWSM 2011 Conference Proceedings, pp. 586–589 (2011)
9. Bakshy, E., Hofman, J.M., Mason, W.A., Watts, D.J.: Everyone's an influencer: quantifying influence on twitter. In: WSDM 2011 Conference Proceedings, pp. 65–74 (2011)
10. Qazvinian, V., Rosengren, E., Radev, D.R., Mei, Q.: Rumor has it: Identifying misinformation in microblogs. EMNLP **2011**, 1589–1599 (2011)
11. Phuvipadawat, S., Murata, T.: Breaking news detection and tracking in twitter. In: WI-IAT 2010 Conference Proceedings, pp. 120–123 (2010)
12. Chen, K., Chen, T., Zheng, G., Jin, O., Yao, E., Yu, Y.: Collaborative personalized tweet recommendation. In: SIGIR 2012 Conference Proceedings, pp. 661–670 (2012)
13. Boyd, D., Golder, S., Lotan, G.: Tweet, tweet, retweet: conversational aspects of retweeting on twitter. In: HICSS 2010 Conference Proceedings, pp. 1–10 (2010)
14. Macskassy, S.A., Michelson, M.: Why do people retweet? anti-homophily wins the day! In: ICWSM 2011 Conference Proceedings, pp. 209–216 (2011)
15. Suh, B., Hong, L., Pirolli, P., et al.: Want to be retweeted? large scale analytics on factors impacting retweet in twitter network. In: SocialCom 2010 Conference Proceedings, pp. 177–184 (2010)
16. Hong, L., Dan, O., Davison, B.D.: Predicting popular messages in twitter. In: WWW 2011, pp. 57–58 (2011)
17. Kupavskii, A., Ostroumova, L., Umnov, A., et al.: Prediction of retweet cascade size over time. In: CIKM 2012 Conference Proceedings, pp. 2335–2338 (2012)
18. Yang, Z., Guo, J., Cai, K., Tang, J., Li, J., Zhang, L., Su, Z.: Understanding retweeting behaviors in social networks. In: CIKM 2010 Conference Proceedings, pp. 1633–1636 (2010)
19. Adams, P.: Grouped: How small groups of friends are the key to influence on the social web. New Riders (2012)
20. Weng, J., Lim, E.P., Jiang, J., et al.: Twitterrank: finding topic-sensitive influential twitterers. In: WSDM 2010 Conference Proceedings, pp. 261–270 (2010)
21. Blei, D.M., Ng, A.Y., Jordan, M.I.: Latent dirichlet allocation. The Journal of machine Learning research **3**, 993–1022 (2003)
22. Gilbert, E., Karahalios, K.: Predicting tie strength with social media. In: SIGCHI 2009 Conference Proceedings, pp. 211–220 (2009)
23. Granovetter, M.: The strength of weak ties. American Journal of Sociology **78**, 1360–1380 (1973)
24. Krackhardt, D.: The strength of strong ties: The importance of philos in organizations. Networks and Organizations: Structure, Form, and Action **216**, 239 (1992)
25. Haythornthwaite, C.: Strong, weak, and latent ties and the impact of new media. The Information Society **18**, 385–401 (2002)
26. Huberman, B.A., Romero, D.M., Wu, F.: Social networks that matter: Twitter under the microscope (2008). arXiv preprint arXiv:0812.1045
27. Marsden, P.V.: Core discussion networks of americans. American Sociological Review, 122–131 (1987)
28. Xiaolin, S., Lada, A., Martin, S.: Networks of strong ties. Physica A: Statistical Mechanics and its Applications **378**, 33–47 (2007)

29. Shen, Y., Yu, J., Dong, K., Nan, K.: Automatic fake followers detection in chinese micro-blogging system. In: Tseng, V.S., Ho, T.B., Zhou, Z.-H., Chen, A.L.P., Kao, H.-Y. (eds.) PAKDD 2014, Part II. LNCS, vol. 8444, pp. 596–607. Springer, Heidelberg (2014)
30. McPherson, M., Smith-Lovin, L., Cook, J.M.: Birds of a feather: Homophily in social networks. Annual Review of Sociology, 415–444 (2001)

Realizing Impact Sourcing by Adaptive Gold Questions: A Socially Responsible Measure for Workers' Trustworthiness

Kinda El Maarry[1(✉)], Ulrich Güntzer[2], and Wolf-Tilo Balke[1(✉)]

[1] IFIS, TU Braunschweig, Braunschweig, Germany
{elmaarry,balke}@ifis.cs.tu-bs.de
[2] Inst. f. Informatik, Universität Tübingen, Tübingen, Germany
ulrich.guentzer@informatik.uni-tuebingen.de

Abstract. In recent years, crowd sourcing has emerged as a good solution for digitizing voluminous tasks. What's more, it offers a *social* solution promising to extend economic opportunities to low-income countries, alleviating the welfare of poor, honest and yet uneducated labor. On the other hand, crowd sourcing's virtual nature and anonymity encourages fraudulent workers to misuse the service for quick and easy monetary gain. This in turn compromises the quality of results, and forces task providers to employ strict control measures like gold questions or majority voting, which may gravely misjudge honest workers with lower skills, ultimately discarding them from the labor pool. Thus, the problem of fairly distinguishing between fraudulent and honest workers lacking educational skills becomes vital for supporting the vision of Impact Sourcing and its social responsibility. We develop a technique with socially responsible gold questions as an objective measure of workers' trustworthiness, rather than a mere discarding mechanism. Our statistical model aligns workers' skill levels and questions' difficulty levels, which then allows adapting the gold questions' difficulty for a fair judgment. Moreover, we illustrate how low-skilled workers' initial payloads, which are usually discarded along with the worker, can be partially recovered for an increased economic gain, and show how low-skilled workers can be seamlessly integrated into high-performing teams. Our experiments prove that about 75% of misjudged workers can be correctly identified and effectively be integrated into teams with high overall result correctness between 70-95%.

Keywords: Crowd sourcing · Impact sourcing · Fraud detection · Quality control

1 Introduction

Crowd sourcing platforms can distribute cognitive tasks requiring human intelligence through digital gateways, which can flexibly tap into huge international workforces. In a nutshell, it creates a win-win opportunity where task providers can cut down their costs through cheaper services, while simultaneously providing economic opportunities to hired workers. Coupled with *Impact Sourcing* it could play a key role in advancing the

(c) Springer International Publishing Switzerland 2015
J. Li and Y. Sun (Eds.): WAIM 2015, LNCS 9098, pp. 17–29, 2015.
DOI: 10.1007/978-3-319-21042-1_2

economic development of low-income countries, alleviating the welfare of less fortunate individuals, as well as connecting them to the global economy. Impact Sourcing, the socially responsible arm of the information technology outsourcing industry [1], specifically aims at employing people at the bottom of the pyramid, who are disadvantaged on an economical, educational and accordingly skill-wise level.

However, the highly distributed nature, virtual and anonymous setup of crowd sourcing platforms, along with the short term task contracts they offer open doors for *fraudulent workers*, who can simply submit randomly guessed answers, in hope of going undetected. The inclusion of such workers of course jeopardies the overall credibility of the returned quality. And with manual checking being both costly and time consuming, this directly invalidates the main gains of crowd sourcing. Hence, task providers are forced to employ strict control measures to exclude such workers, ensure high quality results, and get good return on their investment. However, these measures befall honest, yet low-skilled workers, too. In fact, anecdotal evidence from our own previous work [2] shows that by completely excluding workers from two offending countries, where a high number of fraudulent workers were detected, the overall result correctness instantly saw a 20% increase. Needless to say, this simultaneously excluded many honest workers in those two countries as well.

Indeed, the positive social impact of the Impact Sourcing model is immense, where almost half of the world's population lives on less than $2.50 a day, and 1.8 billion people can't access a formal job[1]. But also with Impact sourcing this huge task force may ultimately fall into a vicious cycle: even with simple task training mechanisms offered by platforms like CrowdFlower, the opportunity provided by crowd sourcing is biased by quality control measures towards educated workers. In fact, quality measures tend to repeatedly exclude uneducated, low-skilled workers. Not giving them a chance at improving their skills leaves them prey for constant exclusion.

Common currently deployed *quality control measures* include gold questions, majority votes, and reputation based systems. Of course, all such control measures are susceptible to the ultimate downside of misjudging honest low-skilled workers. Accordingly, in this paper we develop an objective socially responsible measure of workers' trustworthiness: *adaptive gold questions*. Basically, an initial set of balanced gold questions (i.e. covering all difficulty levels) is used as a mechanism for determining the skill level of a worker rather than a discarding mechanism. Next, a second round of adapted gold questions, whose difficulty levels are within the estimated skill level of the corresponding worker, are injected. The underlying assumption is that, although low-skilled workers may fail the correctness threshold set for the balanced gold questions, since they surpass their own skill level, they should succeed at gold questions, which have been adapted to their lower skill levels. On the other hand, fraudulent workers would also fail such adaptive gold questions, since their responses to both sets of balanced and adaptive gold questions will be random.

To adapt gold questions, our method requires two parameters: workers' skill levels and difficulties of questions. To that end, we make use of psychometric item response theory (IRT) models: in particular, the Rasch Model for estimating these parameters.

---

[1] http://www.impacthub.org/

Our experiments show that around 75% honest misjudged workers can be correctly identified and the payloads that would have been discarded with the worker can be partially recovered i.e. tasks in the payload within a low-skilled worker's ability. Furthermore, we investigate heuristics for forming high-performing skill-based teams, into which low-skilled workers can be later integrated to ensure high quality output.

2   Related Work

The social model of Impact Sourcing was first implemented by *Digital Divide Data (DDD)*[2] back in 2001, and ever since has been adopted by many crowd sourcing platforms such as *Samasource*[3], *RuralShores*[4], or *ImpactHub*[1]. Crowd sourcing provides an accessible solution to both: companies having digital intelligent problems (e.g. web resource tagging [3], completing missing data [4], sentiment analysis [5], text translation [6], information extraction [7], etc.) and underprivileged honest workers lacking high skills. But for actually profiting from this win-win situation, the challenge of identifying fraudulent workers and their compromising contributions must be met.

A rich body of research addresses the quality problem in crowdsourcing. Currently employed solutions include aggregation methods, which rely on redundancy as means to improving the overall quality: By assigning the same task to several workers, the correct answer can be identified through aggregation, e.g. majority voting. Nevertheless, this has been shown to have severe limitations, see e.g. [8]. This was followed by Dawid and Skene [9], who applied an expectation maximization algorithm to consider the responses' quality based on the individual workers. Focusing on such error rates, other approaches emerged such as: a Bayesian version of the expectation maximization algorithm approach [10], a probabilistic approach taking into account both the worker's skill and the difficulty of the task at hand [11], or an even more elaborate algorithm trying to separate unrecoverable error rates from recoverable bias [12].

Another class of solutions focuses on eliminating unethical workers throughout longer time scales. This can be achieved through constantly measuring workers' performance via a reputation-based system (based on a reputation model [13-14], on feedback and overall satisfaction [14], or on deterministic approaches [15], etc.) or through injecting gold questions in the tasks. Except, reliably computing the workers' reputation poses a real challenge, and as we will show in section 3, both techniques are susceptible to the ultimate downside of misjudging honest low-skilled workers. In contrast, we apply gold questions as a socially responsible measure of workers' trustworthiness to measure their skill level rather than as a discarding mechanism.

Furthermore, monetary incentives as means of quality control have been investigated. But the implementation of such an approach proves to be tricky, where low paid jobs yield sloppy work, while high paid jobs attract unethical workers [16].

It is important to note how tightly coupled our work is with the IRT Paradigm [17] in psychometrics, which enables us to focus on the workers' capabilities. We employ

---

[2] http://www.digitaldividedata.com/about/

[3] http://www.samasource.org/

[4] http://ruralshores.com/about.html

the Rasch model [18] to estimate the tasks' difficulty and   workers' skill. This allows us to address the principal concern of Impact Sourcing: distinguishing honest low-skilled workers from unethical workers. Perhaps most similar to our work is the model presented in [11], which is also based on the IRT paradigm: GLAD – a generative model of labels, abilities and difficulties – iteratively estimates the maximum likelihood of the worker's skill, question's difficulty, as well as the worker's correctness probability computed by EM (Expectation-Maximization approach). GLAD's robustness wavers though when faced with unethical workers, especially when they constitute more than 30% of the task force [19]. In contrast, we focus on detecting the workers' skill level to adapt future gold questions to be injected, which then enables us to identify with sufficient robustness honest workers who are merely low-skilled.

Other research focusing on estimating one or more parameters include, Dawid and Skene [9], who considered the worker's skill and utilized confusion matrices as an improved redundancy technique form. The downfall as pointed out and addressed by Ipeirotis [20]   is the underestimation of the workers' quality, who consistently give incorrect results. Considering two parameters, both the workers' abilities as well as the inference of correct answers was investigated in [21]. Except, the difficulty of the task at hand, which in turn influences the workers' perceived skill level is neglected.

3   Motivational Crowd Sourcing in a Laboratory-Based Study

For self-containment, we briefly detail in this section one of our earlier experiments in [22]. To acquire ground truth, we set up a small-scale laboratory experiment, with a total of 18 volunteers. In this paper we formulate our Human Intelligent Tasks (HITs) over an American standardized test for college admission: the Graduate Record Examination (GRE) crawled dataset (http://gre.graduateshotline.com), namely the verbal practice questions section. The task then is to select the correct definition out of 4 choices for a given word. Given a set of 20 multiple choice questions, volunteers were asked to answer questions twice. In the first round, they should just randomly select answers while in the second round, they should consider the questions and answer them to the best of their knowledge. Accordingly, the dataset can be divided into honest and unethical workers.

Figure 1 sorts all workers' answers according to the respective total number of correct answers achieved over 20 questions. Although no worker got all 20 answers correct, it comes as no surprise that truthful answers (58.6%) tend to be more correct than random answers (40%). Furthermore, even though the dataset is in no way biased, random responses at times produced better overall results. Consider the top 10 workers getting the most correct answers in figure 1. In a reputation based system, the worker at rank 5 (scoring 15 correct answers) would be given a higher reputation score than workers on ranks 6 to 9 (scoring 14 correct answers). Yet here, 3 workers at least tried to answer correctly.

Furthermore, with the common 70% correctness threshold set, gold questions would eliminate 61% honest workers (i.e. 11 workers) and 88% of the unethical workers (i.e. 16 workers). Though gold questions are more biased to penalize unethical workers, still the bias is small, and a significant number of honest workers are penalized too.

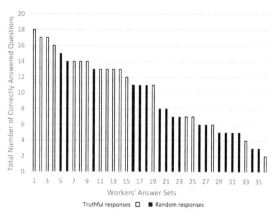

Fig. 1. Truthful versus random responses

4   Identifying Low-Skilled Workers

As shown, gold questions tend to misjudge honest low-skilled workers and can be bypassed by unethical workers. In this section, we provide a short overview of the underlying statistical *Rasch Model* (RM), which is used to align workers' skill levels and questions' difficulty levels to adapt the gold questions' difficulty for a fairer judgment and a socially responsible measure that can identify low-skilled workers.

4.1   The Rasch Model

The intrinsic nature of crowdsourcing involves many human factors. This in turn directed our attention to psychometrics − the science assessing individual's capabilities, aptitudes and intelligence − and it's IRT classes, namely, the Rasch model (RM). Basically, RM computes the probability P_{ij} that the response of a worker $\omega_i \in W$ to a given task $t_j \in \mathbb{P}$ is correct as a function of both: 1) his/her ability θ_{ω_i}, and 2) the difficulty of the task β_{t_j}. Assuming a binary setting, where a worker's response $x_{ij} \in \{0,1\}$ is known (where 0 indicates an incorrect response and 1 a correct response), RM's dichotomous case can be applied. Simply put, both the RM's parameters: a worker's ability θ and a task's difficulty β are depicted as latent variables, whose difference yields the correctness probability P.

Definition 1: (Rasch Model for Dichotomous Items) given a set of workers W= $\{\omega_1, \omega_2, ..., \omega_n\}$, where $|W| = n$ and a HIT $\mathbb{P} = \{t_1, t_2, ..., t_m\}$, where $|\mathbb{P}| = m$. Assume $\omega_i \in W$ and $t_j \in \mathbb{P}$, then the correctness Probability P_{ij} can be given as follows:

$$P_{ij} = P_{ij}(x_{ij} = 1) = \frac{exp(\theta_{\omega_i} - \beta_{t_j})}{1 + exp(\theta_{\omega_i} - \beta_{t_j})}$$

This can also be reformulated, such that the distance between θ_{ω_i} and β_{t_j} is given by the logarithm of the odds ratio, also known as the log odd unit *logit*.

$$log\left(\frac{P_{ij}}{1 - P_{ij}}\right) = \theta_{\omega_i} - \beta_{t_j}$$

The difficulty of a question with a logit value of 0 is average, where a negative logit value implies an easy β_{t_j} and a low θ_{ω_i} and vice versa. Accordingly, the correctness probability of a worker's response is high when his ability exceeds the corresponding task's difficulty. A special vital feature of RM is its emphasis on the "objective" measurement of (θ, β) [23]. That is, measurement of both θ and β should be independent respectively of $\mathbb{P}$ and W.

4.2   Adapting the Gold Questions Based on RM's alignment

Initially, a balanced set of gold questions $\mathbb{G}^B$ are injected in an initial payload $\mathbb{P}_a$ to which worker ω is assigned to. However, failing the correctness threshold $\mathbb{C} > 70\%$ (i.e. worker fails on more than 30% of the gold questions), doesn't instantly eliminate ω. Instead, based on RM's skill level estimation θ_ω, an adapted set of gold questions $\mathbb{G}^A$ are formulated by aligning their difficulty $\mathbb{G}^A_{\beta_{t_j}}$ to the corresponding worker's θ_ω, and injected within a second payload $\mathbb{P}_b$. Surpassing the correctness threshold $\mathbb{C}$ on $\mathbb{G}^A$ indicates that worker ω is indeed honest, though not up to the initial standards. As an example consider the following result from one of our experiments.

Example 1 (Correctness Threshold versus Adapted Gold Questions) assuming a correctness threshold $\mathbb{C} = 70\%$, three workers $\omega_1, \omega_2, \& \omega_3$ are assigned to initial payload $\mathbb{P}_a$ comprising $\mathbb{G}^B$ with difficulty levels β_{t_j} ranging between [-1.04, 1.8]. Logit values of β are interpreted accordingly: 0 is average, $\beta < 0$ implies easiness, & $\beta > 0$ implies difficulty. Given that $\omega_1^\mathbb{C}$ is the correctness threshold achieved by ω_1, the following correctness thresholds were achieved:
$$-\omega_1^\mathbb{C} = 87.5\% > \mathbb{C}\,(= 70\%) \quad -\omega_2^\mathbb{C} = 50\% \;\; < \mathbb{C}\,(= 70) \quad -\omega_3^\mathbb{C} = 37.5\% < \mathbb{C}\,(= 70\%)$$
Accordingly, workers ω_2 and ω_3 would be eliminated in a usual correctness threshold setup. In contrast, following our approach, we compute instead the workers' ability based on $\mathbb{G}^B$ upon which we formulate two $\mathbb{G}^A$, such that $\mathbb{G}^{A_{\omega_2}}_\beta \leq \theta_{\omega_2}$ and $\mathbb{G}^{A_{\omega_3}}_\beta \leq \theta_{\omega_3}$. Next ω_2, and ω_3 are assigned a second payload $\mathbb{P}_b$ comprising the respective $\mathbb{G}^A$. They scored the following correctness thresholds:
$$-\,\omega_2^\mathbb{C} = 37.5\% < \mathbb{C}\,(= 70\%) \quad -\omega_3^\mathbb{C} = 100\% < \mathbb{C}\,(= 70\%)$$
Accordingly, ω_3 is identified as a low-skilled ethical worker to be retained, unlike ω_2.

5   Gains of Recovering Low-Skilled Workers

Impact sourcing is realized through recovering low-skilled workers, who would've been otherwise treated as fraudulent and unfairly discarded. In this section, we list empirically derived heuristics for integrating low-skilled in high performing teams and illustrate how low-skilled workers' earlier payloads can be partially recovered.

5.1   High Performing Team Combinations

Following experimental results in section 6.2, three workers proved to be best as a team-size baseline. Based on a labor pool of 30 workers, 66% out of all the possible team combinations ($^{30}C_3 = 4060$ teams) produced high correctness quality (70-95%) upon aggregating their results through skill-weighted majority vote. By analyzing the teams constituting this 66%, heuristics for formulating high performing teams were empirically found. As shown below, the heuristics range from including two highly-skilled workers along with one average or low-skilled worker like H_1, H_2. Two low-skilled workers along with one highly-skilled worker like H_3, H_4. A combination of unskilled, average and highly skilled workers like H_6, or average to highly skilled workers like H_5.

Heuristics 1-6: (Heuristics for formulating High Performing Team) given a team $\mathcal{T} = \{\omega_1, \omega_2, \omega_3\}$, comprising a combination of three workers with the respective skill levels $\theta =$ vels $\theta = \{\theta_{\omega_1}, \theta_{\omega_2}, \theta_{\omega_3}\}$. Logit values of θ are interpreted accordingly: 0 is average, $\theta < 0$ implies low skill level, & $\theta > 0$ implies high skill level. Through combining low-skilled with higher-skilled workers in the following team combinations, high correctness quality percentage results $\mathbb{Q}$ can be attained through skill-weighted majority vote.

- H_1: If $\left(1 \leq \theta_{\omega_i} < 2.5\right) \wedge \left(\theta_{\omega_j} < 0.6\right)$, then $65 \leq \mathbb{Q} \leq 95$,

  where $P(80 \leq \mathbb{Q} \leq 90) = 0.77$, $i = 1, 2$ and $j = 3$
- H_2: If $\left(\theta_{\omega_i} \geq 0.5\right) \wedge (1 \leq \theta_{\omega_j} \leq 2.5)$, then $80 \leq \mathbb{Q} \leq 85$, where $i = 1, 2$ and $j = 3$
- H_3: If $(-1 \leq \theta_{\omega_i} < 0) \wedge (1 \leq \theta_{\omega_j} \leq 2.5)$, then $70 \leq \mathbb{Q} \leq 85$, where $i = 1, 2$ and $j = 3$
- H_4: If $(-2.9 \leq \theta_{\omega_i} < -1) \wedge (\theta_{\omega_j} > 2.5)$, then $55 \leq \mathbb{Q} \leq 80$,

  where $P(70 \leq \mathbb{Q} \leq 80) = 0.66$, $i = 1, 2$ and $j = 3$
- H_5: If$(\theta_{\omega_i} \geq 0.5)$, then $70 \leq \mathbb{Q} \leq 80$, where $i = 1, 2, 3$
- H_6: If $\left(\theta_{\omega_1} < 0\right) \wedge (\theta_{\omega_2} \geq 0.5) \wedge (\theta_{\omega_3} > 2)$, then $70 \leq \mathbb{Q} \leq 90$,

  where $P(75 \leq \mathbb{Q} \leq 85) = 0.78$

5.2   Recovering Partial Payloads

During the process of identifying low-skilled workers, that is, before they are as-signed to form high contributing teams, low-skilled workers would've already been assigned two payloads 1) $\mathbb{P}_a$: the initial payload worker ω is assigned to, comprising balanced $\mathbb{G}^B$. Failing $\mathbb{C} > 70\%$ at this stage doesn't lead to an instant elimination, but to 2) $\mathbb{P}_b$: the second payload comprising the adapted $\mathbb{G}^A$ as per RM's puted θ_ω. This time, failing $\mathbb{C} > 70\%$ leads to elimination. Succeeding however, implies that worker ω is low-skilled and is to be henceforward enrolled to form high performing teams, which ensures high quality throughput. Rather than discarding $\mathbb{P}_a$ & $\mathbb{P}_b$, we can attain high quality results by recovering those tasks in the payloads, whose difficulty levels are within the worker's skill level.

Definition 2: (Partial Recoverable Payloads) assume a low-skilled worker ω, with computed RM's skill level θ_ω, and two payloads $\mathbb{P}_a = \{t_1^a, t_2^a, ..., t_m^a\}$ and $\mathbb{P}_b = \{t_1^b, t_2^b, ..., t_m^b\}$, where $|\mathbb{P}_a| = |\mathbb{P}_b| = m$, and have corresponding difficulty levels $\mathbb{P}_\beta^a = \{\beta_{t_1^a}, \beta_{t_2^a} ..., \beta_{t_m^a}\}$ and $\mathbb{P}_\beta^b = \{\beta_{t_1^b}, \beta_{t_2^b} ..., \beta_{t_m^b}\}$. Then the recoverable payload is:
$$\mathbb{R}_\omega^{\mathbb{P}} = \{t \in \mathbb{P}_* \mid \beta_t \leq \theta_\omega\}, \text{ where } \mathbb{P}_* = \mathbb{P}_a \cup \mathbb{P}_b, |\mathbb{R}_\omega| < 2m$$

In order to identify the recoverable tasks within a payload, their difficulty level should be computed. However, RM requires the corresponding ground truth in order to estimate the β parameter. To that end, we aim at synthesizing a reliable ground truth for the payloads' tasks, which would then serve as input to RM. We aggregate the responses of the low-skilled workers along with two other workers, such that these three workers' combination adhere to the above Heuristics 1-6 for forming a high performing team. Ultimately the skill-weighted Majority vote produce a reliable synthesized ground truth which RM uses to estimate the tasks' difficulty level. Our experiments show that the synthesized ground truth's correctness quality is always higher than 70%. We provide a description in Algorithm 1 below.

Algorithm 1. Recovering Partial Payloads

```
Input:
•  H  : HIT's list of questions object q, with attributes:1)ID: q.ID,
      2)difficulty: q.difficulty, 3)synthesized ground truth: q.GT
•  ℑ  : high performing team consisting of 3 workers (ω₁,ω₂,ω₃),where ω₁
      is a low-skilled worker
•  ℑᶜ : corresponding skill levels  of ℑ → (θω₁,θω₂,θω₃)
•  RM: matrix holding list of responses of each worker in ℑ
Output:
     ▪ PL  : List of questions recovered for ω₁
 1: begin:
 2:    for each q in H
 3:      q.GT = computeGroundTruthBySkillWeightedMajortyVote(RM, ℑᶜ,q)
 4:      DL  = computeQuestionsDifficultLevelyByRaschModel(H,RM)
 5:      ODL = orderQuestionsAscendinglyByDifficultyLevel(DL)
 6:    for each q in ODL
 7:        if(θω₁ > q.difficulty)
 8:            add q to PL
 9: end
```

---

6   Experimental Results

In this section we evaluate the efficiency of adaptive gold questions in identifying low-skilled workers through laboratory and real crowdsourcing experiments. The open source eRm package for the application of IRT models in R is utilized [24], First, we investigate the percentage of low-skilled honest workers that can be correctly detected by each of $\mathbb{G}^B$ and $\mathbb{G}^A$. Next, we investigate the quality of the synthesized ground-truth from the skill-weighted Majority vote, upon which RM can estimate the task's difficulty levels, eventually allowing us to identify which parts of payloads $\mathbb{P}_a$ and $\mathbb{P}_b$ can be recovered for the correctly identified low-skilled workers. Moreover, we empirically investigate heuristics for forming high performing teams into which low-skilled workers can be later assigned to. Lastly we test our measure in a real crowdsourcing experiment.

6.1   Identifying Low-Skilled Workers

Based on the laboratory experiment's ground truth dataset in section 3, we use the data generated from the second round, which corresponds to honest workers. This allows us to investigate how many honest workers our measure can correctly identify.

As shown in figure 2, with a correctness threshold $\mathbb{C}=70\%$ set, the initial payload with $\mathbb{G}^B$ retained 44.44% ethical workers (i.e. 8 out of 18). The second payload comprising $\mathbb{G}^A$ retained 50% of the previously discarded low-skilled ethical workers. That is, 72% of the honest workers have been detected after both payloads. In fact, the identified low-skilled workers get on average 90.6% of the $\mathbb{G}^A$ correctly i.e. exceeding even the original 70% correctness threshold $\mathbb{C}$ with a tangible margin. On the other hand, those ethical workers who were discarded even after $\mathbb{G}^A$ had lower skill levels than the easiest questions in $\mathbb{P}$, which justifies their exclusion.

Similarly, a laboratory based experiment comprising 30 volunteers, supports the previous findings. The initial payload with $\mathbb{G}^B$ retained 33.3% of the honest workers (i.e. 20 honest workers are discarded), while the second payload comprising $\mathbb{G}^A$ retained 65% of the previously discarded low-skilled workers (13 out of 20 discarded ethical workers were correctly retained). That is, 76% honest workers have been identified instead of 33.3%.

6.2   Investigating Crowd-Synthesized Ground-Truth Quality

Next, we investigate the highest crowd-synthesized ground-truth that can be attained through skill-based majority vote. A high ground-truth quality must be insured since RM base its tasks' difficulty level estimates upon it. That is, poor quality would lead to bad β estimations, which would in turn lead to wrong identification of the recoverable sections of $\mathbb{P}_a$ and $\mathbb{P}_b$. Based on the 30 honest volunteer laboratory experiment, we investigate different team combinations and search for those team combinations producing the highest ground-truth quality.

Initially, we start by all the possible combination of three-sized teams (i.e. $^{30}C_3 = 4060$.) As shown in figure 3, many combinations: 2,671 teams i.e. ≈66%, achieve high correctness quality (70-95%). Further experiments with team combinations of 4 workers show a slight improvement, where 19,726 teams achieve correctness quality ranging between (70-95%), and 4 teams reaching 100% .i.e. ≈72% of all possible team combinations: $^{30}C_4 = 27,405$.) On the other hand, teams of size 2 perform badly, and none reaches 95% quality.

It is clear from figure 3 that certain team combinations work exceedingly better than others. Accordingly, in figure 3 we zoom in only on team combinations of qualified workers (i.e. low-skilled workers that have been identified by $\mathbb{G}^A$ and highly-skilled workers who were identified earlier by $\mathbb{G}^B$). Analyzing the different skill-based team combinations producing the required high quality results (70-95%) yielded the heuristics of creating high-quality skill-based team combinations, as listed in section 5.1. Figure 3 depicts the Ground truth quality achieved by high performing team combinations. This yielded 718 possible team combinations, achieving on average 78% accuracy, which ranges up to 95%. Only the output of such team combinations are accordingly to be used when recovering payloads and when low-skill workers are to be integrated to form high-performing teams.

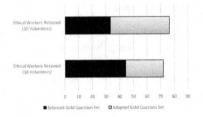

Fig. 2. Percentage of ethical workers retained after $\mathbb{G}^B$ and $\mathbb{G}^A$

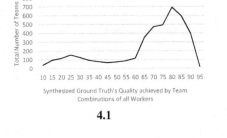

4.1

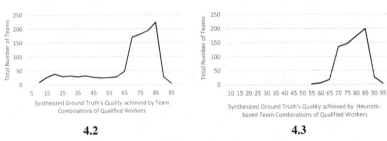

4.2   **4.3**

Fig. 3. Ground truth quality achieved by different team combinations

6.3   Partially Recovering Low-Skilled Workers' Payloads

Based on the previous experiment's findings, we check how well we can identify the recoverable sections of $\mathbb{P}_a$ and $\mathbb{P}_b$ based on RM's β estimates and the quality of the synthesized ground truth. From the 30 honest volunteer laboratory experiment, a random subset of 10 honest low-skilled workers are taken and a set of all possible high performing team combinations were created. For each worker, we compute the aggregate percentage of the recoverable payloads' size and quality over all the possible high performing team combinations this worker formulated.

As seen in figure 4, on average 68% of the payloads can be recovered (i.e. around 13 question from each of $\mathbb{P}_a$ and $\mathbb{P}_b$,). Moreover, the average correctness quality is 76%, which is even higher than the required correctness threshold. This corresponds to 6.50$ savings when recovering the initial and second payloads for each of the 10 workers, given that each payload has 20 questions and costs 50 cents.

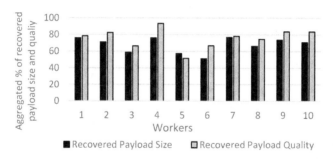

Fig. 4. Size and Quality of Recovered Payloads

6.4   Real Crowd Sourcing Experiment

We evaluate the efficiency of our measure through a real crowdsourcing experiment, which was ran on the generic CrowdFlower crowd sourcing platform. A total of 41 workers were assigned Hits comprising 10 payload questions and 4 gold questions, giving a total of 574 judgments and costing 20.5$, where each HIT costs 35 cents. A correctness threshold $\mathbb{C} = 70\%$ would discard around 30% of the workers (i.e. 12 workers). In contrast, our measure, assigned those 12 workers to a second payload $\mathbb{P}_b$ comprising adapted $\mathbb{G}^A$. This yielded a total of 168 judgments, costing 4.2$. In total, 25% of the workers were identified as low-skilled workers (i.e. 3 workers.)

Unlike the laboratory-based experiment, the real crowd sourcing experiment has no ground truth (i.e. number of low-skilled workers and number of fraudulent workers), accordingly we measure the efficiency of how well these workers were correctly identified by checking the quality of their partial recovered payloads, since these payload tasks are those within their real skill level. On average 50% of both payloads $\mathbb{P}_a$ and $\mathbb{P}_b$ were recovered with an average correctness of 80%, which surpasses even the correctness threshold. This corresponds to 3 payloads (i.e. 1.5$). The small savings reflect nothing more than the number of detected low-skilled workers, whose percentage in this experiment could have been small and lesser than the fraudulent workers.

7   Summary and Future Work

In this paper, we support *Impact Sourcing* by developing a socially responsible measure: adaptive gold questions. Our laboratory-based experiment attests that current employed quality control measures like gold questions or reputation based systems tend to misjudge low-skilled workers and eventually discard them from the labor pool. In contrast, we show how gold questions that are adapted to the corresponding workers' ability can identify low-skilled workers, consequently saving them from the vicious elimination cycle and allow them to work within their skill levels. This can be achieved by utilizing the Rasch Model, which estimates and aligns both the workers' skill level and the gold questions' difficulty level. Furthermore, we show how initial payloads could be partially recovered to reclaim some of the arguable economic loses. Through empirical results, we defined heuristics for building high performing teams.

Following these heuristics, low-skilled workers can be effectively integrated to produce reliable results (70-95%) through skill-weighted majority vote.

Nevertheless, retaining a database of workers and dynamically creating such high performing teams might not always be feasible. Therefore, the next step would be to expand our model's adaptivity to encompass not only the gold questions, but to adapt as well the entire payload to suit each workers' ability, which would boost the overall quality and promote a more efficient assignment of tasks.

References

1. "Digital Jobs Africa: The Rockefeller Foundation," [Online]. Available: http://www. rockefellerfoundation.org/our-work/current-work/digital-jobs-africa/impact-sourcing.
2. Selke, J., Lofi, C., Balke, W.-T.: Pushing the boundaries of crowd-enabled databases with query-driven schema expansion. In: 38th Int. Conf. VLDB, pp. 538-549 (2012)
3. Finin, T., Murnane, W., Karandikar, A., Keller, N., Martineau, J., Dredze, M.: Annotating named entities in twitter data with crowdsourcing. In: CSLDAMT 2010 Proc. of the NAACL HLT 2010 Workshop on Creating Speech and Language Data with Amazon's Mechanical Turk, pp. 80–88 (2010)
4. Lofi, C., El Maarry, K., Balke, W.-T.: Skyline queries in crowd-enabled databases. In: EDBT/ICDT Joint Conf., Proc. of the 16th Int. Conf. on Extending Database Technology (2013)
5. Kouloumpis, E., Wilson, T., Moore, J.: Twitter sentiment analysis: the good the bad and the OMG!. In: International AAAI Conf. on Weblogs& Social Media, pp. 538–541 (2011)
6. Callison-Burch, C.: Fast, cheap, and creative: evaluating translation quality using Amazon's Mechanical Turk. In: EMNLP 2009: Proc. of the 2009 Conf. on Empirical Methods in Natural Language Processing, vol. 1, no. 1, pp. 286–295 (2009)
7. Lofi, C., Selke, J., Balke, W.-T.: Information Extraction Meets Crowdsourcing: A Promising Couple. Proc. of the VLDB Endowment 5(6), 538–549 (2012). 23, 2012
8. Kuncheva, L.I., Whitaker, C.J., Shipp, C.A., Duin, R.P.W.: Limits on the majority vote accuracy in classifier fusion. Journal: Pattern Analysis and Applications -- PAA 6(1), 22–31 (2003)
9. Dawid, A.P., Skene, A.M.: Maximum likelihood estimation of observer error-rates using the EM algorithm. Journal of Applied Statistics. 28, 20–28 (1979)
10. Raykar, V.C., Yu, S., Zhao, L.H., Valadez, G.H., Florin, C., Bogoni, L., Moy, L.: Learning From Crowds. The Journal of Machine Learning Research 11, 1297–1322 (2010)
11. Whitehill, J., Ruvolo, P., Wu, T., Bergsma, J., Movellan, J.: Whose Vote Should Count More: Optimal Integration of Labels from Labelers of Unknown Expertise. Proc. of NIPS 22(1), 1–9 (2009)
12. Ipeirotis, P.G., Provost, F., Wang, J.: Quality management on amazon mechanical turk. In: Proc. of ACM SIGKDD Workshop on Human Computation, pp. 0–3 (2010)
13. El Maarry, K., Balke, W.-T., Cho, H., Hwang, S., Baba, Y.: Skill ontology-based model for quality assurance in crowdsourcing. In: UnCrowd 2014: DASFAA Workshop on Uncertain and Crowdsourced Data, Bali, Indonesia, (2014)
14. Ignjatovic, A., Foo, N., Lee, C.T.L.C.T.: An analytic approach to reputation ranking of participants in online transactions. In: 2008 IEEE/WIC/ACM Int. Conf. Web Intell. Intell. Agent Technol. vol. 1 (2008)
15. Noorian, Z., Ulieru, M.: The State of the Art in Trust and Reputation Systems: A Framework for Comparison. Journal of theoretical and applied electronic commerce research 5(2) (2010)

16. Kazai, G.: In search of quality in crowdsourcing for search engine evaluation. In: Clough, P., Foley, C., Gurrin, C., Jones, G.J., Kraaij, W., Lee, H., Mudoch, V. (eds.) ECIR 2011. LNCS, vol. 6611, pp. 165–176. Springer, Heidelberg (2011)
17. Traub, R.E.: Applications of item response theory to practical testing problems. Book's Publisher: Erlbaum Associates **5**, 539–543 (1980)
18. Rasch, G.: Probabilistic Models for Some Intelligence and Attainment Tests. Book's Publisher: Nielsen & Lydiche (1960)
19. Hung, N.Q.V., Tam, N.T., Tran, L.N., Aberer, K.: An evaluation of aggregation techniques in crowdsourcing. In: WISE (2013)
20. Wang, J., Ipeirotis, P.G.. Provost, F.: Managing crowdsourced workers. In: Winter Conf. on Business Intelligence (2011)
21. Batchelder, W.H., Romney, A.K.: Test theory without an answer key. Journal Psychometrika **53**(1), 71–92 (1988)
22. El Maarry, K., Balke, W.-T.: Retaining rough diamonds: towards a fairer elimination of low-skilled workers. In: 20th Int. Conf. on Database Systems for Advanced Applications (DASFAA), Hanoi, Vietnam (2015)
23. Karabatsos, G.: A critique of Rasch residual fit statistics. Journal of Applied Measures. **1**(2), 152–176 (2000)
24. Mair, P.: Extended Rasch Modeling: The eRm Package for the Application of IRT Models in R. Journal of Statistical Software **20**(9), 1–20 (2007)

A Coalition Formation Game Theory-Based Approach for Detecting Communities in Multi-relational Networks

Lihua Zhou[1(✉)], Peizhong Yang[1], Kevin Lü[2], Zidong Zhang[1], and Hongmei Chen[1(✉)]

[1] Department of Computer Science and Engineering,
Yunnan University, Kunming 650091, China
{lhzhou,hmchen}@ynu.edu.cn, {285342456,272270289}@qq.com
[2] Brunel University, Uxbridge UB8 3PH, UK
kevin.lu.brunel@gmail.com

Abstract. Community detection is a very important task in social network analysis. Most existing community detection algorithms are designed for single-relational networks. However, in the real world, social networks are mostly multi-relational. In this paper, we propose a coalition formation game theory-based approach to detect communities in multi-relational social networks. We define the multi-relational communities as the shared communities over multiple single-relational graphs, and model community detection as a coalition formation game process in which actors in a social network are modeled as rational players trying to improve group's utilities by cooperating with other players to form coalitions. Each player is allowed to join multiple coalitions and coalitions with fewer players can merge into a larger coalition as long as the merge operation could improve the utilities of coalitions merged. We then use a *greedy agglomerative manner* to identify communities. Experimental results and performance studies verify the effectiveness of our approach.

Keywords: Social network · Community detection · Coalition formation game · Multi-relational network

1 Introduction

In accordance with the development of network communication, discovering densely connected actor communities from social networks has become an important research issue in social network analysis and has been receiving a great deal of attention in recent years [1,2,3]. Community detection is helpful to improve user-oriented services [4], visualize the structures of networks [5], response emergency and disaster [6], analysis importance of nodes [7], and other applications.

At present, most existing community detection algorithms are designed to find densely connected groups of nodes from only one single graph in which nodes represent actors and edges indicate only a relatively homogenous relationship (such as friendship) amongst actors. However, in the real world, social networks are mostly multi-relational, i.e., persons or institutions are related through various kinds of relationship types [8], such as friendships, kinships, colleague relationships, and hobby

(c) Springer International Publishing Switzerland 2015
J. Li and Y. Sun (Eds.): WAIM 2015, LNCS 9098, pp. 30–41, 2015.
DOI: 10.1007/978-3-319-21042-1_3

relationships. These various kinds of relationship types usually jointly affect people's social activities [9]. One way to model multi-relational social networks is to use multiple single-relational graphs, each reflecting interactions amongst actors in one kind of relationship type. Of course, a multi-relational social network can also be modeled by a union graph that is an aggregate representation of all single-relational graphs. Although existing community detection algorithms for single-relational networks can be used independently in each single-relational graphs or the union graph, the communities obtained may overlook much information that would be crucial to fully understand phenomena taking place in networks. Thus, it is necessary to develop community detection algorithms that are suitable for multi-relational social networks.

Unfortunately, the inclusion of multiple relationship types between nodes complicates the design of network algorithms [10]. In a single-relational network, with all edges being "equal," the executing algorithm just need to be concerned with the existence of an edge in a graph. In a multi-relational social network, on the other hand, the inter-dependencies and feedbacks between multiple relational interactions are significant [11], community detection algorithms must take the existence of an edge in multiple graphs into account in order to obtain meaningful results.

In this paper, we define the multi-relational communities as the components of a partitioning with respect to the set of all nodes in a multi-relational network. This partitioning is a shared community structure underneath different interactions rather than any one of single-relational graphs or the union graph, so it represents a more global structure and can be easily understood. To detect multi-relational communities, we proposed an approach based on coalition formation game theory [12] that studies the cooperative behaviors of groups of rational players and analyze the formation of coalitional structures through players' interaction. A coalition refers to a group of players who cooperate to each other to form coalitions for improving the group's gains, while the gains are limited by the costs for forming these coalitions. In social network environments, behaviors of actors are not independent [13], and joining a community provides one with tremendous benefits, such as members feel rewarded in some way for their participation in the community, and gaining honor and status for being members [14]. Therefore, every actor shall have the incentive to join communities. However, in the real-world scenario each actor not only receives benefit from the communities it belongs to but also needs to pay certain cost to maintain its membership in the communities [15]. As a consequence of this fact, we believe that coalition formation game theories are promising tools that could be used for detecting communities.

In the approach proposed in this paper, we first model the process of multi-relational community detection as a coalition formation game, in which actors in a social network are modeled as rational players trying to achieve and improve group's utilities by cooperating with other players to form coalitions. Coalitions with fewer players can merge into a larger coalition as long as the merge operation can contribute to improve the utilities of coalitions merged. A multi-relational community structure is defined as a collection of stable coalitions, i.e. coalitions that can not further improve its utility by merging with other coalitions. The process of merging coalitions actually illustrates the process of forming the hierarchy structures amongst coalitions. Meanwhile, each player is allowed to join multiple coalitions, which could capture and reflect the concept of "overlapping communities".

Then, we introduce the utility function with respect to a multi-relational network for each coalition that is the combination of a gain function and a cost function. The gain function measures the degree of the interaction amongst the players inside a coalition, while the cost function instead represents the degree of the interaction between the players of the coalition and the rest of the network.

Next, we develop a greedy agglomerative manner to identify multi-relational communities, which starts from the nodes as separate coalitions (singletons); coalitions are iteratively merged to improve group's utilities with respect to a multi-relational network until no pairs of coalitions are merged. This greedy agglomerative manner does not require a priori knowledge on the number and size of the communities, and it matches the real-world scenario, in which communities are formed gradually from bottom to top.

Finally, we conduct experiments on synthetic networks to quantitatively evaluate proposed algorithm. We compare our multi-relational community structure with the community structures from the single-relational graphs. The experimental results validate the effectiveness of our method.

In summarize, the main contributions of this paper are as follows:

- The multi-relational communities are defined as the shared communities underneath different interactions rather than any one of single-relational graphs or the union graph, so they represent a more global structure and can be easily understood.
- A coalition formation game theory-based approach for identifying multi-relational communities is presented, in which a utility function for modeling the benefit and cost of each coalition from forming this coalition is introduced.
- A greedy agglomerative manner is proposed to identify communities. The proposed manner does not require a priori knowledge on the number and size of communities.
- The proposed approach has been implemented. Experimental results and performance studies verify the effectiveness of our approach.

The rest of this paper is organized as follows: section 2 introduces related work; section 3 presents a coalition formation game theory-based approach for detecting multi-relational communities. The experimental results are presented in section 4, and section 5 concludes this paper.

2   Related Work

So far, community detection in single-relational networks has been extensively studied and many approaches have been proposed [16]. Of which, the modularity-based method proposed by Newman and Girvan [17] is a well known algorithm. To optimize the modularity, Blondel et al. [18] designed a hierarchical greedy algorithm, named the *Louvain Method* that allows detecting communities quickly and efficiently with enlightening results in large networks. Aynaud and Guillaume [19] modified the Louvain Method to detect multi-step communities in evolving networks.

To detect communities in multi-relational social networks, Cai et al. [9] proposed a regression-based algorithm to learn the optimal relation weights and then they combined various relations linearly to produce a single-relational network and utilized threshold cut as the optimization objective for community detection. Ströele et al. [8] used clustering techniques with maximum flow measurement to identify the social structure and research communities in a scientific social network with four different relationship types. Wu et al. [4] proposed a co-ranking framework to determine the weights of various relation types and actors simultaneously, and then they combined the probability distributions of relations linearly to produce a single-relational network and presented a Gaussian mixture model with neighbor knowledge to discover overlapping communities. Tang et al. [21] proposed four strategies to integrate the interaction information presented in different relations, and then utilized the spectral clustering to discover communities. A common way adopted by above works is to transform a multi-relational network into a single-relational network, and then use single-relational network algorithms to detect communities.

It is worthy to mention that Lin et al. [22], Zhang et al. [23] and Li et al. [24] also studied community detection problem in multi-relational networks that contain more than one typed entities, such as users, tags, photos, comments, and stories, but the multi-relational networks in this study are limited to contain multiple typed relations but just one typed entities.

Game theories [25,26] have been used to solve community detection problems. Chen et al. [15] considered the community formation as the result of individual actors' rational behaviors and a community structure as equilibrium of a game. In the process of game, each actor is allowed to select one or multiple communities to join or leave based on its own utility measurement. Alvari et al. [27] considered the formation of communities as an iterative game in which each actor is regarded as a player trying to be in the communities with members such that they are structurally equivalent. Lung et al. [28] formulated the community detection problem from a game theory point of view and solved this problem by using a crowding based Differential Evolution algorithm. Hajibagheri et al. [29] used a framework based on Information Diffusion Model and Shapley Value Concept to address the community detection problem. Zhou et al. [30] used the Shapley Value to evaluate contributions of actors to the closeness of connection and a community is defined as a coalition in which there is no one member receives lower Shapley Value than that it receives from other coalitions. Communities can be detected by iteratively increasing the size of groups until further increase leads to lower Shapley Values for the members. However, all these methods are developed for single-relational networks.

3   A Coalition Formation Game Theory-Based Approach for Detecting Communities in Multi-relational Social Networks

In this paper, a multi-relational network is represented by multiple single-relational graphs, each of them representing the interactions amongst actors in a given relationship. To identify communities in multi-relational social networks, we propose a coalition formation game theory-based approach in which a coalition refers to a group of actors who cooperate to each other to form coalitions for improving the group's gains, while the gains are limited by the costs for forming these coalitions. Coalitions with fewer players can merge into a larger coalition as long as the merge operation can contribute to improve the utilities of coalitions merged. The cost makes coalition formation games are generally not superadditive, hence the grand coalition (the coalition of all players) is seldom the optimal structure. The process of merging coalitions actually illustrates the process of forming the hierarchy structures amongst communities. Meanwhile, each player is allowed to join multiple coalitions, which could capture and reflect the concept of "overlapping communities".

Given a multi-relational social network, the objective of detecting communities is to detect and identify coalitions over multiple single-relational graphs such that these coalitions can not further improve their utilities with respect to the multi-relational network by merging with other coalitions.

In the next, we first present relevant notations and definitions, and then we introduce the algorithm for detecting multi-relational communities.

3.1   Definitions

Let the multi-relational network G on a set of relations $R = \{1, 2, ..., r\}$ be $G = \{G_1, G_2, ..., G_r\}$, where r be the number of relationship types, $G_i = (V_i, E_i)$ be an undirected graph with $|V_i|$ nodes (actors) and $|E_i|$ edges (interactions), representing the interactions amongst actors in the relation $i (i \in R)$, i be the relation index. Let A_i be an adjacency matrix of G_i with $A_i(x, y) = 1$ if $(x, y) \in E_i$ for any pair of nodes $x, y \in V_i$ and 0 otherwise in the relation $i (i \in R)$. Let $V = \bigcup_{i \in R} V_i$, the set of all nodes in the multi-relational network G.

Let S_k denote a subset of V ($S_k \subseteq V$) and S_k be called a coalition, meanwhile let Γ denote a coalition structure (a collection of coalitions), i.e. $\Gamma = \{S_1, S_2, ..., S_s\}$. Let $e_i(S_k)$ be the number of edges amongst nodes inside S_k and $d_i(S_k)$ be the sum of degree of nodes inside S_k in $G_i = (V_i, E_i)$, i.e.

$$e_i(S_k) = \frac{1}{2} \sum_{x, y \in S_k, V_i} A_i(x, y) \quad , \quad d_i(S_k) = \sum_{x \in S_k, V_i} \sum_{y \in V_i} A_i(x, y) \quad . \quad \text{For any coalition}$$

$S_k, S_l \subseteq V$, let $e_i(S_k, S_l)$ be the number of edges connecting nodes of the coalition S_k to the nodes of the coalition S_l in $G_i = (V_i, E_i)$. Let S_{kl} be a super-coalition of S_k and S_l in a merge operation of S_k with S_l.

Depending on the context, an element in V may either be called as an actor, player or a node, and a subset of V may either be called as a group or a coalition.

Definition 1. Utility function of a coalition with respect to a single-relational network. Let S_k be a coalition of Γ, then the utility function of S_k with respect to $G_i = (V_i, E_i)$, $v_i(S_k)$, is defined by the following Equation (1):

$$v_i(S_k) = \frac{2e_i(S_k)}{d_i(S_k)} - \frac{1}{\sqrt{|E_i|}} \left(\frac{d_i(S_k)}{2|E_i|} \right)^2. \tag{1}$$

In Equation (1), the first term and the second term are called as the gain function and the cost function of S_k with respect to $G_i = (V_i, E_i)$, respectively. The gain function is the ratio of edges inside S_k over the total degree of the nodes inside S_k in $G_i = (V_i, E_i)$; the cost function instead represents the ratio of the total degree in S_k over the total degree in the network $G_i = (V_i, E_i)$. The larger gain function value means that there are more interactions amongst the actors inside S_k in $G_i = (V_i, E_i)$, while the larger cost function value means that there are more interaction between the actors of the coalition S_k and the rest of the network $G_i = (V_i, E_i)$. Eq. (1) means that forming a coalition brings gains to its members, but the gains are limited by a cost for forming the coalition in $G_i = (V_i, E_i)$.

Definition 2. Utility function of a coalition with respect to a multi-relational network. Let S_k be a coalition of Γ, $v_i(S_k)$ the utility function of S_k with respect to $G_i = (V_i, E_i)$, $i = 1, 2, ..., r$, then the utility function of S_k with respect to G, $v(S_k)$, is defined by the Equation (2).

$$v(S_k) = \sum_{i=1}^{r} v_i(S_k) = \sum_{i=1}^{r} \frac{2e_i(S_k)}{d_i(S_k)} - \sum_{i=1}^{r} \frac{1}{\sqrt{|E_i|}} \left(\frac{d_i(S_k)}{2|E_i|} \right)^2. \tag{2}$$

The first term and the second term in Equation (2) are called as the gain function and the cost function of S_k with respect to G respectively. They are the sum of the gain function and the cost function of S_k with respect to $G_i = (V_i, E_i)$, $i = 1, 2, ..., r$. $v_i(S_k)$ is defined on a single-relational graph only considering the nodes and edges in this single-relational graph, while the $v(S_k)$ is defined on multiple single-relational graphs.

Definition 3. Stable coalition. A coalition S_k is regarded as a stable coalition if S_k can not further improve its utility by merging with other coalitions, i.e. $\forall S_l \neq S_k$, $v(S_k + S_l) < v(S_k)$ and $\forall S_l \subseteq S_k, v(S_k) > v(S_l)$. Specially, S_k is called as the grand coalition [12] if $S_k = V$, i.e, the coalition of all the actors, while S_k is called as a trivial coalition if S_k just consist of a single node, i.e, $S_k = \{x\}, x \in V$.

Definition 4. Increment of utility of a coalition. Let $S_{kl} = S_k + S_l$, a super coalition obtained by merging coalition S_k with S_l, then the increment of utility of coalition S_k with respect to S_{kl} is defined by $\Delta v(S_k, S_{kl}) = v(S_{kl}) - v(S_k)$.

There are two conditions must be satisfied for merging S_k and S_l, i.e. $\exists e_i(S_k, S_l) \neq 0$, $i = 1,2,...,r$, and $\Delta v(S_k, S_{kl}) > 0$ & $\Delta v(S_l, S_{kl}) > 0$. The condition $\exists e_i(S_k, S_l) \neq 0$, $i = 1,2,...,r$ implies that two coalitions without an edge between them in $G = \{G_1, G_2,..., G_r\}$ can not merge into a larger coalition. This is natural because this merge operation cannot contribute to improve the closeness of connection. So, whether a coalition is merged with others can be decided by looking only at its neighbors (coalitions that have links between them), without an exhaustive search over the entire network. The condition $\Delta v(S_k, S_{kl}) > 0$ & $\Delta v(S_l, S_{kl}) > 0$ implies that the utilities of S_k and S_l should be improved through the merge operation. It ensures that a coalition formed by a merge operation has greater utility than that of its subsets. The unilateral meet of two inequalities shows that two coalitions fail to reach an agreement to cooperate, for example, the case of $\Delta v(S_k, S_{kl}) > 0$ but $\Delta v(S_l, S_{kl}) < 0$ suggests that S_k intends to cooperate with S_l but S_l may not agree.

Definition 5. Stable coalition structure. A collection of coalitions $\Gamma = \{S_1, S_2,..., S_s\}$ is regarded as a stable coalition structure if $\forall S_k \in \Gamma, \max(\max_{S_{kl}} \Delta v(S_k, S_{kl}), 0) = 0$ holds.

A stable coalition structure can be regarded as a kind of equilibrium state of coalitions, in which no group of individuals has an interest in performing a merge operation any further. In this paper, a stable coalition structure Γ is called as a multi-relational community structure of G, and a coalition of Γ is called as a multi-relational community of G.

Given a multi-relational network $G = \{G_1, G_2,..., G_r\}$, detecting multi-relational communities means finding a stable coalition structure Γ in which no group of actors has an interest in performing a merge operation any further.

3.2   An Algorithm for Detecting Multi-relational Communities

In this study, we develop a greedy agglomerative manner to identify multi-relational communities. The main idea of the greedy agglomerative manner is to start from the nodes as separate coalitions (singletons), coalitions that can result the highest utility increment are iteratively merged into a larger coalition to improve groups' utilities until no such merge operation can be performed any further. Now, the number of coalitions represents the number of communities in G. Hence, the number and size of the communities are obtained automatically rather than specified in advance. The pseudo-code for the greedy agglomerative algorithm is given in *MRgame* algorithm.

MRgame algorithm:

Input: a multi-relational network $G = \{G_1, G_2, ..., G_r\}$

Output: the community structure of G

Variables:

CoaSetold : The set of coalitions before merge operation

CoaSetnew : The set of coalitions after merge operation

CooSetMap : The map of *CoaSetold* , i.e. the copy of *CoaSetold*

CooSps : A cooperative sponsor, i.e. a coalition with maximal utility in *CoaSetold*

CooCaSet : The set of cooperative candidates, i.e. coalitions that there is at least a link between the coalition and *CooSps*

CooCas* : A best cooperative candidate, i.e. such a coalition in *CooCaSet* that the cooperation of the coalition with *CooSps* can bring about maximal increment of utility

Steps:

1. $CoaSetnew = \{\{1\}, \{2\}, ..., \{|V|\}\}$;

2. Do

3. $CoaSetold = CoaSetnew$;

4. $CoaSetMap = CoaSetold$;

5. $CoaSetnew = \phi$;

6.   while $CoaSetMap \neq \phi$

7. $CooSps = \arg \max\limits_{S \in CoaSetMap} v(S)$;

8. $CoaSetMap = CoaSetMap - \{CooSps\}$;

9. $CooCaSet(CooSps) = \{S \mid \exists e_i(CooSps, S) \neq 0, S \in CoaSetold, i = 1, 2, ..., r\}$;

10.   while $CooCaSet(CooSps) \neq \phi$

11. $CooCas^* = \arg \max\limits_{CooCas \in CooCaSet(CooSps)} v(CooSps + CooCas)$;

12.   if $v(CooSps + CooCas^*) > v(CooSps)$ &

                   $v(CooSps + CooCas^*) > v(CooCas^*)$

13. $CooSps = CooSps + CooCas^*$;

14. $CoaSetMap = CoaSetMap - \{CooCas^*\}$;

15. $CoaCaSet(CooSps) = CoaCaSet(CooSps) - \{CooCas^*\}$;

                   $+ (CooCaSet(CooCas^*) - \{CooSps\})$

16.   else $CoaCaSet(CooSps) = CoaCaSet(CooSps) - \{CooCas^*\}$;

17.   end if

18.   end while

19. $CoaSetnew = CoaSetnew + \{CooSps\}$;

20.   end while

21. while $CoaSetnew \neq CoaSetold$

22. Output $CoaSetnew$

In a single graph, the time complexity of *MRgame* algorithm is $O(|V|\log|V|)$ at worst case. Note that, $|V|-1$ iterations are an upper bound and the algorithm will terminate as soon as no pair of coalitions would be merged any further. It is possible that the algorithm ends up before the grand coalition forms. But with respect to a multi-relational network, *MRgame* searches neighbors for each node over multiple single-relational graphs, thus the time complexity of *MRgame* algorithm is $O(|r \times V|\log|r \times V|)$ at worst case.

4   Experimental Validation

In this section, we demonstrate the effectiveness of our approach for detecting multi-relational communities in multi-relational social networks. Typically, a real-world network does not provide the ground truth information about the membership of actors, so we resort to purpose-built synthetic data to conduct some controlled experiments. The synthetic network contains $|V|$ nodes which roughly form s communities containing $|S_1|,...,|S_s|$ nodes respectively. Furthermore, r different relations are constructed to look at the clustering structure in different angles. For each type of relations, nodes with same membership connect randomly with each other following a intra-group interaction probability p_{in}, while nodes with different membership connect randomly with each other following a inter-group interaction probability p_{out}. The interaction probability p_{in} (p_{out}) may differ with respect to groups at distinct relations. For the sake of simplicity, we set same p_{in} (p_{out}) for all groups. Figure 1 (a)~(c) shows one example of the generated 3-relational network under $|V|=100$, $s=3$, $p_{in}=0.7$, $p_{out}=0.1$, where $S_1=\{1,...,33\}, S_2=\{34,...,66\}$, $S_3=\{34,...,100\}$. Figure 1 (d) (union graph) is a single relation network obtained by enumerating all edges of G_1, G_2 and G_3. Clearly, different dimensions demonstrate different interaction patterns, and the interactions amongst nodes in the union graph become denser.

Since a priori community structure (i.e. ground truth $\{S_1, S_2, S_3\}$) are known, we then adopt *Normalized mutual information (NMI)* [31,32] to evaluate quantitatively the performance of the *MRgame* algorithm. Figure 2 (a) shows the overall comparison on the synthetic dataset of Figure 1. In Figure 2 (a), the aggregated bars on G_1 to G_3 correspond to the results on 3 different relationship types, "Union-graph" corresponds to the result on the union graph, and "Multi-graphs" corresponds to the result over multiple single-relational graphs. As can be seen, *NMI* on the multiple single-relational graphs outperforms any of single relations and the union graph. That implies that community structure over multiple single-relational graphs is clearer than that on any original single-relational graphs and the union graph.

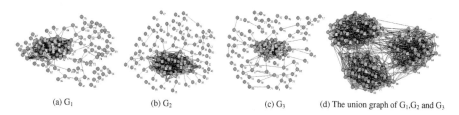

(a) G_1 (b) G_2 (c) G_3 (d) The union graph of G_1,G_2 and G_3

Fig. 1. A 3-relational networks and their union graph

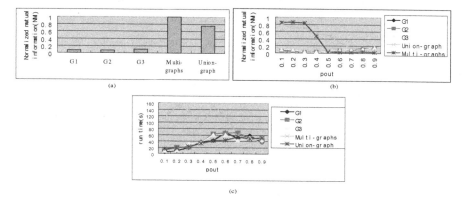

Fig. 2. (a) The *NMI* values between the community structures detected by *MRgame* and the structure of $\{S_1,S_2,S_3\}$ in different graphs, where p_{in}=0.7, p_{out}=0.1; (b) *NMI* values under $|V|$=1000, s=3, p_{in}=0.9 and different p_{out}; (c) the run times of the *MRgame* algorithm under $|V|$=1000, s=3, p_{in}=0.9 and different p_{out}.

Figure 2 (b) presents the *NMI* values between the priori community structures and the community structures detected by *MRgame* in different networks produced under $|V| = 1000$, $s = 3$, $p_{in} = 0.9$ and different p_{out}. As can be seen, *NMI* on the multiple single-relational graphs outperforms any of single relations and the union graph when $p_{out} < 0.4$. Also, *NMIs* decrease with the increments of p_{out}. Results in networks under other parameters, such as different p_{in}, are omitted by the limitation of space.

Figure 2 (c) presents the run times of the *MRgame* under the same parameters with Figure 2 (b). As can be seen that *MRgame* runs more times in multi-graphs than in three single-relational graphs and the union graph when $p_{out} < 0.7$, the reason for this is that more communities are detected and many nodes are overlapping when $p_{out} < 0.7$.

Acknowledgments. This work is supported by the National Natural Science Foundation of China under Grant No.61262069, No. 61472346, Program for Young and Middle-aged Skeleton Teachers, Yunnan University, and Program for Innovation Research Team in Yunnan University (Grant No. XT412011).

References

1. Francesco, F., Clara, P.: An evolutionary multiobjective approach for community discovery in dynamic networks. IEEE Transactions on Knowledge and Data Engineering **26**(8), 1838–1852 (2014)
2. Zhou, L., Lü, K.: Detecting communities with different sizes for social network analysis. The Computer Journal (2014). doi:10.1093/comjnl/bxu087
3. Xin, Y., Yang, J., Xie, Z.Q.: A semantic overlapping community detection algorithm based on field sampling. Expert Systems with Applications **42**, 366–375 (2015)
4. Yuan, W., Guan, D., Lee, Y.-K., Lee, S., Hur, S.J.: Improved trust-aware recommender system using small-worldness of trust networks. Knowledge-Based Systems **23**(3), 232–238 (2010)
5. Wu, P., Li, S.K.: Social network analysis layout algorithm under ontology model. Journal of Software **6**(7), 1321–1328 (2011)
6. Wang, D., Lin, Y.-R., Bagrow, J.P.: Social networks in emergency response. In: Alhajj, R., Rokne, J. (eds.) Encyclopedia of Social Network Analysis and Mining, vol. 1, pp. 1904–1914 (2014)
7. Li, G.P., Pan, Z.S., Xiao, B., Huang, L.W.: Community discovery and importance analysis in social network. Intelligent Data Analysis **18**(3), 495–510 (2014)
8. Ströele, V., Zimbrão, G., Souza, J.M.: Group and link analysis of multi-relational scientific social networks. Journal of Systems and Software **86**(7), 1819–1830 (2013)
9. Cai, D., Shao, Z., He, X., Yan, X., Han, J.: Community mining from multi-relational networks. In: Jorge, A.M., Torgo, L., Brazdil, P.B., Camacho, R., Gama, J. (eds.) PKDD 2005. LNCS (LNAI), vol. 3721, pp. 445–452. Springer, Heidelberg (2005)
10. Rodriguez, M., Shinavier, J.: Exposing multi-relational networks to single relational network analysis algorithms. Journal of Informetrics **4**(1), 29–41 (2010)
11. Szell, M., Lambiotte, R., Thurner, S.: Multirelational organization of large-scale social networks in an online world. Proceedings of the National Academy of Sciences of the United States of America **107**(31), 13636–13641 (2010)
12. Saad, W., Han, Z., Debbah, M., Hjørungnes, A., Basar, T.: Coalitional game theory for communication networks: a tutorial. IEEE Signal Processing Magazine **26**(5), 77–97 (2009)
13. Zacharias, G.L., MacMillan, J., Hemel, S.B.V. (eds.): Behavioral modeling and simulation: from individuals to societies. The National Academies Press, Washington, DC (2008)
14. Sarason, S.B.: The Psychological Sense of Community: Prospects for a Community Psychology. Jossey-Bass, San Francisco (1974)
15. Chen, W., Liu, Z., Sun, X., Wang, Y.: A Game-theoretic framework to identify overlapping communities in social networks. Data Mining and Knowledge Discovery **21**(2), 224–240 (2010)
16. Fortunato, S.: Community detection in graphs. Physics Reports **486**, 75–174 (2010)
17. Newman, M.E.J., Girvan, M.: Finding and evaluating community structure in networks. Physical Review E **69**, 026113 (2004)
18. Blondel, V.D., Guillaume, J.L., Lambiotte, R., Lefebvre, E.: Fast unfolding of communities in large networks. Journal of Statistical Mechanics: Theory and Experiment **10**, P10008 (2008)
19. Aynaud, T., Guillaume J.-L.: Multi-step community detection and hierarchical time segmentation in evolving networks. In: Proceedings of the fifth SNA-KDD Workshop on Social Network Mining and Analysis, in conjunction with the 17th ACM SIGKDD International Conference on Knowledge Discovery and Data Mining (KDD 2011), San Diego, CA, pp. 21–24, August 2011

20. Wu, Z., Yin, W., Cao, J., Xu, G., Cuzzocrea, A.: Community detection in multi-relational social networks. In: Lin, X., Manolopoulos, Y., Srivastava, D., Huang, G. (eds.) WISE 2013, Part II. LNCS, vol. 8181, pp. 43–56. Springer, Heidelberg (2013)

21. Tang, L., Wang, X., Liu, H.: Community detection via heterogeneous interaction analysis. Data Mining Knowledge Discovery 25(1), 1–33 (2012)

22. Lin, Y.-R., Choudhury, M.D., Sundaram, H., Kelliher, A.: Discovering Multi-Relational Structure in Social Media Streams. ACM Transactions on Multimedia Computing, Communications and Applications 8(1), 1–28 (2012)

23. Zhang, Z., Li, Q., Zeng, D., Gao, H.: User community discovery from multi-relational networks. Decision Support Systems 54(2), 870–879 (2013)

24. Li, X.T., Ng, M.K., Ye, Y.M.: MultiComm: finding community structure in multi-dimensional networks. IEEE Transactions on Knowledge and Data Engineering 26(4), 929–941 (2014)

25. Nash, J.F.: Non-cooperative games. Annals of Mathematics 54(2), 286–295 (1951)

26. Zlotkin, G., Rosenschein J.: Coalition cryptography and stability mechanisms for coalition formation in task oriented domains. In: Proceedings of The Twelfth National Conference on Artificial Intelligence, Seattle, Washington, August 1–4, pp. 432–437. The AAAI Press, Menlo Park (1994)

27. Alvari, H., Hashemi, S., Hamzeh, A.: Detecting overlapping communities in social networks by game theory and structural equivalence concept. In: Deng, H., Miao, D., Lei, J., Wang, F.L. (eds.) AICI 2011, Part II. LNCS, vol. 7003, pp. 620–630. Springer, Heidelberg (2011)

28. Lung, R.I., Gog, A., Chira, C.: A game theoretic approach to community detection in social networks. In: Pelta, D.A., Krasnogor, N., Dumitrescu, D., Chira, C., Lung, R. (eds.) NICSO 2011. SCI, vol. 387, pp. 121–131. Springer, Heidelberg (2011)

29. Hajibagheri, A., Alvari, H., Hamzeh, A., Hashemi, A.: Social networks community detection using the shapley value. In: 16th CSI International Symposium on Artificial Intelligence and Signal Processing (AISwww.lw20.comP), Shiraz, Iran, May 2–3, pp. 222–227 (2012)

30. Zhou, L., Cheng, C., Lü, K., Chen, H.: Using coalitional games to detect communities in social networks. In: Wang, J., Xiong, H., Ishikawa, Y., Xu, J., Zhou, J. (eds.) WAIM 2013. LNCS, vol. 7923, pp. 326–331. Springer, Heidelberg (2013)

31. Danon, L.: Danony, Díaz-Guilera, A., Duch, J., Arenas, A.: Comparing community structure identification. Journal of Statistical Mechanics: Theory and Experiment 9, P09008 (2005)

32. Lancichinetti, A., Fortunato, S.: Benchmarks for testing community detection algorithms on directed and weighted graphs with overlapping communities. Physical Review E 80(1), 16118 (2009)

Individual Influence Maximization via Link Recommendation

Guowei Ma[1], Qi Liu[1]([✉]), Enhong Chen[1], and Biao Xiang[2]

[1] University of Science and Technology of China, Hefei, Anhui, China
gwma@mail.ustc.edu.cn, {qiliuql,cheneh}@ustc.edu.cn
[2] Microsoft, Search Technology Center Asia (STCA), Beijing, China
bixian@microsoft.com

Abstract. Recent years have witnessed the increasing interest in exploiting social influence in social networks for many applications, such as viral marketing. Most of the existing research focused on identifying a subset of influential individuals with the maximum influence spread. However, in the real-world scenarios, many individuals also care about the influence of herself and want to improve it. In this paper, we consider such a problem that maximizing a target individual's influence by recommending new links. Specifically, if a given individual/node makes new links with our recommended nodes then she will get the maximum influence gain. Along this line, we formulate this link recommendation problem as an optimization problem and propose the corresponding objective function. As it is intractable to obtain the optimal solution, we propose greedy algorithms with a performance guarantee by exploiting the submodular property. Furthermore, we study the optimization problem under a specific influence propagation model (i.e., Linear model) and propose a much faster algorithm (*uBound*), which can handle large scale networks without sacrificing accuracy. Finally, the experimental results validate the effectiveness and efficiency of our proposed algorithms.

1 Introduction

Social network platforms, such as Twitter and Facebook, play an important and fundamental role for the spread of influence, information, or innovations. These diffusion processes are useful in a number of real-world applications, for instance, the social influence propagation phenomenon could be exploited for better viral marketing [1]. To this end, both modeling the influence propagation process and identifying the influential individuals/nodes in social networks have been hot topics in recent years [2].

Indeed, researchers have proposed several influence models to describe the dynamic of influence propagation process, such as Independent Cascade (IC) model [3], Linear Threshold (LT) model [4], a stochastic information flow model [5] and the linear social influence model (Linear) [6]. Meanwhile, other researchers focus on learning the real or reasonable influence propagation probability between two individuals in the influence models [7,8]. Based on the influence propagation models and the influence propagation probabilities, influence

© Springer International Publishing Switzerland 2015
J. Li and Y. Sun (Eds.): WAIM 2015, LNCS 9098, pp. 42–56, 2015.
DOI: 10.1007/978-3-319-21042-1_4

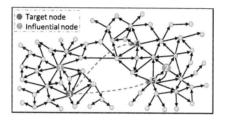

Fig. 1. A toy example

maximization (IM) is the problem of identifying a subset with K influential nodes in a social network so that their aggregated influence in the network is maximized. Since influence maximization is a fundamental problem in viral marketing, various aspects of it have been studied extensively in the last decade [2,9,10]. For instance, Eftekhar et al. [10] studied influence propagation at group scale, where they aimed at identifying the influential groups instead of a subset of individuals.

However, most of existing works about IM focus on identifying a small subset of individuals or groups in a network so that their aggregated influence spread is maximized. In the real-world scenarios, an individual also cares about her own influence and wants to improve it by making new links. Formally, if a given target node (e.g., a person or a company) in a social network wants to maximize its influence by making several new links (i.e., target node could spread its influence through these links), which nodes should it link with? In this situation, linking it with the most influential nodes or the nodes with largest degree, may not lead to the maximum influence gain, since we have to consider the topology of the target node and the overlap of influence spread between the target node and the selected nodes. Let us take the network in Fig.1 as a toy example. If node 3 is the given target node and we want to improve its influence by recommending two new links for node 3. Suppose that node set $\{1, 2\}$ are the most influential node set found by IM method (e.g, by CELF [11]). Actually, the total influence of node 3 after linking with nodes $\{1, 2\}$ is less than that with nodes $\{4, 5\}$. The reason may be that there is much overlap of the influence spread between nodes 3 and 1.

Though similar link recommendation problems have been studied in the literature (e.g., adding new links or strengthening the weaken social links to boost the information spread across the entire network [12]), the problem of eliminating the influence overlap to maximize the target node's influence via link recommendation remains pretty much open. Actually, there are two challenges to solve this problem efficiently: First, how to design a rational measure to eliminate the influence overlap between nodes; Second, because the computation of influence spread is very time-consuming, it is urgent to propose an efficient algorithm which can sharply reduce the times of influence spread estimations. To address these challenges, in this paper, we provide a focused study on the problem of maximizing the target individual's own influence by recommending new links for this individual (i.e., individual influence maximization via link recommendation). Our contributions could be summarized as follows:

- We formulate this individual influence maximization-oriented link recommendation problem as an optimization problem, and define the corresponding objective function, which can be generally applied to different influence propagation models.
- To solve the intractable optimization problem effectively, we propose a *greedy* algorithm with a performance guarantee. One step further, we present another algorithm *lazy* for scaling up this simple *greedy*. Both algorithms can be used in general influence models, such as IC and LT.
- We leverage the properties of influence spread estimations under the specific Linear model, and propose a much faster recommendation algorithm *uBound*, which can handle large scale networks without sacrificing accuracy.
- We conduct extensive experiments on four real world datasets and the results demonstrate the effectiveness and efficiency of our proposed algorithms.

2   Related Work

Influence Propagation and Maximization. Researches proposed several models for describing the influence propagation process. IC model [3] and LT model [4] are two widely used ones. However, both of them require time-consuming Monto Carlo simulations to estimate the influence spread, some researchers designed more efficient (or tractable) models, e.g., the stochastic information flow model [5] and the linear social influence (Linear) model [6]. Since learning the influence propagation process is beyond the scope of this paper, we use these existing influence models for illustration.

The IM problem can be traced back to Domingos and Richardson [1,13]. Kempe et al. [14] first formulated it as a discrete optimization problem, demonstrated it as NP-hard and presented a greedy approximation algorithm with provable performance guarantee. From then on, researchers proposed many computationally efficient algorithms, such as CELF [11], PMIA [15], IPA [16] and TIM [17], by exploiting specific aspects of the graph structure or the influence model. Some researchers also consider other aspects of the IM problem [9,18]. For instance, Guo et al. [18] studied local influence maximization, aiming to find the top-K local influential nodes on the target node. However, to the best of our knowledge, few attention has been paid to the problem of maximizing the target node's own influence via link recommendation.

Recommendations in Social Networks. The user-to-user recommendation in social networks is an important task for many social network sites like Twitter, Google+ and Facebook, for the purpose of guiding user discover potential friends [19–21] or improving the connectivity of the network [12,22]. Researchers have proposed a number of recommendation algorithms to recommend potential friends to users in a social platform, such as the Friend-of-Friend(FoF) algorithm [19] and other interest-based or profile-based algorithms [20,21]. Some of these works also consider the influence propagation effect, such as selecting a set of "influential" users for a new user [23] or a new product [24], like solving the cold-start problem in recommender systems.

In addition, some works in the area of *network/graph augmentation* also try to add links in the network for improving some quality of the graph [12,22]. For instance, Tian et al. [22] suggested users to re-connect their old friends and strengthen the existing weak social ties in order to improve the social network connectivity; Chaoji et al. [12] recommended an edge set in order to increase the connectivity among users and boost the content spread in the entire social network. However, these related works pay more attention to the entire social network rather than the target individual's own influence.

3   Individual Influence Maximization

Preliminaries. Let the directed graph $G(V, E, T)$ represents an influence network, where $V = \{1, 2, ..., n\}$ are n nodes in graph and E stores all the influence links(edges) between nodes. $T = [t_{ij}]_{n*n}$ is a given propagation probability matrix. For each edge $(j, i) \in E$, $t_{ij} \in (0, 1)$ denotes the influence propagation probability from node i to node j. For any edge $(j, i) \notin E$, $t_{ij} = 0$. G is assumed to be directed as influence propagation is directed in the most general case. Given this graph, the influence spread $\mathbf{f}_i$ for each node $i \in V$ can be computed by the influence propagation models (e.g., IC [3], LT [4] and Linear [6]). Specifically, $\mathbf{f}_i = [f_{i \to 1}, f_{i \to 2}, ..., f_{i \to n}]'$, an $n \times 1$ vector, denotes the influence of node i on each node in the network. Thus, the total influence spread of node i in network equals to the sum of influence of node i on other nodes, namely $f_{i \to V} = \sum_{j \in V} f_{i \to j}$. Indeed, $f_{i \to V}$ is the expected number of the nodes that will be influenced by node i.

3.1   Problem Statement and Formulation

In a real-world network, such as Twitter, nodes represent users, and edges represent their links/connections. If a target user wants to improve her own influence, she should make new influence links[1] with other users, especially the influential ones, then the information she posts will be read and followed by more users (e.g., by retweet). Since making new links with other nodes may require money or time, we also associate a nonnegative cost $c(j)$ with each node j. That is, the cost of linking to node j is $c(j)$ if a target node makes a new link with j. The less the cost is, the easier to create the link for the target node. We denote the total cost of the target node for making new links with a subset $\mathbf{S}$ as $c(\mathbf{S}) = \sum_{j \in \mathbf{S}} c(j)$. Hence, the problem of individual influence maximization is to find a subset $\mathbf{S}$ such that if the target node t makes new links with nodes in

---

[1] In this paper, we use the expressions of "influence link "and "link" without distinction.

S, the t's influence gain is maximum, and $c(\mathbf{S})$ does not exceed a specific budget B. Now this problem could be formulated as an optimization problem:

$$\arg\max_{\mathbf{S}} \{f^{\mathbf{S}}_{t\to V} - f_{t\to V}\} \text{ , subject to } c(\mathbf{S}) \leq B, \tag{1}$$

where $f^{\mathbf{S}}_{t\to V} - f_{t\to V}$ is the influence gain of the target node t after linking with nodes in set **S**. Notice that, we assume that the other parts of network structure stay unchanged before t makes links with the nodes in **S**. To reduce complexity, in this paper, we consider $c(j) = 1$ for each $j \in V$, i.e., every new link shares the same cost. Hence, the cost $c(\mathbf{S})$ equals to the number of nodes in **S**, namely $c(\mathbf{S}) = |\mathbf{S}|$. Let $\mathcal{F}(\mathbf{S}) = \{f^{\mathbf{S}}_{t\to V} - f_{t\to V}\}$ and $K = B$, we can rewrite Eq. (1) as below.

$$\arg\max_{\mathbf{S}} \mathcal{F}(\mathbf{S}) = \{f^{\mathbf{S}}_{t\to V} - f_{t\to V}\} \text{ , s.t. } |\mathbf{S}| \leq K. \tag{2}$$

In summary, the individual influence maximization problem is formalized as recommending a subset **S** with K nodes such that node t can achieve the maximum influence gain by making new links with the nodes in **S** (i.e, adding new edges $(j, t), j \in \mathbf{S}$).

3.2   Definition of the Objective Function

The key of the above optimization problem is to design an appropriate objective function $\mathcal{F}(\mathbf{S})$ to eliminate the influence overlap (the first challenge given in Introduction) when adding **S** to link the target individual. For introducing our definition of $\mathcal{F}(\mathbf{S})$, we start with a single link from node t to node c.

Definition 1. *If a target node t makes a new link with a candidate node c, we define $\mathcal{F}(\mathbf{S}) = \mathcal{F}(\{c\})$ as :*

$$\begin{aligned}
\mathcal{F}(\{c\}) &= f^{\{c\}}_{t\to V} - f_{t\to V} \\
&= \lambda_c \cdot (1 - f_{t\to c}) \cdot \sum_{i\in V}(1 - f_{t\to i})f_{c\to i},
\end{aligned}$$

where $\lambda_c \in (0, 1)$ is a hyper parameter to model the real-world social influence propagation process.

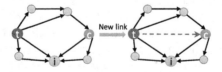

Fig. 2. A Simple Example

Definition Explanation. Let us take a simple example. Suppose we want to improve the target node t's influence in Fig. 2, and thus we should estimate the total influence gain of t after making a new link (the red dashed line) with a candidate node c. We first show how to estimate the influence gain of t on any node $i \in V$. Before making the new link with c (Left part of Fig. 2), the target t has an influence on node i ($f_{t \to i}$) and node c ($f_{t \to c}$) respectively, and meanwhile, the node c also has an influence on node i ($f_{c \to i}$). When t makes a new link with c (Right part of Fig. 2), we define that the influence of t on node c has increased by $\lambda_c(1 - f_{t \to c})$. Now, let's explain this definition: Suppose t always influences (or actives) c successfully ($f_{t \to c}^{\{c\}} = 1$), then the influence of t on node c will be increased by ($f_{t \to c}^{\{c\}} - f_{t \to c}) = (1 - f_{t \to c})$. However, this assumption is a little unrealistic. Thus, for better modeling the real-world influence propagation process, we introduce $\lambda_c \in (0, 1)$ to weaken the influence gain of node t on c, and get $\lambda_c(1 - f_{t \to c})$. [2] One step further, we can represent that the influence of t on i through c has increased by $\lambda_c(1 - f_{t \to c})f_{c \to i}$. Hence, the influence gain of t on i after making a link with c is: $f_{t \to i}^{\{c\}} - f_{t \to i} = (1 - f_{t \to i}) \cdot \lambda_c \cdot (1 - f_{t \to c}) \cdot f_{c \to i}$. Then, we can get the total influence gain of node t on the entire network: $f_{t \to V}^{\{c\}} - f_{t \to V} = \sum_{i \in V}(f_{t \to i}^{\{c\}} - f_{t \to i}) = \sum_{i \in V}(1 - f_{t \to i}) \cdot \lambda_c(1 - f_{t \to c})f_{c \to i} = \lambda_c(1 - f_{t \to c})\sum_{i \in V}(1 - f_{t \to i})f_{c \to i}$.

From this example, we could get the implication of the Definition 1. Though we do not show more rigorous justification for this function, the extensive experimental results show that the nodes selected by this function can really obtain much real influence gain of a given target node, which illustrate that this function is rational and effective.

One step further, we introduce the following definition of the objective function $\mathcal{F}(\mathbf{S})$, i.e, the influence gain of a target node when it makes new links with nodes in $\mathbf{S}$.

Definition 2. *If a target node t makes new links with the nodes in $\mathbf{S}$, we define $\mathcal{F}(\mathbf{S})$:*

$$\mathcal{F}(\mathbf{S}) = f_{t \to V}^{\mathbf{S}} - f_{t \to V} = \sum_{c \in \mathbf{S}} \mathcal{F}(\{c\})$$

$$= \sum_{c \in \mathbf{S}} \lambda_c(1 - f_{t \to c}) \cdot \sum_{i=1}^{n}(1 - f_{t \to i})f_{c \to i}$$

We could demonstrate that the function $\mathcal{F}(\mathbf{S})$ satisfies the properties below:

1. $\mathcal{F}(\emptyset) = 0$, i.e., we cannot improve the influence of the target node without making any new link.
2. $\mathcal{F}(\mathbf{S})$ is nonnegative and monotonically increasing. It is obvious that making new links can not reduce the influence of the target node.
3. $\mathcal{F}(\mathbf{S})$ is *submodular*. That is, $\mathcal{F}(\mathbf{S})$ satisfies the "diminishing returns" property.

---

[2] Notice that, in real-world applications, λ_c's value could be determined based on the specific influence models and the characteristics of nodes c and t in the social network.

3.3 Greedy Strategy

Indeed, maximizing submodular function in general is NP-hard [25], and thus it is intractable to obtain the optimal solution of the problem we formulated. However, for a nonnegative monotone submodular function, such as $\mathcal{F}(\mathbf{S})$, the greedy strategy, a common used heuristic, approximates the optimum within a factor of $(1 - 1/e)$ [26].

The simple *greedy* algorithm starts with the empty set $\mathbf{S} = \emptyset$, and requires about n times influence spread estimation in each iteration to select one node (with the maximum influence margin) to join $\mathbf{S}$. Thus, *greedy* requires about $(n \cdot K)$ times influence spread estimations, where $K = |S|$. As each influence spread estimation calculated by influence models (e.g., IC, LT) is very time-consuming, *greedy* is quite slow.

Scaling Up. Here, we exploit the submodularity of $\mathcal{F}(\mathbf{S})$ and adopt the lazy-forward strategy [11] for scaling up the simple algorithm *greedy* . Specifically, based on the fact that the influence gain of node t after making a link with node c in the current iteration cannot be larger than its marginal influence gain in the previous iteration, we propose the algorithm *lazy* without sacrificing any accuracy. Algorithm 1 shows the details about the algorithm *lazy*. Because *lazy* just requires n times influence spread estimations in the initial iteration for calculating the upper bound of influence gain of each candidate node i, it requires totally $(n + \theta \cdot K)$ times influence spread estimations, where $\theta \ll n$ is the expected number of influence spread estimations in each iteration.

Algorithm 1. The *lazy* Algorithm

Input: $G(V, E, T)$, a given target node t, K

Output: $\mathbf{S}$ with K nodes

1   initialize $\mathbf{S} = \emptyset$;
2   **for** *each node* $i \in V$ **do**
3   calculate $\mathcal{F}(\{i\}) = f_{t \to V}^{\{i\}} - f_{t \to V}$;
4     $flag_i = |\mathbf{S}|$;   // here, $|\mathbf{S}| = 0$
5   // $flag_i$ indicates that $\mathcal{F}(\{i\})$ is
6   // calculated in the $|\mathbf{S}|$ iteration
7   **while** $|\mathbf{S}| < K$ **do**
8   s = Find the greatest $\mathcal{F}(\{s\})$ in $\mathcal{F}$;
9   **if** $flag_s == |\mathbf{S}|$ **then**
10      $\mathbf{S} = \mathbf{S} \cup s$;
11      $\mathcal{F}(s) = 0$;
12   **else**
13   recalculate $\mathcal{F}(s) = f_{t \to V}^{\mathbf{S} \cup s} - f_{t \to V}^{\mathbf{S}}$;
14      $flag_s = |\mathbf{S}|$;
15 **Return** $\mathbf{S}$;

---

4 Optimization Under the Linear Model

To address the challenge of inefficiency, we further explore this problem on a specific influence model, the linear social influence (Linear) model [6]. Specifically, the reasons could be summarized as: (1) Linear model is tractable and efficient; (2) Linear has close relations with the traditional influence models. For instance, it can approximate the non-linear stochastic model [5], and the linear approximation method for the IC model [27] is a special case of Linear. In the following, we first review the Linear model.

Review. Given a directed graph $G(V, E, T)$, Linear model [6] is defined as below.

Definition 3. *Define the influence of node i on j as*

$$f_{i \to i} = \alpha_i, \quad \alpha_i \in (0, 1] \tag{3}$$

$$f_{i \to j} = d_j \sum_{k \in N_j} t_{kj} f_{i \to k}, \quad \text{for } j \neq i \tag{4}$$

where $N_j = \{u \in V | (u, j) \in E\}$, α_i is the self-confidence of node i, which represents the prior constraint of node i for spreading the information. The parameter $d_j \in (0, 1]$ is the damping coefficient for the influence propagation.

Under the Linear model, there is an upper bound to measure a node's influence [6]:

$$f_{i \to V} = \sum_{j=1}^{n} f_{i \to j} \leq \alpha_i \cdot (I - DT)_{i \cdot}^{-1} e \tag{5}$$

where, I is an n-by-n identity matrix, $D = diag(d_1, d_2, ..., d_n)$ is a diagonal matrix, e is an $n \times 1$ vector consisting all 1s, $(I - DT)_{i \cdot}^{-1}$ denotes the i-th row of matrix $(I - DT)^{-1}$.

Optimization with Upper Bounds. In this part, we further exploit the properties of the influence computation in Linear model and demonstrate that if target node t makes a new link with an arbitrary candidate node i, the influence gain cannot be greater than the upper bound, $\lambda_i \alpha_i \cdot (I - DT)_{i \cdot}^{-1} e$.

Theorem 1. *(Upper bound) If a given target node t makes a new link with node $i \in (V \setminus \{t\})$, then the influence gain of node t satisfies the equation:*

$$\mathcal{F}(\{i\}) = f_{t \to V}^{\{i\}} - f_{t \to V} \leq \lambda_i \cdot \alpha_i \cdot (I - DT)_{i \cdot}^{-1} e$$

Proof. We first prove that $\mathcal{F}(\{i\}) = f_{t \to V}^{\{i\}} - f_{t \to V} \leq \lambda_i \cdot f_{i \to V}$. According to the influence gain definition,

$$\mathcal{F}(\{i\}) = f_{t \to V}^{\{i\}} - f_{t \to V} \tag{6}$$

$$= \lambda_i (1 - f_{t \to i}) \sum_{k=1}^{n} (f_{i \to k} \cdot [1 - f_{t \to k}]) \tag{7}$$

$$\leq \lambda_i (1 - f_{t \to i}) \cdot \sum_{k=1}^{n} f_{i \to k} \tag{8}$$

$$\leq \lambda_i \sum_{k=1}^{n} f_{i \to k} = \lambda_i \cdot f_{i \to V} \tag{9}$$

Both Eqs. (8) and (9) hold because $f_{i \to j} \in [0, 1]$. Combining Eqs. (5) with (9), we have proved that $\mathcal{F}(\{i\}) = f_{t \to V}^{\{i\}} - f_{t \to V} \leq \lambda_i \alpha_i (I - DT)_{i \cdot}^{-1} e$. □

Here, we can rewrite Theorem 1 into vector: $[\mathcal{F}(\{1\}), \mathcal{F}(\{2\}), ..., \mathcal{F}(\{n\})]' \leq diag(\lambda_1, \lambda_2, ..., \lambda_n) \cdot diag(\alpha_1, \alpha_2, ..., \alpha_n) \cdot (I - DT)^{-1} e$. As $(I - DT)$ is a strictly diagonally dominant matrix, $(I - DT)^{-1} e$ can be quickly calculated through

Gauss-Seidel method in $O(|E|)$ time. We use these upper bounds to replace the influence gain estimations in the initial iteration of algorithm *lazy*, and then propose the corresponding *uBound* algorithm without sacrificing any accuracy. Algorithm 2 shows the details about *uBound*. According to the analysis above, we know that *uBound* requires only $(1+\eta \cdot K)$ times influence spread estimations, where $\eta \ll n$ is the expected number of influence spread estimations and it is related to the tightness of the upper bound. In contrast, *lazy* (Algorithm 1) requires $(n + \theta \cdot K)$ times influence spread estimations.

Algorithm 2. The *uBound* Algorithm

Input: $G(V, E, T)$, a given target node t, K

Output: **S** with K nodes

1 initialize $\mathbf{S} = \emptyset$;
2 Compute the upper bound vector $\mathbb{U} = diag(\lambda_1, \lambda_2, ..., \lambda_n) \cdot diag(\alpha_1, \alpha_2, ..., \alpha_n) \cdot (I - DT)^{-1} \cdot \mathbf{e}$ in $O(|E|)$ time;
3 **for** *each node* $i \in V$ **do**
4    $\quad \mathcal{F}(i) = \mathbb{U}_i$;
5    $\quad flag_i = 0$; // here, $|\mathbf{S}| = 0$

6 **while** $|\mathbf{S}| < K$ **do**
7    $\quad s = $ Find the greatest $\mathcal{F}(s)$ in $\mathcal{F}$;
8    $\quad$ **if** $flag_s == |\mathbf{S}|$ **then**
9      $\quad\quad \mathbf{S} = \mathbf{S} \cup s$;
10     $\quad\quad \mathcal{F}(s) = 0$;
11   $\quad$ **else**
12     $\quad\quad$ recalculate $\mathcal{F}(s) = f_{t \to V}^{\mathbf{S} \cup s} - f_{t \to V}^{\mathbf{S}}$;
13     $\quad\quad flag_s = |\mathbf{S}|$;

14 **Return S**;

---

5   Experiments

Experimental Setup. The experiments are conducted on four real-world datasets with different sizes. (a)Wiki-Vote, a who-votes-on-whom network at Wikipedia where nodes are users and an edge(j, i) represents that user j voted on user i; (b)Weibo, a social media network in China, where nodes are the users and edges are their followships. We crawled this data from weibo.com[3] at March 2013 and then sampled a small network which only contains the verified users for filtering the zoombie accounts; (c)Cit-HepPh, an Arxiv High Energy Physics paper citation network where nodes represent papers and an edge(j, i) represents that paper j cites paper i. Both Cit-HeppPh and Wiki-Vote are downloaded from SNAP[4]; (d)Twitter, another social media network. We downloaded this data from Social Computing Data Repository at ASU[5]. Table 1 show the detailed dataset information.

Table 1. Experimental Datasets

Name	Wiki-Vote	Weibo	cit-HepPh	Twitter
Nodes	7,115	7,378	34,546	11,316,811
Edges	103,689	39,759	421,578	85,331,845

---

[3] http://www.weibo.com/
[4] http://snap.stanford.edu/data/
[5] http://socialcomputing.asu.edu/datasets/Twitter

Influence Model and Propagation Probability. We validate our discoveries under the IC, LT and Linear models, as widely used in the literature [6,11,14,18, 28]. For each network, we transform it into a directed influence graph $G(V, E, T)$. Specifically, if there is an edge (j, i) in the original network, we add an influence link $(i, j) \in E$ [6] in G and then assign the corresponding influence propagation probability $t_{ij} = 1/indegree(j)$. For LT [14], each node j chooses a *threshold* θ_j uniformly at random from $[0, 1]$, and the Monte Carlo simulation times are set to be 20,000 for both IC and LT. For Linear model, we use the same damping coefficient for all nodes similar to Xiang et al. [6,28] (i.e., $d_i = 0.85$ for $i \in V$), and we set $\alpha_i = 1$ assuming that each initial node shares the same prior influence probability. Note that, for simplicity, we manually set the $\lambda_c = 0.85$ for all the nodes in objective function $\mathcal{F}(\mathbf{S})$.

We implemented the algorithms in Java and conducted the following experiments on Windows 64-bit OS with 2.20GHz Intel Core i3-2330M and 16GB memory.

5.1   Real Influence Gain Comparison

We first demonstrate that our objective function is rational and effective, i.e., the node set $\mathbf{S}$ recommended based on our $\mathcal{F}(\mathbf{S})$ can help a target node t make more influence gain than the benchmark methods. Specifically, for a given target node t, we first calculate its original influence $f_{t \to V}$. Secondly, we let t make new links with the nodes in $\mathbf{S}$ recommended by different methods, and recalculate the t's new influence $f_{t \to V}^{\mathbf{S}}$ [7]. Finally, we get the t's real influence gain $\mathcal{F}(\mathbf{S}) = f_{t \to V}^{\mathbf{S}} - f_{t \to V}$. Thus, the performance of each method is evaluated by the influence gain it could provide to the target node, i.e., the larger influence gain, the better the method is. In the following, we call our method as ISIM (**I**ndividual **S**ocial **I**nfluence **M**aximization) and the results are based on the *lazy* algorithm. For comparison, we choose several benchmark methods:

- **Random.** Let the target node make links with K nodes that are selected randomly.
- **OutDeg.** Let the target node link to the top K nodes with the largest out-degree.
- **LongDist.** The recommended K nodes are the farthest ones from the target node, i.e., those have the fewest influence overlap with the target node. Here, the distance is measured by Random Walk with Restart [29].
- **PageRank.** Recommend the nodes with top K ranked PageRank values [30].
- **HighestInf (Highest Influence).** Let the target node make links with the top-K nodes with highest influence. This method is also competitive because the largest influential nodes can improve the target node's influence sharply. However, this method does not consider the influence overlap.

---

[6] For example, if user j follows user i in Twitter, then i influences j.

[7] After linking t to nodes in $\mathbf{S}$, the indegree of node $c \in \mathbf{S}$ pluses one. The influence propagation probability of each edge $(u, c) \in E$ will be updated by $1/new\_indegree(c)$.

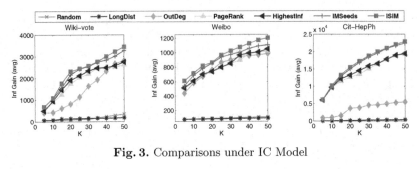

Fig. 3. Comparisons under IC Model

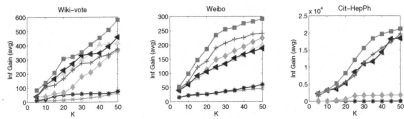

Fig. 4. Comparisons under LT Model

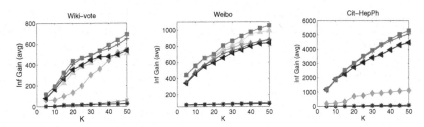

Fig. 5. Comparisons under Linear Model

- **IMSeeds.** It selects and recommends the most influential node set **S** by using the CELF algorithm [11] for traditional social influence maximization problem. This method could alleviate the influence overlap between the nodes in **S**. However, it does not consider the influence overlap between the target node and those in **S**.

On each dataset, we run the above selection algorithms on the randomly chosen 100 target nodes from different out-degree ranges, and then we compute and compare the average influence gain (with the size of the recommendation set $|\mathbf{S}| = 5, 10, ..., 50$) for each algorithm. We compare the effectiveness of each algorithm under IC, LT and Linear model, respectively. Figs. 3, 4 and 5 show the corresponding results. Actually, similar results could be seen from all figures. That is, the node set **S** selected by our ISIM could help the target node to get more real influence gain than the benchmarks; the node set **S** recommended by IMSeeds cannot always guarantee the best performance. What's more, we only show the results on the three data sets, because IMSeeds (i.e., CELF) cannot obtain a result within feasible time on the Twitter data.

5.2 Time Complexity Analysis

In the part, we compare the efficiency of our proposed algorithms (*greedy*, *lazy* and *uBound*) for our ISIM method under Linear model from two aspects: *the number of influence spread estimations* and the *running time* of the algorithms.

Table 2. Numbers of Influence Spread Estimation

Datasets	Alg.	5	10	15	20	25	30	35	40	45
Wiki-Vote	greedy	35208	70390	105547	140679	175786	210868	245925	280957	315964
	lazy	7124	7139	7171	7195	7220	7250	7306	7331	7374
	uBound	19	53	84	119	157	204	265	297	352
Weibo	greedy	36836	73646	110431	147191	183926	220636	257321	293981	330616
	lazy	7414	7446	7505	7572	7638	7738	7838	7917	8047
	uBound	50	98	180	278	356	494	622	726	868
Cit-HepPh	lazy	34554	34566	34586	34608	34642	34660	34691	34710	34744
	uBound	17	37	64	89	137	161	194	221	264
Twitter	uBound	16	43	71	99	136	166	212	253	286

Table 2 shows the expected numbers of influence spread estimations when selecting different K seeds using different algorithms. The results illustrate that *greedy* needs the largest number of influence spread estimations. Compared to *lazy*, the expected number of influence spread estimations of *uBound* at top $K = 45$ is reduced at a rate of 95.2%, 89.2%, 99.2% on the three datasets(ie, Wiki-Vote, Weibo, Cit-HepPh), respectively. The reason is that *lazy* requires n times influence spread estimations in the initial iteration to establish the upper bounds of the marginal influence, while *uBound* requires only one time. Correspondingly, Fig. 6 shows the real running time of different algorithms when selecting K seeds on different datasets. From the results, we know that the simple *greedy* algorithm is very time-consuming as the number K increases. That is because *greedy* requires about $(n \cdot K)$ times influence spread estimations. What's more, we can observe that *uBound* is much faster than *lazy*. Actually, *uBound* is so efficient that it can handle the Twitter data, a large scale network with tens of millions of nodes, and the running time is growing linearly as the the number K increases.

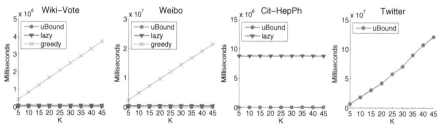

Fig. 6. Real Runtime Comparisons

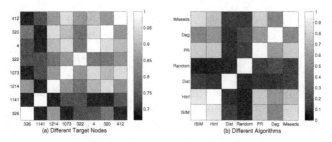

Fig. 7. Jaccard Index Comparison

5.3 Case Study

We finally use a case study to illustrate the necessity of designing individualized link recommendation algorithms. Fig. 7(a) shows the Jaccard index of the 25 nodes recommended by ISIM for 8 target nodes which are randomly selected (distinguished by node ID) from Wiki-Vote. This figure illustrates that the nodes recommended for different target nodes are different, and this is because the method ISIM considers the target node's personalized information, such as the topology structure of the target node. Similarly, Fig. 7(b) shows the Jaccard index of different node sets recommended by different methods. This figure illustrates that the nodes selected by different algorithms are also quite different. Meanwhile, the more similar with our proposed method(i.e., ISIM), the more effective of the algorithm (combining the results in Figs. 3, 4 and 5).

6 Discussion

In this part, we mainly discuss the limitations and possible extensions of this study. For better illustration, in this paper we only deal with individual influence maximization by designing general algorithms, and our solutions could be further improved in the future. First, it is better to include different costs for the link connection (i.e., the $c(\mathbf{S})$ in Eq. (1)) instead of treating them equally. Meanwhile, more reasonable settings for parameters λ_c or threshold θ are required when we know about more prior knowledge or real propagation action logs (like using the data-driven approach for threshold θ settings [31]). What's more, our assumption that the rest of the network stays unchanged during the link connection may be relaxed. Secondly, it is also better to study individual influence maximization and social influence modeling from the observed data rather than the simple simulation. For one thing, the information diffusion process may be affected by some other factors, e.g., information topic and homophily [32]. For another, as is only exploratory in nature, the conclusions of the simulation studies often have a great deviation to the actual propagation data. Thirdly, this study only focuses on the one target individual's influence, and one possible extension is to add links for improving the influence spread of several individuals simultaneously, where the competitions or cooperations between each target individual may be a big challenge. Last but not least, like the *uBound* algorithm under Linear model, we would like to find out the upper bounds under other influence models (e.g., IC, LT) and propose the corresponding scalable algorithms.

7 Conclusion

In this paper, we studied the problem of maximizing individual's own influence by recommending new links. We first formulated it as an optimization problem and designed a rational objective function. Then we proposed three algorithms to solve this intractable problem; especially the *uBound* algorithm with $O(1+\eta \cdot K)$ time complexity could handle large scale network. The experiments have shown encouraging results, and we hope this study could lead to more future work.

Acknowledgments. This research was partially supported by grants from the National Science Foundation for Distinguished Young Scholars of China (Grant No. 61325010), the National High Technology Research and Development Program of China (Grant No. 2014AA015203), the Natural Science Foundation of China (Grant No. 61403358) and the Anhui Provincial Natural Science Foundation (Grant No. 1408085QF110). Qi Liu acknowledges the support of the Youth Innovation Promotion Association, CAS.

References

1. Richardson, M., Domingos, P.: Mining knowledge-sharing sites for viral marketing. In: KDD (2002)
2. Chen, W., Lakshmanan, L.V., Castillo, C.: Information and influence propagation in social networks. Synthesis Lectures on Data Management **5**(4), 1–177 (2013)
3. Goldenberg, J., Libai, B., Muller, E.: Talk of the network: A complex systems look at the underlying process of word-of-mouth. Marketing letters **12**(3), 211–223 (2001)
4. Granovetter, M.: Threshold models of collective behavior. American journal of sociology, 1420–1443 (1978)
5. Aggarwal, C., Khan, A., Yan, X.: On flow authority discovery in social networks. In: SDM, pp. 522–533 (2011)
6. Xiang, B., Liu, Q., Chen, E., Xiong, H., Zheng, Y., Yang, Y.: Pagerank with priors: an influence propagation perspective. In: IJCAI (2013)
7. Goyal, A., Bonchi, F., Lakshmanan, L.V.: Learning influence probabilities in social networks. In: WSDM (2010)
8. Saito, K., Ohara, K., Yamagishi, Y., Kimura, M., Motoda, H.: Learning diffusion probability based on node attributes in social networks. In: Kryszkiewicz, M., Rybinski, H., Skowron, A., Raś, Z.W. (eds.) ISMIS 2011. LNCS, vol. 6804, pp. 153–162. Springer, Heidelberg (2011)
9. He, X., Kempe, D.: Stability of influence maximization. In: SIGKDD, pp. 1256–1265 (2014)
10. Eftekhar, M., Ganjali, Y., Koudas, N.: Information cascade at group scale. In: KDD (2013)
11. Leskovec, J., Krause, A., Guestrin, C., Faloutsos, C., VanBriesen, J., Glance, N.: Cost-effective outbreak detection in networks. In: KDD (2007)

12. Chaoji, V., Ranu, S., Rastogi, R., Bhatt, R.: Recommendations to boost content spread in social networks. In: WWW (2012)
13. Domingos, P., Richardson, M.: Mining the network value of customers. In: KDD (2001)
14. Kempe, D., Kleinberg, J., Tardos, É.: Maximizing the spread of influence through a social network. In: KDD (2003)
15. Chen, W., Wang, C., Wang, Y.: Scalable influence maximization for prevalent viral marketing in large-scale social networks. In: KDD (2010)
16. Kim, J., Kim, S.K., Yu, H.: Scalable and parallelizable processing of influence maximization for large-scale social networks? In: ICDE, pp. 266–277. IEEE (2013)
17. Tang, Y., Xiao, X., Shi, Y.: Influence maximization: near-optimal time complexity meets practical efficiency. In: SIGMOD, pp. 75–86. ACM (2014)
18. Guo, J., Zhang, P., Zhou, C., Cao, Y., Guo, L.: Personalized influence maximization on social networks. In: CIKM (2013)
19. Facebook official blog: People You May Know. https://www.facebook.com/notes/facebook/people-you-may-know/15610312130
20. Chen, J., Geyer, W., Dugan, C., Muller, M., Guy, I.: Make new friends, but keep the old: recommending people on social networking sites. In: SIGCHI (2009)
21. Guy, I., Ronen, I., Wilcox, E.: Do you know?: recommending people to invite into your social network. In: IUI (2009)
22. Tian, Y., He, Q., Zhao, Q., Liu, X., Lee, W.c.: Boosting social network connectivity with link revival. In: CIKM (2010)
23. Saito, K., Kimura, M., Ohara, K., Motoda, H.: Which targets to contact first to maximize influence over social network. In: Greenberg, A.M., Kennedy, W.G., Bos, N.D. (eds.) SBP 2013. LNCS, vol. 7812, pp. 359–367. Springer, Heidelberg (2013)
24. Goyal, A., Lakshmanan, L.V.S.: Recmax: exploiting recommender systems for fun and profit. In: KDD (2012)
25. Khuller, S., Moss, A., Naor, J.: The budgeted maximum coverage problem. Inf. Process. Lett. **70**(1), 39–45 (1999)
26. Krause, A., Golovin, D.: Submodular function maximization. Tractability: Practical Approaches to Hard Problems **3**, 19 (2012)
27. Yang, Y., Chen, E., Liu, Q., Xiang, B., Xu, T., Shad, S.A.: On approximation of real-world influence spread. In: Flach, P.A., De Bie, T., Cristianini, N. (eds.) ECML PKDD 2012, Part II. LNCS, vol. 7524, pp. 548–564. Springer, Heidelberg (2012)
28. Liu, Q., Xiang, B., Zhang, L., Chen, E., Tan, C., Chen, J.: Linear computation for independent social influence. In: ICDM (2013)
29. Tong, H., Faloutsos, C., Pan, J.Y.: Random walk with restart: fast solutions and applications. In: KIS (2008)
30. Page, L., Brin, S., Motwani, R., Winograd, T.: The pagerank citation ranking: bringing order to the web (1999)
31. Goyal, A., Bonchi, F., Lakshmanan, L.V.: A data-based approach to social influence maximization. Proceedings of the VLDB Endowment **5**(1), 73–84 (2011)
32. Anagnostopoulos, A., Kumar, R., Mahdian, M.: Influence and correlation in social networks. In: KDD, pp. 7–15. ACM (2008)

Community Detection Based on Minimum-Cut Graph Partitioning

Yashen Wang[✉], Heyan Huang, Chong Feng, and Zhirun Liu

Beijing Engineering Research Center of High Volume Language Information Processing
and Cloud Computing Applications, School of Computer,
Beijing Institute of Technology, Beijing, China
{yswang,hhy63,fengchong,zrliu}@bit.edu.cn

Abstract. One of the most useful measurements of community detection quality is the modularity, which evaluates how a given division deviates from an expected random graph. This article demonstrates that (i) modularity maximization can be transformed into versions of the standard minimum-cut graph partitioning, and (ii) normalized version of modularity maximization is identical to normalized cut graph partitioning. Meanwhile, we innovatively combine the modularity theory with popular statistical inference method in two aspects: (i) transforming such statistical model into null model in modularity maximization; (ii) adapting the objective function of statistical inference method for our optimization. Based on the demonstrations above, this paper proposes an efficient algorithm for community detection by adapting the Laplacian spectral partitioning algorithm. The experiments, in both real-world and synthetic networks, show that both the quality and the running time of the proposed algorithm rival the previous best algorithms.

Keywords: Community detection · Modularity · Minimum-cut

1 Introduction

Network could represent data entities and the relationship among them. A property that seems common to many networks is community structure, the division of the network vertices into communities in which the connections of vertices are dense, but between which the connections are sparser [1, 2]. Community detection has been proven to be valuable in a wide range of domains [3, 4]. To solve the problem of community detection, we first get insight into the definition of community. There are several classes of global community definitions [5]:

(i)   The most intuitive definition relies on the number of edges lying between communities, referred to as *cut size* (or *edge cut*), which derives minimum-cut   graph partitioning/clustering problem aiming at producing division with smallest cut size;

(ii)   Another one is based on the widely used concept of *modularity*. Modularity evaluates how a given division deviates from an expected random graph (null model), and modularity-based algorithm [6, 14] has become one of the most popular algorithms for detecting community because of its reasonable accuracy and feasibility.

(c) Springer International Publishing Switzerland 2015
J. Li and Y. Sun (Eds.): WAIM 2015, LNCS 9098, pp. 57–69, 2015.
DOI: 10.1007/978-3-319-21042-1_5

The core of our investigation is to explore the relationship between problem of *modularity maximization* and problem of *minimum-cut graph partitioning* which minimizes the cut size. [7] focused on the maximization of the likelihood of generating a particular network, and showed that inference methods could be mapped to standard minimum-cut problem. Our research is inspired by it, however, we focus on the maximization of modularity, and we demonstrate that (i) the modularity maximization could be transformed into minimum-cut graph partitioning, and (ii) normalized modularity maximization is identical to normalized cut graph partitioning. Thus, many existing heuristics for graph partitioning could be adopted to solve our community detection problem [13, 15]. As an example, we apply the Laplacian spectral partitioning algorithm [15], which is simple and fast, to derive a community detection algorithm competitive with the best currently available algorithms in terms of both speed and quality of results.

The other contribution of this paper is that we innovatively combine the modularity theory with popular statistical inference method. Research about community detection recently has focused particularly on statistical inference methods (stochastic block model and its degree-corrected counterpart, etc.), which assume some underlying probabilistic model that generates the community structure and estimates model's parameters [9, 11]. We try to exploit the line lying between modularity-based algorithm and inference model-based algorithm:

(i)   Statistical inference models could be transformed into the *null models* in modularity maximization problem based on the analysis in [11];

(ii)   The core idea of the statistical inference model is to generate a suitable *objective function* for community detection with maximizing the profile likelihood. So, we adapt such objective function for our optimization.

Based on the rigorous demonstrations, the implementation of the proposed algorithm is intuitive and simple: we first get the candidate set of community structure solutions by implementing Laplacian spectral partitioning algorithm; then we compute objective function of degree-corrected model for each candidate, and chose the one with the highest score as final community detection result. The experimental results guarantee the feasibility and robustness of the proposed demonstration. To the best of our knowledge, this is the first clear relationship among (i) modularity-based theory, (ii) statistical inference models, and (iii) traditional graph clustering/partitioning-based algorithms.

2   Related Work

There have been significant interests in algorithms for detecting community structure in networks for decades [1, 2, 3, 4, 5, 7]. In [1], Newman describes an algorithm based on a characteristic of clustering quality called modularity, a measure that takes into account the number of in-community edges and the expected number of such edges. [6] proposed a greedy modularity-based algorithm CNM, one of the most widely used methods recently. It iteratively selects and merges the best pair of vertices, which has the largest modularity gain, until no pairs improve the modularity. Then several algorithms were proposed based on other theories [10, 12]. Recently, interest has particularly focused on statistical inference methods, such as stochastic block model [9] and its degree-corrected counterpart [11]. Moreover, community detection is closely related to graph partitioning, although there are differences between them [5]. Graph partitioning, dividing the vertices of a graph into some communities of given size (not necessarily with equal size) such that the number of edges between groups is minimized, has been well studied thoroughly by researchers [13, 15]. So far, there is few study on the relationship between modularity maximization and minimum-cut partitioning. One early work on modifying the objective function of modularity maximization is [16], which introduced the cluster size normalized much like the Ratio-Cut spectral clustering. But the modified modularity is not exactly the graph Laplacian.

3 Degree-Corrected Stochastic Block Model

Original stochastic block model ignores variation in vertex degree, making it unsuitable for applications to real-world networks [9]. For such reason, degree-corrected stochastic block model adds a parameter θ_i for each vertex i, which controls expected degree of vertex i, making it accommodate arbitrary degree distributions.

In this paper, let V and E be sets of n vertices and m edges of the graph $G(V, E)$ respectively, and use K to denote the number of communities. A_{ij} denotes an element of the adjacency matrix of $G(V, E)$, conventionally equal to the number of edges between vertices i and j. Assuming there are no self-edges in the network, so $A_{ii} = 0$. And ω_{rs} is an element of a $K * K$ symmetric matrix of parameters, controlling edges between communities. Following [11], we let expected value of A_{ij} be $\theta_i \theta_j \omega_{g_i g_j}$, where g_i is the community to which vertex i belongs. The likelihood of generating $G(V, E)$ is

$$P(G|\theta, \omega, g) = \prod_{i<j} \frac{(\theta_i \theta_j \omega_{g_i g_j})^{A_{ij}}}{A_{ij}!} exp\left(-\theta_i \theta_j \omega_{g_i g_j}\right) \qquad (1)$$

Parameter θ is arbitrary within a multiplicative constant absorbed into ω, and their normalization can be fixed by imposing the constraint $\sum_i \theta_i \delta_{g_i s} = 1$ for each community s ($\delta_{g_i s}$ is the *Kronecker Delta* [11]). Then the maximum likelihood values of parameters θ_i and ω_{rs} are given by

$$\hat{\theta}_i = \frac{k_i}{g_i}, \text{ and } \hat{\omega}_{rs} = m_{rs}, \qquad (2)$$

where k_i is the degree of vertex i, and $_s$ is the total number of degrees in community s. As the number of edges connecting community r and s, m_{rs} is defined as

$$m_{rs} = \sum_{i<j} A_{ij} \delta_{g_i, r} \delta_{g_j, s} \, . \qquad (3)$$

Substituting these values of θ and ω into Eq. (1) with little manipulation gives the *profile likelihood*. In most cases, it is simpler to maximize the *logarithm of the profile likelihood* rather than itself, because the maximum is in the same place. Neglecting overall constants, the logarithm is given by

$$\mathcal{L}(G|g) = \sum_{r<s} m_{rs} \log \frac{m_{rs}}{rs} \, . \qquad (4)$$

This log likelihood depends on community structure only. Since the block model is generally used as a source of flexible and challenging benchmark for evaluation of community detection algorithms, the maximum of Eq.(4) with respect to the community structure could tell us the most suitable community structure.

4 Community Detection Based on Minimum-Cut Graph Partitioning

[1] defines a measurement for quality of community structure, *modularity*, which is widely used. And most recent researches about community detection focus on the problem of *modularity maximization*. In this section, we will demonstrate: *how the problem of (normalized) modularity maximization could be transformed into the problem of (normalized) minimum-cut graph partitioning.*

4.1 Preliminaries

Given
  (i)   a graph $G(V,E)$ with n vertices and m edges,
  (ii) a partition $\mathcal{P}$ of K communities $\mathcal{P} = (P_1, \dots, P_K)$, and
  (iii) a null model (random graph distribution) $\mathcal{G}$ on V,
  Modularity $Q(\mathcal{P}, G, \mathcal{G})$ is defined as

$$Q(\mathcal{P}, G, \mathcal{G}) = \sum_{i=1}^{K} (|E(P_i)| - |E(P_i, \mathcal{G})|) , \qquad (5)$$

wherein $| \dots |$ is the size of the corresponding set, $E(P_i)$ is the set of edges that join vertices in community P_i, and $|E(P_i, \mathcal{G})|$ is the expected number of such edges in the null model $\mathcal{G}$. If the number of inner-community edges is same as that in the null model $\mathcal{G}$, $Q(\mathcal{P}, G, \mathcal{G})$ will be zero. And a nonzero value evaluates deviation from randomness.

There are several choices of the null model $\mathcal{G}$, such as Erdös-Renyi model [17] and Chung-Lu model [18]. The original stochastic block models have an internal relation with random graph model proposed by Erdös and Renyi [11]. However, this model produces highly unrealistic networks. Specifically, it produces networks with Poisson degree distributions, in stark contrast to most real networks. To avoid this problem, modularity is usually defined using the model proposed by Chung and Lu, which fixes the expected degree sequence to be the same as that of the observed network. In this case, the objective function of modularity could also be rewritten as

$$Q = \sum_{i=1}^{K} \left(\frac{m_{ii}}{2m} - \left(\frac{i}{2m} \right)^2 \right) \qquad (6)$$

wherein $2m = \sum_{ij} A_{ij} > 0$ is the sum of the degrees of all vertices in the network, which is a constant. If the edges of a network is randomly assigned, then on average,

$$A_{ij} = \frac{k_i k_j}{2m} \quad \text{and} \quad m_{rs} \approx \frac{rs}{2m}. \qquad (7)$$

This choice of null model is found to give significantly better results than the former one (original uniform model), because it allows for the fact that vertices with high degree are, more likely to be connected than those with high degree. So, we get better results with incorporating this observation into stochastic block model, and we get degree-corrected stochastic block model [11].

Based on the analysis above, we wish to find compact community, meaning that we need to maximize the $|E(P_i)|$ in Eq. (5) (or m_{ii} in Eq. (6)). But since $|E(P_i, \mathcal{G})|$ is the statistically expected value we should maximize $(|E(P_i)| - |E(P_i, \mathcal{G})|)$ instead of $|E(P_i, \mathcal{G})|$. For the same reason, we should maximize $(m_{ii} - \frac{(i)^2}{2m})$ instead of m_{ii}. Summing over all the communities, the objective function of modularity maximization is to maximize Eq. (5) or Eq. (6).

4.2   Demonstration 1: Mapping from Modularity Maximization to Minimum-Cut Graph Partitioning

This article treats the previous two statistical inference models as potential choices of $\mathcal{G}$ by a small amount of manipulation, which will be discussed latter, respectively.

If high value of modularity $Q(\mathcal{P}, G, \mathcal{G})$ corresponds to good division of a community structure, the purpose of algorithm is to find such division with the global maximum modularity, as follows:

$$\underset{\mathcal{P}}{max} \sum_{i=1}^{K}(|E(P_i)| - |E(P_i, \mathcal{G})|) \tag{8}$$

To simplify the description, we define the following quantity

$$|cut(\mathcal{P})| = m - \sum_{i=1}^{K}|E(P_i)| \tag{9}$$

as the *cut size* of the partition $\mathcal{P}$. And minimum-cut partitioning problem optimizes

$$\underset{\mathcal{P}}{min}|cut(\mathcal{P})|. \tag{10}$$

Theorem 1. *The modularity maximization of Eq. (8) could be transformed into minimum-cut partitioning of Eq. (10).*

Proof. Objective function of modularity maximization (Eq. (8)) could be written as

$$\underset{\mathcal{P}}{max} \sum_{i=1}^{K}(|E(P_i)| - |E(P_i, \mathcal{G})|)$$
$$= -\underset{\mathcal{P}}{min}(-\sum_{i=1}^{K}|E(P_i)| + \sum_{i=1}^{K}|E(P_i, \mathcal{G})|)$$
$$= -\underset{\mathcal{P}}{min}(m - \sum_{i=1}^{K}|E(P_i)| - m + \sum_{i=1}^{K}|E(P_i, \mathcal{G})|) \tag{11}$$

Substituting Eq. (9) to Eq. (11), we get

$$-\underset{\mathcal{P}}{min}(|cut(\mathcal{P})| - m + \sum_{i=1}^{K}|E(P_i, \mathcal{G})|). \tag{12}$$

Eq. (12) is similar to the objective function of minimum-cut partitioning, except for the item $\sum_{i=1}^{k}|E(P_i, \mathcal{G})|$ which depends on the null model $\mathcal{G}$ and community structure. We first discuss the bisection condition (i.e., $K = 2$), and later we will illustrate how the proposed algorithm could be extended to the condition of multi-communities.

Thus, Eq. (12) could be rewritten as

$$-\min_{\mathcal{P}}(|cut(\mathcal{P})| - m + \Sigma_{i=1}^2 |E(P_i, \mathcal{G})|) \tag{13}$$

Here, for the choices of $\mathcal{G}$ as discussed above, we will investigate standard stochastic block model and its degree-corrected counterpart respectively.

Standard Stochastic Block Model. We rewrite Eq. (13) as

$$-\min_{\mathcal{P}}(|cut(\mathcal{P})| - m + |E(\mathcal{G})| - |E(\mathcal{G})| + \Sigma_{i=1}^2 |E(P_i, \mathcal{G})|) , \tag{14}$$

where $|E(\mathcal{G})|$ is the expected number of all edges in $\mathcal{G}$. Neglecting constants and terms independent of the community structure, we get

$$-\min_{\mathcal{P}}(|cut(\mathcal{P})| + |E(\mathcal{G})| - (|E(\mathcal{G})| - \Sigma_{i=1}^2 |E(P_i, \mathcal{G})|)) . \tag{15}$$

If we use w_{12} to denote the expected number of edges between community 1 and 2,

$$|E(\mathcal{G})| - \Sigma_{i=1}^2 |E(P_i, \mathcal{G})| = \Sigma_{i<j}(1 - \delta_{g_i g_j}) w_{12} . \tag{16}$$

Considering

$$\Sigma_{i<j}(1 - \delta_{g_i g_j}) = n_1 n_2 , \tag{17}$$

where n_1 and n_2 are numbers of vertices in communities 1 and 2 respectively,

$$\Sigma_{i<j}(1 - \delta_{g_i g_j}) w_{12} = n_1 n_2 w_{12} . \tag{18}$$

Thus the modularity maximization could be rewritten as:

$$\max_{\mathcal{P}} \Sigma_{i=1}^2 (|E(P_i)| - |E(P_i, \mathcal{G})|) = -\min_{\mathcal{P}}(|cut(\mathcal{P})| + |E(\mathcal{G})| - n_1 n_2 w_{12}) . \tag{19}$$

Instead of problem (8), we could address the problem of computing

$$\underset{\mathcal{P}}{argmin}(|cut(\mathcal{P})| + |E(\mathcal{G})| - n_1 n_2 w_{12}) . \tag{20}$$

So the maximization of modularity corresponds to the minimization of the cut size of the partition with an additional penalty term $(|E(\mathcal{G})| - n_1 n_2 w_{12})$ that favors communities of equal size. Different graph distribution $\mathcal{G}$ leads to different value of w_{12} and $|E(\mathcal{G})|$. So the quantity of the formula above varies when we apply various $\mathcal{G}$.

Degree-Corrected Stochastic Block Model. Dropping constant m in Eq. (13), then

$$-\min_{\mathcal{P}}(|cut(\mathcal{P})| + \Sigma_{i=1}^2 |E(P_i, \mathcal{G})|) . \tag{21}$$

In degree-corrected stochastic block model,

$$\Sigma_{i=1}^2 |E(P_i, \mathcal{G})| = (w_{11} \Sigma_{ij} \delta_{g_i,1} \delta_{g_j,1} \theta_i \theta_j + w_{22} \Sigma_{ij} \delta_{g_i,2} \delta_{g_j,2} \theta_i \theta_j)/2 , \tag{22}$$

where w_{11} and w_{22} represent the expected numbers of edges connecting vertices in community 1 and 2, respectively. And,

$$w_{11} \Sigma_{ij} \delta_{g_i,1} \delta_{g_j,1} \theta_i \theta_j = w_{11} \Sigma_i \delta_{g_i,1} \theta_i \Sigma_j \delta_{g_j,1} \theta_j . \tag{23}$$

Considering the constraint $\sum_i \theta_i \delta_{g_{is}} = 1$ for community s,

$$w_{11} \sum_{ij} \delta_{g_{i,1}} \delta_{g_{j,1}} \theta_i \theta_j = w_{11} \sum_i \delta_{g_{i,1}} \theta_i = w_{11} \ . \tag{24}$$

Similarly,

$$w_{22} \sum_{ij} \delta_{g_{i,2}} \delta_{g_{j,2}} \theta_i \theta_j = w_{22} \ . \tag{25}$$

Instead of problem (8), we now could address the problem of computing

$$\underset{\mathcal{P}}{argmin}(|cut(\mathcal{P})| + (w_{11} + w_{22})/2) \ . \tag{26}$$

Eq. (26) is very similar to Eq. (20). Namely, the original problem of modularity maximization could be converted into the minimization of the cut size with an additional term. Taking both Eq. (20) and Eq. (26) into account, this article minimizes the formula over community structure first for fixed $\mathcal{G}$, and then over $\mathcal{G}$ at the end.

In the above additional terms, w_{11}, w_{22}, w_{12} and $|E(\mathcal{G})|$ are unknown, and rely on $\mathcal{G}$ only. To get around such terms, we implement a limited minimization, in which $\mathcal{G}$ is given and the community sizes n_1 and n_2 are held fixed as some values we choose. Hence discarding the constant terms $(|E(\mathcal{G})| - n_1 n_2 w_{12})$ (in Eq. (20)) and $(w_{11} + w_{22})/2$ (in Eq. (26)), the optimization is equal to standard minimum-cut graph partitioning problem.

4.3   Demonstration 2: Mapping from Normalized Modularity Maximization to Normalized Minimum-Cut Graph Partitioning

In this section, we show for that a *normalized* version of modularity maximization is identical to normalized cut spectral partitioning, whose objective function is

$$C_n = \sum_{r<s}(\frac{m_{rs}}{r} + \frac{m_{rs}}{s}). \tag{27}$$

We then define normalized modularity maximization as

$$Q_n = \sum_{i=1}^{K} \frac{1}{i}(\frac{m_{ii}}{2m} - (\frac{i}{2m})^2). \tag{28}$$

Theorem 2. *The normalized modularity maximization of Eq. (28) is identical to normalized cut partitioning of Eq. (27).*

Proof. Eq.(28) could be written as

$$Q_n = \frac{1}{2m} \sum_{i=1}^{K}(\frac{m_{ii}}{i} - \frac{i}{2m}). \tag{29}$$

Considering $\sum_{i=1}^{K} i = 2m$,

$$Q_n = \frac{1}{2m}(\sum_{i=1}^{K} \frac{m_{ii}}{i} - 1). \tag{30}$$

Meanwhile, Eq. (27) could also be written as

$$C_n = \sum_{i=1}^{K} \frac{\sum_{j \neq i} m_{ij}}{i} = \sum_{i=1}^{K} \frac{i - m_{ii}}{i} = K - \sum_{i=1}^{K} \frac{m_{ii}}{i}. \tag{31}$$

Thus,

$$C_n = (K - 1) - 2m * Q_n. \tag{32}$$

Therefore, minimization of C_n is identical to maximization of Q_n. In other words, minimizing normalized cut is identical to maximizing normalized modularity.

4.4   The Implementation of Community Detection Based on Minimum-Cut Graph Partitioning

Based on the analysis above, we utilize Laplacian partitioning algorithm as [7], which is shown effective in practice ($O(m)$), although there really exist other heuristics [12, 15]. The choice of n_1 and n_2 could be done with a greedy search strategy [6], and there are $O(n)$ possible choices (ranging from putting all vertices in only one community to two communities with equal size, and everything in between). All the choices construct a *candidate set*. Then for each choice, which is standard minimum-cut partitioning problem, we utilize Laplacian spectral bisection algorithm, and get corresponding *candidate solution* (i.e., candidate community structure). Since the final community structure should maximized the log of the profile likelihood (Eq. (4)), we calculate this quality for each candidate solution, and the one with the highest score is the maximum-modularity community structure. We recursively partition a current community into two sub-coummunities using the second eigenvector of the generalized Laplacian matrix, $\mathbf{Lv} = \lambda \mathbf{Dv}$. Having calculated such eigenvector, one divides the network into communities of the required size n_1 and n_2 by inspecting the vector elements and assigning the n_1 vertices with the largest elements to community 1 and the rest to community 2. A nice feature of this algorithm is that, only in a single calculation, it provides us entire one-parameter family of minimum-cut divisions of the network [7].

This is repeated until the desired number of communities is reached [12], and one issue here is how to choose the next community to split. We choose the community P_k with smallest average similarity, $\sum_{ij} \delta_{g_i,k} \delta_{g_j,k} A_{ij} / |P_k|^2$, which is the loosest cluster. And if there exists no division of a community that gives any positive contribution to the modularity of the entire network, and it should be left alone.

5   Experiments and Results

In this section, the proposed algorithm (denoted as **MC**) was tested on both *real-world* networks and *synthetic* (i.e., computer generated) networks in terms of accuracy and sensitivity. This article mainly utilizes normalized mutual information (NMI) [19] as evaluation, and the experimental results showed the proposed algorithm worked well.

5.1   Experiments on Real-World Networks

We tested our algorithm on a range of real-world networks, and it produced community structures consistent with prior knowledge in all case. Here, we only describe three networks in detail with the limitation of space.

The first example is the "karate club" network [20], which represents the interactions amongst the 34 members of a karate club at an American university, in Fig. 1(a). In Fig. 1(b), we show another social network of a community of 62 bottlenose dolphins living in Doubtful Sound, New Zealand, established by observation of statistically significant frequent association [21]. The network first splits naturally into two large groups, and the larger one then splits into three smaller subgroups.

In the third real-world example, the proposed algorithm was applied to the American political blogs network [22]. This network is composed of American politics blogs and the links between them around the time of the 2004 presidential election, which classified the first 758 blogs as left-leaning and the remaining 732 as right-leaning. This experiment was implemented as a contrast to the similar experiment described in [11]. As shown in Table 1, applying our algorithm to this network of all the 1,490 vertices, the detected community structure has a NMI of only 0.5005. However the NMI increases to 0.7204 when the experiment only examines the largest connected component by dropping 225 isolated vertices. This result is similar to [11], which gets NMI of 0.72 in 1,222 main vertices. **MC** represents the proposed algorithm in the following table. We obtain similar results for other real-world networks, although we do not provide the plots for lack of space.

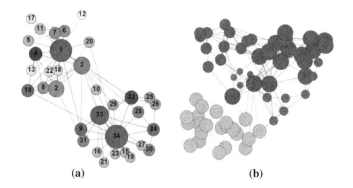

(a)   **(b)**

Fig. 1. Experiments on (a) karate club network and (b) bottlenose dolphin network

Table 1. Experiment results of the American political blog network

	Remained Vertices	Removed Vertices	NMI
MC	1490	0	0.5005
MC	1265	225	**0.7204**
Ref.[11]	1222	268	0.72

5.2   Experiments on Synthetic Networks

For direct comparison with benchmarks or other algorithms, we now turn to applications of the proposed algorithm to synthetic networks, and such networks are generated by using degree-corrected stochastic model (DCM) [11] (which are published online[1]) and Lancichinetti-Fortunato-Radicchi (LFR) benchmark [23].

---

[1] http://hlipca.org/index.php/2014-12-09-02-58-51/2014-12-09-02-59-44/44-communitydetectionmc

Experiments on DCM Synthetic Networks. Following [11], this article chooses ω_{rs} which has the particular form $\omega_{rs} = \lambda\omega_{rs}^{planted} + (1 - \lambda)\omega_{rs}^{random}$ to generate synthetic networks. In this form, the parameter λ is used to interpolate linearly between the values $\omega_{rs}^{planted}$ (a planted network where the communities are completely separated) and ω_{rs}^{random} (a fully random network with no community structure). In [24], ω_{rs}^{random} is defined as the expected value of ω_{rs} in a random graph with fixed expected degree, i.e. $\omega_{rs}^{random} = \frac{rs}{2m}$. For the value of $\omega_{rs}^{planted}$, this article chooses a simple form as $\omega_{rs}^{planted} = \begin{bmatrix} 1 & 0 \\ 0 & 2 \end{bmatrix}$ to create community structure.

In the first group of experiments, each generated network has 100,000 vertices with two communities of equal size and equal average expected degree of 20. After we have chosen: (i) the group structure, (ii) expected degrees of all vertices (θ_i in Eq. (2)), and (iii) the value of ω_{rs}, the generation of graph is implemented as follows: we first draw a Poisson-distributed number of edges for each pair of communities r and s with mean ω_{rs}, and then assign each end of an edge to an arbitrary vertex i according to θ_i. The proposed algorithm (**MC**), **SBM** and **DCM** were respectively applied to the experimental networks. For **SBM** and **DCM**, we choose the best result from 10 independent runs, initialized either by random (with the suffix **R**) or planted assignment (with the suffix **P**).

As shown in Fig. 2(a), the proposed algorithm, with excellent stability, outperforms other models as λ increases from zero, although none of them finds any really significant result because the network in this sense is nearly random. Note that, **MC** calculates objective score (Eq.(4)) for each candidate community structure, but only the one with the highest score is chosen as the final result. So there exists a *gap* between the NMI of the final result and the highest NMI among candidates: when λ is close to zero, the gap is a little bigger, and decreases with λ increasing. With λ beyond the mid-value, the curves of **MC** and **DCM** nearly coincide. The uncorrected model (**SBM**), as expected, finds no significant community structure even at λ=0.5.

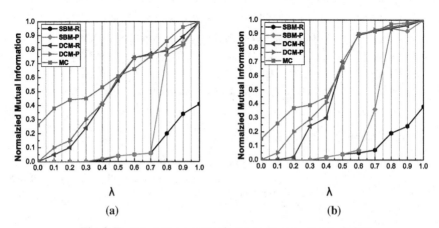

Fig. 2. Experiments on synthetic networks generated by DCM

Then we turn to another group of experiments on DCM synthetic networks, which are closer to the practical networks (wherein community 1 follows power-law degree distribution with exponent $\gamma = 2.8$, and other settings remain unchanged). Figure 2(b) shows that **MC** works well even for small λ. That means we could reveal acceptable result even on nearly random graph, because it can classify most of the vertices simply based on their degrees. In contrast, the **DCM** has accuracy near 0 when $\lambda \leq 0.2$, however NMI of **DCM** increases rapidly, especially **DCM-P**. When each edge has nearly the same probability of being drown from the random model and from the planted structure (i.e., $0.5 \leq \lambda \leq 0.7$), the accuracy of **DCM** increases faster. Moreover, the broad degree distribution in community 1 makes **SBM** perform much more badly here than in the former experiment. Comparing the two synthetic network experiments above, we could conclude that the proposed algorithm performs better in practical network.

Experiments on LFR Synthetic Networks. LFR is a realistic benchmark for community detection, accounting for the heterogeneity of both degree and community size. We test effect of flowing main parameters for network generation on diverse algorithms: (a) mixing parameter, λ; (b) number of vertices, N; (c) average degree of vertices, $\langle k \rangle$.

Firstly, we make comparisons with [10] by using the same parameters with two different network sizes (1,000 and 5,000 vertices), average degree of 20, maximum degree of 50, degree exponent of -2 and community exponent of -1. The results of **MC** are shown in Fig. 3(a), wherein, notations **S** and **B** denote "small" (10-50 vertices) and "big" (20-100 vertices) community sizes, respectively. The results show that **MC** performs as well as, even better than, [10], and there is also a sharp ascent of NMI as λ increases from 0.3 to 0.4. Secondly, we compare the proposed algorithm with other algorithms, such as **CNM**, **SBM** and **DCM**. To test scalability, we generate a family of networks of increasing sizes, and Fig. 3(b) and Fig. 3(c) present the NMI of competing algorithms respect to N and $\langle k \rangle$. We could indicate the number of vertices has only limited effect on the precision, whereas the average degree appears to have a strong impact (Almost all the competing algorithms suffer considerable loss in NMI, and most of them reach the peak when $\langle k \rangle$ is about 24.83).

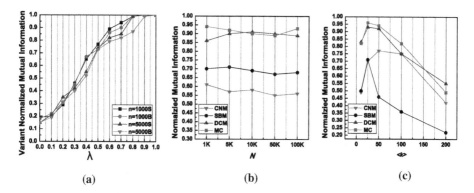

Fig. 3. Experiments on LFR standard benchmark networks

6   Conclusion

Different kinds of algorithms about community detection, such as (i) modularity-based algorithms, (ii) statistical inference methods and (iii) graph clustering/partitioning-based algorithms, have been proposed. This article is an innovative attempt, aiming at exploring the potential connecting lines among them.

We propose a new algorithm for detecting community structure from topology of the network, based on the demonstration that the problem of (normalized) modularity maximization could be transformed into versions of (normalized) minimum-cut graph partitioning. The experimental results showed that the proposed algorithm has generally a better quality than the best existing algorithms. Especially, it performed well in the case of strong degree imbalances as in power-law distributions. Besides unprecedented scale, another important property of social network nowadays is dynamic nature. So, further research will focus more on online or dynamic community detection problem.

Acknowledgement. The work described in this article was supported by the National Basic Research Program of China (973 Program, Grant No. 2013CB329605) and National Natural Science Foundation of China (Grant No. 61201351).

References

1. Newman, M.E.J., Girvan, M.: Finding and Evaluating Community Structure in Networks. Physical Review E. **69**(2), 026113 (2004)
2. Fortunato, S.: Community detection in graphs. Physics Reports. **486**(3), 75–174 (2010)
3. Zhou, T.C., Ma, H., Lyu, M.R., King, I.: User-rec: a user recommendation framework in social tagging systems. In: Proceedings of 24th AAAI Conference on Artificial Intelligence, pp. 1486–1491. AAAI Press, Atlanta (2010)
4. Weng, J., Lee, B.S.: Event detection in twitter. In: Proceedings of 5th International AAAI Conference on Weblogs and Social Media, pp. 401–408. AAAI Press, Barcelona (2011)
5. Papadopoulos, S., Kompatsiaris, Y., Vakali, A., Spyridonos, P.: Community detection in Social Media. Data Mining and Knowledge Discovery **24**(3), 515–554 (2011)
6. Clauset, A., Newman, M., Moore, C.: Finding community structure in very large networks. Physical Review E. **70**(6), 066111 (2004)
7. Newman, M.E.J.: Community detection and graph partitioning. EPL (Europhysics Letters). **103**(2), 28003 (2013)
8. Karypis, G., Kumar, V.: A Fast and High Quality Multilevel Scheme for Partitioning Irregular Graphs. SIAM Journal on Scientific Computing **20**(1), 359–392 (1998)
9. Condon, A., Karp, R.M.: Algorithms for Graph Partitioning on the Planted Partition Model. Random Structures and Algorithms. **18**(2), 116–140 (2001)
10. Lancichinetti, A., Fortunato, S.: Community detection algorithms: A comparative analysis. Physical Review E. **80**, 056117 (2009)
11. Karrer, B., Newman, M.E.J.: Stochastic blockmodels and community structure in networks. Physical Review E. **83**(1), 016107 (2011)
12. Newman, M.E.J.: Finding community structure in networks using the eigenvectors of matrices. Physical Review E. **74**(3), 036104 (2006)

13. Fjallstrom, P.: Algorithms for Graph Partitioning: A Survey. Linkoping Electronic Articles in Computer and Information Science. **3**(10) (1998)
14. Newman, M.E.: Modularity and community structure in networks. Proceedings of the National Academy of Sciences **103**(23), 8577–8582 (2006)
15. Fiedler, M.: Algebraic Connectivity of Graphs. Czechoslovak Mathematical Journal **23**(98), 298–305 (1973)
16. White, S., Smyth, P.: A spectral approach to find communities in graphs. Proceedings of SIAM Conf. on Data Mining **5**, 76–84 (2005)
17. Erdös, P., Renyi, A.: On Random Graphs. Publ. Math. **6**, 290–297 (1959)
18. Chung, F., Lu, L.: Connected Components in Random Graphs with Given Expected Degree Sequences. Annals of Combinatorics. **6**(2), 125–145 (2002)
19. Fred, A.L.N., Jain, A.K.: Robust data clustering. In: Proceedings of 2003 IEEE Computer Society Conference on Computer Vision and Pattern Recognition II, pp. 128–133. IEEE Press, Madison, Wisconsin, USA (2003)
20. Zachary, W.W.: An Information Flow Model for Conflict and Fission in Small Groups. Journal of Anthropological Research. **33**(4), 452–473 (1977)
21. Lusseau, D., Schneider, K., Boisseau, O.J., Haase, P., Slooten, E., Dawson, S.M.: The bottlenose dolphin community of Doubtful Sound features a large proportion of long-lasting associations. Behavioral Ecology and Sociobiology. **54**, 396–405 (2003)
22. Adamic, L.A., Glance, N.: The political blogosphere and the 2004 U.S. election. In: Proceedings of 3rd International Workshop on Link Discovery, pp. 36–43. ACM Press, New York (2005)
23. Lancichinetti, A., Fortunato, S., Radicchi, F.: Benchmark graphs for testing community detection algorithms. Physical Review E. **78**, 046110 (2008)
24. Chung, F., Lu, L.: The average distances in random graphs with given expected degrees. Proceedings of the National Academy of Sciences of the United States of America **99**(25), 15879–15882 (2002)

Coherent Topic Hierarchy: A Strategy for Topic Evolutionary Analysis on Microblog Feeds

Jiahui Zhu[1], Xuhui Li[2,3], Min Peng[2,4(✉)], Jiajia Huang[2], Tieyun Qian[2], Jimin Huang[2], Jiping Liu[2], Ri Hong[2], and Pinglan Liu[2]

[1] State Key Lab of Software Engineering, School of Computer, Wuhan University, Wuhan, China
[2] School of Computer, Wuhan University, Wuhan, China
pengm@whu.edu.cn
[3] School of Information Management, Wuhan University, Wuhan, China
[4] Shenzhen Research Institute, Wuhan University, Wuhan, China

Abstract. Topic evolutionary analysis on microblog feeds can help reveal users' interests and public concerns in a global perspective. However, it is not easy to capture the evolutionary patterns since the semantic coherence is usually difficult to be expressed and the timeline structure is always intractable to be organized. In this paper, we propose a novel strategy, in which a coherent topic hierarchy is designed to deal with these challenges. First, we incorporate the sparse biterm topic model to extract some coherent topics from microblog feeds. Then the topology of these topics is constructed by the basic Bayesian rose tree combined with topic similarity. Finally, we devise a cross-tree random walk with restart model to bond each pair of sequential trees into a timeline hierarchy. Experimental results on microblog datasets demonstrate that the coherent topic hierarchy is capable of providing meaningful topic interpretations, achieving high clustering performance, as well as presenting motivated patterns for topic evolutionary analysis.

Keywords: Coherent topic hierarchy · Topic evolution · Microblog feed · Bayesian rose tree

1   Introduction

The prosperity of the microblogging services at all levels of the social life has been witnessed in the past few years. This kind of services, such as Twitter and Sina Weibo, are becoming the preferred online platforms for people to express their experiences, opinions, thoughts, etc. Therefore, microblog feeds, i.e., the textual aspects of user generated contents, gradually act as the carriers of topics which somehow reflect public concerns at that time. With time elapsing, these topics often evolve in both word presentation and occurrence intensity, leading

This research is supported by the Natural Science Foundation of China(No.61472291, No.61272110, No.61272275, No.71420107026, No.164659), and the China Postdoctoral Science Foundation under contract No.2014M562070.

J. Li and Y. Sun (Eds.): WAIM 2015, LNCS 9098, pp. 70–82, 2015.
DOI: 10.1007/978-3-319-21042-1_6

to complex relations between every two sequential time points. Modeling the topic evolution can be of great significance to track user preferences or event stages, thus providing helpful guidance in personalized recommendation, opinion summarization and emergency detection.

LDA-based [1] probabilistic topic models, combined with temporal information or transition patterns, are usually considered as the mainstream technologies of solving the topic evolution due to their advantages of high robustness and fitness. These variants of LDA [2, 3] not only strive to extract the topics over the word distribution, but also reconcile the topics to their evolutionary threads. Apart from the topic models, classical data mining methods, including clustering [4], frequent patterns mining [5] and automatic summarization [6] are also quite suitable for the topic evolutionary analysis. These non-parametric or less parametric methods are overall more efficient and motivated than the topic models which are always involved in intractable training and inferring.

However, there are still some challenges to analyze topic evolution based on all of the above methods. The first one is how to effectively extract a set of semantically coherent topics from the microblog feeds which have the characteristics of short text, erratic form and voluminous amount. The topics released by this kind of LDA variants [2, 3] are named as *latent topics*, with a paucity of concentration and interpretation. These incoherent topics are not appropriate for the evolutionary analysis as they may entangle users in the semantic confusion. The second challenge is how to efficiently construct the evolutionary structure. The results of the methods based on classical data mining are not always intuitive enough since they pay little attention to the topology of the topics. Recent works [7, 8] attempt to clarify that the relations of different topics can be modeled into the structure of the hierarchy, namely *topic hierarchy*. This type of hierarchical structure sketches the topic evolutionary patterns both horizontally and vertically, resulting in a more succinct presentation compared with a welter of topic threads as in [6].

In this paper, in order to effectively overcome the challenges described above, we design a coherent topic hierarchy (CTH) to improve the coherence of topics as well as the presentation of evolution. First, the *biterm topic model* (BTM) [9] is incorporated to discover some coherent topics from the microblog feeds at each time point. Then the simBRT, which is a *topic similarity enhanced* version of the basic *Bayesian rose tree* (BRT) [10], is developed to figure out the relations of the topics at different time points. Finally, a *cross-tree random walk with restart* (CT-RWR) model is established to effectively generate the whole topic hierarchy along the timeline. Our contributions are listed as follows:

(1) We design a coherent topic hierarchy to cope with the topic evolutionary analysis. In this hierarchical strategy, both the topic coherence and the relation structure are carefully considered.

(2) We smooth the basic Bayesian rose tree to better accommodate the real-world topics topology of the microblog feeds. Therefore, the topic similarity is of great utilization during the tree construction.

(3) We formulate a CT-RWR model to measure the relations between each pair of sequential trees, so as to maintain a hierarchical structure along the timeline.

2   Related Work

Topic Model based Methods: In order to analyze the evolution of topics, researchers were assiduous to devise probabilistic topic models with temporal information. Some of the fundamental works include DTM [2], TOT [11], and OLDA [3]. To further discover some essential evolutionary patterns, the evolutionary hierarchical Dirichlet processes (EvoHDP) [12] was proposed. However, the EvoHDP requires a cascaded Gibbs sampling scheme to infer the model parameters, which may prohibit the explicit comprehension. More efficient models, such as TM-LDA [13], focus on formulating the topic transition through matrix decomposition, while neglecting the interpretation of its topics. All the topic models above are incoherent topic models, which fail to present meaningful topics.

Coherent Topic Models: To enhance the interpretation of the topics, coherent topic models [14] are developed to refine the LDA's topics in a more meaningful way. Usually, the coherent topic models are linked to the knowledge bases [15], and the domain information is helpful to improve their semantic coherence. Coherent topic models are also known as focused topic models, which concentrate on the posterior sparsity of the document-topic and topic-word distributions. Presenting the document-topic and topic-word distributions sparsely, like the DsparseTM in [16], can sufficiently enhance the distinctiveness of the topics. The biterm topic model [9] is also a coherent topic model that is designed to better fit the short text cases. In this paper, we employ the BTM to extract some coherent topics.

Topic Hierarchy: Hierarchical structure can vividly reveal the relations between topics and their traces along timeline [17]. Topic hierarchy is not easily constructed when encountered with multi-source texts [18]. Recently, Bayesian rose tree [10] has attracted great attention in text mining [7] due to its high fitness and smoothness. It can organize the relations of documents or topics into a multi-branch tree through likelihood maximization. Though the interpretable characteristics of coherent topic models and the fitness of Bayesian rose tree are quite appropriate for modeling the topic concepts, previous works still concern little about their cooperation.

3   Coherent Topic Hierarchy

The main objective of our work is to construct a coherent topic hierarchy along the timeline for topic evolutionary analysis. Generally, the construction of our CTH can be divided into three parts:

(1) *Coherent topics extraction*: Including the coherent topic model BTM and sparsification of topic distributions.

(2) *Topic tree construction*: Including the topic topology construction based on the combination of BRT and topic distribution similarity.

(3) *Timeline topic hierarchy construction*: Including the construction of the topic hierarchy along the timeline by our CT-RWR.

The overall procedures of the CTH construction are shown in Figure 1.

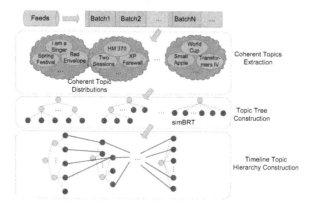

Fig. 1. The whole procedures of the CTH construction

3.1   Coherent Topics Extraction

In our cases, the whole microblog feeds are regarded as a stream $S = \{D_1, D_2, ..., D_t, ...\}$, where $D_t = \{d_1^t, d_2^t, ..., d_M^t\}$ is a feeds collection (feeds batch) at time point t, d_M^t is the Mth feed in D_t. After text modeling with vector space of vocabularies, the BTM can be employed to extract the coherent topics.

In BTM, all the feeds in a batch are considered to have K topics, so BTM first extracts K topics from the feeds batch D_t by using the document-level word co-occurrence patterns, i.e., biterms. These K topics are the topic-word distributions, denoted as ϕ_t, each of which is a multinational distribution over the vocabulary with Dirichlet prior β. Then for each biterm in the feeds batch, topics are assigned by collapsed Gibbs sampling [9] to form the topic assignment sequence of all the feeds. Finally, the feed-topic distribution θ_t with Dirichlet prior α can be inferred by counting the proportion of the biterms' topic assignments indirectly. By performing the BTM, we can derive the feed-topic distribution $\theta_t \in \mathbb{R}^{M \times K}$ and the topic-word distribution $\phi_t \in \mathbb{R}^{K \times V}$ at time point t, where M is the number of the feeds and V is the size of the vocabulary.

To further enhance the topic coherence, the sparsity of the topic-word distribution should be ensured. In real-world microblog stream, it is also reasonable to assume that each topic is only related with a limited number of words, rather than the whole vocabulary [16]. Hence we define the sparsity of each topic as sp ($sp \ll V$), that means each topic only cares about no more than sp words in this vocabulary. Therefore, we improve the BTM via considering the sparsity of word distributions of its topics to make it more appropriate in generating coherent topics from microblog stream. The procedures of the topics sparsification are: 1) For each topic in ϕ_t, select the sp most relevant words with the highest probabilities as its topic descriptive words; 2) sum up the probabilities over these sp words, denoted as p_{sum}; 3) normalize each of these sp probabilities with p_{sum}, while set the rest irrelevant probabilities as zeroes. We denote these sparse topic-word distributions as ς_t, namely, the coherent topics.

3.2   Topic Tree Construction

In this section, we organize the coherent topics at each time point to a topic tree. As the merge or split patterns of topics in real-world applications are not always *one-on-two*, so the traditional topic hierarchy based on binary tree may fail to figure out the intrinsic topology. To overcome this problem, we take the advantage of the Bayesian rose tree [10], a kind of multi-branch tree which allows *one-to-many* relations between topics. To further reveal the real topic topology, we take the topic similarity into account. We name our tree as simBRT.

Basic BRT. Usually, the BRT is constructed in a greedy aggregation. For a BRT at time point t, all the coherent topics $\varsigma_t = \{z_1^t, z_2^t, ..., z_K^t\}$ are its leaves, the main problem is how to organize them into a multi-branch tree to better fit the text data. At first, each topic z_i^t is regarded as an individual tree on its own, namely, $T_i^t = \{z_i^t\}$. Then in each iteration, two trees are selected to greedily consist a new tree T_m by one of the three following basic operations:

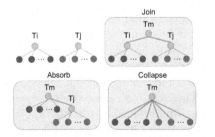

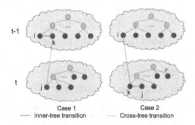

Fig. 2. Three basic operations of BRT   **Fig. 3.** Two types of cross-tree transition patterns: BBA (Case 1) and BAA (Case 2)

(1) Join: $T_m^t = \{T_i^t, T_j^t\}$, so T_m has two branches.
(2) Absorb: $T_m^t = \{ch(T_i^t), T_j^t\}$, so T_m has $|T_i^t| + 1$ branches.
(3) Collapse: $T_m^t = \{ch(T_i^t), ch(T_j^t)\}$, so T_m has $|T_i^t| + |T_j^t|$ branches.

Here $ch(\cdot)$ denotes the children set of a tree. The join is the traditional operation as in binary tree, while the absorb and collapse operations cater to the multi-branch tree. These three operations are shown in Figure 2. In each iteration, the combining objective is to maximize the following ratio of probability:

$$p(\varsigma_m^t | T_m^t) / \big(p(\varsigma_i^t | T_i^t) p(\varsigma_j^t | T_j^t)\big) \qquad (1)$$

where $\varsigma_m^t = \varsigma_i^t \cup \varsigma_j^t$ are the coherent topics under tree structure T_m^t, and $p(\varsigma_m^t | T_m^t)$ is the likelihood of topics ς_m^t under T_m^t. Previous works [7, 10] have addressed that $p(\varsigma_m^t | T_m^t)$ can be calculated through a dynamic programming paradigm:

$$p(\varsigma_m^t | T_m^t) = \pi_{T_m^t} f(\varsigma_m^t) + (1 - \pi_{T_m^t}) \prod_{T_i^t \in ch(T_m^t)} p(\varsigma_i^t | T_i^t) \qquad (2)$$

where $f(\varsigma_m^t)$ is the marginal probability of ς_m^t, which can be modeled by the DCM distribution [7], $ch(T_m^t)$ is the children set of T_m^t, and $\pi_{T_m^t}$ is the prior probability that all the topics in T_m^t are kept in the same partition, $\pi_{T_m^t}$ is defined as:

$$\pi_{T_m^t} = 1 - (1 - \gamma)^{n_{T_m^t} - 1} \qquad (3)$$

where $n_{T_m^t} = |ch(T_m^t)|$, and $0 \le \gamma \le 1$ is the partition granularity.

simBRT. Eq.1 has provided high fitness to the tree construction, but in our scenario, the tree nodes are the topic distributions, rather than the document vectors in previous works. To simultaneously achieve high smoothness, we must also take the topic similarity into consideration. To this end, we refactor Eq.1 by adding the similarities of the topic distributions into join, absorb, or collapse operations.

As the topics are the distributions over the vocabularies, so the Kullback-Leibler divergence can be used to measure the similarity between every two topics. For topic z_i and z_j, their similarity is defined as:

$$topic\_sim(z_i || z_j) = 1 / ((KLD(z_i || z_j) + KLD(z_j || z_i))/2 + 1) \qquad (4)$$

where $KLD(z_i || z_j) = \sum_{k=1}^{V} \phi_{ik} \log (\phi_{ik}/\phi_{jk})$ is the KL divergence between these two topics. To obtain the topic similarity in tree construction, the most important step is to define the weighted topic distribution in each operation.

(1) For the join operation, where the new tree $T_m^t = \{T_i^t, T_j^t\}$, is combined by T_i^t and T_j^t. Hence, their weighted topic distribution is defined as:

$$WT = \frac{avg(\varsigma_i^t)p(\varsigma_i^t | T_i^t) + avg(\varsigma_j^t)p(\varsigma_j^t | T_j^t)}{p(\varsigma_i^t | T_i^t) + p(\varsigma_j^t | T_j^t)} \qquad (5)$$

(2) For the absorb operation, where the new tree T_m^t is comprised by T_j^t and all the children of T_i^t. Then the weighted topic distribution is defined as:

$$WT = \frac{avg(\varsigma_j^t)p(\varsigma_j^t | T_j^t) + \sum_{Ta \in ch(T_i^t)} avg(\varsigma_{Ta})p(\varsigma_{Ta} | Ta)}{p(\varsigma_j^t | T_j^t) + \sum_{Ta \in ch(T_i^t)} p(\varsigma_{Ta} | Ta)} \qquad (6)$$

(3) For the collapse operation, where the new tree T_m^t is comprised by all the children of T_i^t and T_j^t. So the weighted topic distribution is defined as:

$$WT = \frac{\sum_{Ta \in ch(T_i^t)} avg(\varsigma_{Ta})p(\varsigma_{Ta} | Ta) + \sum_{Tb \in ch(T_j^t)} avg(\varsigma_{Tb})p(\varsigma_{Tb} | Tb)}{\sum_{Ta \in ch(T_i^t)} p(\varsigma_{Ta} | Ta) + \sum_{Tb \in ch(T_j^t)} p(\varsigma_{Tb} | Tb)} \qquad (7)$$

Specifically, the final merged topic distribution under T_m^t is $avg(\varsigma_m^t)$. Then the topic similarity between the weighted topic WT and the final merged topic $avg(\varsigma_m^t)$ is added into the primitive objective function in Eq.1. Thus, Eq.1 can be rewritten as:

$$topic\_sim(WT, avg(\varsigma_m^t)) \cdot p(\varsigma_m^t|T_m^t) / \left(p(\varsigma_i^t|T_i^t)p(\varsigma_j^t|T_j^t) \right) \qquad (8)$$

Therefore, in our scenario, to construct the most reasonable tree to interpret the real topology of the topics is equivalent to maximize Eq.8 with join, absorb, or collapse operation at each step. In this way, we can derive a similarity enhanced Bayesian rose tree of the coherent topics at time point t.

3.3   Timeline Topic Hierarchy Construction

In Section 3.2, we have structured a tree of all the coherent topics at each time point. To further analyze the evolution of the topics along the timeline, we need to bond each pair of sequential trees with topic transition metrics.

Sequential Trees Modeling with CT-RWR. A traditional way of measuring the relations of topics from two sequential time points is to directly compute the KL divergence of each pair of topics, as in [5]. However, in this case, the tree structure of the topics is neglected. In order to make full use of the tree structure, we not only take the KL divergence, but also take the *likelihood of the First Common Ancestor* (lFCA) as the similarity metrics.

For two sequential simBRTs at time $t-1$ and t, we model the topic evolution based on the restart version of random walk (RWR) [19]. As our RWR model is settled between two sequential trees, we name it cross-tree RWR. In traditional RWR, topic z_i evolves to other topics with probabilities $p_{ik}(k = 1, 2, ...K, i \neq k)$, and evolves to itself with probability p_{ii}. Usually, the evolution yields to a steady state after several times of walking. Here, for all the topics in ς^{t-1} and all the topics in ς^t, we have two types of similarities: *inner-tree similarity* and *cross-tree similarity*. The inner-tree similarity within either ς^{t-1} or ς^t is defined by the lFCA of each pair of topics:

$$W_{ij}^{t-1} = lFCA(\varsigma_i^{t-1}, \varsigma_j^{t-1}), W_{ij}^t = lFCA(\varsigma_i^t, \varsigma_j^t) \qquad (9)$$

For the cross-tree similarity between ς^{t-1} and ς^t, we resort to KL divergence[1]:

$$W_{ij}^* = topic\_sim(\varsigma_i^{t-1}, \varsigma_j^t) \qquad (10)$$

Then the transition probability matrices are as follows:

$$\mathbf{P}^{t-1} = (\mathbf{D}^{t-1})^{-1}\mathbf{W}^{t-1}, \mathbf{P}^t = (\mathbf{D}^t)^{-1}\mathbf{W}^t, \mathbf{P}^* = (\mathbf{D}^*)^{-1}\mathbf{W}^* \qquad (11)$$

where $\mathbf{D}^{t-1}$, $\mathbf{D}^t$ and $\mathbf{D}^*$ are the degree matrices respectively. According to the RWR model, the inner-tree state probability matrices $\mathbf{R}^{t-1}$ and $\mathbf{R}^t$ can be formulated by the iterative ways as in [19]. Consequently, both of these two state

---

[1] Attention that the vocabularies of the topic distributions in $t-1$ and t are different.

probability matrices will converge after several steps. The converged matrices can be calculated as:

$$\mathbf{R}^{t-1} = (1-\mu)(\mathbf{I} - \mu\mathbf{P}^{t-1})^{-1}, \mathbf{R}^t = (1-\eta)(\mathbf{I} - \eta\mathbf{P}^t)^{-1} \qquad (12)$$

where μ and η are prior probabilities that the topic will not evolve to itself.

Cross-tree Relations Organization. Note that the key in our timeline topic hierarchy construction is the cross-tree transition probability matrix $\mathbf{P}^*$, which seems not so easy to be deduced (like Eq.11) because of its dynamic characteristic. In our cases, the inner-tree transition probability matrices $\mathbf{P}^{t-1}$ and $\mathbf{P}^t$ are stable, as we have constructed the simBRT to reveal their intrinsic relations. As for the cross-tree transition probability matrix $\mathbf{P}^*$, its stability is not guaranteed. For the topic ς_i^{t-1} in time point $t-1$ and the topic ς_j^t in time point t, their transition may be influenced not only by the similarity of this pair of topics, but also some other topics that are implicitly related to ς_i^{t-1} or ς_j^t.

Overall, the $\mathbf{P}^*$ in CT-RWR model is quite complicated due to the various cross-tree transition patterns. After taking into full account of the transition rules, we focus on two types of cross-tree transition patterns: *Before->Before->After* (BBA) and *Before->After->After* (BAA), as shown in Figure 3. Then the stable transition probability, denoted as $\mathbf{SP}^*$, can be calculated as:

$$\mathbf{SP}^* = \delta\mathbf{R}^{t-1}\mathbf{P}^* + (1-\delta)\mathbf{P}^*\mathbf{R}^t \qquad (13)$$

where δ is the prior probability of the transition pattern BBA. $\mathbf{R}^{t-1}\mathbf{P}^*$ involves all the relations that come from the topics in time point $t-1$, while $\mathbf{P}^*\mathbf{R}^t$ involves all the relations that come from the topics in time point t.

The matrix $\mathbf{SP}^*$ reflects the stable transition probability (i.e., the intrinsic relevance) of the topics between two sequential batches both in topic-word distributions and tree structures. Additionally, we give out two thresholds ω and ε to burn out the weak relations. The relations are selected by two rules: 1) All the relations with strength above ω will be reserved, while all the relations with strength below ε will be discarded. 2) In other cases, for each topic in ς^t, only the topic with the highest strength from ς^{t-1} will be related.

By implementing CT-RWR to all the pair of sequential simBRTs, we finally capture the whole topic hierarchy along the timeline.

4   Experiments

To evaluate the effectiveness of our CTH, we carry out our experiments on both Sina Weibo feeds and Twitter feeds. The Sina Weibo feeds are gathered by our web crawler[2] when given the search keywords, e.g., "*Two Sessions*", "*MH 370*". We set 65 keywords, each of which represents a global topic. The dataset spans from Jan. 1 to Sep. 15 in 2014, with totally about 6.6 million feeds labelled

---

[2] http://sc.whu.edu.cn/

by these 65 keywords. All the data is divided into 257 batches, each of which contains all the feeds produced in one day. The Twitter feeds dataset is provided by [8], including 12 global topics and approximately 3 million feeds from Oct. 1, 2012 to Jan. 4, 2014. The Twitter feeds are divided into 231 batches, two days per batch.

Preprocessing work involves feeds batch filtering, word segmentation, stop words removal, vector space modeling, etc. Here we use Jieba[3] for Chinese word segmentation. All the algorithms are implemented in Python except the BTM[4].

4.1 Evaluation of Topic Coherence

To evaluate the coherence of topics is difficult since there is no such thing as standard "coherence". In [16], the point-wise mutual information (PMI) has been adopted as a metric to measure the semantic coherence of topics. Therefore, we also take the PMI as the major criterion. Here, we compute the PMI of the topics extracted by the sparse BTM (in our CTH) against the LDA (as a baseline) in both the Sina Weibo and Twitter dataset. Additionally, the topic-word distribution sparsity sp is set to 5. The results are shown in Figure 4.

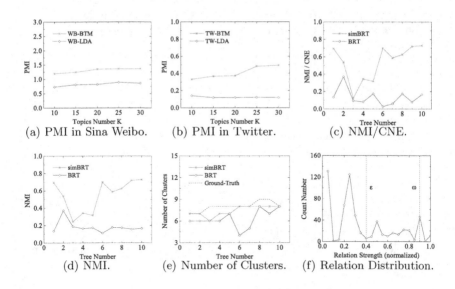

Fig. 4. Evaluation Results

From Figure 4(a) and Figure 4(b), we can see that the PMI scores of the BTM are higher than the LDA in both datasets, indicating the better coherence of the topics produced by the BTM. Meanwhile, the PMI scores of the Sina Weibo topics are expressly higher than those of the Twitter topics. This is partly

---

[3] https://github.com/fxsjy/jieba
[4] https://github.com/xiaohuiyan/BTM

because that the Twitter feeds contains more noisy words or phrases and the word feature space of their topics is larger than that of the Sina Weibo topics. Compared with the smaller word feature space, the larger one is usually not so workable to reveal topics with semantic coherence.

4.2   Evaluation of simBRT

To directly evaluate the effectiveness of our simBRT is also an open ended question. However, the tree structure somehow reflects a clustering result of the topics. Hence we conduct our experiments on the normalized mutual information (NMI) and cluster number error (CNE) as used in [7]. The NMI reveals the similarity of topics within the same cluster, while the CNE measures the deviation of the cluster number between the simBRT and the ground-truth. A higher NMI and a lower CNE indicate a better performance, so we use the NMI/CNE to present the soundness of our simBRT. The results of 10 sequential batches (10 trees) from the Sina Weibo dataset are also displayed in Figure 4. We employ the basic BRT without topic similarity enhancement as a baseline. Here, the topics number K is set to 20 to provide an over-complete topics set.

Figure 4(c) suggests that in most time points, the NMI/CNE of our simBRT is higher than that of the BRT. This can be sure that the simBRT integrates the topic distribution similarity while the basic BRT only depends on the tree structure. Figure 4(d) displays the NMI of each topic tree, where the NMI is highly consistent with the NMI/CNE. Figure 4(e) describes the number of clusters of each topic tree. In our scenario, the clusters number is judged by the number of branches in the second level (root is the first level). Also, we can see that the clusters number of our simBRT is more approximate to the real topology, nearly half of the total cases is strictly equal to the ground-truth, while the BRT is not so well-behaved.

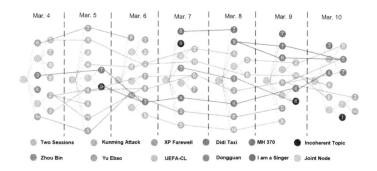

Fig. 5. Timeline topic hierarchy from Mar. 4 to Mar. 10, 2014

4.3 Timeline Topic Hierarchy Presentation

We manipulate each pair of sequential trees with the CT-RWR from Mar. 4, 2014 to Mar. 10, 2014 (totally a week) in Sina Weibo dataset. Here, the topic number K is set to 10 as to give a more succinct structure. The prior probabilities μ and η in CT-RWR are all set as 0.8. Besides, we set the prior probability δ of the cross-tree transition pattern BBA to 0.5, as to share equal probability with BAA. As for the burning thresholds ω and ε for screening out the weak relations, we set them as 0.9 and 0.4 (approximate values after normalized by the range of the relation scores) respectively according to their count distribution, as displayed in Figure 4(f). The coherent topic hierarchy is shown in Figure 5.

In Figure 5, each simBRT contains ten coherent topics and each pair of sequential simBRTs is bonded with colorful edges. Note that each color is related to a certain global topic, e.g., all the green nodes are the coherent topics of the global topic *"Kunming Attack"*. Focusing on the topic tree at Mar. 4, as we can see, all the light blue nodes are put together, so are the green ones. This means that the topics linked to the same global topic can be accurately aggregated together by the simBRT. But this is not always true when the topic tree is too broad (in Mar. 5). Besides, all the simBRTs during this week contain lots of branches at the second level. This is partly because that the topic distinctiveness of the BTM is favorable. And this also suggests that the simBRT can still aggregate the similar topics in such a strictly discriminative environment.

Table 1. Topic keywords evolution of global topics *"Two Sessions"* and *"MH 370"*

Date	Two Sessions	MH 370
Mar. 4	CPPCC committee reporter	
Mar. 5	committee deputy NPC	
Mar. 6	committee nation Li Keqiang	
Mar. 7	government work report	
Mar. 8	nation committee deputy	Kunming MH 370 China
Mar. 9	China committee government	MH 370 Kunming lost
Mar. 10	NPC deputy people	MH 370 Kunming event

According to the whole topic hierarchy, we can explicitly pick out a thread of a global topic. For example, we can see that during this week, the global topic *"Two Sessions"* is popular, thus a complete thread (the light blue thread in Figure 5) "1-1-4-10-6-10-8" runs through all the timeline. Besides, we can also grasp the keywords evolution of this topic, for example, the keywords mentioned in the first day of *"Two Sessions"* are *"CPPCC+committee+reporter"*, but in the next day, the keywords are *"committee+deputy+NPC"*, as listed in the middle column of Table 1 (topic keywords of thread "1-1-4-10-6-10-8"). What's more, we can also discover some emergent topics in this topic hierarchy. For example, in the simBRT at Mar. 8, we can see a red node labelled with "9" that is not related to a previous node and the red color is the first come. In this situation, a new global topic occurs. This global topic is the *"MH 370"*. In general, the coherent topic hierarchy is rather beneficial for evolutionary analysis of topics.

5 Conclusions

In this paper, we present a hierarchical strategy by constructing a timeline coherent topic hierarchy to perform the topic evolutionary analysis on microblog feeds. In our CTH, the semantic coherence is highly ensured by the sparse BTM, and the topic relation is soundly enhanced by the topic distribution similarity based on the Bayesian rose tree. Particularly, the cross-tree relation is well modeled by the CT-RWR, which creates an efficient and precise hierarchical connection along timeline. Our experimental results show that the coherence of topics in our CTH is higher than those of the LDA. Meanwhile, the simBRT outperforms the basic BRT in topics clustering. Besides, the timeline coherent topic hierarchy is quite reasonable for topic evolutionary analysis. In the future, we will refine the CTH to generate more coherent topics and elaborate the CT-RWR to fit for big data cases, so as to consolidate its scalability in real-world applications.

References

1. Blei, D., Ng, A., Jordan, M.: Latent Dirichlet allocation. The Journal of Machine Learning Research **3**, 993–1022 (2003)
2. Blei, D., Lafferty, J.: Dynamic topic models. In: ICML 2006, pp. 113–120. ACM (2006)
3. AlSumait, L., Barbar, D., Domeniconi, C.: Online lda: adaptive topic models for mining text streams with applications to topic detection and tracking. In: ICDM 2008, pp. 3–12. IEEE (2008)
4. Long, R., Wang, H., Chen, Y., Jin, O., Yu, Y.: Towards effective event detection, tracking and summarization on microblog data. In: Wang, H., Li, S., Oyama, S., Hu, X., Qian, T. (eds.) WAIM 2011. LNCS, vol. 6897, pp. 652–663. Springer, Heidelberg (2011)
5. Yang, X., Ghoting, A., Ruan, Y., et al.: A framework for summarizing and analyzing twitter feeds. In: KDD 2012, pp. 370–378. ACM (2012)
6. Shou, L., Wang, Z., Chen, K., et al.: Sumblr: continuous summarization of evolving tweet streams. In: SIGIR 2013, pp. 533–542. ACM (2013)
7. Wang, X., Liu, S., Song, Y., et al.: Mining evolutionary multi-branch trees from text streams. In: KDD 2013, pp. 722–730. ACM (2013)
8. Zhu, X., Ming, Z., Hao, Y., et al.: Customized organization of social media contents using focused topic hierarchy. In: CIKM 2014, pp. 1509–1518. ACM (2014)
9. Yan, X., Guo, J., Lan, Y., et al.: A biterm topic model for short texts. In: WWW 2013, pp. 1445–1456. ACM (2013)
10. Blundell, C., Teh, Y., Heller, K.: Bayesian rose trees. In: UAI 2010 (2010). arXiv:1203.3468
11. Wang, X., McCallum, A.: Topics over time: a non-Markov continuous-time model of topical trends. In: KDD 2006, pp. 424–433. ACM (2006)
12. Zhang, J., Song, Y., Zhang, C., et al.: Evolutionary hierarchical Dirichlet processes for multiple correlated time-varying corpora. In: KDD 2010, pp. 1079–1088. ACM (2010)

13. Wang, Y., Agichtein, E., Benzi, M.: Tm-lda: efficient online modeling of latent topic transitions in social media. In: KDD 2012, pp. 123–131. ACM (2012)
14. Chang, J., Gerrish, S., Wang, C., et al.: Reading tea leaves: how humans interpret topic models. In: NIPS 2009, pp. 288–296. MIT Press (2009)
15. Chen, Z., Mukherjee, A., Liu, B., et al.: Discovering coherent topics using general knowledge. In: CIKM 2013, pp. 209–218. ACM (2013)
16. Lin, T., Tian, W., Mei, Q., et al.: The dual-sparse topic model: mining focused topics and focused terms in short text. In: WWW 2014, pp. 539–550. ACM (2014)
17. Lin, C., Lin, C., Li, J., et al.: Generating event storylines from microblogs. In: CIKM 2012, pp. 175–184. ACM (2012)
18. Zhu, X., Ming, Z., Zhu, X., et al.: Topic hierarchy construction for the organization of multi-source user generated contents. In: SIGIR 2013, pp. 233–242. ACM (2013)
19. Tong, H., Faloutsos, C., Pan, J.: Fast random walk with restart and its applications. In: ICDM 2006, pp. 613–622. IEEE (2006)

Age Detection for Chinese Users in Weibo

Li Chen, Tieyun Qian$^{(\boxtimes)}$, Fei Wang, Zhenni You,
Qingxi Peng, and Ming Zhong

State Key Laboratory of Software Engineering, Wuhan University, Wuhan, China
{whulichen,feiw14,whuznyou}@163.com,{qty,clock}@whu.edu.cn,
pengqingxi@gmail.com

Abstract. Age is one of the most important attributes in one user's profile. Age detection has many applications like personalized search, targeted advertisement and recommendation. Current research has uncovered the relationship between the use of western language and social identities to some extents. However, the age detection problem for Chinese users is so far unexplored. Due to the cultural and societal difference, some well known features in English may not be applicable to the Chinese users. For example, while the frequency of capitalized letter in English has proved to be a good feature, Chinese users do not have such patterns. Moreover, Chinese has its own characteristics such as rich emoticons, complex syntax and unique lexicon structures. Hence age detection for Chinese users is a new big challenge.

In this paper, we present our age detection study on a corpus of microblogs from 3200 users in Sina Weibo. We construct three types of Chinese language patterns, including stylistic, lexical, and syntactic features, and then investigate their effects on age prediction. We find a number of interesting language patterns: (1) there is a significant topic divergence among Chinese people in various age groups, (2) the young people are open and easy to accept new slangs from the internet or foreign languages, and (3) the young adult people exhibit distinguished syntactic structures from all other people. Our best result reaches an accuracy of 88% when classifying users into four age groups.

Keywords: Age detection · Chinese users · Feature selection · Feature combination

1 Introduction

In recent years, there is an increasingly research interest in user profiling in social media [4, 7–9, 11–13, 16]. This task is about predicting a user's demographics from his/her writing, mainly including gender, age, occupation, education, and political orientation. Among which, gender and age are two natural attributes and attract the most research attention. Compared to the problem of gender classification [1–3, 10, 14, 20], age detection is much more difficult and less examined. One reason is that it is hard to collect the labeled age data. As the age information is more sensitive and personal, many users tend to hide it in their profile.

© Springer International Publishing Switzerland 2015
J. Li and Y. Sun (Eds.): WAIM 2015, LNCS 9098, pp. 83–95, 2015.
DOI: 10.1007/978-3-319-21042-1_7

The other reason is the lack of distinct features for age classification. In contrast, there are a bunch of gender features. For example, female users are generally more emotional than male users, and thus many sentimental words can be used to as the identifier of gender. Both these result in less studies on detecting age than on gender. Several pioneering work aims to reveal the relationship between users' age and the language use [4,5,11–13,16–19]. However, existing researches are all based on western languages, i.e., English and Dutch. The age detection problem for Chinese users is so far unexplored.

Due to the cultural and societal difference, some well known features in English may not be applicable to the Chinese users. For example, while the frequency of capitalized letter in English has proven to be a good feature, Chinese users do not have such patterns. More importantly, Chinese has its own characteristics such as rich stylistic expressions, complex syntax and unique lexicon structures. There are about 161 western emoticons. In contrast, there are more than 2900 emoticons commonly used in the main social media in China. In addition, the document written in Chinese needs word segmentation to perform lexical analysis. Considering the fact that there are a number of informal use of language expression, the accuracy of word segmentation will be lowered down and this may affect the performance of word based model. Finally, the syntax structure for Chinese is also quite different from that for English. Will all these points lead to new challenges or chances for the age prediction task for Chinese users? What kind of features is of the most importance to this problem? To what extent can we identify a person's age group given his/her records in social media?

In this paper, we present our study to classify people into age categories. Since this is the first attempt to detecting ages of Chinese users, we have to build our own labeled data. For this purpose, we collect and annotate a corpus of microblogs from 3200 users in Sina Weibo, which is one of the biggest social networking sites in China. We then treat this as a supervised classification problem and derive models in three ways: 1) extracting stylistic features including emoticon, punctuation, and acronym; 2) using lexical based unigram features; and 3) using syntactic part-of-speech (POS) unigram structures. The word and POS unigram features are investigated to examine their effectiveness in Chinese in spirit of a fair comparison with their counterpart in English and Dutch [11–13,15]. It should be also noted that the stylistic features used in [17], [4], and [5] are different from those used here in that they are treated as tokens as word unigrams rather than only a total occurence, given the fact that we extract a large number of novel stylistic features from Chinese microblogs.

The contributions of this paper are as follows:

1. We present a first-of-a-kind age detection problem in Chinese social media. We collect and carefully build an annotated data set. We will publish this data set later for research use.
2. We construct a set of dictionaries, i.e., the stylistic, lexicon, and syntactic feature list. Then we systematically study their effects on age detection.

3. We find a significant topic divergence among Chinese people in various age groups. Furthermore, while old people tend to use conventional expressions, young and young adult people exhibit unique stylistic and syntactic structures, respectively.

Our study will provide new insights into the relationship between Chinese language and its users among different ages. It also has a great number of potential applications such as personalized search, targeted advertisement and recommendation.

The rest of this paper is organized as follows. Section 2 reviews related work. Section 3 presents the overall framework. Section 4 introduces three types of features. Section 5 provides experimental results. Section 6 analyzes some age-related interesting language patterns. Finally, Section 7 concludes the paper.

2   Related Work

With the rise of social media, user profiling has received considerable attention in recent years. There are a number of user identities in one user's profile like gender, age, occupation, education, and political orientation. The overall framework for these tasks are all similar. We will review the literature in age detection in this section, organized by the age group, the feature set, and the classification method.

2.1   Age Group

Age prediction in most of existing studies is approached as a binary or multi-class classification problem. Rao et al. classify users into two major demographic pools, users who are below or above 30 [16]. Rosenthal and McKeown experimented with varying the binary split in order to find Pre- and Post-Social media generations [17]. With the research objective of detecting adults posing as adolescents, Peersman el at. set up the binary classification experiment for 16- vs. 16+, 16- vs. 18+, and 16- vs. 25+ [15]. When being posed as a multi-class problem, persons are categorized according to their life stages. Some of them are related to generations [4,19], i.e., 10s, 20s, 30s, 40s, and higher, or multiple generations [11,13], i.e., 20-, 20-40, and 40+. Others are based on life events (social age) and physical maturity (biological age). For example, Kabbur et al. use a biological age: kid (3..12 years), teen (13..17 years), young-adult (18..34 years), adult (35..49 years) and old (50+ years) [6]. Rosenthal and McKeown also use a three social age spans [17], namely, 18..22, 28..32, and 38..42 in their experiment. More recently, researches also start to predict age as a continuous variable and predicting lifestages [11–13].

In general, there are a number of ways to categorize people into age groups. Nevertheless, it is still problematic in finding clear boundaries. We adopt the three age groups in [17] and add one group for old people, which is close to the social age distribution in China.

2.2   Feature Set

Finding good features has long been a major interest for age detection. The content based features have been proved to be useful for detecting the age of users. The widely used ones are social-lingusitic features [12,16], character n-gram [15,19], word n-gram [11–13,15,17,19], pos n-gram [12], function words [5], and various stylistic features such as the number of punctuation and exclamation marks, slang words, or sentence length [4,5,17]. Besides the above features extracted from the texts, other features may also be used depending on whether their data sources contain such information. For example, the webpage structure is employed in [6], and the network structure and communication behavior are used in [16,17].

While most of the studies are interested in comparing the effects of individual feature set, a few works pay attention to the combination of features [17]. We also note that feature selection is less examined in author profiling unlike that in text classification. The reason can be due to that the removal of common features may incur information loss in stylistic features. One exception is that in [15] which used χ^2 to do feature selection. Their results show that the best performance is achieved by using the largest informative feature set.

We extract the stylistic, lexical and syntactic features from Sina weibo, which are specific to Chinese language. We do not use structure and behavior features since we find they perform very poor in our preliminary experiment. Besides, we do feature selection and feature combination to evaluate their impacts.

2.3   Classification Method

A number of machine learning approaches have been explored to solve the age grading problem, including support vector machine (SVM) [15,16], Naïve Bayes [4], logistic regression [11,17]. When age is treated as a continuous variable, linear regression is adopted to train the model [1,11,12]. There are also a few literatures that study to combining the classifier built from different feature sets. Gressel et al. used a random forest classifier to ensemble learning approaches [5]. Rao et al. employed a stacked model to do simple classifier stacking [16].

The classification model is not the focus of this research. In our study, we choose to use SVM as our classifier.

3   Overall Framework

We follow a supervised learning framework for age detection. We look at four age groups containing a 5-year gap, i.e., young (18..22), young adult(28..32), adult(38..42), old(48..52). Each group is a category or class. A model is trained using features that are extracted from the contents of training users with annotated age spans, and then the model is used to predict the age group of test users. The overall framework is shown in Figure 1.

Since we are the first one studying this problem, there is no available data for this use. It takes us great efforts in building the corpus. Our data set consists of weibo downloaded from Sina, which is the largest social media in China. We

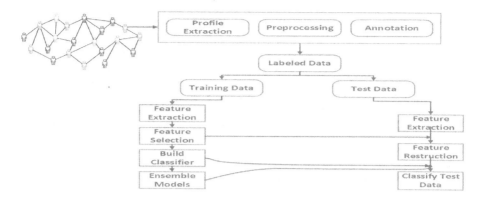

Fig. 1. The overall framework for age detection

use weibo as our testbed since the website provides the users the opportunity to post their age on their profile. We take advantage of this feature by downloading weibo where the user chooses to make this metadata publicly available. We downloaded 319,542 weibos containing age. However, the raw data is extremely imbalanced and contains a large number of noises. The dominant weibo users are those born between 1980 and 1990. There are more than 240,000 people in this span and less than 2,000 people aged over 50-year-old. So we randomly sample 1000 users for each group. We then place the following restrictions to pre-process the corpus:

1. The age in the profile should be valid (in an interval of [10,80]).
2. The account should represent an actual person, e.g. not an organization.
3. The account should have followers less than 1,000 to ensure that he/she is not a celebrity.
4. The account should have sufficient weibos (at least 10).

After the above filtering and pre-processing procedure, we then employ two students to manually check on a sample set of 300 profiles. The age labels are found to be correct for over 95% of them. Finally, we construct a corpus with 3200 users in 4 age classes, each having 800 users. Note that one user's weibos are merged into one document and then it is treated as an instance in each class.

4   Features

We extracted three types of features from users' weibos. They are treated as pseudo-words and their frequencies are counted to form various feature vectors for one user in terms of tf.idf values.

4.1   Stylistic Features

The stylistic features are extracted from the weibo. We use three types of stylistic features including emoticon, punctuation, and non-Chinese characters.

Chinese users are very active in using emoticons. To meet the requirements of users, many IMEs (Input Method Editor) and social networking tools have created their own emoticon set. We integrate the emoticons from Sina and several popular IMEs like Sogo and build our emoticon dictionary, which contains 2919 emoticons. The punctuations are extracted by the same way. There are 227 punctuations in total. The acronym list consists of English characters and digits. It includes English words such as "good", "you", "love", internet slangs such as "hoho", "MM", and abbreviations such as "OMG", "CEO". The feature set size for acronym is 29289.

The usage of stylistic features are different from those in previous studies in that we use them as tokens as word unigrams rather than the total number of occurrences.

4.2 Lexical Features

We represent each user's tweets by a vector of word frequencies. We use the ICTCLAS tool[1] to segment each tweet into words. The vocabulary size for word unigram in our experiment is *158910*. We do not remove stop word as in text categorization. This is because some of the stop words are actually discriminative.

4.3 Syntactic Features

Since the ICTCLAS software also outputs the POS tags for each word, we then use these tags as syntactic features. Figure 2 shows the syntactic structures of two sample sentences. One is correct, and the other is wrongly segmented.

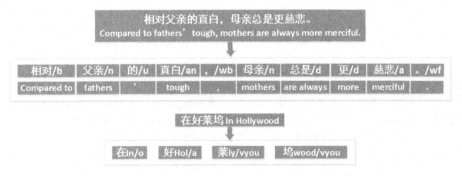

Fig. 2. Syntactic structure of two sample sentences

After segmentation, each word has a POS tag. For example, "father" is tagged as a "*n*" standing for "noun", and "always" to be a "*d*" for "adverb". The total number of syntactic features is 94.

---

[1] http://ictclas.nlpir.org/

4.4   Feature Selection and Combination

We use the χ^2 statistic as a metric to select features [21]. The χ^2 statistic can be used to measure the lack of independence of a token t and a class label c. It is defined as: $\chi(t,c) = \frac{N*(AD-CB)^2}{(A+B)(A+C)(B+D)(C+D)}$, where N is the total number of documents (users in our case), A is the number of times t and c co-occur, B is the number of times t occurs without c, C is the number of times c occurs without t, and D is the the number of times neither t nor c occurs.

We also do feature combinations by 1) merging different types of features into one long vector, and 2) ensembling the classifiers built from different types of features.

5   Experimental Evaluation

In this section, we evaluate the proposed approach. We first introduce the experiment setup, and then present the results using different settings.

5.1   Experiment Setup

All our experiments use the $SVM^{multiclass}$ classifier[2] with default parameter settings. The data are randomly split into two parts: 80% for training and the rest 20% for test. The results are averaged over the 5-fold cross validations. We report the classification accuracy as the evaluation metric.

5.2   Experimental Results

Below we will show our experimental results.

Effects of Feature Selection

In order to compare the effects of feature selection, we compute the χ^2 value for all features in each class of training data, and then sort the features in the descending order of their χ^2 statistic. For training and test data, we only keep features ranked high of the χ^2 list and discard those in the tail. In Figure 3, we show the effects of feature selection with a decreasing ratio of features kept. Note that a 100% setting means using all the original features, i.e., no feature selection is done.

We have the following observations from Figure 3.

– Feature selection has very positive effects on age detection for Chinese users. It can greatly improve the accuracy for all kinds of features. The accuracy increases very fast when removing the most ambiguous features and then goes steady. This finding is new and contradicts to those in existing study in Western language [15]. In the following, we will use 50% as our default setting.

---

[2] http://www.cs.cornell.edu/People/tj/svm_light/svm_multiclass.html

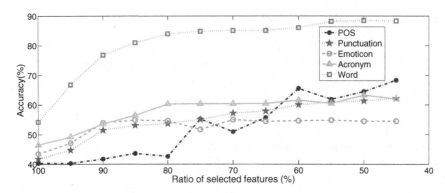

Fig. 3. The effects of feature selection

– The lexical features perform the best among all features, whether or not applying the feature selection procedure. The performances for punctuation and emoticon are generally in the middle. The accuracy for POS fluctuates when the ratio changes. At the start point, its performance is the worst. However, it increases very fast and finally outperforms all others kinds of features except the word unigram.

Effects of Feature Combinations

The effects of feature combinations are shown in Table 1. For easy comparison, we also show the accuracy for single feature on the right side.

Table 1. Effects of feature combination

Combined Features	Acc(longvec)	Acc(ensemble)	Single Feature	Acc
Emotion_Punt_Acronym	65.49	65.56	Acronym	63.31
Emotion_Punt_Acronym_Word	86.81	85.19	Punt	61.34
Emotion_Punt_Acronym_POS	68.66	68.31	Emoticon	54.47
Word_POS	88.72	88.47	POS	64.53
ALL	87.12	85.44	Word	88.50

From Table 1, we can see that:

– Merging features into a long vector is a bit better than ensembling the outputs of classifiers from different features. In the following, we will use feature mergence as the default setting for feature combination.
– While other features benefit from the mergence or ensemble, the accuracy for word unigram is lower down when it is combined with stylisitic features. Even if the combination of word and POS improves the performance a little, the accuracy for all features is lower down when stylistic features join. This indicates that the complicated stylistic structure is not good itself for the combination process.

Effects of the Size of Training Data

We evaluate the effects of number of training data. Figure 4 shows the results.

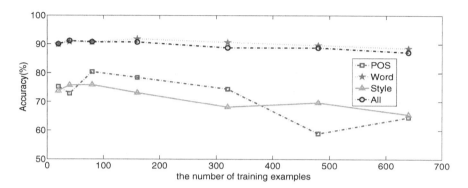

Fig. 4. The effects of training set size

In Fig. 4, we see that the accuracy for word and all features are less sensitive to the number of training examples as the stylistic (style) and syntactic (POS) features do, showing that the training set size has fewer impacts on the features whose dictionary is large. We also notice that most of the classifiers get the best result when using about 160 training examples. This can be due to the fact that the labeled set contains more noises when more users are added.

One Age Group vs. the Other Group

In Table 2, we shows the results for binary classification, i.e., using two age spans as two classes.

Table 2. Results for binary classification

	C1 vs. C2	C1 vs. C3	C1 vs. C4	C2 vs. C3	C2 vs. C4	C3 vs. C4
POS	50.13	65.50	61.44	64.63	67.25	49.94
Punt	64.00	74.44	72.25	65.94	72.06	50.69
Emoticon	54.63	56.38	68.00	54.38	65.00	60.81
Acronym	72.44	80.06	82.50	62.38	70.81	56.38
Word	64.94	81.19	88.13	69.69	75.06	50.56

In general, C1(18..22) *VS.* C4(48..52) get the best result. This is easy to understand because they have the biggest age difference. However, the highest accuracy for POS is achieved on C2(28..32) *VS.* C4(48..52).This is consistent with our observation that the young adults show specific syntactic patterns, as we will will show in Fig. 7. In addition, the accuracy for C3(18..22) *VS.* C4(28..32) is the worst, showing that the adults and old people share more common language patterns than others do.

6   Age Related Language Patterns

In this section we analyze some interesting language patterns.

6.1   Lexical Patterns

We find significant topic divergence among the four age groups in spite of the segmentation errors. Figure 5 shows the top 10 discrimitive word unigrams for each group. From the table, we have the following important notes:

1. The young people seem to be polarized. On one hand, they are interested in those close to their life such as "service" or "experience". On the other hand, the political affairs are still their major topics. For example, "China" and "people" appear as two of the top 10 words in this class.
2. The young adults focus mainly on their daily life. The top three words are about their work status, and none of the top 10 words is related with public affairs. The reason may be that the folks in this age just start to work and many of them are going to get or already got married. They do not have enough spare time.
3. The adult people are the main stream in society. Their interests totally devote to their career. As can be seen, all the top 10 patterns are on the business or marketing.
4. The old people pay more attentions on authorities and abstractions. Eight of ten words are related to the identity which reflects one's social status such as "civilian", "cadre", "chairman", and "scholar".

Young	Young Adult	Adult	Old
公司 *company*	加班 *work overtime*	行业 *industry*	百姓 *civilian*
服务 *service*	上班 *go to work*	营销 *advertise*	干部 *cadre*
体验 *experience*	下班 *off work*	商业 *commerce*	书记 *party secretary*
中国 *China*	哎 *sigh*	消费 *consumption*	中央 *government*
工作 *work*	果然 *as expected*	案例 *case*	主席 *chairman*
进行 *ongoing*	貌似 *look like*	推广 *promotion*	学者 *scholar*
北京 *Beijing*	伤不起 *like a dream*	互联网 *internet*	现代 *modern*
人民 *people*	哈哈 *haha*	业务 *business*	发表 *publish*
及 *and*	肿么 *what*	运营 *operation*	教授 *professor*
功能 *function*	婚礼 *wedding*	智慧 *wisdom*	主任 *director*

Fig. 5. Top 10 discrimitive word unigrams

6.2   Stylistic Patterns

The most discriminative acronyms for age groups are shown in Table 3. Besides the topic divergence as in lexical patterns, we find that the young people are active in using foreign or new Internet slangs. For example, "get" usually means one masters a new skill and "QAQ" stands for a sad expression. Please also note

Table 3. Discriminative acronyms for age groups

Young	Young Adult	Adult	Old
get	ing	CEO	bull
po	mark	App	it
LOL	TM	GDP	like
QAQ	ps	pptx	is
come	pose	HR	on

that "LOL" is a game name in Chinese in most of the cases instead of "laugh out loudly" in English. Meanwhile, we find that other three groups all use acronym in a regular way.

The emoticon patterns are shown in Figure 6. We do not observe obvious patterns for old people. This indicates that the old ones are conservative and tend to use conventional emoticons. In contrast, the young people are willing to accept new and vivid emoticons such as "doge" and "shy". Furthermore, while young people are fond of use emoticons for self-expression, young adults are more concerned with the others. For example, they use "V5" and "handsome" to compliment or praise other people.

Young	Young Adult	Adult	Old
[doge]	[hehe]	[cake]	
[oh yeah]	[V5]		
[love]	[money]		
[laugh to tears]	[snow]		
[shy]	[handsome]		

Fig. 6. Emoticon patterns

Token	Syntax		Examples
e	叹词	*exclamation*	哦、嗨、啊
al	形容词性惯用语	*adjective phrase*	不得了、出神入化、神清气爽
tg	时间词性语素	*time related phrase*	为止、及、来着
y	语气词	*interjection*	么、呢、吧
qv	动量词	*special purpose verb*	次、回、趟

Fig. 7. Special syntactic patterns for young adults

6.3 Syntactic Patterns

While the other three groups are similar in using of syntactic structures, we find the young adults show special POS patterns. The most distinctive patterns for this group are shown in Figure 7. We notice the top two patterns are exclamation and adjective phrases, and the others are about time and unit. We believe this corresponds to the special topic for this group. Remember that they are overwhelmingly related to daily life.

7 Conclusion

In this paper, we investigate the age detection problem for Chinese users. We construct an annotated corpus using rule based filtering and manual check. We extract three types of features to represent users, namely, stylistic, lexical, and syntactic features. We find the word unigrams are the most discriminative features in detecting age, and the syntactic and stylistic features are also informative. The improvement of feature selection is evaluated as significant on all types of features, which contradicts to existing study in Western languages. Our research also discloses a number of interesting language patterns specific to a particular age group. The results do provide us important insights in terms of analyzing the relationship between the Chinese language and their users at various ages.

Our current study focuses on classifying Chinese users into social age group. In the future, we plan to extend our work by detecting age as a continuous variable and predicting users' life stages. In addition, as a first attempt, we only explore three basic types of features. The impacts of other features need further investigation. Finally, our current experiment only involves users' own information. Our next work will study how the users in social network are affected by their neighbors.

Acknowledgments. The work described in this paper has been supported in part by the NSFC projects (61272275, 61232002, 61202036, 61272110, and U1135005), the 111 project(B07037), and SRFDP (20120141120013).

References

1. Bergsma, S., Durme, B.V.: Using conceptual class attributes to characterize social media users. In: Proc. of ACL, pp. 710–720 (2013)
2. Cheng, N., Chen, X., Chandramouli, R., Subbalakshmi, K.P.: Gender identification from e-mails. In: CIDM, pp. 154–158 (2009)
3. Garera, N., Yarowsky, D.: Modeling latent biographic attributes in conversational genres. In: Proc. of ACL and IJCNLP, pp. 710–718 (2009)
4. Goswami, S., Sarkar, S., Rustagi, M.: Stylometric analysis of bloggers' age and gender. In: Proc. of ICWSM, pp. 214–217 (2009)
5. Gressel, G., Hrudya, P., Surendran, K., Thara, S., Aravind, A., Poornachandran, P.: Ensemble learning approach for author profiling. In: PAN at CLEF (2014)
6. Kabbur, S., Han, E.H., Karypis, G.: Content-based methods for predicting web-site demographic attributes. In: Proc. of ICDM (2010)
7. Kosinski, M., Stillwell, D., Graepel, T.: Private traits and attributes are predictable from digital records of human behavior. PNAS **110**, 5802–5805 (2013)

8. Li, J., Ritter, A., Hovy, E.: Weakly supervised user profile extraction from twitter. In: Proc. of ACL, pp. 165–174 (2014)
9. Mislove, A., Viswanath, B., Gummadi, P.K., Druschel, P.: You are who you know: inferring user profiles in online social networks. In: Proc. of WSDM, pp. 251–260 (2010)
10. Mukherjee, A., Liu, B.: Improving gender classification of blog authors. In: Proc. of EMNLP, pp. 207–217 (2010)
11. Nguyen, D., Gravel, R., Trieschnigg, D., Meder, T.: "how old do you think i am?": A study of language and age in twitter. In: Proc. of ICWSM, pp. 439–448 (2013)
12. Nguyen, D., Smith, N.A., Rosé, C.P.: Author age prediction from text using linear regression. In: Proc. of the 5th ACL-HLT Workshop, pp. 115–123 (2011)
13. Nguyen, D., Trieschnigg, D., Doğruöz, A.S., Grave, R., Theune, M., Meder, T., de Jong, F.: Why gender and age prediction from tweets is hard: lessons from a crowdsourcing experiment. In: Proc. of COLING, pp. 1950–1961 (2014)
14. Otterbacher, J.: Inferring gender of movie reviewers: exploiting writing style, content and metadata. In: Proc. of CIKM, pp. 369–378 (2010)
15. Peersman, C., Daelemans, W., Vaerenbergh, L.V.: Predicting age and gender in online social networks. In: Proc. of SMUC, pp. 37–44 (2011)
16. Rao, D., Yarowsky, D., Shreevats, A., Gupta, M.: Classifying latent user attributes in twitter. In: Proc. of SMUC, pp. 37–44 (2010)
17. Rosenthal, S., McKeown, K.: Age prediction in blogs: a study of style, content, and online behavior in pre- and post-social media generations. In: Proc. of ACL, pp. 763–772 (2011)
18. Schler, J., Koppel, M., Argamon, S., Pennebaker, J.W.: Effects of age and gender on blogging. In: Proc. of AAAI Spring Symposium on Computational Approaches for Analyzing Weblogs, pp. 199–205 (2005)
19. Tam, J., Martell., C.H.: Age detection in chat. In: Proc. of ICSC, pp. 33–39 (2009)
20. Xiao, C., Zhou, F., Wu, Y.: Predicting audience gender in online content-sharing social networks. JASIST **64**, 1284–1297 (2013)
21. Yang, Y., Pedersen, J.O.: A comparative study on feature selection in text categorization. In: Proc. of ICML, pp. 412–420 (1997)

Batch Mode Active Learning for Networked Data with Optimal Subset Selection

Haihui Xu[1], Pengpeng Zhao[1(✉)], Victor S. Sheng[2], Guanfeng Liu[1], Lei Zhao[1], Jian Wu[1], and Zhiming Cui[1]

[1] School of Computer Science and Technology, Soochow University,
Suzhou 215006, People's Republic of China
szhhxu@163.com, {ppzhao,gfliu,zhaol,jianwu,szzmcui}@suda.edu.cn
[2] Computer Science Department, University of Central Arkansas, Conway, USA
ssheng@uca.edu

Abstract. Active learning has increasingly become an important paradigm for classification of networked data, where instances are connected with a set of links to form a network. In this paper, we propose a novel batch mode active learning method for networked data (BMALNeT). Our novel active learning method selects the best subset of instances from the unlabeled set based on the correlation matrix that we construct from the dedicated informativeness evaluation of each unlabeled instance. To evaluate the informativeness of each unlabeled instance accurately, we simultaneously exploit content information and the network structure to capture the uncertainty and representativeness of each instance and the disparity between any two instances. Compared with state-of-the-art methods, our experimental results on three real-world datasets demonstrate the effectiveness of our proposed method.

Keywords: Active learning · Batch mode · Correlation matrix · Optimal subset

1 Introduction

With a large amount of data produced by social services, like Twitter and Sina Weibo, much effort has been devoted to processing and understanding networked data. However, obtaining the labels for these data is usually an expensive and time consuming process [1]. One promising approach of reducing the cost of labeling is active learning [2]. The task of active learning is to determine which instances should be selected to query for labels such that a learned classifier could achieve a good performance with requiring as few labels as possible.

A lot of social network systems have been widely used. These systems enable us to obtain a huge amount of networked data. The instances of networked data are connected by links. For example, users in a social network may share friendships. These friendships form a network. In this case, traditional active learning approaches are not appropriate, because they assumes that data instances are

© Springer International Publishing Switzerland 2015
J. Li and Y. Sun (Eds.): WAIM 2015, LNCS 9098, pp. 96–108, 2015.
DOI: 10.1007/978-3-319-21042-1_8

independent and identically distributed. Recent research articles [3][4] exploit the explicit network structure or links in the data instances. They have shown that selectively querying for labels based on data links can significantly improve classification performance. According to the strategies of evaluating data instances to query for labels, existing active learning methods on networked data can be roughly divided into two categories: individual assessment or single mode based, and set assessment or batch mode based. The first category methods select one instance at a time to query for its label. The second category methods intend to select an optimal subset , by levering the inter-correlations to estimate the utility value of a set of instances. The methods belonging to the first category is inefficient to retrain or update the classification model, especially for networked data, because of information redundancy exists among the selected instances in different iterations. There are a few methods [4][5] that solve the problem in a batch mode way, but these batch mode methods are in essence in a single mode. They selected instances with a greedy algorithm, which selects one instance with the maximum utility value each time, iteratively forms a batch instances. The greedy selection can be regarded as a local optimum approach.

We expect batch mode methods can effectively select a batch instances to request labels, such that we can build a good classification model as soon as possible. With this aim in mind, we propose a new active learning framework for networked data. We utilize the network structure as representativeness criteria inside a graph-based metric—betweenness centrality to measure instance informativeness. To capture instance correlations, we combine instance content uncertainty with links representativeness and disparity to form a matrix, where each element denotes a correlation of instances indexed by its corresponding row and column. Instead of using a greedy algorithm, we solve the problem in a global way. We transform this problem to a semi-defined programming (SDP) problem, which selects a subset of instances out of the entire unlabeled set, such that the selected subset of instances have the maximum utility value. Compared to existing methods, the contributions of this paper are threefold:

- Instance selection in a global optimum way. Individuals with the highest informative do not necessarily form an optimal subset. we select an optimal subset instances in a global way.
- A new utility measure for evaluating an instance utility value. To consider the instance correlation, we combine content information and link information in a heuristic weighted strategy to form an item in the correlation matrix, coupling with disparity to capture instance level correlations.
- A new general framework for networked data active learning. We present a novel method by considering multi-instances as a whole to simultaneously make use of their content information and their link information.

The remainder of the paper is organized as follows. Section 2 reviews the related work. The preliminaries are introduced in Section 3. The proposed approach is detailed in Section 4, followed by experiment results in Section 5. In Section 6, we conclude our paper.

2 Related Work

Depending on query strategies, existing active learning methods are generally divided into three categories. The first category is based on uncertainty sampling [6].The second category is based on query-by-committee (QBC) [7], which selects the instances that is considered to be the one the committee disagrees the most. The third category is based on expected error reduction [3], which selects the instance where the expected classification error can be reduced the largest.

Recent researches extend to classify related examples like networked data by exploiting relational dependencies or link information between data instances, which have been shown to be effective to improve the classification model performance under inter-related conditions [8]. Due to the success of taking the data instance dependencies (e.g., instance structure, relationship) into consideration, many graph-based active learning methods have been proposed to address the problem of classifying networked data [3][4][9][10]. We review such single mode based approaches first, and then batch mode based ones. Macskaasy et al. [3] employed graph-based metrics to assess the informativeness of data instances with an empirical risk minimization technique. Fang et al. [11] introduced a social regularization term in the instance utility measure to capture its network structure. Xia hu et al. [9] proposed an active learning method for networked texts in microblogging, which utilizes network metrics and community detection to take advantage of network information.

These aforementioned methods belong to the single mode based approaches, which select one instance to query the oracle for its label. There exist several batch mode methods. Shi et al. [4] proposed a unified active learning framework by combining both link and content information. They defined three criteria to measure the informativeness of a set of data instances.Yang et al. [5] focused on the setting of using the probabilistic graphical model to model the networked data. They used a greedy algorithm to select one data instance with the maximum score at a time and chose all the examples one by one. Obviously, this method essentially is a single mode active selection method.

3 Preliminaries

3.1 Problem Definition

In our batch mode active learning, input data are denoted as a graph $G =< V, E >$, where denotes a set of nodes (one node presents one instance in the input data) including labeled instances V_L and unlabeled instances V_U , and E denotes the edges between nodes. Each node $v_i \in V$ is described by a feature vector and a class label $y_i \in Y$, i.e., $v_i =< x_i, y_i >$. $x_i = [x_{i_1}, x_{i2}, x_{i3}, ..., x_{id}]$ is a vector of dimensional features, and Y denotes a set of class labels,i.e.,$Y = [y_1, y_2, y_3..., y_m]$. Each edge $e_{ij} \in E$ describes some relationship between two node v_i and v_j . For example, in a citation network, the nodes are publications. The features of each node include words, and the label of a node may be one of the topics of the papers. The edges denote as the citations.

Definition 1. *Batch Mode Active Learning(BMAL) for Networked Data:*
Given a set of labeled instances $V_L \in G$ with $V_L = \{< x_i, y_i >\}_{i=1}^m$, a large
number of unlabeled instances V_U, a budget B, our goal is to actively select a
subset of unlabeled instances as a whole at each iteration of active learning, and
query the oracle for their labels under the budget. And this subset is the best for
improving the classification model performance. To solve this problem, a general
objective function for selecting instances to request labels is defined as follows.

$$\max_{V_S \subseteq V_U} Q(V_S), \; |V_S| \leq k, \; \sum |V_S| \leq B \tag{1}$$

where k is the size of the selected subset at each iteration, and $\sum |V_S|$ denotes
the total selected instances.

Thus, this problem is an optimization problem. The following task is how to
instantiate the objectives function $Q(V_S)$ and solve it effectively.

4   A New Framework — BMALNeT

In the above formulation, the objective function can be well instantiated in
different ways. This definition of active learning for networked data has been
widely used in the literature [12][4][13]. We now use a correlation matrix M
to formulate the function. It is formulated as a quadratic integer programming
problem as follows:

$$\max_{X} \; X^T M X$$

$$\text{Subject to}: \sum_{i=1, x_i \in X}^{n} x_i = k, \; x_i \in \{0, 1\} \tag{2}$$

where X is an n-dimensional column vector and n is the size of unlabeled set
V_U. The constraint k defines the size of the subset for labeling at each iteration
of active learning. $x_i = 1$ means that the instance x_i is selected for labeling and
$x_i = 0$ otherwise. In the following, we elaborate how to build the correlation
matrix and how to actively select an optimal subset respectively.

4.1   Correlation Matrix Construction

To build a correlation matrix $M \in R^{n \times n}$, where n is the number of instances
in the unlabeled set V_U, we separate elements in M into two parts. Specifically,
assuming that $C_{i,i}$ denotes the utility value of the uncertainty and representativeness of an instance x_i, and $I_{i,j}, i \neq j$ denotes the disparity between two
instances x_i and x_j, the correlation matrix M is constructed as follows:

$$M_{i,j} = \begin{cases} C_{i,j}, & if \; i = j \\ I_{i,j}, & if \; i \leq j \end{cases} \tag{3}$$

Note that the utility value of an instance integrates the uncertainty and the representativeness of an instance and the disparity. The detail explanations are as follows.

Given the classification model θ_L trained on the labeled set V_L, the uncertainty of an instance $U(x_i)$ is defined as the conditional entropy of its label variable as follows:

$$U(x_i) = -\sum P(y|x_i, \theta_L) log P(y|x_i, \theta_L) \tag{4}$$

This uncertainty measure captures the utility value of the candidate instance with respect to the labeled instances. The larger the conditional entropy, the larger the uncertainty of an instance has.

As we know, representativeness-based active learning methods choose instances which can well represent the overall distribution of unlabeled data. For the networked data, we intend to select representative nodes to capture topological patterns of the entire network. Many methods have been proposed to capture particular features of the network topology in social networks. These methods quantify the network structure with various metrics [14]. In our method, we use one of the widely used graph-based metric, Betweenness Centrality [15], to select representative nodes in a network. Betweenness is one of the most prominent measures of centrality. In a network, a node with a greater betweenness centrality has a more important role between nodes communication. The betweenness centrality is defined as follows:

$$C_B(v_i) = \sum_{v_s \neq v_i \neq v_t \in V} \frac{\sigma_{st}(v_i)}{\sigma_{st}} \tag{5}$$

where σ_{st} is the number of shortest paths between nodes v_s and v_t in the graph ((s, t) denotes as a path), and $\sigma_{st}(v_i)$ is the number of (s, t)-paths that go through the node v_i. We need to compute all-pairs shortest-paths to measure the centrality for all nodes. However, we notice that there are efficient ways for computing this [16].

Given the uncertainty measure and the betweenness centrality measure defined above, we develop a combination framework to integrate the strengths of both content information and link information. Specifically, we combine the two measures in a general form as follows:

$$C_{i,j} = \alpha U + (1 - \alpha) C_B \tag{6}$$

where $0 \leq \alpha \leq 1$ is a tradeoff controlling parameter over the two terms .

Diversity discriminates the differences between instances when taking the sample redundancy into consideration. We use the disparity[17] between each pair of instances to capture the difference between instances such that an optimal subset can contain instances with high disparity and low redundancy. To calculate the disparity of each pair of instances, we employ two types of distance measures, the prediction distance and the feature distance.

The prediction distance intends to compare the prediction dissimilarity of a classifier on two instances. The purpose is to assess the behavioral difference between the pair of instances with respect to the classifier. We estimate the prediction distance based on their prediction probabilities. Given a classifier θ_L, the estimated class membership distribution of the classifier for an unlabeled instance is denoted by $p_x = \{p(y_1|x, \theta_L), p(y_2|x, \theta_L), ..., p(y_m|x, \theta_L)\}$. For a pair of instances x_i and x_j, their prediction difference with respect to this classifier over all class labels $Y = [y_1, y_2, y_3..., y_m]$ is denoted by

$$P_{i,j} = \sum_{l=1}^{m} \mid p(y_l|x_i, \theta_L) - p(y_l|x_j, \theta_L) \mid \tag{7}$$

The feature distance intends to capture the disparity of a pair of instances in the feature space. Given an instance $x_i = [x_{i_1}, x_{i2}, x_{i3}, ..., x_{id}]$, which is also a vector of d-dimensional features of the node v_i , the feature distance between x_i and x_j is calculated as follows:

$$F_{i,j} = \sqrt{\sum_{\lambda=1}^{d} (x_{i\lambda} - x_{j\lambda})^2} \tag{8}$$

Because the prediction distance and the feature distance represent the difference between instances x_i and x_j from different perspectives, we simultaneously the behaviors of two consider instances and their distance in the feature space. We define the final disparity between x_i and x_j as the product of the two distances as follows:

$$I_{i,j} = P_{i,j} \times F_{i,j} \tag{9}$$

Assuming that each of the prediction distance and the feature distance assesses the instance distribution from one dimension, the product therefore assesses the joint distribution from both dimensions and is a better way of assessing the instance disparity.

4.2   Optimal Instance Subset Selection

In this section, we will discuss our solution of selecting an optimal subset of instances. This problem of finding the optimal subset of instances is actually a standard 0–1 optimization problem, which generally is NP-hard. Our solution is based on semi-definite programming. In order to find a optimal solution, we first change the original problem to a max-cut with a size k (MC-k) problem [18]. whose objective is to partition an edge-weighted graph containing N vertices into two parts, with one of which containing k vertices, such that the total weight of edges across the cut is maximum. To transform the original problem into the MC-k problem, we change variable x_i as follows:

$$x_i = \frac{t_i + 1}{2} \tag{10}$$

As $x_i \in \{0,1\}$, $t_i \in \{-1,1\}$. Replacing x_i in (2) using in (10), we have

$$\max_t \frac{1}{4}(e+t)^T M(e+t) \tag{11}$$

$$Subject\ to: \ (e+t)^T I(e+t) = 4k; \ t_i \in \{-1,1\}$$

where t is an n-dimensional vector with values of either 1 or -1 and e is the same-sized column vector with all values being 1 . The original cardinality constraint is rewritten as a quadratic form, where I is an identity matrix. To convet the transformed objective function in (11) and its cardinality constraints into a quadratic form, we expand the vector $t = (t_1, t_2, ..., t_n)$ into an extended from $t = (t_0, t_1, t_2, ..., t_n)$ with $t_0 = 1$ and construct a new matrix $\eta \in R^{(n+1) \times (n+1)}$ as follows:

$$\eta = \begin{pmatrix} e^T M e & e^T M \\ M e & M \end{pmatrix} \tag{12}$$

Also, we can apply the same extension to the cardinality constraints and build a new constraint matrix $\tau \in R^{(n+1) \times (n+1)}$ as follows:

$$\tau = \begin{pmatrix} n & e^T \\ e & 1 \end{pmatrix} \tag{13}$$

As a result, the original instance-selection problem in (2) is transformed into an MC-k problem as follows:

$$\max_t t^T \eta t \tag{14}$$

$$Subject\ to: \ t^T \tau t = 4k; \ t_i \in \{-1,1\}, t_0 = 1, \forall I \neq 0$$

To solve (14), we let $T = t \times t^T$, where $T \in R^{(n+1) \times (n+1)}$, so an SDP form is:

$$\max_t \eta \bullet T \tag{15}$$

$$Subject\ to: \ \tau \bullet T = 4k; \ t_i \in \{-1,1\}, t_0 = 1, \forall I \neq 0$$

where $\bullet$ reprents the dot product.We integrate the constraints on a binary variable t_i. Because t_i has only two possible values (1 or -1), together with the constraint $t_0 = 1$, all the diagonal terms in T are all 1. Accordingly, the constraints $t_i \in \{-1,1\}$ can be expressed as $Diag(T) = I$, where I is an $(n+1)$-dimensional identity matrix. Therefore, the SDP relaxation of (14) is denoted as follows:

$$\max_t \eta \bullet T$$

$$(16)$$

$$Subject\ to: \quad \tau \bullet T = 4k; \quad Diag(T) = I; T \geq 0$$

where $T \geqslant 0$ means that the symmetric matrix T is positive semi-definite. Following the SDP problem formulation defined in (16), we can employ publicly available open source packages to solve this problem. In our experiments, we use a semi-definite programming algorithm [19], which is based on an interior point method to find solutions.The pseudo-code of the framework of our proposed active learning method is shown in Algorithm 1.

Algorithm 1. *BMALNeT*: Batch Mode Active Learning for Networked Data With Optimal Subset Selection

---

Input: An entire data set V, a budget B , the batch size k
Output: Labeled instance V_L with B instances
1: Initialize V_L with a small random instances
2: Unlabeled data $V_U \leftarrow$ V\V_L
3: **while** *Labeled instances* $|V_L| \leq B$ **do**
4:     Utilize the labeled instances to train a classifier model
5:     Calculate instance uncertainty U based on Eq.(4)
6:     Calculate betweenness centrality C_B based on Eq.(5)
7:     Calculate instance disparity I based on Eq.(9)
8:     Build instance-correlation matrix M
9:     Utilize $SDPA$ to select a subset $\triangle$ with k instances
10:     $V_L \leftarrow V_L \cup \triangle$; $V_U \leftarrow V_U \backslash \triangle$
11:     Labeled instances $\leftarrow$ labeled instances $+ k$;
12: **end while**

In line 1, we randomly initialize some instances as the labeled dataset, so the unlabeled data are the remaining ones, as line 2 describes. In line 3 to 12, *BMALNeT* proceeds in iterations until the budget B is exhausted. In each iteration, we select the optimal instance subset based on correlation matrix from the unlabeled data, then update the labeled instances and the unlabeled instances.

5   Experiments

In this section, we empirically evaluate the performance of our proposed active learning strategy on three datasets, in comparison with the state-of-the-art methods.

5.1   Datasets

In order to validate the performance of our proposed algorithm,we conducted extensive experiments on are three real-world datasets – Cora,CiteSeer and

WebKB[20]. In the three datasets, instances are corresponding to documents, which are represented by a set of features vectors, and the network structure of each dataset is provided by the citations between different documents. The general statistic description of each dataset is shown in Table 1.

Table 1. General Characteristics of Datasets

Dataset	Cora	CiteSeer	WebKB
♯ of Instances	2708	3312	877
♯ of Links	5429	4723	1608
♯ of Classes	7	6	5

5.2 Experimental Setup

In the experiment, each dataset is divided into an unlabeled set and a testing set. We use LibSVM [21] to train a SVM based on the labeled set, and use it to classify the instances in the testing set. In order to evaluate the effectiveness of our proposed active learning framework, we compare it with following active learning methods:

Random: this method randomly selects instances from the unlabeled set to query for labels.

Uncertainty: this method selects the instances with the least prediction margin between the two most probable class labels given by the classification model.

ALFNET [12]: this method clusters the nodes of a graph into several groups, and then randomly samples nodes from each cluster.

ActNeT [9]: this method presents two selection strategies to exploit network structure to select the most informative instances to query for their labels.

To make fair comparisons, all of the methods actively select k instances to request their labels in each iteration of the active learning process. Batch mode active learning methods actively select k instances each time, and single mode active learning methods gradually select k instances for each iteration. We perform 10-fold cross-validation to obtain the average classification accuracy.

5.3 Experimental Results

We use the average classification accuracy as the assessment criteria. There are two important parameters involved in our experiments: the tradeoff controlling parameter α in Equation (6) and the batch size k. We first set the tradeoff controlling parameter $\alpha = 0.5$, and the batch size $k = 5$ for each dataset. Under this setting, our experimental results are shown in Figure 1. From Figure 1, we can see that our proposed method—BMALNeT always achieves higher classification accuracy at the fastest rate on all three datasets. It consistently performs the best on the three datasets, followed by ActNeT. ActNeT consistently performs better than ALFNET, and ALFNET consistently performs better than Uncertainty. Random consistently is the worst. Our proposed method performs better ActNeT and ALFNET, because it takes the redundancy of instances into consideration. However, both ALFNET and ActNeT do not consider the redundancy among instances. The Uncertainty method performs worse than our method,

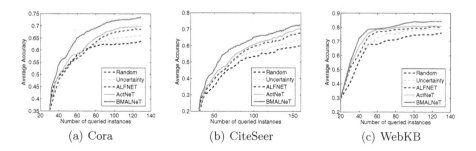

(a) Cora   (b) CiteSeer   (c) WebKB

Fig. 1. Average Accuracy on Three Datasets

ActNeT, and ALFNET, because it does not take advantage of link information. All these methods are expected better than Random.

A. Results under Different Tradeoff Controlling Parameters

As our method has two parameters, we further investigate its performance under different parameter settings. In this section, we set the batch size to a fixed value $k = 5$ for each dataset, and vary the tradeoff controlling parameter α from 0.25 to 0.5 to 0.75. The experimental results are shown in Figure 2. From Figure 2, we can see that our proposed method is slightly impacted by the tradeoff controlling parameter α. When $\alpha = 0.25$, our proposed method has the best performance on the dataset Cora and WebKB. However, it has the best performance on the dataset CiteSeer when $\alpha = 0.75$. We attribute this variant to the ratio of the number of instances to the number of links in a dataset. Notice that α is a tradeoff controlling parameter between content uncertainty and link information, which is directly related to the ratio of the number of instances and the number of links in a dataset. The ratio of the dataset CiteSeer is larger than the ratios of the dataset Cora and WebKB. That is, its link information has a relative high utility.

B. Results under Different Batch Sizes

In this section, we further investigate the performance of our proposed method under different batch sizes. Again, in our experiment, we set the tradeoff controlling parameter to a fixed value $\alpha = 0.5$ for each dataset, and vary the batch size k from 5 to 10 to 15. Our experimental results are shown in Figure 3. From Figure 3, we can see that the batch size has slightly impacted the performance of our proposed method. Our proposed method performs slightly better when the batch size decreases from 15 to 10 to 5 on Cora and WebKB datasets. On the CiteSeer dataset, our proposed method performs the best when the batch size

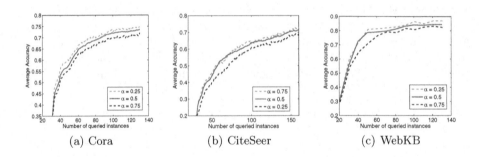

(a) Cora　　　　　　　(b) CiteSeer　　　　　　(c) WebKB

Fig. 2. Average Accuracy under Different α on Three Datasets

is set to be 10. We attribute this to the characteristics of each dataset, such as the number of instances, and the number of links, and the number of features. However, the impact of the batch size is marginal to our proposed method.

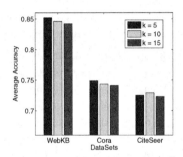

Fig. 3. Average Accuracy under Different Batch Sizes on Three Datasets

6   Conclusion

In this paper, we have proposed a new batch mode active learning method for networked data levering optimal subset selection. In our novel method, we evaluate the value of each instance based on content uncertainty, link information and disparity. Based on the value of each instance, a correlation matrix is constructed. Then, we transform the batch mode active learning selection problem into a semi-definite programming problem, select a subset of instances with an optimal utility value. The use of the correlation matrix and the SDP algorithm ensure that the solution of the subset is global optimal. Our experimental results on three real-world datasets show that our proposed approach outperforms the state-of-the-art methods.

Acknowledgments. This work was partially supported by Chinese NSFC project (61170020, 61402311, 61440053), and the US National Science Foundation (IIS-1115417).

References

1. Baldridge, J., Osborne, M.: Active learning and the total cost of annotation. In: EMNLP 2004, A meeting of SIGDAT, pp. 9–16 (2004)
2. Cohn, D.A., Ghahramani, Z., Jordan, M.I.: Active learning with statistical models. J. Artif. Intell. Res. (JAIR) **4**, 129–145 (1996)
3. Macskassy, S.A.: Using graph-based metrics with empirical risk minimization to speed up active learning on networked data. In: KDDM 2009, pp. 597–606. ACM (2009)
4. Shi, L., Zhao, Y., Tang, J.: Batch mode active learning for networked data. ACM Transactions on Intelligent Systems and Technology (TIST) **3**(2), 33 (2012)
5. Yang, Z., Tang, J., Xu, B., Xing, C.: Active learning for networked data based on non-progressive diffusion model. In: Proceedings of the 7th ACM International Conference on Web Search and Data Mining, pp. 363–372. ACM (2014)
6. Joshi, A.J., Porikli, F., Papanikolopoulos, N.: Multi-class active learning for image classification. In: CVPR 2009, pp. 2372–2379. IEEE (2009)
7. Melville, P., Mooney, R.J.: Diverse ensembles for active learning. In: Proceedings of the Twenty-First International Conference on Machine Learning, p. 74. ACM (2004)
8. Jensen, D., Neville, J., Gallagher, B.: Why collective inference improves relational classification. In: KDDM 2004, pp. 593–598. ACM (2004)
9. Hu, X., Tang, J., Gao, H., Liu, H.: Actnet: Active learning for networked texts in microblogging. In: SDM, pp. 306–314. SIAM (2013)
10. Cesa-Bianchi, N., Gentile, C., Vitale, F., Zappella, G.: Active learning on trees and graphs. arXiv preprint arXiv:1301.5112 (2013)
11. Fang, M., Yin, J., Zhang, C., Zhu, X., Fang, M., Yin, J., Zhu, X., Zhang, C.: Active class discovery and learning for networked data. In: SDM, pp. 315–323. SIAM (2013)
12. Bilgic, M., Mihalkova, L., Getoor, L.: Active learning for networked data. In: ICML 2010, pp. 79–86 (2010)
13. Zhuang, H., Tang, J., Tang, W., Lou, T., Chin, A., Wang, X.: Actively learning to infer social ties. Data Mining and Knowledge Discovery **25**(2), 270–297 (2012)
14. Newman, M.: Networks: an introduction. Oxford University Press (2010)
15. Freeman, L.C.: A set of measures of centrality based on betweenness. Sociometry, 35–41 (1977)
16. Brandes, U.: On variants of shortest-path betweenness centrality and their generic computation. Social Networks **30**(2), 136–145 (2008)
17. Fu, Y., Zhu, X., Elmagarmid, A.K.: Active learning with optimal instance subset selection. IEEE Transactions on Cybernetics **43**(2), 464–475 (2013)
18. Goemans, M.X., Williamson, D.P.: Improved approximation algorithms for maximum cut and satisfiability problems using semidefinite programming. Journal of the ACM (JACM) **42**(6), 1115–1145 (1995)

19. Fujisawa, K., Kojima, M., Nakata, K.: Sdpa (semidefinite programming algorithm) user manual-version 4.10. Department of Mathematical and Computing Science, Tokyo Institute of Technology, Research Report, Tokyo (1998)
20. Sen, P., Namata, G.M., Bilgic, M., Getoor, L., Gallagher, B., Eliassi-Rad, T.: Collective classification in network data. AI Magazine **29**(3), 93–106 (2008)
21. Chang, C.C., Lin, C.J.: Libsvm: a library for support vector machines. ACM Transactions on Intelligent Systems and Technology (TIST) 2(3), 27 (2011)

Information and Knowledge

A Question Answering System Built on Domain Knowledge Base

Yicheng Liu[1]($\boxtimes$), Yu Hao[1], Xiaoyan Zhu[1], and Jiao Li[2]

[1] Tsinghua National Laboratory of Intelligent Technology and Systems (LITS)
Department of Computer Science and Technology, Tsinghua University,
Beijing 100084, China
liuchen250.student@gmail.com, {haoyu,zxy-dcs}@tsinghua.edu.cn
[2] Institute of Medical Information and Library Chinese Academy of Medical Sciences,
Beijing 100020, China
li.jiao@imicams.ac.cn

Abstract. Interactive Question Answering (QA) system is capable of answering users' questions with managing/understanding the dialogs between human and computer. With the increasing amount of online information, it is highly needed to answer users' concerns on a specific domain such as health-related questions. In this paper, we proposed a general framework for domain-specific interactive question answering systems which takes advance of domain knowledge bases. First, a semantic parser is generated to parse users' questions to the corresponding logical forms on basis of the knowledge base structure. Second, the logical forms are translated into query language to further harvest answers from the knowledge base. Moreover, our framework is featured with automatic dialog strategy development which relies on manual intervention in traditional interactive QA systems. For evaluation purpose, we applied our framework to a Chinese interactive QA system development, and used a health-related knowledge base as domain scenario. It shows promising results in parsing complex questions and holding long history dialog.

1 Introduction

Question answering is a long-standing problem. Nowadays, with the development of intelligent devices, many business question answering systems are in great use in our everyday life. Siri, a widely-used business question answering system, can handle many types of questions such as "How is the weather like today?" and "Call [*somebody*]". Although these business question answering systems perform well in such short and single questions, they have the following two major limitations: First, these systems can only deal with simple or pattern-fixed questions. Once a complex question is raised, they will lead user to a search engine instead of providing any answers. Second, dialog management has received less attention in real-world QA systems. They treat input questions as stand-alone ones while questions are often related to dialogs.

To provide a desirable domain question answering system, we make it working on a wide variety of domain questions and being able to hold the dialog

© Springer International Publishing Switzerland 2015
J. Li and Y. Sun (Eds.): WAIM 2015, LNCS 9098, pp. 111–122, 2015.
DOI: 10.1007/978-3-319-21042-1_9

history. In this paper, we developed a general framework for question answering systems which takes only a domain knowledge base as input. For a given domain knowledge base, a semantic parser is generated from it to parse the user questions to logical forms, and answer candidates are extracted from the knowledge base by the queries translated from the logical forms. Dialog management, a key component of the framework, collects all user-computer conversations and generates dialog strategies for more accurate answer prompt.

Accurate question analysis and smart dialog management are two challenging technique issues in our framework development. Where, a good question analyzer is expected to parse entities and their relations from users questions accurately, And it should have a good capability in extension to handle complex questions such as those questions with reasoning. A smart dialog manager is expected to use the conversation contexts for quick identification of users concerns when handling the ambiguous questions.

In summary, we made the following contributions in this study:

- We proposed a novel framework for question answering system with dialog management
- We developed a method to build a domain-specific semantic parser from a domain knowledge base
- We developed a general dialog management method with capability of managing the logic form of the dialog history

The rest of this paper is organized as follows: Section 2 reviews related work. Section 3 provides an overview of our proposed framework. Section 4 and Section 5 describe more details on semantic parsing and dialog management in our framework. Section 6 analyzes the performance of a health domain QA system built in our framework. Section 7 concludes.

2   Related Work

2.1   Semantic Parsing

Semantic parsing targets on parsing natural language questions into logical forms. There are usually two major levels for a semantic parser. At the lexical level, a semantic parser maps natural language phrases to logical predicates in knowledge base. In a limited domain, a lexicon can be learned from annotated corpus [1,2]. At the compositional level, the predicates are combined into a logic form of the question. Pattern based approaches use dependency pattern [3] or question pattern generated from paraphrase[4]. Rule-based approaches requires combination rules which are specified manually [5,6] or learned from annotated logical forms [7].

2.2   Dialog Management

The main task of dialog manager is to give the user reasonable responses according to dialog status. Early approaches [8,9] use finite-state based methods or task based methods to manage the dialog. In these approaches, tasks are described as frames or topic trees which have to be written manually. Latest approaches [10,11] target on learning the dialog model automatically. In our approach, we implement general strategies by introducing a history graph and domain-independent inference algorithms and then leave the high-level strategies to the business logic which will be implemented system-specific.

3   Setup

3.1   Knowledge Base

The main input of our framework is a knowledge base $\mathcal{K} = (E, R, C, \Delta)$, in which E represents the set of entities, R the relations, C the categories and Δ the knowledge instances which are triples of two entities and the relations between them. In our approach, we require that every entity in $\mathcal{K}$ is typed and labeled. Figure 1 shows the schema and some sample instances of our knowledge base.

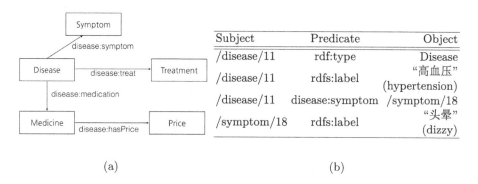

Subject	Predicate	Object
/disease/11	rdf:type	Disease
/disease/11	rdfs:label	"高血压" (hypertension)
/disease/11	disease:symptom	/symptom/18
/symptom/18	rdfs:label	"头晕" (dizzy)

(a)   (b)

Fig. 1. Sample knowledge base. (a) shows the schema of the the knowledge base and (b) shows some instances.

3.2   Logical Form

Following the study of [12], we use another graph QG_q called **question graph** to represent the semantic of a given question q. Every node (or edge) in QG_q represents an entity (or relation) mentioned in q and it must match with an concept in the underlying knowledge base. QG_q also has to meet the type constraints of the ontology: every two nodes n and n' in QG_q can be connected by an edge e only if $type_n/type_{n'}$ satisfies the domain/range of r.

3.3   Problem Statement

Given a knowledge base $\mathcal{K}$, our goal is to generate a semantic parser which can map natural language questions to their logical forms which are semantically equivalent to them. And within every session, we maintain the dialog history in our framework. When we find that there are missing or ambiguous conditions in a question, we use the history to complete or disambiguate it before generating the knowledge base query.

3.4   Framework

The architecture of our framework is shown as Figure 2. A question q in natural language is fed into the following pipeline:

Firstly, semantic parser maps q to its logical form (i.e., a question graph). Then in graph injection step, history graph and other constraints from business logic are injected if the question graph is unsolvable with its conditions. After that query generation translates the logical form to SPARQL query. Knowledge base executes the query and returns the answer as an answer graph. Finally, the answer graph is pushed back into the history graph and all nodes in it are re-weighted for future injection. Business logic translates the answer graph into a natural language sentence with some rules. Business logic also executes the domain specific rules in dialog management, e.g. asking for more information when there are too many candidate answers , handling the user's response after a system-ask round or providing other constraints for the next round of dialog.

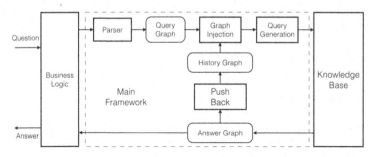

Fig. 2. Framework

4   Semantic Parsing

The main goals of semantic parsing are finding the best logical form for a natural language question and generating the knowledge base query. In this approach, we generate a context-free grammar from the knowledge base (Section 4.1). By using a context-free grammar, we can parse questions with reasoning which are incapable for pattern-based methods that are widely used in real-world dialog systems. With the grammar, we can generate a derivation d_q from an input question q. Then the derivation d_q is converted to the question graph QG_q then a SPARQL query which can be executed by knowledge base (Section 4.2). Figure 3 shows an example parsing of a question contains an one-step reasoning.

4.1   Parser Generation

In this section, we discuss the construction of a context-free grammar $\mathcal{G}$ from $\mathcal{K}$. With this grammar, we can generate derivations for questions. Figure 3a shows the derivation of a question with an one-step reasoning.

Entities and Categories. For each entity $e \in E_{\mathcal{K}}$, it generates production rules $\{URI_e \rightarrow l \mid \forall l \in label(e)\}$. And for each category $c \in C_{\mathcal{K}}$, it generates a set of production rules as $\{URI_c \rightarrow URI_e \mid \forall e : type_e = c\}$. A category also has labels, which are phrases in natural language that represent an object in this category. For example, the category `Medicine` may be represented by "medicine" or "drug". So for a category $c \in C_{\mathcal{K}}$, $\{URI_c \rightarrow l \mid \forall l \in label(c)\}$ is added to $\mathcal{G}$. In Figure 3a, for example, `Type:Disease` produces the non-terminal character `URI:Cold`, `URI:Cold` produces the terminal character "感冒 (Cold)" and `Type:Medicine` produces its label phrase "药 (Medicine)".

Relations. For a binary relation $r \in R_{\mathcal{K}}$, if label of r occurs in the sentence it combines two part which are adjacent to it and generates a higher-level semantic category. Table 1 shows all production rules generated for a given relation r.

Table 1. Rules generated for relation r

Normal	$domain(r) \rightarrow domain(r)\ label(r)\ range(r)$
	$range(r) \rightarrow domain(r)\ label(r)\ range(r)$
Bridging	$domain(r) \rightarrow domain(r)\ range(r)$
	$range(r) \rightarrow domain(r)\ range(r)$
Slot	$domain(r) \rightarrow domain(r)\ label(r)$
	$range(r) \rightarrow domain(r)\ label(r)$
	$domain(r) \rightarrow label(r)\ range(r)$
	$range(r) \rightarrow label(r)\ range(r)$

As shown in Table 1, we first generate two normal rules for r: Note that $domain(r)$ and $range(r)$ are both categories in $C_{\mathcal{K}}$. Since we cannot know which one between $domain(r)$ and $range(r)$ is the semantic subject, so rules for both $domain(r)$ and $range(r)$ are generated. This strategy will not cause disambiguation in derivation because higher level reduction will accept the right category.

But in a natural language question, there are many cases where predicates are expressed weakly or implicitly [6]. For example, the relation is omitted in "高血压的药 (Medicines for hypertension)". Since it is impossible to extract relations by phrases in these questions, we add **bridging rules** of relations to $\mathcal{G}$.

Besides, our framework works as a dialog system so questions may have missing components which are actually implicit references to previous conversation. For example, in a medical diagnose system, user may input the question "吃什么药 (What medicine should be taken)?" after he is told that he has a cold. Relation `disease:medication` appears in the question but one of its argument `Disease` is missing. **Slot rules** to $\mathcal{G}$ to handle these incomplete questions.

Lexicon Extension. In grammar $\mathcal{G}$, we use labels of concepts to map natural language phrases to knowledge base concepts. To avoid ambiguity, we only use entities' labels and aliases in knowledge base. But to cover various utterances, we extend the lexicon of relations by alignment of a large text corpus to the knowledge base. Intuitively, a phrase and a relation align if they share many of the same entity arguments.

We use 25 million questions from Baidu Zhidao[1] and replace the name phrases in questions by simple string matching with knowledge base entities. For those questions which have exactly 2 entities in it, we add the entity pair to the context set of the text between them. For example, the sentence is "感冒用阿司匹林 (Cold use aspirin)", (Cold, Aspirin) is added to $Context($"用 (use)"$)$. The context of a relation r is defined by the triples in knowledge base, $Context(r) = \{(s, o)|(s, r, o) \in \mathcal{K}\}$. Then we calculate the overlap between the contexts of phrases and relations. The phrase p is added to the label set of relation r if $|Context(p) \cap Context(r)| > \delta$, where δ is a manual-specified threshold.

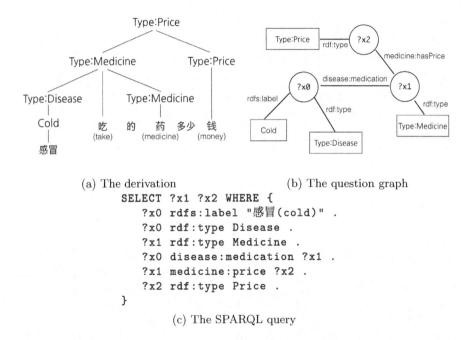

(a) The derivation   (b) The question graph

```
SELECT ?x1 ?x2 WHERE {
    ?x0 rdfs:label "感冒(cold)" .
    ?x0 rdf:type Disease .
    ?x1 rdf:type Medicine .
    ?x0 disease:medication ?x1 .
    ?x1 medicine:price ?x2 .
    ?x2 rdf:type Price .
}
```

(c) The SPARQL query

Fig. 3. An example parse result of the question "感冒吃的药多少钱 (The price of the medicines for cold)"

4.2 Question Graph and SPARQL Construction

We apply our parser on a question and the derivation which has the most coverage on question words is selected as the best derivation. With this derivation,

---

[1] http://zhidao.baidu.com, a Chinese question answering community

a question graph that represents the semantic of the question can be generated. Specificlly, we make a bottom-up traverse on the derivation to generate the question graph.

For an entity e and its rule $URI_e \rightarrow label(e)$, a graph containing one blank node with a label is created. And for the category rule $URI_c \rightarrow URI_e$, a type property is added to the blank node in graph. Until now there is only one node in each graph, so the **subject node** of the graph is the blank node itself. A subject node is the semantical center of the graph, it will be connected with other graphs in future.

A relation rule combines two sub-trees in derivation. When a rule generated from relation r is applied, we connect the graphs generated by the sub-trees with the relation r on their subject nodes. A slot rule have only one sub-tree, so we will generate a blank node which marked as "slot" to the omitted side of the relation. The subject node of the new graph will be chosen from two old ones whose type matches the left-hand side of the rule.

When we get to the root of the derivation, we get a connected question graph which is semantically equivalent to the question. Figure 3b shows the question graph generated from the derivation shown in Figure 3a.

The question graph must be converted to a SPARQL query to be executed. In final step, every edge in question graph and two nodes it connects will be mapped to one triple condition in the SPARQL query. Blank nodes are mapped to variables while nodes with URIs are added as is. Figure 3c shows the SPARQL query generated from the question graph shown in Figure 3b.

5   Dialog Management

In a dialog system, users' questions are related to the contexts, so a major challenge of dialog system is to infer implicit references in questions.

In this paper, we use a inference method based on the history graph. Since the history graph is built totally from question graphs and answer graphs, the inference algorithms are domain-independent. This framework can be used directly in any domain without extra efforts on transferring.

5.1   Answer Graph

The results returned by knowledge base contains all possible solutions in which the variables are solved to entities in knowledge base. Since variables in SPARQL query are generated from blank node in question graph, we replace every blank node in the question graph with its corresponding solution to generate a new graph we call **answer graph**. Note that there might be more than one group of solution for a query. One answer graph will be generated for each group of solution.

5.2 History Graph and Push Back

History graph maintains the history of the dialog. Every answer graph will be merged into history graph, and we call this procedure **push back**.

Timeliness is important in the history, latest contexts are likely to be referenced while far contexts should be forgotten. In history graph, we add an extra property *weight* for every concept (entities and relations). Concepts that are just mentioned in answer graph will be given the highest weight (e.g. 1) while the weight of concepts not mentioned will decline by a coefficient. This *weight* will be used as the priority in graph injection step.

5.3 Graph Injection

After at least one round of dialog, the history graph gets non-empty and can be used to infer implicit references or missing conditions in following questions.

> Q1: 感冒有什么症状 (What's the symptoms of cold)?
> A1: 头晕, 发烧 (Dizzy, Fever)
> Q2: 怎么治疗?(What is the treatment)
> A2: ...
> Q3: 高血压呢?(How about hypertension)
> A3: ...

(a) Sample Dialog

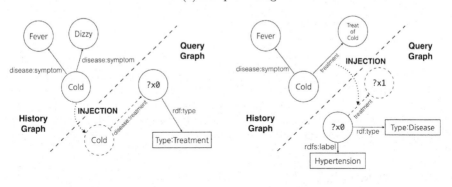

(b) Graph Injection for Q2   (c) Graph Injection for Q3

Fig. 4. Example of graph injection

For the question graph has only one node b, these questions have no enough conditions to be solved. If the single node b does not have a label (Q2 in Figure 4a, its question graph shown as 4b), we pick a relation $\tilde{r}$ whose domain (or range) matches $type_b$ with highest weight. Then we choose an entity $\tilde{e}$ that matches the range (or domain) of $\tilde{r}$ from the history graph to build a triple in question graph. In the worst case, there is no available $\tilde{r}$ meets the type requirement, we only pick an entity $\tilde{e}$ and traverse all relations in knowledge base $\mathcal{K}$ and choose one relation whose domain and range match $type_{\tilde{e}}$ and $type_b$ (or vice verse). If b has a label (Q3 in Figure 4a, question graph shown as 4c), user may want to change a condition. So we pick a relation $\tilde{r}$ whose domain (or range) matches $type_b$ with highest weight. Then we complete the question graph by adding b to domain (or range) of $\tilde{r}$ and creating a blank node of range (or domain) of $\tilde{r}$. Now the question graph have a complete triple as the condition.

For questions with slot nodes (discussed in Section 4.1), we just pick up an entity which matches the type constraints of the slot node with the highest weight and replace the slot node in question graph with it.

After graph injection, question graph should be able to generate a solvable query. Note that this graph injection algorithm needs neither hand-written frames or scripts nor extra knowledge about the task of the system.

System-specific business logic also work through graph injection. For example in a health diagnose system, when statements given by patient are not enough, the system can ask patient for more statements and send the answer of last round to the dialog manager as candidates. In graph injection step, answer candidates are injected to question graph as a filter. This interface is also useful in deal with user profile since user profile can also be represented as a profile graph and injected to the question graph.

6   Evaluation

In this section we describe our experimental evaluation. For the semantic parser, natural language questions sampled from Internet are used as the input. We measure the precision of our semantic parser by compare the question graphs generated by the parser with the manually annotated question graph. For the dialog management, we test the behavior of it by some test cases.

6.1   Datasets and Setup

As the test bed for evaluating our framework, we use a medical domain knowledge base in which entities are labeled in Chinese. It contains 16155 instances with 1093 diseases, 236 symptoms and 903 medicines.

To evaluate the performance of the semantic parser, we randomly sample 132 natural language questions from Baidu Zhidao. These questions are all about the categories and relations covered by our knowledge base, including disease, symptom, medicine, treatment and so on. The equivalent conjunctive SPARQL queries written manually are used as the golden standard of the semantic parser.

And for the dialog management, we designed some test cases to examine whether the actions taken by the framework are correct.

6.2   Results and Discussion

Semantic Parsing. Table 2 shows the results of semantic parsing on the test questions.

Table 2. The result of semantic parsing on 132 questions

	Correct	ME	UCR	WP	Out-of-Grammar
Non-extend	48%	28%	12%	4%	8%
Lexicon Extend	56%	28%	4%	4%	8%

In Table 2 we can see 56% (74 questions) are parsed correctly. Note that lexicon extension for relations significantly improves the performance by 8%.

28% (37 questions) are not parsed because the missing entities (ME). Entities are labeled by their scientific name in knowledge base while people usually using the popular names of them. There are also some misspelled names or traditional chinese medicine, which are totally not contained in our knowledge base.

4% (5 questions) have unrecognized categories or relations (UCR). For example, category Medicine has label "medicine" so it can parse the questions like "感冒吃什么药 (What medicines to take for cold)" but will failure on "感冒有什么秘方 (What's the recipe for cold)".

4% (5 questions) are wrongly parsed (WP). Those caused by the ambitious parsing in derivations. For example, "克拉霉素对红眼病有治疗作用吗 (The cure effect to pinkeye by clarithromycin)", our parser combined the "cure" and "pinkeye" firstly and then the "clarithromycin". These questions are described in an inverted order. This linguistic phenomenon is not covered by our rules so a wrong derivation is generated.

8% (11 questions) have structures that are not covered by the grammar (Out-of-Grammar). Grammar are generated strictly by the schema of knowledge base. But people often use implicit reasoning when they think the it is self-evident. For example, user may ask "头疼吃什么药 (What are the medicines for headache)". Headache is a Type:Symptom and it has no property medication. Type:Disease should be the bridge between Type:Symptom and Type:Medicine, but is omitted in the given question. Questions with aggregation functions are also not covered in our grammar.

Dialog Management. In this section, we will use some test cases to evaluate our dialog management algorithm.

The first case in Figure 5 shows a typical scenario that user's question contains implicit references to the previous dialog: Q2 asks about medicines in A2; Q3 has the same question as Q1 but the subject changed; Q4 shares the same subject with Q3. Note that when Q2 is raised, there are two entities that can be injected to question graph ("Reserpine" and "Digoxin"). They both have the highest weight so both of them are injected. When Q4 is raised, there are also two entities that can be injected to question graph ("hypertension" and "diabetes"). As "diabetes" is mentioned recently so it has higher weight than "hypertension" and is chosen to be injected.

Figure 6 shows a case when business logic takes the control. In the diagnose mode, users give their symptoms and the system tells the possible diseases. When there are too many results, the business logic will not show a long list but ask

Q1: 高血压吃什么药?(What medicine to take for hypertension)
A1: 利血平, 地高辛 (Reserpine, Digoxin)
Q2: 多少钱?(The price)
A2: 利血平 10 元, 地高辛 20 元 (10RMB for Reserpine, 20RMB for Digoxin)
Q3: 糖尿病呢?(How about diabetes)
A3: 胰岛素, 维生素 A(Insulin, Vitamin A)
Q4: 去哪个科室?(Which department to go)
A4: 内分泌科 (Department of Endocrinology)

Fig. 5. Example of inferring implicit references

for further information and save the answer of this round (A1) for the candidate of next round (Q2). When there are just several answers left (A2), the business logic will pick up a decidable symptom and give yes-no questions for choosing.

Q1: 我腹痛还发热是什么病?(Stomachache and fever means what disease)
A1: 您可能患有感冒、胃炎、麻疹、腹膜炎。您还有哪些症状?
(Maybe cold, gastritis, measles, peritonitis. Any other symptoms?)
Q2: 腹泻 (Diarrhea)
A2: 您可能患有感冒、胃炎。请问您消化不良嘛?
(Maybe cold, gastritis. Do you have dyspepsia?)
Q3: 是 (Yes)
A3: 您可能患有胃炎 (You may have gastritis)

Fig. 6. Example of business logic

7   Conclusions and Future Work

We presented a general framework for domain-specific question answering systems with dialog management. With this framework, one can generate a domain-specific dialog system with only a domain knowledge base as input. In the framework, we designed a logical form to represent the semantic of questions and the history of conversations. And we proposed a method to generate a semantic parser from the knowledge base and extended it with web corpus. Then we achieve a general dialog management algorithm about inferring implicit references in questions. Our experiments showed that our framework has good performance in handling complex questions and dealing with multi-round dialog.

For the semantic parsing part, our future work includes (1) adding more question patterns in our grammar to handle more linguistic phenomenon such as aggregation; (2) collecting more labels or alias for entities and categories to increase the coverage of our parser; (3) creating more rules from the knowledge base to handle variety of questions. For dialog management, we will design more interface for business logic to implement more complex dialog control strategies that are useful in business systems.

References

1. Kwiatkowski, T., Zettlemoyer, L., Goldwater, S., Steedman, M.: Lexical generalization in ccg grammar induction for semantic parsing. In: Proceedings of the Conference on Empirical Methods in Natural Language Processing, pp. 1512–1523. Association for Computational Linguistics (2011)
2. Liang, P., Jordan, M.I., Klein, D.: Learning dependency-based compositional semantics. Computational Linguistics **39**, 389–446 (2013)
3. Yahya, M., Berberich, K., Elbassuoni, S., Ramanath, M., Tresp, V., Weikum, G.: Natural language questions for the web of data. In: Proceedings of the 2012 Joint Conference on Empirical Methods in Natural Language Processing and Computational Natural Language Learning, pp. 379–390. Association for Computational Linguistics (2012)
4. Berant, J., Liang, P.: Semantic parsing via paraphrasing. In: Proceedings of ACL (2014)
5. Krishnamurthy, J., Mitchell, T.M.: Weakly supervised training of semantic parsers. In: Proceedings of the 2012 Joint Conference on Empirical Methods in Natural Language Processing and Computational Natural Language Learning, pp. 754–765. Association for Computational Linguistics (2012)
6. Berant, J., Chou, A., Frostig, R., Liang, P.: Semantic parsing on freebase from question-answer pairs. In: EMNLP, pp. 1533–1544 (2013)
7. Cai, Q., Yates, A.: Large-scale semantic parsing via schema matching and lexicon extension. In: ACL (1), Citeseer, pp. 423–433 (2013)
8. Wu, X., Zheng, F., Xu, M.: Topic forest: A plan-based dialog management structure. In: Proceedings of the 2001 IEEE International Conference on Acoustics, Speech, and Signal Processing (ICASSP 2001) vol. 1, pp. 617–620. IEEE (2001)
9. Rudnicky, A., Xu, W.: An agenda-based dialog management architecture for spoken language systems. In: IEEE Automatic Speech Recognition and Understanding Workshop, pp. 1–337 (1999)
10. Doshi, F., Roy, N.: Efficient model learning for dialog management. In: 2007 2nd ACM/IEEE International Conference on Human-Robot Interaction (HRI), pp. 65–72. IEEE (2007)
11. Young, S., Schatzmann, J., Weilhammer, K., Ye, H.: The hidden information state approach to dialog management. In: IEEE International Conference on Acoustics, Speech and Signal Processing, ICASSP 2007, vol. 4, pp. IV-149. IEEE (2007)
12. Shekarpour, S., Ngonga Ngomo, A.C., Auer, S.: Question answering on interlinked data. In: Proceedings of the 22nd International Conference on World Wide Web, International World Wide Web Conferences Steering Committee, pp. 1145–1156 (2013)

Fast and Accurate Computation of Role Similarity via Vertex Centrality

Longjie Li[1]([✉]), Lvjian Qian[1], Victor E. Lee[2], Mingwei Leng[3],
Mei Chen[1,4], and Xiaoyun Chen[1]

[1] School of Information Science and Engineering,
Lanzhou University, Lanzhou 730000, China
{ljli,mchen,chenxy}@lzu.edu.cn
[2] Department of Mathematics and Computer Science,
John Carroll University, University Heights, OH 44118, USA
[3] School of Mathematics and Computer Science,
Shangrao Normal University, Shangrao 334001, China
[4] School of Electronic and Information Engineering,
Lanhou Jiaotong University, Lanzhou 730070, China

Abstract. There is growing evidence that vertex similarity based on structural context is the basis of many link mining applications in complex networks. As a special case of vertex similarity, role similarity which measures the similarity between two vertices according to their roles in a network can facilitate the search for peer vertices. In RoleSim, graph automorphism is encapsulated into the role similarity measure. As a real-valued role similarity, RoleSim shows good interpretative power in experiments. However, RoleSim is not sufficient for some applications since it is very time-consuming and may assign unreasonable similarities in some cases. In this paper, we present CentSim, a novel role similarity metric which obeys all axiomatic properties for role similarity. CentSim can quickly calculate the role similarity between any two vertices by directly comparing their corresponding centralities. The experimental results demonstrate that CentSim achieves best performance in terms of efficiency and effectiveness compared with the state-of-the-art.

Keywords: Complex network · Vertex similarity · Role similarity · Vertex centrality · Similarity metric

1 Introduction

Nowadays, networked data, *e.g.*, social network and web page, is proliferating and attracting a growing interest among researchers. In sociology, individuals

---

L. Li—L. Li was supported by the Fundamental Research Funds for the Central Universities (No. lzujbky-2014-47).

M. Chen—M. Chen was supported by the Gansu Provincial Natural Science Fund (No. 145RJZA194) and the Fundamental Research Fund for the Gansu Universities (No. 214151).

J. Li and Y. Sun (Eds.): WAIM 2015, LNCS 9098, pp. 123–134, 2015.
DOI: 10.1007/978-3-319-21042-1_10

are often assigned "social roles", such as a father, a doctor, or a professor. In the past, role studies have primarily been the interest of sociologists on offline social networks [1,17]. Recent studies have found that roles also appear in many other type of networks, such as biological networks [13], web graphs [14], and techno-logical networks [15]. On the one hand, role discovery is indeed an important task for general graph mining and exploratory analysis since it is useful in many real applications [20–22]. On the other hand, measuring role-based similarity between any two vertices is also a key question in studying the roles in a network system [7]. One reason is role similarity can help to predict vertex functionality within their domains. For instance, in a protein-protein interaction network, proteins with similar roles usually serve similar metabolic functions. Thus, if the func-tion of one protein is known, all other proteins having the similar role would be predicted to have similar function [5].

Despite its significance, the problem of role similarity has received little attention. From the viewpoint of a network, automorphic vertices have equivalent surroundings and hence share the same role. In [7], graph automorphism is encap-sulated into the role similarity measure: *two automorphically equivalent vertices share the same role and have maximal role similarity*. Take the network shown in Fig. 1a as an example. Clearly, vertices b and c are automorphically equiva-lent, thus they share the same role and should have the maximal role similarity. Although vertices d and e are not automorphically equivalent, they have the very similar surroundings and hence should have higher role similarity. To estimate how role-similar two vertices are, a real-valued role similarity measure, called RoleSim [7], was proposed. For vertex-pairs (b, c) and (d, e), RoleSim can successfully assign their similarities, *i.e.*, $RoleSim(b, c) = 1$ and $RoleSim(d, e) = 0.589$. For other two vertex-pairs (a, b) and (a, e), $RoleSim(a, b) < RoleSim(a, e)$. This result is acceptable and reasonable.

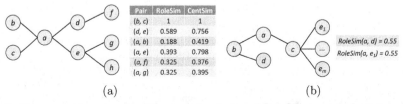

(a)   (b)

Fig. 1. Two example networks. The damping factor of RoleSim is set to be 0.1 and the initialization of RoleSim is *ALL-1* scheme [7]. The Degree, PageRank and Closeness centralities are used in CentSim, and their weights are equally set to 1.

However, RoleSim assigns the same similarity score to vertex-pairs (a, f) and (a, g) in Fig. 1a. That seems to be unreasonable. Furthermore, in Fig. 1b, RoleSim always thinks that vertex-pairs (a, d) and (a, e_1) have the same role similarity regardless of the number of neighbors of vertex c. Although this situation should not be deemed as a failure of RoleSim, there is clearly room to improve its accuracy and sensitivity. In addition, the very serious problem in RoleSim is it is a time-consuming method. Its time complexity is $O(kN^2d')$ where k is the number of iterations, N is the number of vertices in a network, and d' is the average of $(d_u \times d_v) \times \log(d_u \times d_v)$ of all vertex-pairs (u, v). d_u is the degree of vertex u. In [8], a scalable algorithm for RoleSim, namely IcebergRoleSim, was presented to speed the computation of RoleSim while the cost is to prune the vertex-pairs whose similarities are lower than a given threshold.

In this paper, we propose CentSim (centrality-based similarity measure), a new real-valued role similarity measure to quickly and accurately compute the role similarity of any vertex-pairs. In CentSim, we employ the centralities of vertices to calculate their role similarities. When measuring two vertices' role similarity, CentSim just compares several of their centralities, while RoleSim investigates all neighbor-pairs of the two vertices. Thus, CentSim can quickly calculate any vertex-pairs' role similarities. Furthermore, CentSim obeys all the axiomatic role similarity properties [7] and hence is an admissible role similarity metric. To show the performance of CentSim, we review the network shown in Fig. 1a. Vertices b, c have the same role and d, e have the very similar surroundings, CentSim can assign reasonable similarities to (b, c) and (d, e), respectively. That is $CentSim(b, c) = 1$ and $CentSim(d, e) = 0.756$. For vertex-pairs (a, f) and (a, g), $CentSim(a, f) = 0.376$ and $CentSim(a, g) = 0.395$. For the network shown in Fig. 1b, CentSim always assigns a smaller similarity to (a, d) than that to (a, e_1) for any value of $m > 1$. Compared to the similarities assigned by RoleSim, these results are more reasonable.

The rest of this paper is organized as follows. Section 2 gives a brief introduction of related work, and Section 3 contains some preliminaries for this work. In Section 4, we detail the proposed role similarity measure and prove it is an admissible role similarity metric. Section 5 demonstrates the experimental results. Finally, the conclusion of this paper is presented in Section 6.

2   Related Work

To date, many link-based similarities have been proposed. Among them, SimRank [6] is a well-known one, which is based on the intuition that *two vertices are similar if they are linked by similar vertices*. The computation of SimRank is iterative. In each iteration, SimRank updates the similarity score between two different vertices according to the average similarity of all their neighbor pairs in the previous iteration. The idea of SimRank seems to be solid and elegant; however, it may assign inaccurate or even counter-intuitive similarity scores [12] as well as undesirably introduces the "zero-similarity" issue [25, 26]. With the help of *maximal weighted matching* of neighbor pairs, MatchSim [12] overcomes the counter-intuitive results of SimRank. In order to remedy the "zero-similarity" issue, SimRank* [25] introduces a new strategy to find more paths that are largely overlooked by SimRank. Similarly, E-Rank [26] deals with the meetings of two vertices that walk along any length paths and also solves the "zero-similar" issue.

PageSim [11] is a quite different link-based similarity measure from SimRank. Motivated by the propagating mechanism of PageRank [19] and simultaneously employing the PageRank scores as vertices' features, PageSim propagates PageRank score of each vertex to other vertices via links, and then represents each vertex by a feature vector. The PageSim score of one vertex-pair is derived by comparing their feature vectors.

Role similarity, which measures the similarity of vertices based on their roles, is a special case in the link-based similarity problem. RoleSim [7], an admissible role similarity metric, was proposed to evaluate how role-similar two vertices are. Two main properties make RoleSim a role similarity metric. The first is when updating the similarity between two different vertices, RoleSim adopts the maximal weighted matching between their neighbors. And the second one is the initialization of RoleSim is admissible.

3 Preliminaries

In this section, we give the necessary background and notations before we discuss role similarity further.

3.1 Role Similarity Properties

A social network or other complex network is defined as an undirected graph $G(V, E)$ where V is the vertex set and E represents the edge set. For a given vertex u in graph G, the set of its neighbors is denoted as $N(u)$ and the degree of u is the number of its neighbors, denoted as d_u, $d_u = |N(u)|$.

Given a graph $G(V, E)$ and two vertices $u, v \in V$, an *automorphism* of G is a *permutation* σ of V such that $(u, v) \in E$ iff $(\sigma(u), \sigma(v)) \in E$. If $u = \sigma(v)$, then vertices u and v are *automorphically equivalent*, denoted as $u \equiv v$.

To theoretically depict the role similarity measure, Jin *et al.* [7] formulated a series of axiomatic properties that all role similarity measures should obey.

Definition 1 (Axiomatic Role Similarity Properties). *Let $G(V, E)$ be a graph and $s(u, v)$ be the similarity score between any two vertices $u, v \in V$. Five axiomatic properties of role similarity are developed as follows:*

1) Range: $s(u, v) \in [0, 1]$.
2) Symmetry: $s(u, v) = s(v, u)$.
3) Automorphism confirmation: If $u \equiv v$, $s(u, v) = 1$.
4) Transitive similarity: If $u \equiv v, x \equiv y$, then $s(u, x) = s(u, y) = s(v, x) = s(v, y)$.
5) Triangle inequality: $d(u, x) \leq d(u, v) + d(v, x)$, where $d(u, v) = 1 - s(u, v)$.

If $s(u, v)$ obeys the first four properties, it is an **admissible role similarity measure**. If $s(u, v)$ satisfies all five properties, it is called an **admissible role similarity metric**.

3.2   Centrality

Centrality is a general measure of vertex activity in a network and can be calculated by several metrics. The most popular ones are *degree, closeness, betweenness* and *eigenvector centrality* [18]. These measures determine the relative importance of a vertex within a network, particularly a social network. In a social network, vertices with larger values of centrality measures are powerful vertices and occupy the critical positions [4].

The degree centrality represents the local importance of a vertex. Generally, a vertex with higher degree is inclined to have a greater ability of local influence than others, or to be closer to the center of a network. Closeness measures the reachability of a vertex to other vertices. Formally, it is the average length of all shortest paths from a given vertex to all others in a network. Higher closeness value of a vertex indicates more vertices can be reached with shorter paths, which fits the human intuition of "centrally located." The betweenness of a vertex can commonly be interpreted as the frequency that this vertex lies on the shortest paths between any two vertices. A vertex with high betweenness usually occupies a critical position which connects two different regions and controls the information flow between different communities. Eigenvector centrality measures the influence or importance of a vertex. The basic idea of eigenvector centrality is that the influence of a vertex is recursively defined by the influence of its neighbors. *PageRank* [19] can be treated as a variant of eigenvector centrality.

4   CentSim: A Novel Role Similarity Metric

In this section, we describe the proposed new role similarity measure, CentSim.

4.1   Definition of CentSim

The basic idea of our similarity measure comes from two aspects. The first one is the role of a vertex is deeply influenced by its position in a network. And the second one is centrality is a general measure of how the position of a vertex is within a network [18]. Thus, centrality can be elected as a favorable tool to evaluate role similarity. Consequently, in CentSim, we employ vertex's centralities to calculate their role similarities. The formal computation of CentSim is given in Definition 2.

Definition 2 (CentSim). *Given a graph G(V,E) and two vertices $u, v \in V$, the CentSim score between u and v is defined as:*

$$CentSim(u,v) = \frac{\sum_{i=1}^{l} w_i \theta_i(u,v)}{\sum_{i=1}^{l} w_i} \qquad (1)$$

where l is the number of different centralities adopted in CentSim. Coefficient $w_i > 0$ is the weight of centrality c_i. $\theta_i(u,v)$ is defined as:

$$\theta_i(u,v) = \frac{\min(c_i(u), c_i(v))}{\max(c_i(u), c_i(v))} \qquad (2)$$

where $c_i(u)$ is the value of the centrality c_i of u. In Equation 2, we define $\frac{0}{0} = 1$.

From Definition 2, we can see that the core of the computation of CentSim score is to compare the centrality values of vertices. Suppose the centrality values of vertices are obtained in advance. Clearly, the CentSim score between any two vertices can be computed straightforwardly. In our implementation, three centralities, *i.e.*, PageRank, Degree and Closeness, are employed. In default, we set the weights of the three centralities equally to 1.

4.2   Admissibility of CentSim

Theorem 1 (Admissibility). *CentSim is an admissible role similarity metric.*

To prove Theorem 1, we can separately prove that CentSim obeys each of the five axiomatic role similarity properties listed in Definition 1. Trivially, CentSim holds true for the Range (property 1) and Symmetry (property 2). For the Transitive similarity (property 4), Jin *et al.* [7] proved that it is implied by the Triangle inequality property. Therefore, in the following, we only need to prove that CentSim satisfies the Automorphism confirmation (property 3) and Triangle inequality (property 5).

Lemma 2. *For any two vertices u, v in graph G, if $u \equiv v$ then $c_i(u) = c_i(v)$.*

Actually, two automorphically equivalent vertices are identical with respect to all graph theoretic properties and hence have the same centrality score on every possible measure (see chapter 12 in [24]).

Proof of Automorphism Confirmation. Since $u \equiv v$, in the light of Lemma 2, we get $\theta_i(u, v) = \frac{\min(c_i(u), c_i(v))}{\max(c_i(u), c_i(v))} = \frac{c_i(u)}{c_i(v)} = 1$. So that

$$CentSim(u, v) = \frac{\sum_{i=1}^{l} w_i \theta_i(u, v)}{\sum_{i=1}^{l} w_i} = \frac{\sum_{i=1}^{l} w_i}{\sum_{i=1}^{l} w_i} = 1. \qquad \square$$

Proof of Triangle Inequality. Given any vertices x, y and z in G, we get

$$d(x, y) + d(y, z) - d(x, z) = 1 + CentSim(x, z) - CentSim(x, y) - CentSim(y, z)$$

$$= 1 + \frac{\sum_{i=1}^{l} w_i \theta_i(x, z)}{\sum_{i=1}^{l} w_i} - \frac{\sum_{i=1}^{l} w_i \theta_i(x, y)}{\sum_{i=1}^{l} w_i} - \frac{\sum_{i=1}^{l} w_i \theta_i(y, z)}{\sum_{i=1}^{l} w_i}$$

$$= \frac{\sum_{i=1}^{l} w_i (1 + \theta_i(x, z) - \theta_i(x, y) - \theta_i(y, z))}{\sum_{i=1}^{l} w_i}$$

Let $c_i(x) = a$, $c_i(z) = b$ and $c_i(y) = c$, then

$$1 + \theta_i(x, z) - \theta_i(x, y) - \theta_i(y, z) = 1 + \frac{\min(a, b)}{\max(a, b)} - \frac{\min(a, c)}{\max(a, c)} - \frac{\min(b, c)}{\max(b, c)}$$

$$= \frac{|a - c|}{\max(a, c)} + \frac{|b - c|}{\max(b, c)} - \frac{|a - b|}{\max(a, b)}$$

Thus, to prove CentSim satisfies triangle inequality, it suffices to show that Inequality 3 holds true.

$$\frac{|a - b|}{\max(a, b)} \leq \frac{|a - c|}{\max(a, c)} + \frac{|b - c|}{\max(b, c)} \tag{3}$$

If a, b and c are positive numbers, the proof of Inequality 3 can be found in [23]. If one or two or all of a, b and c are 0s, Inequality 3 also holds true, since we defined that $\frac{0}{0} = 1$ in Definition 2. Therefore, CentSim satisfies the Triangle inequality property.   □

In conclusion, CentSim obeys all the axiomatic role similarity properties. That is to say, CentSim is an admissible role similarity metric.

4.3   Complexity of CentSim

Suppose the centrality values of vertices are given in advance. Obviously, the time complexity of CentSim is $O(lN^2)(l \ll N)$ where N is the number of vertices in graph G and l is the number of centralities used in CentSim. In fact, l is a constant when CentSim is implemented, thus the time complexity of CentSim is reduced to $O(N^2)$. To obtain the similarity of all vertex-pairs, we need to compute $N(N-1)/2$ similarity scores and record lN centrality values. Therefore, the space complexity of CentSim is $O(lN + N^2)$.

5   Experiments

In this section, we experimentally study the performance of CentSim in terms of efficiency and effectiveness. Five baselines are SimRank [6], SimRank* [25], MatchSim [12], PageSim [11] and RoleSim [7].

5.1   Experimental Setup and Datasets

All experiments are conducted on a machine with AMD Opteron 8347 4 core CPU and 16GB DDR2 memory. The operating system is Suse Linux Enterprise Server 10 SP2. CentSim and five baselines are implemented in C++, while the scores of PageRank, Degree and Closeness centrality are computed by the NetworkX[1] package of Python. In [25], the authors presented an algorithm for computing SimRank* by means of *fine-grained memoization*, namely memo-gSR*. However, in this paper, we only implement the naive algorithm of SimRank*. We set the damping factors $C = 0.8$ for both SimRank and SimRank*, and $\beta = 0.1$ for RoleSim. The initialization of RoleSim is *ALL-1* [7].

For impartial comparison of similarity measures, we utilize four real-world datasets from varying fields as benchmarks, which are PGP [16], Yeast [2], Enron [9] and DBLP[2]. The DBLP is a co-author network derived from 7-year publications (2006-2012) in conferences of SIGMOD, VLDB, ICDE, KDD, ICDM, and

---

[1] http://networkx.github.io/
[2] http://dblp.uni-trier.de/~ley/db/

Table 1. Statistics of the largest connected components of the four networks. N: number of vertex; M: number of edge; $\langle k \rangle$: average degree of vertices; k_{max}: maximal degree of vertices; $\langle d \rangle$: average shortest distance of all vertex-pairs.

	PGP	Yeast	Enron	DBLP
N	10680	2224	33696	5890
M	24316	7049	180811	19845
$\langle k \rangle$	4.554	6.339	10.732	6.739
k_{max}	205	66	1383	157
$\langle d \rangle$	7.486	4.377	4.025	5.782

Table 2. Running time of all measures on four benchmarks (unit: second).

	PGP	Yeast	Enron	DBLP
SimRank	1939	200	108121	984
SimRank*	849	57	20395	327
MatchSim	4614	921	273625	2604
PageSim	4395	93	422835	1271
RoleSim	4890	1003	326754	2698
CentSim	**91**	**5**	**981**	**28**
Centralities	486	13	7204	156

SDM. Each network is treated as an undirected unweighted graph and pruned into its largest connected component. The statistics are listed in Table 1.

5.2 Comparison of Time Performance

This section compares the time performance of CentSim with the five baselines. We perform each measure on the four benchmark datasets to compute the similarities of all vertex-pairs and then count the running time. The results of time are listed in Table 2. From Table 2, we can clearly see that CentSim outperforms the others on all benchmarks. This achievement of CentSim is due to its straightforward computation of similarity. The last row in Table 2 gives the total time of computing the scores of PageRank, Degree and Closeness centrality. As shown in Table 2, even counting the time of computing centralities, CentSim still costs the least time compared with the baselines. Therefore, our CentSim is more efficient than the state-of-the-art on assigning the similarity scores for vertex-pairs.

5.3 Comparison of Accuracy Performance

For the time performance, it is easy to evaluate by tracking the total running time. However, evaluating the performance of accuracy is quite hard, since it is difficult to identify a benchmark in which the real roles of vertices are identified or the role similarities of vertices are known. To delineate roles, two alternatives are utilized in this work: (1) as in [7], we use K-shell [3] as a proxy; (2) we adopt the roles of vertices discovered by the method proposed in [27] as ground-truth.

To quantitatively evaluate the performance of accuracy of a role similarity measure, two criteria are utilized in this paper.

The first criterion comes from the following idea. That is, the higher role similarity score two vertices have, the more likely they are within the same shell or share the same role. To formulate this idea, we compute the **fraction** of top ranked vertex-pairs that are within the same shells or share the same roles.

The accuracy performance evaluated by the fraction based on the two alternatives are shown in Fig. 2 and 3, respectively. Obviously, CentSim achieves the

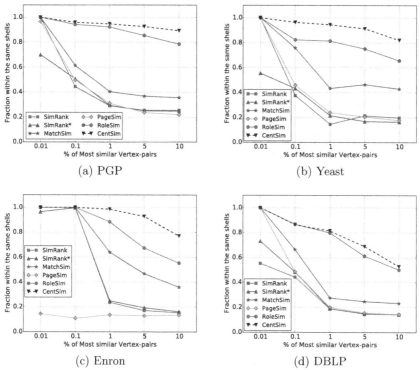

Fig. 2. Fraction of top ranked vertex-pairs that within the same shells

best performance on accuracy while RoleSim obtains the second best. In Fig. 2, on both PGP and Yeast, all measures, except SimRank*, do well for the top 0.01% vertex-pairs. However, CentSim markedly outperforms the baselines when more top-ranked vertex-pairs are considered. On Enron, PageSim unexpectedly achieves the very bad results, while others do well for the top 0.01% and 0.1% vertex-pairs. And furthermore, when the range is expanded, CentSim still does well, while the performance of both SimRank and SimRank* decline significantly. In our viewpoint, two reasons cause the poor accuracy of PageSim on Enron: one is the variety of vertex's degree is large, and the other is that the feature propagating mechanism makes PageSim to assign high similarity scores to vertex-pairs cross-shells. For the top 0.01% vertex-pairs ranked by PageSim, the maximum, average and variance of the difference of degree of vertex-pair are 1382, 12.118 and 1375.711, respectively. On DBLP, CentSim, RoleSim, MatchSim and PageSim do very well for the top 0.01% vertex-pairs; CentSim and RoleSim do well for the top 0.1% and 1% vertex-pairs. But CentSim shows better performance on accuracy than RoleSim for the top 5% and 10% vertex-pairs. In Fig. 3, the results are similar to those of Fig. 2. Due to lack of space, the details are omitted. In one word, experimental results in Fig. 2 and 3 indicate that more similar vertex-pairs ranked by CentSim are more likely to be within the same roles.

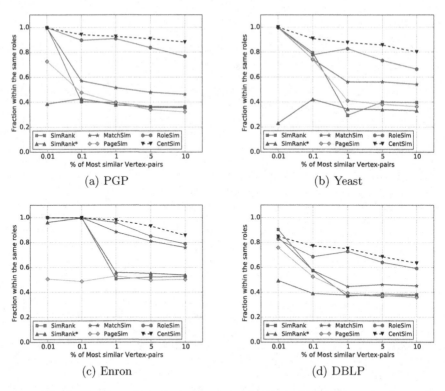

Fig. 3. Fraction of top ranked vertex-pairs that within the same roles

Table 3. Accuracies of similarity measures measured by auc statistic (ground-truth: K-shell). Each number is obtained by averaging over 100 independent realizations.

	PGP	Yeast	Enron	DBLP
SimRank	0.485	0.448	0.491	0.551
SimRank*	0.568	0.466	0.795	0.635
MatchSim	0.527	0.680	0.635	0.618
PageSim	0.428	0.374	0.498	0.489
RoleSim	0.822	0.864	0.877	**0.827**
CentSim	**0.823**	**0.872**	**0.933**	0.776

Table 4. Accuracies of similarity measures measured by auc statistic (ground-truth: role [27]). Each number is obtained by averaging over 100 independent realizations.

	PGP	Yeast	Enron	DBLP
SimRank	0.494	0.478	0.500	0.508
SimRank*	0.580	0.491	0.590	0.608
MatchSim	0.531	0.639	0.588	0.504
PageSim	0.444	0.407	0.499	0.424
RoleSim	0.760	0.755	0.720	0.627
CentSim	**0.820**	**0.801**	**0.795**	**0.699**

The basic opinion of the second criterion is that the role similarity of two vertices within-role (or shell) should be bigger than that of the other vertices cross-role (or shell). So, we group all vertex-pairs into two parts: within-role pairs, P^w, and cross-role pairs, P^c. To quantify the accuracy of similarity measures, we use the **AUC statistic**. It can be interpreted as the probability that a randomly chosen within-role pair (a pair in P^w) is given a higher role similarity score than a randomly chosen cross-role pair (a pair in P^c). In the implementation, among n independent comparisons, if there are n' cases that the within-role pair has the higher score, and n'' cases that the within-role pair and the cross-role pair have the same score, as in the research of link prediction [10], we define AUC in Eq. 4.

$$AUC = \frac{n' + 0.5 \times n''}{n} \tag{4}$$

The accuracy results measured by AUC are shown in Table 3 and 4, respectively. Generally speaking, CentSim can give overall better accuracy than the baselines. After CentSim, RoleSim performs the next best, while the others, particularly PageSim and SimRank, perform far worse. These results show that CentSim has high probability to assign large role similarity scores for vertex-pairs within-role.

In summary, the above experiments conducted in Section 5.2 and 5.3 demonstrate that CentSim is not only more efficient than the state-of-the-art but also outperforms them in accuracy. Therefore, we can conclude that the framework of measuring role similarities of vertices by comparing their centrality scores is competitive. Consequently, CentSim is competent to the task of role similarity.

6   Conclusion

In this paper, we proposed a novel and qualified similarity measure, namely CentSim, to quickly and accurately assign the role similarity between any two vertices of a network. We observe that the role of a vertex is related to its position in a network and centrality generally measures the position of a vertex in a network. Motivated by these two aspects, CentSim computes the role similarity between two vertices by means of comparing their corresponding centralities. Importantly, CentSim is an admissible role similarity metric since it obeys all the axiomatic role similarity properties.

We experimentally evaluate the performance of CentSim in terms of efficiency and effectiveness compared with SimRank, SimRank*, MatchSim, PageSim and RoleSim on four real-world datasets. The experimental results demonstrate that CentSim achieved overall best performance on both time and accuracy compared with the state-of-the-art. Thus, CentSim is a qualified role similarity metric.

References

1. Anderson, C.J., Wasserman, S., Faust, K.: Building stochastic blockmodels. Social Networks **14**(1), 137–161 (1992)
2. Bu, D., Zhao, Y., Cai, L., Xue, H., Zhu, X., Lu, H., Zhang, J., Sun, S., Ling, L., Zhang, N., Li, G., Chen, R.: Topological structure analysis of the protein-protein interaction network in budding yeast. Nucleic Acids Research **31**, 2443–2450 (2003)
3. Carmi, S., Havlin, S., Kirkpatrick, S., Shavitt, Y., Shir, E.: A model of internet topology using k-shell decomposition. PNAS **104**(27), 11150–11154 (2007)
4. Freeman, L.C.: Centrality in social networks: Conceptual clarification. Social Networks **1**(3), 215–239 (1979)

5. Holme, P., Huss, M.: Role-similarity based functional prediction in networked systems: application to the yeast proteome. Journal of the Royal Society Interface **2**(4), 327–333 (2005)

6. Jeh, G., Widom, J.: Simrank: a measure of structural-context similarity. In: SIGKDD 2002, pp. 538–543. ACM (2002)

7. Jin, R., Lee, V.E., Hong, H.: Axiomatic ranking of network role similarity. In: SIGKDD 2011, pp. 922–930. ACM (2011)

8. Jin, R., Lee, V.E., Li, L.: Scalable and axiomatic ranking of network role similarity. TKDD **8**(1), 3 (2014)

9. Klimt, B., Yang, Y.: Introducing the enron corpus. In: CEAS 2004

10. Lü, L., Zhou, T.: Link prediction in complex networks: A survey. Physica A **390**(6), 11501170 (2011)

11. Lin, Z., King, I., Lyu, M.R.: Pagesim: A novel link-based similarity measure for the world wide web. In: WI 2006, pp. 687–693. IEEE (2006)

12. Lin, Z., Lyu, M.R., King, I.: Matchsim: a novel neighbor-based similarity measure with maximum neighborhood matching. In: CIKM 2009, pp. 1613–1616. ACM (2009)

13. Luczkovich, J., Borgatti, S.P., Johnson, J.C., Everett, M.G.: Defining and measuring trophic role similarity in food webs using regular equivalence. Journal of Theoretical Biology **220**(3), 303–321 (2003)

14. Ma, H., King, I., Lyu, M.R.: Mining web graphs for recommendations. TKDE **24**(6), 1051–1064 (2012)

15. Mahadevan, P., Krioukov, D., Fomenkov, M., Dimitropoulos, X., Claffy, K., Vahdat, A.: The internet as-level topology: Three data sources and one definitive metric. ACM SIGCOMM Computer Communication Review 36(1), 17–26 (2006)

16. Ná, M.B., Satorras, R.P., Guilera, A.D., Arenas, A.: Models of social networks based on social distance attachment. Physical Review E **70**, 056122 (2004)

17. Nowicki, K., Snijders, T.A.B.: Estimation and prediction for stochastic blockstructures. Journal of the American Statistical Association **96**(455), 1077–1087 (2001)

18. Oliveira, M.: Gama, J.A.: An overview of social network analysis. Wiley Interdisciplinary Reviews. Data Mining and Knowledge Discovery **2**(2), 99–115 (2012)

19. Page, L., Brin, S., Motwani, R., Winograd, T.: The pagerank citation ranking: Bringing order to the web. In: WWW 1998, pp. 161–172. ACM (1998)

20. Rossi, R.A., Ahmed, N.K.: Role discovery in networks. arXiv preprint arXiv: 1405.7134 (2014)

21. Rossi, R.A., Gallagher, B., Neville, J., Henderson, K.: Modeling dynamic behavior in large evolving graphs. In: WSDM 2013, pp. 667–676. ACM (2013)

22. Ruan, Y., Parthasarathy, S.: Simultaneous detection of communities and roles from large networks. In: COSN 2014, pp. 203–214. ACM (2014)

23. Tiakas, E., Papadopoulos, A., Nanopoulos, A., Manolopoulos, Y., Stojanovic, D., Djordjevic-Kajan, S.: Searching for similar trajectories in spatial networks. Journal of Systems and Software **82**(5), 772–788 (2009)

24. Wasserman, S., Faust, K.: Social Network Analysis: Methods and Applications. Cambridge University Press (1994)

25. Yu, W., Lin, X., Zhang, W., Chang, L., Pei, J.: More is simpler: Effectively and efficiently assessing node-pair similarities based on hyperlinks. PVLDB **7**(1), 13–24 (2013)

26. Zhang, M., He, Z., Hu, H., Wang, W.: E-rank: A structural-based similarity measure in social networks. In: WI-IAT 2012, pp. 415–422. IEEE (2012)

27. Zhuge, H., Zhang, J.: Topological centrality and its e-science applications. Journal of the American Society for Information Science and Technology **61**(9), 1824–1841 (2010)

Multimodal-Based Supervised Learning for Image Search Reranking

Shengnan Zhao, Jun Ma$^{(\boxtimes)}$, and Chaoran Cui

School of Computer Science and Technology,
Shandong University, Jinan 250101, China
{belthshengnan,bruincui}@gmail.com, majun@sdu.edu.cn

Abstract. The aim of image search reranking is to rerank the images obtained by a conventional text-based image search engine to improve the search precision, diversity and so on. Current image reranking methods are often based on a single modality. However, it is hard to find a general modality which can work well for all kinds of queries. This paper proposes a multimodal-based supervised learning for image search reranking. First, for different modalities, different similarity graphs are constructed and different approaches are utilized to calculate the similarity between images on the graph. Exploiting the similarity graphs and the initial list, we integrate the multiple modality into query-independent reranking features, namely PageRank Pseudo Relevance Feedback, Density Feature, Initial Ranking Score Feature, and then fuse them into a 19-dimensional feature vector for each image. After that, the supervised method is employed to learn the weight of each reranking feature. The experiments constructed on the MSRA-MM Dataset demonstrate the improvement in robust and effectiveness of the proposed method.

Keywords: Image search reranking · Supervised reranking · Multimodal learning

1 Introduction

Thanks to the growth of image sharing Web sites, such as Flickr, image retrieval gains its popularity in our daily lives. General search engines exploit associated textual information of images, e.g., the titles, surrounding texts and the URL of the web pages containing the images, to retrieval the images. However, a picture is worth a thousand words, those textual information lack the discriminative power to deliver visually search results. Due to the mismatch between images and their associated textual information, the performance of general search engines is less than satisfactory. To tackle this problem, image search reranking is proposed to reorder the images in the initial list returned by text-based search engine to enhance the performance of search engine.

Though many efforts have been devoted to the research on image reranking, the performance of reranking methods varies a lot when the methods employ with different queries [1]. One reason is that the visual feature which the model applies

© Springer International Publishing Switzerland 2015
J. Li and Y. Sun (Eds.): WAIM 2015, LNCS 9098, pp. 135–147, 2015.
DOI: 10.1007/978-3-319-21042-1_11

is not effective for some queries. Therefore, we employ multimodal features which mean the multiple visual features extracted from images.

Two most popular methods using multimodal features are early fusion and late fusion [2]. Early fusion directly fuses all multimodal features within a long feature vector which suffers from " curse of dimensionality ". For this reason, we employ late fusion which utilizes multimodal features to develop a set of query-independent reranking features and integrates them within a short feature vector. Then a supervised method is used to learn the weight of each feature. The main difficulties of our method lie in what reranking features we should develop and what supervised method we should utilize. In our method, we develop two different features for each modality, namely, PageRank Pseudo Relevance Feedback and Density Feature, and develop Initial Ranking Score Feature for initial list (presented in Section 3.2) and integrate them into 19-dimensional feature vector. We employ Rank SVM (presented in Section 3.1) to learn the weight of each feature.

The main contributions of this paper are described as below:

- We propose a multimodal-based supervised learning for image search reranking.
- To exploit the query-independent features for each modality, we design PageRank Pseudo Relevance Feedback and Density Feature.
- Each modality has its own character, we introduce different strategies for different modality to obtain the similarity between a couple of images.
- We conduct experiments on the MSRA-MM Dataset[3] and the results demonstrate the improvement in robust and effectiveness of the proposed method.

The rest of this paper is organized as follows: In Section 2, we review some related works on image search reranking. In Section 3, we explain the detail of the proposed multimodal-based supervised learning method. In Section 4, we illustrate our experimental results conducted on MSRA-MM Dataset. The conclusion and future work are presented in Section 5.

2   Related Work

The existing image search reranking methods can be divided into two categories: supervised methods and unsupervised methods.

Supervised methods usually first exploit a feature vector. Then they employ a part of data in dataset for training to learn the weight of each feature while the rest of data is for testing. Nowadays, there are various training methods. The most common method that the researchers employed is Rank SVM [4]. The groundtruth they usually use is the relevance between query and corresponding images which is annotated by volunteers. After obtaining the weight of each feature, they test their methods on the rest of data. Finally they rerank the images in the test data by the predicted score which is obtained by the product of the weight vector and feature vector. Different methods vary mostly in which features they use. In [5], the features are exploited from the initial list and contextual information which are based on the whole image. Yang et al. [6] argued that it is too rigid to operate on the whole image, by reason that there may be some noises on the background. Therefore, they proposed a bag-of-objects-retrieval model which is based on the object extracted from the images.

Unsupervised methods rerank images on the strength of textual and visual features. In [7] which is merely based on visual features, Jing et al. first detect visual features of images and calculate relevance between images. Then they construct a similarity graph where they set the images as nodes and the edges between nodes are weighted by the visual similarity. After that, the random walk is applied to the graph until the tolerance converges to a very small number. Through random walk, the nodes spread their visual information to the other images in the whole network. [8] is based on both textual and visual features. Tian et al. treat the textual features which are detected by the initial ranking list as a good baseline to do the image search reranking. With those features, they construct a model with Bayesian framework.

3   Multimodal-Based Supervised Learning

Before presenting the proposed image reranking method, we give some definitions.

Definition 1:   Reranking optimization is to find a new ranking function which is defined as:

$$r = \arg \min_r D(r, \bar{r}) \tag{1}$$

where $\bar{r}$ is the ideal list which sorts the relevance of images in descending order. $D(\cdot)$ is the distance between reranking list r and ideal list $\bar{r}$. It can be modeled using Kullback-Leibler divergence (KL divergence) computed by Eq (2):

$$D_{KL}(r, \bar{r}) = \sum_i P(r_i|q, d) \ln \frac{P(r_i|q, d)}{P(\bar{r}_i|q, d)} \tag{2}$$

where r_i is the ranking number of i-th image in the reranking list r, q is the given query, d refers to the image collection returned by search engines.

Substituting (2) into (1), we obtain:

$$r = \arg \min_r D_{KL}(r, \bar{r}) \tag{3}$$

Based on the theoretical derivation, we get:

$$\min_r D_{KL}(r, \bar{r}) \tag{4}$$

$$\propto \min - \sum_i P(r_i|q, d) \ln P(\bar{r}_i|q, d) + \xi_q \tag{5}$$

$$\propto \max \sum_i P(r_i|q, d) + \xi_q \tag{6}$$

where ξ_q is a constant which can be ignored for reranking and $P(r_i|q,d)$ can be modeled using Supervised Model[5]:

$$r = \arg\max_r \sum_{j=0}^{Z} w_j r \psi_j(q,d) \tag{7}$$

where ψ are the functions which can master of the feature extraction processes and w is the weight for each function. Hence, the solution of reranking optimization Eq (1) can be solved by:

$$r = sort \quad by \quad \sum_{j=0}^{Z} w_j \psi_j(q,d) \tag{8}$$

The performance of Supervised Model varies a lot from query to query. The reason is that the visual feature the model applies is not effective for some of the query. Therefore, in our method, we employ multimodal features which mean the multiple visual features extracted for the images.

Fig. 1 briefly describes the framework of the proposed multimodal-based supervised learning method. First, we utilize a training model to learn the weight of each query-independent feature. The training data is submitted to the Feature Generating Model (FGM). When a query is submitted, FGM presents the query to a search engine which returns a initial list L and a image set D. Then, we detect six visual features which are detailed in Section 3.2 and calculate six different similarity metrics for each image in D based on each visual feature. For calculating query-independent reranking features (i.e., PPRE, DF), we construct six different similarity graphs which nodes are images. The difference between six similarity graphs is the weight of each edge which is measured by the similarity between images. After that, exploiting the similarity graphs and the initial list, the query-independent features (i.e., PPRE, DF, IRS) are calculated in accordance with Eq (10), Eq (11), Eq (14). With the knowledge of features and the groundtruth, the weight for each feature is learned by Rank SVM which is recommended in Section 3.1. In practice, after the query committed to the search engine, a query-independent feature vector is exploited by FGM. With the product of features and weight, the reranking process is conducted and the reranking list is returned. More detail of the proposed method for the practical image search reranking is shown in algorithm 1.

3.1   Rank SVM

Rank SVM [4] turns the ranking problem into a classification problem by using pair-wise classification SVM.

$$min_{w,\varepsilon} \quad \frac{1}{2}\|w\|^2 + C\sum \varepsilon_{x,y}^i$$
$$s.t. \quad \forall q_i, k_x \succ k_y \quad R(x;q_i) - R(y;q_i) \geq 1 - \varepsilon_{x,y}^i \tag{9}$$
$$\forall x,y,i \quad \varepsilon_{x,y}^i \geq 0$$

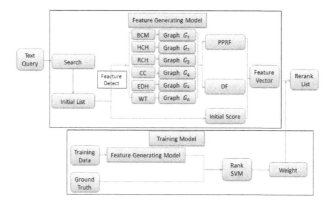

Fig. 1. The framework of the proposed Multimodal-based Supervised Learning for Image Search Reranking

where w is the weight for each feature and C is a trade-off. $k_x \succ k_y$ means image X is ranked before image Y.

Many approaches can be used to solve Rank SVM, such as SMO (Sequential Minimal Optimization) [9] and cutting-plane algorithm [10]. In this paper, we employ the system designed by Thorsten Joachims [11].

3.2   Feature Generation

We employ the visual features presented below. And on this basis, we exploit two types of features, i.e. , PageRank Pseudo Relevance Feedback(PPRF) and Density Feature(DF). PPRF is extracted by assuming top m images in the PageRank list are relevant images while lowest-ranked images are irrelevant. The score of PPRF is calculated based on relevance and irrelevance feedback. DF is extracted from the density of image calculated by Kernel Density Estimation(KDE) [12]. At last, we employ the initial list to develop a initial score.

Visual Features. Visual features are low-level features used in image search reranking. A range of visual features can be extracted from the images: color features, textural features, shape features, spatial relationship features, and the others. In this paper, we utilize the following visual features:

- **Block-wise Color Moment (BCM)** BCM is a simple but effective characteristic of color. First, BCM divides the image into n regions. After that in each region, three color moments (i.e., Mean, Standard Deviation and Skewness) are computed for per channel. So the dimensions of feature vector $d = n \times c \times 3$ where c is the number of channels. In this paper, we use 225D Block-wise Color Moment.
- **HSV Color Histogram (HCH)** HCH is a kind of color histogram on HSV color space. First, HCH divides HSV color space into n bins. Each bin has

Algorithm 1. SMRModel(D, L, f)

Input:

    D : the image set;

    L : the initial list;

    f : reranking function;

Output:

    rerank list r;

1: extract six visual features(i.e., BCM, HCH, RCH, CC, EDG, WT) recommended in Section 3.2 for each image in D;

2: calculate six different similarity metrics based on each visual feature using the methods which are described in Section 3.4;

3: construct six different graphs(i.e., G_1, G_2, G_3, G_4, G_5, G_6) on the basis of different similarity metrics which are computed above;

4: calculate query-independent features(i.e., PPRF, DF) recommended in Section 3.2 for each image on different graphs;

5: calculate initial ranking score(IRS) recommended in Section 3.2 for each image;

6: integrate the features we have gained into a query-independent reranking feature vector (PPRE1, ..., PPRE12, DF1, ..., DF6, IRS);

7: transfer feature vector to rerank function f (Eq. 8) which weight is learned by Rank SVM to get the reranking list r;

8: **return** r;

its own wavelength range of the light spectrum. Through merely counting the number of pixels on the image for each bin, the histogram is obtained. HCH is a representation of the distribution of colors in a HSV image. In this paper, we use 64D HSV Color Histogram.

- **RGB Color Histogram (RCH)** The same as HCH, RCH is a kind of color histogram on RGB color space. RCH can be implemented in exactly the same way as HCH. In this paper, we use 256D RGB Color Histogram.

- **Color Correlogram (CC)** Different from the traditional color histogram, CC builds the histogram by the distance between different colors. The first step of CC is the same as HCH. After that CC choose different distance criterions $d_1, d_2 \cdots d_m$. Then for each distance and each pair of bins, we count the number of pairs whose spatial distance between bins is d_i. So the dimensions of feature vector $d = n \times n \times m$. In this paper, we use 144D Color Correlogram.

- **Edge Distribution Histogram (EDH)** EDH is the textural feature of image. First, EDH detects image edge by Sobel operator, Canny operator, and so on. Based on the image edge we get, the EDH is constructed and normalized. In this paper, we use 75D Edge Distribution Histogram.

- **Wavelet Texture (WT)** WT is a global texture feature. First, WT needs to choose a mother wavelet function. Based on this function, at each level, it decomposes the image into four part. Finally, for each part at each level we calculate mean and standard deviation of the energy distribution to develop the feature vector. In this paper, we use 128D Wavelet Texture.

PageRank Pseudo Relevance Feedback (PPRF). PRF [13] is widely adopted today to do the image retrieval. PRF assumes top-ranked images in the initial list are relevant to the query. Then a majority of PRF methods rerank the images according to the relevance of top ranked images. Due to the very top-ranked results are not always meeting the users intention, so we design the PPRF to convert the image reranking into identifying the "authority" nodes in the similarity graph. First, we employ PageRank [7] to obtain the "authority" of nodes. By ordering the authority of images, the PageRank list is obtained. Then we assume top-ranked images in the PageRank list are relevant to the query while lowest-ranked images are irrelevant. Finally, we use relevance and irrelevance feedback to compute the rerank score for images. Duplicate voting is adapted to do the relevance and irrelevance feedback.

$$PPRF(x_i) = \frac{1}{m} \sum_{j=1}^{m} \frac{dup(x_i, x_j)}{\lg(r(x_j) + 1)} \qquad (10)$$

where m is the number of images we used to feedback. $dup(\cdot)$ is the duplicate detection function elected from [14].

Density Feature (DF). Density Feature is based on the density assumption which means relevant images should have higher density than irrelevant images. It is confirmed by [1]. KDE is employed to calculate the density of image x_i.

$$DF(x_i) = \frac{1}{|\mathcal{N}(x_i)|} \sum_{x_j \in \mathcal{N}(x_i)} K(x_i - x_j) \qquad (11)$$

where $\mathcal{N}(x_i)$ is the neighbors of image x_i. $K(\cdot)$ is the kernel function which is a positive function that integrates to one. A range of kernels are commonly used: uniform, normal, triangular, and others. In this paper, the Gaussian kernel is employed.

Initial Ranking Score Feature (IRS). Initial ranking list is a good baseline for the "best" ranking list. Though noisy, we should discard the dross and select the essence as well. However, most of the search engines only provide the ranking list while the ranking score is unavailable. There are a number of ways to convert the ranking list into the ranking score which are detailed in Section 3.3. In this paper, we pick the equation (14) to do the work mentioned above.

3.3   Initial Ranking Score

Initial ranking score s_i is the ranking score of image X_i in the initial list. The most direct way to get this score is to use the score returned by text-based search engines. However, in most cases, this score is hard to obtain and the search engines just return the initial lists. For these and other reasons, we should transform ranking number to score. Many alternative strategies are proposed by [8] [15]:

- **Ranking Number (RN)** RN employs the difference between the number of total images and the ranking number of the image to obtain the ranking score s_i.

$$s_i = N - r_i \qquad i = 1 \ldots N \tag{12}$$

where N is the number of total images in the initial list and r_i is the ranking number of i-th image in the initial list.
- **Normalized Ranking Number (NRN)** Based on RN, NRN uses the total number N to normalize the RN.

$$s_i = 1 - \frac{r_i}{N} \qquad i = 1 \ldots N \tag{13}$$

- **Initial Relevance Estimation (IRE)** IRE investigates the relationship between ranking score s_i and the ranking number r_i with all queries.

$$s_i = E_{q \in \varrho}[rel(q, r_i)] \qquad i = 1 \ldots N \tag{14}$$

where ϱ means all the queries in the dataset. $E_{q \in \varrho}$ is the expectation on the query set ϱ. $rel(q, r_i)$ is the ground truth of i-th image in the query q which is labeled to be 0, 1 or 2. A heuristic way to obtain $E_{q \in \varrho}$ is to average the relevance score $rel(q, r_i)$ on the query set ϱ.

3.4 Image Similarity

In this paper, we should construct a graph G, whose nodes are the images returned by text-based search engine and edges are established between a couple of similar images. The similarity between images are calculated by:

- **Jensen-Shannon divergence (JSD)** JSD is based on the Kullback-Leibler divergence[16].

$$D_{JSD}(q, t) = \sum_{m=0}^{M-1} \left(h_q[m] \lg \frac{h_q[m]}{o[m]} + h_t[m] \lg \frac{h_t[m]}{o[m]} \right)$$
$$where \qquad o[m] = \frac{h_q[m] + h_t[m]}{2} \tag{15}$$

where $h_q[m]$ is the value of $m - th$ bin in the histograms for image q.
- **Radial basis function kernel (RBF)** The value of RBF merely depends on the distance from the origin. The norm radial function is usually Euclidean distance.

$$D_{RBF}(q, t) = exp(-\frac{\|x_q - x_t\|^2}{2\sigma^2}) \tag{16}$$

where x_q is the feature vector of image q and σ is the scaling parameter.
- **Histogram Intersection Method(HIM)** HIM [17] exploits a average similarity of images.

$$D_{HIM}(q, t) = \frac{1}{M} \sum_{m=0}^{M-1} (1 - \frac{|h_q[m] - h_t[m]|}{max(h_q[m], h_t[m])}) \tag{17}$$

As we mentioned before, many policies can be employed to develop the similarity between images. The different features have different characteristics, and for some features, another policy works better. For the feature of Color Histogram, i.e., HSV Color Histogram and RGB Color Histogram, Histogram Intersection Method (17) is employed while for Color Correlogram, the Jensen-Shannon divergence (15) is used and for the others, we utilize Radial basis function kernel (16).

4   Experiments

In this section, we first describe the experimental settings that we will use in the aforementioned Multimodal-based Supervised learning method. Then we will present the performance of our method compared with several existing methods. Last of all, we will analysis the effect of different initial lists on different methods.

4.1   Experimental Settings

Dataset. In our experiment, we evaluate our method on MSRA-MM dataset. In this dataset, there are 68 representative queries which are selected from the query log of Microsoft Live search and then for each query they collect about 1000 images. So there are a total of 65443 images. For each image, they provide a set of features. In order to evaluate the result, they also provide a vector of relevance score which is construct between the query and the corresponding images returned by text-based search engine. The relevance score is in $\{0, 1, 2\}$. The score $\{0\}$ means that the image is irrelevant to the corresponding query. The score $\{1\}$ means that the image is relevant to the corresponding query. The score $\{2\}$ means that the image is very relevant to the corresponding query.

Evaluate. We employ widely used Normalized Discounted Cumulative Gain ($NDCG$) to evaluate the performance of our method. The $NDCG$ for a given query is calculated as follow:

$$NDCG@n = \frac{DCG@n}{IDCG@n} \tag{18}$$

where

$$DCG@n = rel_1 + \sum_{i=2}^{n} \frac{rel_i}{\log_2 i} \tag{19}$$

The rel_i is the relevance of i-th image in the reranking list. The n means we use top n images to calculate the $NDCG@n$. $IDCG@n$ is a regularization factor which is computed in a ideal way. First, we sort the images on basis of the relevance score, then we employ the equation (19) to compute it. It is the maximum value of $DCG@n$. We utilize $mNDCG@n$(mean of $NDCG@n$ on all queries) to evaluate our methods on all of the queries.

4.2 Performance on All Queries

In this section, to demonstrate the effectiveness of the proposed multimodal-based supervised learning method(SM), we compare it with the methods using merely one modality (i.e., BCM, HCH, RCH, CC, EDH, WT) and three state-of-the-art approaches. The baseline includes Supervised Reranking (SR) [5], Pseudo Relevance Feedback Reranking (PRF) [13], Bayesian Visual Reranking (BVR) [8]. For SR, the Rank SVM [4]is used to train the model to gain the weight of features. For PRF, we treat top-10 images as positive samples to do the relevance feedback. For BVR, according to the suggestion presented by Tian et al. , the local learning regularizer is adopted to learn the regularizer term while the pairwise ranking distance is adopted to measure the ranking distance.

Fig. 2 shows the performance of aforementioned three baseline methods and proposed SM in the light of $mNDCG@20$, $mNDCG@40$ and $mNDCG@60$. From Fig. 2, we can obvious see that SM can outperform the baseline methods. Even though the SM with only one modality, the baseline methods are less effective than it. As position of $mNDCG \ n$ increases, the performance of three baseline methods certainly fall a great deal while the proposed method makes only a small dent. It demonstrates the effectiveness of our method.

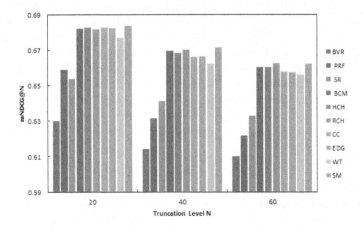

Fig. 2. Performance on All Queries

4.3 Performance on Different Initial lists

We employ three initial lists constructed by [1] and the initial list provided by MSRA-MM dataset to analysis the effect of different initial list. The initial lists include the following components:

- Perfect initial list (Perfect): This list orders the images based on their relevance score. All very relevant images are ordered on the top of list while all the irrelevant images are ordered on the bottom of list.

- Worst initial list (Worst): The same as perfect initial list, images are ordered based on their relevance score. The difference is that all irrelevant images are ordered on the top of list while all very relevant images are ordered on the bottom of list.
- Random initial list (Random): Images are ordered randomly.
- Initial list (Initial): Initial list means the list returned by search engine. In other words, it is the list provided by MSRA-MM dataset.

Fig. 3 compares the three baseline methods and the proposed method on different initial lists. For the result, we can see that PRF and BVR are sensitive to the initial list. A possible reason is that PRF assumes the top-ranked images which are in the initial are pseudo relevant images. If we use the worst initial list to do PRF reranking, most of the pseudo relevant images will be irrelevant images. Based on those images, all the images are reranked. It leads to irrelevant images have a higher rank than relevant images. In the opinion of Bayesian visual reranking, the reranking list should close to the initial list. For this reason, it causes Bayesian visual reranking sensitive to the different initial lists. Supervised multimodal method does a good job on both perfect initial list and worst initial list. In the learning step of our model, we use Rank SVM to learn the weight of each features we construct. When using worst initial list, the weight of features which are about initial list will have a small value. And it gives a large value to the features which are about perfect initial list. In this way, whether using the perfect initial list or not, our method puts up a good performance.

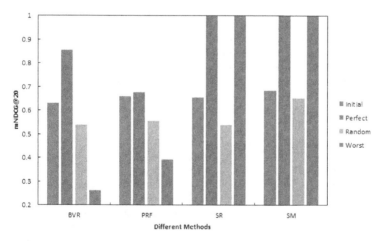

Fig. 3. Performance on Different Initial lists

5   Conclusion

In this paper, we propose a novel model to do image search reranking, multimodal-based supervised learning for image search reranking. This model employs multiple modalities to exploit the query-independent features. Evaluations on MSRA-MM dataset demonstrate the effectiveness and stability of the proposed method.

In the future, we can concentrate on the following respects:

- In this paper, we only design two features for each modality. We can dig through more effective query-independent features.
- The proposed method does not take the diversity into consideration. After reranking, we can append with a step to realize the diversity.

Acknowledgments. This work was supported by Natural Science Foundation of China (61272240, 71402083, 6110315).

References

1. Tian, X., Lu, Y., Yang, L., Tian, Q.: Learning to judge image search results. In: Proceedings of the 19th ACM International Conference on Multimedia, pp. 363–372. ACM (2011)
2. Snoek, C.G.M., Worring, M., Smeulders, A.W.M.: Early versus late fusion in semantic video analysis. In: Proceedings of the 13th Annual ACM International Conference on Multimedia, pp. 399–402. ACM (2005)
3. Wang, M., Yang, L., Hua, X.-S.: Msra-mm: Bridging research and industrial societies for multimedia information retrieval, Microsoft Research Asia. Tech. Rep. (2009)
4. Joachims, T.: Optimizing search engines using clickthrough data. In: Proceedings of the eighth ACM SIGKDD International Conference on Knowledge Discovery and Data Mining, pp. 133–142. ACM (2002)
5. Yang, L., Hanjalic, A.: Supervised reranking for web image search. In: Proceedings of the International Conference on Multimedia, pp. 183–192. ACM (2010)
6. Yang, Y., Yang, L., Wu, G., Li, S.: A bag-of-objects retrieval model for web image search. In: Proceedings of the 20th ACM International Conference on Multimedia, pp. 49–58. ACM (2012)
7. Jing, Y., Baluja, S.: Visualrank: Applying pagerank to large-scale image search. IEEE Transactions on Pattern Analysis and Machine Intelligence **30**(11), 1877–1890 (2008)
8. Tian, X., Yang, Y., Wang, J., Xiuqing, W., Hua, X.-S.: Bayesian visual reranking. IEEE Transactions on Multimedia **13**(4), 639–652 (2011)
9. Platt, J.C.: Fast training of support vector machines using sequential minimal optimization. In: Advances in Kernel Methods, pp. 185–208. MIT Press (1999)
10. Joachims, T.: Training linear svms in linear time. In: Proceedings of the 12th ACM SIGKDD International Conference on Knowledge Discovery and Data Mining, pp. 217–226. ACM (2006)
11. Ranking SVM. http://www.cs.cornell.edu/people/tj/svm_light/svm_rank.html
12. Scott, D.W.: Multivariate density estimation: theory, practice, and visualization, vol. 383. John Wiley & Sons (2009)
13. Yan, R., Hauptmann, A., Jin, R.: Multimedia search with pseudo-relevance feedback. In: Bakker, E.M., Lew, M.S., Huang, T.S., Sebe, N., Zhou, X.S. (eds.) CIVR 2003. LNCS, vol. 2728, pp. 238–247. Springer, Heidelberg (2003)

14. Philbin, J., Chum, O., Isard, M., Sivic, J., Zisserman, A.: Object retrieval with large vocabularies and fast spatial matching. In: IEEE Conference on Computer Vision and Pattern Recognition, CVPR 2007, pp. 1–8. IEEE (2007)
15. Wang, M., Li, H., Tao, D., Ke, L., Xindong, W.: Multimodal graph-based reranking for web image search. IEEE Transactions on Image Processing **21**(11), 4649–4661 (2012)
16. Jensen Shannon divergence. http://en.wikipedia.org/wiki/Jensen-Shannon_divergence
17. Lee, S.M., Xin, J.H., Westland, S.: Evaluation of image similarity by histogram intersection. Color Research and Application **30**(4), 265–274 (2005)

Novel Word Features for Keyword Extraction

Yiqun Chen[1,2], Jian Yin[1(✉)], Weiheng Zhu[3], and Shiding Qiu[3]

[1] Department of Computer Science, Sun Yat-sen University, Guangzhou 510275, China
issjyin@mail.sysu.edu.cn
[2] Department of Computer Science, Guangdong University of Education,
Guangzhou 510303, China
[3] College of Information Science Technology, Jinan University, Guangzhou 510632, China

Abstract. Keyword extraction plays an increasingly crucial role in several texts related researches. Applications that utilize feature word selection include text mining, and information retrieval etc. This paper introduces novel word features for keyword extraction. These new word features are derived according to the background knowledge supplied by patent data. Given a document, to acquire its background knowledge, this paper first generates a query for searching the patent data based on the key facts present in the document. The query is used to find files in patent data that are closely related to the contents of the document. With the patent search result file set, the information of patent inventors, assignees, and citations in each file are used to mining the hidden knowledge and relationship between different patent files. Then the related knowledge is imported to extend the background knowledge base, which would be extracted to derive the novel word features. The newly introduced word features that reflect the document's background knowledge offer valuable indications on individual words' importance in the input document and serve as nice complements to the traditional word features derivable from explicit information of a document. The keyword extraction problem can then be regarded as a classification problem and the Support Vector Machine (SVM) is used to extract the keywords. Experiments have been done using two different data sets. The results show our method improves the performance of keyword extraction.

Keywords: Keyword extraction · Patent data · Information retrieval

1 Introduction

Keywords offer an important clue for people to grasp the topic of the document, and help people easily understand what a document describes, saving a great amount of time reading the whole text. Being increasingly exposed to more and more information on the Internet, people of today have to be more selective about what to read and it's impossible for people to mark keywords for every document manually. Consequently, automatic keyword extraction is in high demand. Meanwhile, it is also fundamental to many other natural language processing applications, such as information retrieval, text clustering, text categorization and so on.

When people mark the documents manually, they try to understand the key facts present in the documents according to their background knowledge, and then pick out

J. Li and Y. Sun (Eds.): WAIM 2015, LNCS 9098, pp. 148–160, 2015.
DOI: 10.1007/978-3-319-21042-1_12

the most important content words as keywords. Our keyword extraction method tries to provide the background knowledge for the target document, which make our method can 'understand' the document as human being, then 'identify' the most important words in the document.

A straightforward solution for building the background knowledge is to make use of massive data. Patent data is the record of science and technology, which has recognized authority. In recently, patent data has become the main source of science and technology in the world. The United States Patent and Trademark Office publish patent data set (named as US Patent) [1] includes all patent granted in United States Patent. Each patent is presented in well XML format, including patent title, number, inventors, assignees, abstract, description and citations, etc. Each patent file is a rigorous description of science and technology, including abundant knowledge. It can be a very good support for background knowledge building.

The key contribution of our work is the introduction on how to make use of a universal knowledge base, patent data, to construct the background knowledge for the target document. To the best of our knowledge, this paper is the first attempt to use patent data as the external knowledge repository.

The second contribution of our work is the introduction of novel word features for keyword extraction. We built a document's background knowledge derived from patent data. To retrieve background knowledge related to the target document, we first form a patent data query according to key facts present in the document. We then search the patent data. Once a set of patent files on the document's background knowledge is obtained, we can mine the hidden knowledge by using the relationship between patent inventors, assignees and citations for semantic relatedness. We then derive four word features based on the related patent, related inventors, related assignees and patent citations link in the retrieved files. Our approach utilizes all four types of information during the keyword selection process. The newly introduced word features that reflect the document's background knowledge offer valuable indications on individual words' importance in the input document and serve as nice complements to the traditional word features derivable from explicit information of a document. The results of our experiments confirm the effectiveness of our new word features for keyword extraction by considering the background knowledge of a document.

2   Related Work

There are a number of classical approaches to extracting keywords. Traditional keyword extraction methods only use information explicitly contained in a document based on lexical and syntactic features, such as the word frequency and word positions etc. However, these methods were generally found to lead to poor results, and consequently other methods were explored.

The research community investigated supervised learning methods whereby training data is used to provide syntactic and lexical features for keyword extraction. In KEA [2] system, the candidate terms are represented using three features: TFxIDF, distance (the number of words that precede the first occurrence of the term, divided by the number of words in the document), key phrase-frequency (the number of times a candidate term occurs as a key term in the training documents). The classifier is trained using the naive Bayes learning algorithm. Thus, KEA analyses only simple statistical properties of candidate terms and considers each candidate term independently from the other terms in the document.

An alternative class of approaches applies graph-based semantic relatedness measures for extracting keywords. The example is TextRank [3]: a document is modeled as a graph of candidate terms in which edges represent a measure of term relatedness. Some graph analysis technique is used to select key terms. Usually the graph is analyzed using a graph-based ranking algorithm (such as PageRank [4], HITS [5], etc.) to range candidate terms. Top N terms with the highest scores are selected as keywords.

Last but not least, in the past few years, different works have leveraged the structured knowledge from Wikipedia to extract keywords. The new methods have been proposed that extend classical methods with new features computed over Wikipedia corpus. For instance, Wikify! [6] introduces keyword feature of a candidate term that is defined as the number of Wikipedia articles in which the term appears and is marked up as a link divided by the total number of Wikipedia articles where the term appears. This feature can be interpreted as probability that the candidate term is selected as a key term in a Wikipedia article as according to the Wikipedia editorial rules only key terms should be used as links to other articles. Maui [7] is the successor of the KEA system, and it extends the feature set used by KEA with several Wikipedia-based features: a) Wikipedia keyphraseness: likelihood of a term in being a link in Wikipedia; b) semantic relatedness: semantic score derived from Wikipedia associated with each candidate keyword; and c) inverse Wikipedia linkage, which is the normalized number of pages that link to a Wikipedia page. The semantic relatedness feature is calculated by linking each candidate keyword to a Wikipedia article, then comparing – by means of a Wikipedia-based semantic measure – a given candidate keyword to all the other candidates. The final value corresponds to the total of all the pairwise comparisons. Paper [8] exploits the titles of Wikipedia articles as well of the graph of Wikipedia categories, and paper [9] utilize Wikipedia as a thesaurus for candidate keyword selection. The paper [10] uses not only information explicitly contained in the document such as word frequency, position, and length but also the background knowledge of the document, which is acquired from Wikipedia via analyzing the inlink, outlink, category, and infobox information of the document's related articles in Wikipedia. Paper [11] compares the state-of-the-art systems, and none of them make use of patent data.

The information retrieval research community is increasingly making use of the information richness in knowledge sources for improving the effectiveness of Web search applications. Wikipedia has been intensively used recently to provide background knowledge for natural language processing while the patent data hasn't been noticed. To the best of our knowledge, this paper is the first attempt for using patent data as background knowledge depository on the task of keyword extraction. Furthermore, we differ from previous approaches [6,7,10] in that we not only using the text content of the patent files but also mining the hidden knowledge by using the relationship between patent inventors, assignees and citations for semantic relatedness.

3   Methodology

Patent data is the main carrier of science and technology progress and innovation, which provides up-to-date and extensive coverage for a broad range of domains. Every Patent file is a strictly verified document, which has considerable authority. Patent file provides information of inventors, assignees, title, abstract, patent classification, description and citations etc., as showed in Fig 1. The information provided in patent file is good for patent understanding and searching. Also can be a good support for related domain's background knowledge.

In this section we discuss the characters of patent data, and discover the relationship between different patent files to build a background knowledge base for the input document. Novel word features derived from patent data that reflect the document's background knowledge are described in section 3.4. Section 3.5 introduce how the keyword extraction problem can be regarded as a classification problem and how the Support Vector Machine (SVM) is used to extract the keywords.

```
   US-08621894-B2.xml

  1       <?xml version="1.0" encoding="UTF-8"?>
  2       <!DOCTYPE us-patent-grant SYSTEM "us-patent-grant-v44-2013-05-16.dtd" [ ]>
  3       <us-patent-grant lang="EN" dtd-version="v4.4 2013-05-16" file="US08621894-2
          date-produced="20131224" date-publ="20140107">
  4       <us-bibliographic-data-grant>
  5       <publication-reference>
 13       <application-reference appl-type="utility">
 20       <us-application-series-code>12</us-application-series-code>
 21       <us-term-of-grant>
 24       <classifications-ipcr>
 41       <classification-national>
 50       <invention-title id="d2e53">Sleep support surface that includes a layer wit
 51       <us-references-cited>
422       <number-of-claims>13</number-of-claims>
423       <us-exemplary-claim>1</us-exemplary-claim>
424       <us-field-of-classification-search>
450       <figures>
454       <us-related-documents>
471       <us-parties>
512       <assignees>
525       <examiners>
536       </us-bibliographic-data-grant>
537       <abstract id="abstract">
540       <drawings id="DRAWINGS">
554       <description id="description">
599       <us-claim-statement>What is claimed is:</us-claim-statement>
600       <claims id="claims">
651       </us-patent-grant>
652
```

Fig. 1. Patent file example

3.1   Process

Our patent based document background knowledge acquisition approach consists of following steps to acquired the novel words features and extract the keywords for the input document:

1) **Pre-processing.** Given a document, we first generate a patent inquiry query for retrieving the document's background knowledge through searching the patent data.

2) Knowledge obtains. We execute the patent query to obtain a set of patent search result files for the input document. After that, we discover the semantic relatedness between patent inventors, assignees and citations, and then extend the background knowledge.

3) Feature extraction. The input is a bag of words in a document. We make use of both local context information and global context information from background knowledge of a word in the document to define its features (see section 3.4 for a detailed definition of the features). The output is the feature vectors, and each vector corresponds to a word.

4) Learning and Extraction. We construct a SVM model that can identify the keyword. In the SVM model, we view a word as an example, the words labeled with 'keyword' as positive examples, and the other words as negative examples. Each word has the features value defined in step 3). We use the labeled data to train the SVM model in advance. Note that we consider only single words as candidates for training. In extraction, the input is a document. We employ the preprocessing and the feature extraction on it, and obtain a set of feature vectors. We then predict whether or not a word is a keyword by using the SVM model from above. The multi-word keywords will then be eventually reconstructed. Finally, we get the extracted keywords for that document.

The key issue here is how to obtain the background knowledge and how to define features for effectively performing the extraction task.

3.2   Pre-processing: Generating a Patent Search Query

To generate a query to find the most relevant background knowledge of a document via searching the patent data, we construct the query based on the key facts carried in the input document. This is done through applying a modified version of the TextRank [3] to select the important content words from the input document. In the original TextRank algorithm, pairwise sentence similarity is based on the word matching degree of two sentences. In the modification method, the way the pairwise sentence similarity is calculated using semantics of words than through mere counting of the number of overlapping characters in the spellings of two words. In our pre-processing, Qtag [12] is used to identify the part of speech of the words. Then we measure sentence similarity through word semantic relativeness analysis [13] using WordNet [14]. Finally we get a few key sentences selected from the input document through the above process. We then perform stop word removal and word stemming over all the words in these key sentences. The remaining words constitute our patent search query.

3.3   Knowledge Obtains: Searching the Patent Data

To obtain the background knowledge for the input document, we call on the full text search engine, Lucene [15], to retrieve files from the patent data that are related to the input document's key contents. We defined two levels background knowledge for the input document.

Using the search query for the input document generated in section 3.2, we get the search results, which are returned as a ranked list of patent files and their corresponding relativeness to the search query (query relativeness scores). We denote the set of retrieved patent files (duplicated files are discard) as $\prod$; the r-th patent file in the search file set $\prod$ as p_r; the query relativeness score of the patent file p_r as $z(p_r)$.

It is agreed the title and abstract elaborately reflect the content of a document in brief. Considering the efficiency, we use only the title and abstract of p_r to construct the first level background knowledge base $BK(p_r)$. We remove the stop words and duplicated words inside, and then perform word stemming. Finally, the remaining words constitute the first level background knowledge base $BK(p_r)$. The words in $BK(p_r)$ are the important content words about patent p_r. The number of words in $BK(p_r)$ is denoted as $|BK(p_r)|$.

In each file p_r, the information of patent's inventors, patent assignee and patent citations is presented. It is obviously that the same patent's inventor concentrate on the same/related research field. While the patent assignees also, in generally, concentrate on the same business area. The patent citation provides the most related patents, which is a great help for understanding the current patent file. Both these information provide additional related information to help the readers better understand the topic(s) discussed in the current patent file, they can be the second level background knowledge for the input document.

To make use of the inventor information in our patent search result set $\prod$, for a patent file p_r, we first extract all the inventors of it. For every inventor of p_r, we search out all his/her patent files. Patent files of every inventor together form a patent files set IL (duplicated files is discard). Then similarly, the words from title and abstract of each patent file in set IL are extracted to construct the inventor knowledge base $IK(p_r)$ as described above. The number of words in $IK(p_r)$ is denoted as $|IK(p_r)|$. $IK(p_r)$ supplies the additional background knowledge from inventors of p_r.

Similarly, we derive the assignee background knowledge base $AK(p_r)$ and citation background knowledge base $CK(p_r)$ following the above procedure. Thus $AK(p_r)$ supplies another kind of additional background knowledge from assignees for p_r while $CK(p_r)$ supplies additional background knowledge about related works for p_r.

In summary, via searching the patent data, we find the most relevant patent file set $\prod$, which is the first level background knowledge. For every file p_r in $\prod$, we use the words in the title and abstract to construct $BK(p_r)$, which serve as the most relevant background knowledge of the input document. We then get the related information following the similar procedure to construct $IK(p_r)$, $AK(p_r)$, $CK(p_r)$. By this way, the hidden knowledge and relationship between different patents files is used to build the second level background knowledge. The second level background knowledge is helpful for understanding the patent file p_r. As the patent file p_r is the background knowledge about the input document, the second level background knowledge is also helpful for understanding the input document.

3.4   Feature Extraction: Novel Word Features

In the following, we will first introduce some novel word features for keyword extraction by making use of background knowledge from patent data. In the same time, traditional word features based on lexical and syntactic features are also used as supplement.

1)   Word Background Knowledge Feature

For every word x_i in an input document, we derive a word background knowledge feature $f(x_i)$ using the first level background knowledge base as follows:

$$f(x_i) = \frac{\sum_{p_r \in \Pi} [z(p_r) * \sum_{k \in BK(p_r)} \sigma_1(x_i, k)]}{\sum_{p_r \in \Pi} z(p_r) * |BK(p_r)|} \tag{1}$$

In the above, $z(p_r)$ is the query relativeness score of the patent file p_r, $BK(p_r)$ is the first level background knowledge base of patent file p_r, k is the word in $BK(p_r)$, $\sigma_1(x_i, k)$ is the pairwise word semantic similarity calculating [13] by wordNet [14]. $|BK(p_r)|$ is the number of words in $BK(p_r)$. By the above definition, the more semantically similar a word x_i is to words in $BK(p_r)$, the larger would be x_i's background feature value, $f(x_i)$.

2)   Word Inventor Background Feature

For every word x_i in an input document, we derive a word inventor background feature $fi(x_i)$ using the second level background knowledge base $IK(p_r)$ as follows:

$$fi(x_i) = \frac{\sum_{p_r \in \Pi} [z(p_r) * \sum_{k \in IK(p_r)} \sigma_1(x_i, k)]}{\sum_{p_r \in \Pi} z(p_r) * |IK(p_r)|} \tag{2}$$

In the above, $z(p_r)$ is the query relativeness score of the patent file p_r, $IK(p_r)$ is the inventor background knowledge base of patent file p_r, k is the word in $IK(p_r)$, $\sigma_1(x_i, k)$ is the pairwise word semantic similarity calculating by wordNet. $|IK(p_r)|$ is the number of patent file in set $IK(p_r)$. By the above definition, the more semantically similar a word x_i is to words in $IK(p_r)$, the larger would be x_i's inventor background feature value, $fi(x_i)$.

3) Word Assignee Background Feature and Word Citation Background Feature

Similarly, for every word x_i in an input document, we can derive the word assignee background feature and word citation background feature using the assignee information tion $AK(p_r)$ and citation information $CK(p_r)$ of each patent file p_r in the input document's patent search result file set $\prod$. The definition is very similar to the way we compute the word inventor background feature in the above.

4) Common Word Features.

The comment word features directly derivable from the input document including word frequency feature, specific name feature, word position features, relative word length feature, and conclusion sentence feature are also used as supplement.

3.5   Learning and Extraction

Once all the word features introduced in the above are derived, we can then apply machine learning based approach to extract the keywords. We formalize keyword extraction as a classification problem. We take 'keyword' and 'not keyword' as classes, words manually labeled with the two classes in the 'training data' as training examples, words in the 'testing data' as test examples, so that keyword extraction can be automatically accomplished by predicting the class of each test example.

We perform keyword extraction in two passes of processing: learning and keyword extraction. In learning, we construct the classification model that can predict the class of each word. In the classification model, we view each word as an example. For each example, we define a set of background knowledge features and the document global context features, totally 9 word features, as defined in section 3.4 and assign a label represents whether the word is a keyword or not as showed in table 1. Since SVM is a good way to using features without discussions, and our method is totally based on the feature it self, it is good to avoid the over fitting situations. We make use of LIBSVM [16] as the classification model to obtained the classifier with small generalization errors. The labeled data is used to train the classification models in advance. One attempts to find an optimal separating hyper-plane that maximally separates the two classes of training examples (more precisely, maximizes the margin between the two classes of examples). The hyper-plane corresponds to a classifier.

In keyword extraction, the input is a document. We extract a bag of words from that document. Then, for each word in that document, we employ the learned classification models to predict whether it is 'keyword' or 'not keyword'. We next view the words that are predicted as 'keyword' as keywords. All lexical units selected as potential keywords in the above are marked in the document, and sequences of adjacent keywords are collapsed into a multi-word keyword. For instance, in the text 'It is essential for the success of grid computing', if both 'grid' and 'comput' are selected as potential keywords, since they are adjacent, they are collapsed into one single keyword 'grid comput'. Note that we perform word stemming over all the words for simply.

Table 1. Words with 9 features for classification

Word	Feature1	...	Feature 9	isKeyword
W1	V11	...	V19	0/1
W2	V21	...	V29	0/1
...	...	...	...	0/1
Wn	Vn1	...	Vn9	0/1

4   Experimentation

4.1   Data Sets and Evaluation Measures

Patent Data as Background Knowledge Depository. We download the US Patent [1] data from 2003 to 2013 as the background knowledge depository.

To evaluate the performance of our approach for keyword extraction, we first construct our own keyword extraction ground truth data set through collecting 300 patent files from US Patent dataset in 2013~2014 as the Patent Dataset. We also use the SemEval [17] Dataset for experiment.

Patent Dataset. We choose 300 patent files (with the same classification number) published in 2013~2014 as the patent dataset. It is well agreed that the title has a similar role to the key phrases. They are both elaborated to reflect the content of a document. Therefore, words in the titles are often appropriate to be keywords. However, one can only include a couple of most important words in the title prudently due to the limitation of the title length, even though many other keywords are all pivotal to the understanding of the document. On the other hand, some words maybe put in title to attract the readers but they are not exactly reflecting the content of the document. Therefore, we perform stop word removal and word stemming over all the words in title. The remaining words constitute one set of keyword answer. We then asked 10 master students in our computer science department to extract keywords manually from these files. Each patent file was analyzed by four students. Students were asked to extract 5 to 10 keywords from every patent file assigned to them. Thus we get five sets of answers for each patent file, one from title and four from students. If a word appears at least three times in the five sets of answer, we agree that it is the keyword of the document. After carrying out this manual keyword extraction process, we constructed a data set consisting of 300 files and their corresponding keywords. The average number of the key phrases is 7.2 per document. During the evaluation, the dataset was dived into two parts: 200 patent files were treated as training data while the rest 100 ones served for the final evaluation.

SemEval [16] Dataset. The SemEval dataset is a standard benchmark in the keyword extraction field. It is comprised of 244 scientific articles, usually composed of 6 to 8 pages. The articles cover 4 research areas of the ACM classification related to Computer Science, including the Computer-Communication Networks (C), Information Storage and Retrieval (H), Artificial Intelligence (I) and Computer Applications (J).

Human annotators who assigned a set of keywords to each document carried out the annotation of the gold standard. Besides the annotators' keywords, the gold standard also considers keywords originally assigned by the authors of the papers. On average, the annotators provided 75% of the keywords and the authors provide the 25%. During the evaluation, the corpus was divided into two parts corresponding to the training and testing stages: 144 articles (2265 keywords) were dedicated to training and the other 100 articles (1443 keywords) served for the final evaluation.

Evaluation Measures for Keyword Extraction. In all the experiments, we employ widely used Precision, Recall and F-Measure to evaluate the performance of different approach for keyword extraction. The F-Measure is a standard method that is used to analyze the performance of the keyword extraction methods. The higher the F-Measure, the better will be the performance of extraction. These standard measures to address the performance are given by:

$$\Pr ecision, P = \frac{Terms\_found\_and\_correct}{Keywords\_extracted} \tag{3}$$

$$\mathrm{Re} call, R = \frac{Terms\_found\_and\_correct}{Total\_number\_of\_Keywords} \tag{4}$$

$$F - Measure, F = \frac{2PR}{(P+R)} \tag{5}$$

4.2   Experimental Results

We evaluated the performances of our keyword extraction methods on two datasets. This section conducts two sets of experiment: 1) to evaluate the performance of our algorithm on patent dataset; 2) to compare different query generation methods on our algorithm's overall keyword extraction performance; 3) to compare with other well known state-of-art keyword extraction approaches on SemEval dataset.

Keyword Extraction on Patent Dataset. It is the first time that patent data being used as background knowledge serving for keyword extraction. To evaluate the performance of our approach, we compare the keywords of a document identified by our algorithm to the keywords of the document as labeled in the ground truth dataset in section 4.1. We also implement several existing keyword extraction methods including TFxIDF [18], TextRank [3] and [10](we name it as AAAI10) to compare the performance of these peer methods and our algorithm. Different parameter setting can influence the performance of the classifier. We change the value of parameter c and gamma for LIBSVM. The result is shown in Table 2, where P stands for precision, R stands for recall and F stands for F-Measure. We choose C=512 and gamma=8 as the best result. Different systems' result shown in Table 3 confirms the advantage of our method for keyword extraction.

We also consider the influence of different query generation methods to the overall performance of our method for keyword extraction. To this aim, we implemented two different query generation methods and experimentally compared the overall keyword extraction performance when employing each query generation method in the first step of our algorithm respectively. According to the results reported in Table 4, we can see that the modified TextRank method, currently employed in our algorithm, optimizes the performance of our keyword extraction approach.

Table 2. Comparison of parameters setting for c and gamma (%)

gamma C		8	16	32	64
256	P	62.3	37.7	43.5	44.8
	R	52.2	44.4	44.4	44.3
	F	46.3	47.9	43.9	44.5
512	P	**62.1**	49.9	44	43.9
	R	**41.1**	45.9	46.4	42.9
	F	**49.5**	47.8	45.2	43.4
1024	P	57.2	46.9	41.8	44
	R	41.2	45.8	45.3	42.4
	F	46.3	46.3	46.3	46.3
2048	P	57.7	45.2	41.5	43.2
	R	42.7	46.1	45.2	43.5
	F	46.3	46.3	46.3	46.3

Table 3. Performance comparison between different methods on patent dataset (%)

Algorithm	Precision	Recall	F-Measure
TF x IDF	28	38.1	32.3
TextRank	34.1	37.7	35.9
AAAI10	45.6	51.3	48.3
Our method	62.1	41.1	49.5

Table 4. Comparison of different query generation methods on our algorithm's overall keyword extraction performance (%)

Query Generation Method	Precision	Recall	F-Measure
TextRank	33.2	39.8	36.2
Modified TextRank	62.1	41.1	49.5

Experiments on SemEval Dataset. Paper [11] evaluates five semantic annotators and compares them to two state-of-the-art keyword extractors, namely KP-miner [19] and Maui [20]. Experiments show that the top three systems on SemEval dataset are Alchemy Keyword(Alch_key) [21], KP-Miner and Maui. The experiment for top 15 keywords extraction is the best. The top 15 keywords were obtained based on the confidence scores returned by the systems. We compare our method with the top three systems' best results in paper [11]. For the SemEval dataset, we used both authors and readers selected keywords as a gold standard. Table 5 indicates the results for keywords extracted by the systems on testing data provided by SemEval.

In the experiments on SemEval data, all the systems are only able to achieve less than 30% F1-score for the keyword extraction task, which is much lower than the experiments over patent data. It is because we concentrate on the same class patent file during the training and testing. The classification model is trained on the same domain. While the SemEval data provide four classes of scientific articles, each domain has only about 20~30 articles. For scientific articles, choices of words vary between different domains. Especially, keywords are usually made up of specialized words, most of which are unique to the domain. Therefore, it influences the performance of keyword extraction.

Table 5. Performance comparison between different methods on SemEval dataset(%)

Algorithm	Precision	Recall	F- Measure
KP-Miner	25.87	26.47	26.16
Maui	20.67	21.15	20.9
Alch_Key	21.08	21.35	21.21
Our method	22.81	35.77	27.86

5   Conclusion and Future Work

In this paper, we address the problem of automatic keyword extraction using novel word features. We investigate the patent data to serve for the problem of keyword extraction. We propose some novel word features for keyword extraction.

These new word features are derived through background knowledge of the input document. The background knowledge is acquired via first querying patent data, and then mining the hidden knowledge by exploring the inventor, assignees, and citation information of the patent data search result file set. Novel word features are derived from this background knowledge. Experimental results have proved that using these novel word features, we can achieve superior performance in keyword extraction to other state-of-the-art approaches.

In future work we would like to investigate the following issues: 1) to make further improvement on the efficiency, 2) to investigate the effect of including flat text from the patent file across the (current) title and abstract only approach, and 3) to extend our approach to document classification and other text mining applications.

Acknowledgments. This work is supported by the National Natural Science Foundation of China (61033010, 61272065,61370186), Natural Science Foundation of Guangdong Province (S2011020001182, S2012010009311), Research Foundation of Science and Technology Plan Project in Guangdong Province (2011B040200007, 2012A010701013).

References

1. USPTO Bulk Downloads. Patent Grant Full Text. The United States: Patent and Trademark Office. http://www.google.com/googlebooks/uspto-patents-grants-text.html
2. Witten, I.H., Paynter, G.W., Frank, E., Gutwin, C., Nevill-Manning, C.G.: Kea: Practical automatic keyphrase extraction. In: Proceedings of DL (1999)
3. Mihalcea, R., Tarau, P.: Textrank: Bringing order into texts. In: Proceeding of Conference on Empirical Methods in Natural Language, pp. 404–411 (2004)
4. Page, L., Brin, S., Motwani, R., et al.: The pagerank citation ranking: Bringing order to the web. Technical report of Stanford Digital Library
5. Kleinberg, J.: Hubs, Authorities, and Communities. Cornell University (1999)
6. Mihalcea R., Csomai A.: Wikify!: linking documents to encyclopedic knowledge. In: Proceedings of the 16th ACM Conference on Information and Knowledge Management, pp. 233–242 (2007)
7. Medelyan, O., Frank, E., Witten, I.H.: Human-competitive tagging using automatic keyphrase extraction. In: Proceedings of EMNLP (2009)
8. Qureshi, M.A., O'Riordan, C., Pasi, G.: Short-text domain specific key terms/phrases extraction using an n-gram model with wikipedia. In: Proceedings of ACM CIKM (2012)
9. Joorabchi, A., Mahdi, A.E.: Automatic subject metadata generation for scientific documents using wikipedia and genetic algorithms. In: EKAW 2012. LNCS (LNAI), vol. 7603, pp. 32–41. Springer, Heidelberg (2012)
10. Songhua, X., Yang, S., Lau, F.C.M.: Keyword Extraction and Headline Generation Using Novel Word Features. In: Proceedings of the 24th Conference on Artificial Intelligence, AAAI, Atlanta, Georgia, USA, July 11-15, pp. 1461–1466 (2011)
11. Jean-Louis, L., Zouaq, A., Gagnon, M., Ensan, F.: An Assessment of Online Semantic Annotators for the Keyword Extraction Task. In: Pham, D.-N., Park, S.-B. (eds.) PRICAI 2014. LNCS, vol. 8862, pp. 548–560. Springer, Heidelberg (2014)
12. QTag. http://web.bham.ac.uk/O.Mason/software/tagger/
13. Pedersen, T., Patwardhan, S., Michelizzi, J.: Wordnet: similarity – measuring the relatedness of concepts. In: Proceedings of the 19th National Conference on Artificial Intelligence (AAAI 2004), pp. 1024–1025 (2004)
14. Miller, G.A.: Wordnet: a Lexical Database for English. Communications of the ACM (1995)
15. Lucene. http://lucene.apache.org/
16. LIBSVM – A Library for Support Vector Machines. http://www.csie.ntu.edu.tw/~cjlin/libsvm/
17. Kim, S.N., Medelyan, O., Kan, M.Y., Baldwin, T.: Semeval-2010 task 5: Automatic keyphrase extraction from scientific articles. In: Proceedings of SemEval (2010)
18. Salton, G., Buckley, C.: Term-weighting approaches in automatic text retrieval. Technical report, Ithaca, NY, USA (1987)
19. El-Beltagy, S.R., Rafea, A.: Kp-miner: A keyphrase extraction system for english and Arabic documents. Information Systems (2009)
20. Medelyan, O., Frank, E., Witten, I.H.: Human-competitive tagging using automatic keyphrase extraction. In: Proceedings of EMNLP (2009)
21. Alchemyapi. http://www.alchemyapi.com/api/keyword-extraction/

A Multi-round Global Performance Evaluation Method for Interactive Image Retrieval

Jiyi Li[✉]

Department of Social Informatics, Kyoto University,
Yoshida-Honmachi, Sakyo-ku, Kyoto 606-8501, Japan
garfieldpigljy@gmail.com

Abstract. In interactive image retrieval systems, from the image search results, a user can select an image and click to view its similar or related images until he reaches the targets. Existing evaluation approaches for image retrieval methods only focus on local performance of single-round search results on some selected samples. We propose a novel approach to evaluate their performance in the scenario of interactive image retrieval. It provides a global evaluation considering multi-round user interactions and the whole image collection. We model the interactive image search behaviors as navigation on an information network constructed by the image collection by using images as graph nodes. We leverage the properties of this constructed image information network to propose our evaluation metrics. We use a public image dataset and three image retrieval methods to show the usage of our evaluation approach.

Keywords: Image retrieval · Information network · Evaluation

1 Introduction

In interactive image retrieval systems, when users search images in an image collection, they first input some queries, which can be keywords or images, to get some initial image search results. In contrast to web page search, in image retrieval, it may be not easy for users to reach the targets in the direct search results of the queries. Some reasons can lead to their failures on reaching the targets in these initial search results. For example, users do not exactly describe their specific targets in the queries because it is a difficult or boring task; the targets of users may be still not clear when they start their searches, because they are just seeking the information they are interested in; the performance of image retrieval methods are still not good enough to return results matching the query topics or detect user intentions for personalized search results. Therefore, many image retrieval systems provide interfaces of user interactions, e.g., relevance feedback, to gather additional information for refining the search results.

Because it may be boring for users to input too many additional actions, some recent systems have succinct and convenient interfaces. They allow users to click

© Springer International Publishing Switzerland 2015
J. Li and Y. Sun (Eds.): WAIM 2015, LNCS 9098, pp. 161–168, 2015.
DOI: 10.1007/978-3-319-21042-1_13

Fig. 1. Example of Interactive Image Retrieval

images to view other similar or related images until they reach the targets. Figure 1 shows an example with Google image search[1], after a user inputs "Kyoto" as the search keyword, Google returns the initial image search results. There is a link of "search by image" for each image in the search results. The user can click these links to reach further results until he finds satisfied results. After several rounds of clicks and views, this user finally finds his target image which is about the cherry-blossom at Arashiyama in Kyoto in spring.

Existing evaluation approaches of image retrieval methods in the scenario of interactive image retrieval provide local evaluation by selecting some samples and using single-round evaluation with initial search results and first-round user interaction on the initial search results. On one hand, they does not evaluate the performance of results after multi-round interactions. On the other hand, whether the selected samples can represent the whole information space of the data collection which may have tens of thousands of or millions of images, and whether this local performance is equivalent to the global performance are unknown. Therefore, to evaluate the performance of image retrieval methods in the interactive scenario, besides single-round local evaluation approach, we also need to propose a multi-round global evaluation approach by considering the images in the whole data collection and multi-round interactions in a search session.

We find that the image search behavior with user interactions in a search session can be regarded as navigation on the information space of the image collection. A query image or the first clicked image in the search results of a query keyword is the start of this navigation. The end of this navigation is some images which may contain the final target images. The user may find the targets or fail to find the targets in current search session and give up at the end of this navigation. In a search session, there is a route from the start image to the target image. The user navigates the images in the image information network through this route. Figure 2 shows an example of the navigation perspective of

---

[1] http://images.google.com

Fig. 2. Example of Image Navigation

interactive image retrieval. In this search session, there is a route from the start image to the target image. Full lines mark the route, and dotted lines show other images that are not clicked in the search results. A user navigates the images in the image information network through this route.

Therefore, we propose an approach to construct an image information network for an image collection based on a specific image retrieval method. The nodes in the network are images, and the edges in the network depend on the image retrieval method which generates top k search results of an image from the image collection. We utilize the properties of this information network to propose our multi-round global evaluation metrics. This information network can also be utilized for various applications such as image re-ranking, user intention detection and personalized target prediction in image retrieval. In this paper we concentrate on using it for the evaluation on the performance of an image retrieval method on an image collection. The length of shortest path of two images on information network can show the difficulty of reaching the target image from the start image in ideal.

The contributions of this paper are as follows.

– We propose an approach for evaluating multi-round global performance of an image retrieval method on the whole data collection. This evaluation approach is a useful supplement to existing single-round local evaluation approaches. Note that our work does not focus on the evaluation of relevance feedback methods. Our work focuses on the evaluation of image search methods in the scenario of multi-round user interactions.
– We provide a novel model with navigation perspective for interactive image retrieval. We construct information network for a given image collection based on a specific image retrieval method to reflect user navigation behavior on images in a search session.

The remainder of this paper is organized as follows. We list some related work in Section 2. In section 3, we introduce how to construct the image information network and propose our evaluation method. We report the experimental results in Section 4 and give a conclusion in Section 5.

2   Related Work

On the hand, Ref. [1] lists many works on image retrieval. Many of them focus on the issue of one query and its direct image search results without further interactions. In their evaluation, they select a subset with some labeled samples.

On the other hand, Relevance feedback has been widely used in image re-ranking in supervised scenarios and is an important solution in interactive image retrieval. Rui et al. [3] and Porkaew et al. [4] proposed query reweighting, query reformulation, query modification, query point movement and query expansion approaches. Ref. [5] only requires the user to click one image in the interaction to minimize the effort of selecting relevance feedback instances. Our model of interactive image retrieval assumes a succinct and convenient user interaction interface in which users click an image in the results to search by this image. The evaluation approaches used in interactive image retrieval, e.g., Ref. [3–5], mainly consider only one round user interaction; they select a small number of samples and take local evaluation in the data collection.

In contrast to the single-round local evaluation used in related work, our proposed evaluation approach based on the image information network can provide an ideal multi-round global evaluation.

3   Multi-round Global Performance Evaluation

In this section, we first describe our model of interactive image retrieval and then introduce the approach of how to construct the information network for a given image collection based on a specific image retrieval method. After that we propose our evaluation approach based on this information network.

3.1   Interactive Image Retrieval

Our model of interactive image retrieval is as follows. We define the images as a_i in a given image collection C. In a search session, the queries that user input can be images or keywords. In the case of query by image, the start image is the query; in the case of query by keyword, the start image is the clicked image in the initial search results. This definition of start image can simplify the constructed information network with only one type of nodes to facilitate the analysis on the graph for various applications. Each node in this graph represents an image. There is no node representing other types of objects such as text.

It may be boring for users to input too many additional actions in user inter-actions, and users may prefer succinct and convenient interfaces. Therefore, in our scenario of interactive image retrieval, in each round of user interaction, we simplify that a user selects and clicks only one image. Such one-click inter-face is used in recent image search engine like Google image search. After the user clicks, a specific image retrieval method F is used to return top k image search results by using this clicked image as query. This image retrieval method can use either visual content information or textual context information of the clicked image. There are several rounds of user interactions in a search session.

The images clicked in all rounds construct a navigation sequence on the graph which is a path from the start image to the end images. Note that end images can be target images or not. It depends on whether the search sessions success or fail.

The definition of search session is somewhat different from existing work in the areas like web page search. We assume that when a user modifies his original query, the current search session is regarded as finished, no matter whether the user reaches his targets. There are no jumps from one image node to another image node between which there is no edge, because such jump is regarded as a new session. In this way, the navigation behavior on the image information network in a search session is continuous by following the edges on the graph. All the simplifications in this model can smooth the analysis on the graph when using it for applications.

3.2   Information Network Construction

The steps of image information network construction for an image collection C based on the search results of an image retrieval method F are as follows. This automatic process doesn't require users' participation.

- 1. Create a node for each image a_i in the image collection C.
- 2. For each image a_i, compute its top k image search results A_i from the image collection C by using the image retrieval method F. For some images, some methods can only generate less than k similar images, e.g., the images with few textual information and text based image retrieval methods.
- 3. Create an edge from a_i to each image a_j in A_i. This image information network is thus a directed graph. Different image retrieval methods generate different search results and thus lead to different information networks.

We assume that users only view and click the images from the top k ranked images in the search engine results page (SERP). Therefore, the out-degrees in this constructed information network is at most k. This constrain is able to control the edges of the graph and the time cost when computing the graph properties for analysis.

The rank information or similarity between a_i and each image a_j in A_i can be stored as the weights of the edges. Because it is not yet necessary in this paper, we still use same weights (equal to 1) for all edges.

3.3   Performance Evaluation Method

The image information network can be used in various applications. In this section, we specially discuss the issue about how to evaluate the performance of an image retrieval method on an image collection using this information network. In the communities of graph theory and social network analysis, many metrics have been proposed for analyzing various properties of graph, e.g., average shortest path, diameter, centrality, and so on.

We select two metrics which can reflect the difficulty of navigation from one start image to another target image. One is the length of average shortest path of two images on the information network which shows the average navigation difficulty in ideal; the other is the diameter of the graph which shows the upper limit of steps in a navigation in ideal.

Because we use top k search results to construct the information network and thus the out-degrees in this network are at most k, the cut-off size parameter k becomes an important factor that influences the network properties. Therefore, the evaluation metrics we propose include this parameter k. They are Average Shortest Path at top k results (**ASP@k**) and Diameter at top k results (**Dia@k**).

The evaluation approach we propose reflect the multi-round global performance of an image retrieval method by using all rounds of user interactions in a search session and considering the images in the whole data collection. Our evaluation approach is an evaluation approach in the ideal case on the image collection for the following several reasons.

- First, users may not select the ideal shortest path to reach the targets. West et al. [6] made an analysis of human wayfinding in entity information network. They find that human wayfinding is different from shortest paths but is still very efficient. How to assist users to select the paths close to ideal ones is one of our future work.
- Second, not all start and target image pairs included in the evaluation metrics are meaningful enough. Although we assume that users intention in a query session can change and the start and target images can be not similar or related to each other, there are still start and target pairs that rarely appear in the real world search.
- Third, these evaluation metrics assume that users do not give up the search sessions, do unlimited rounds and do not click the back button until they reach the targets.

Because it is an ideal evaluation, this evaluation approach is not proposed to instead the exiting evaluation methods. It provides the evaluation on another aspect of the performance and is a useful supplement to existing single-round local evaluation approaches.

Because the graph construction and property analysis are time consuming for huge collections, in practice, we can only use a subset of the collection to speed up the construction. Such subset can contain tens of thousands of images and is still much larger than a selected samples used in exiting evaluation approaches.

4   Experiment

4.1   Experimental Settings

The dataset we use for our experiments is MIRFlickr 25000 [2] which is constructed by images and tags downloaded from Flickr. It contains 25,000 images and provides the raw tags of these images on Flickr. We use two kinds of visual features as two different content based image retrieval methods respectively. One is color feature. The feature descriptor is a 1024-Dimension color histogram on the HSV color space. The other one is the SIFT feature with the bags of words model and 1000 visual words in the codebook.

With each kind of feature, we generate the top k similar images of an image by computing the distance of visual feature vectors. The distance between image a_i and a_j is computed using a Pearson correlation distance $s(a_i, a_j)$ defined as

$$\mathcal{H}'_i(x) = \mathcal{H}_i(x) - \frac{\sum_y \mathcal{H}_i(y)}{\mathcal{N}},$$

$$s(a_i, a_j) = \frac{\sum_x (\mathcal{H}'_i(x) * \mathcal{H}'_j(x))}{\sqrt{(\sum_y \mathcal{H}'_i(y)^2) * (\sum_y \mathcal{H}'_j(y)^2)}},$$

where $\mathcal{H}_i$ and $\mathcal{H}_j$ are feature vectors. $\mathcal{N}$ is the size of the feature vector.

We also use a textual based image retrieval method which searches textual related images to the query image. The textual information in the dataset is social tag. We use the tag list of an image as its textual feature vector, and generate the top k relevant images of an image by computing the Ochiai coefficient of textual feature vectors. The distance between image a_i and a_j is defined as

$$t(a_i, a_j) = \frac{|\mathcal{T}_i \cap \mathcal{T}_j|}{\sqrt{|\mathcal{T}_i| * |\mathcal{T}_j|}},$$

where $\mathcal{T}_i$ and $\mathcal{T}_j$ are textual feature vectors, and $|\cdot|$ is the number of tags. The images without textual information have no related images.

4.2   Experimental Results

After we construct the information networks with this dataset and the three image retrieval methods, we evaluate the performance of these methods on the metrics of ASP@k and Dia@k. Table 1 lists the experimental results.

Table 1. Performance of Different Image Retrieval Methods

Method	Metric	k 10	25	50	Metric	k 10	25	50
HSV	ASP@k	12.7	7.25	5.52	Dia@k	67	29	19
SIFT	ASP@k	11.2	6.68	5.02	Dia@k	51	24	24
TEXT	ASP@k	10.3	5.37	3.95	Dia@k	28	14	8

Because content based methods and textual based methods search images based on different purposes, i.e, visual similar images and textual related images. We use two content based methods to show how to make comparison with our evaluation metrics. In these two content based methods based on different visual features, for a given parameter k, the method using SIFT feature always has smaller average shortest path and diameter (except the diameter on $k = 50$) than the method using HSV feature. We therefore claim that this SIFT feature based image retrieval method has better global performance than the HSV feature based image retrieval method for this data collection. The reason may be that in this image collection, more images are better to be represented by SIFT feature when searching for visual similar images.

In addition, Table 1 also shows that as the k increases, both length of average shortest path and diameter decrease. It is as expected because when more edges are added in the graph, it is easier to move from one node to another node. $ASP@k$ can also be used as reference for evaluating the proper value of k for an image retrieval method, which means the number of images returned in the top search results, e.g., the maximum k with $ASP@k \leq 6$. It means that in average any two images "know" each other within six degrees. Although there is no proof that the six degrees of separation theory [8] is proper here, it can be a candidate criteria for tuning parameter k.

5   Conclusion

In this paper, we model the image search behavior with user interactions as navigation on a information network of image collection. We propose a global evaluation approach on the performance of multi-round user interactions on the whole image collection by using the properties of this image information network. For the future work, we will make further verification to this evaluation approach.

References

1. Datta, R., Joshi, D., Li, J., Wang, J.Z.: Image retrieval: Ideas, influences, and trends of the new age. ACM Computing Surveys (CSUR) **40**(2), 1–60 (2008)
2. Huiskes, M.J., Lew, M.S.: The MIR Flickr Retrieval Evaluation. In: Proceedings of the 1st ACM International Conference on Multimedia Information Retrieval (MIR 2008), pp. 39–43. ACM, New York (2008)
3. Rui, Y., Huang, T.S., Ortega, M., Mehrotra, S.: Relevance feedback: a power tool for interactive content-based image retrieval. IEEE Trans. on CSVT. **8**(5), 644–655 (1998)
4. Porkaew, K., Mehrotra, S., Ortega, M.: Query reformulation for content based multimedia retrieval in MARS. In: Proceedings of the IEEE International Conference on Multimedia Computing and Systems (ICMCS 1999), vol. 2, pp. 747–751. IEEE Computer Society, Washington, DC (1999)
5. Tang, X.O., Liu, K., Cui, J.Y., Wen, F., Wang, X.G.: IntentSearch: Capturing User Intention for One-Click Internet Image Search. IEEE Trans. on PAMI **34**(7), 1342–1353 (2012)
6. West, R., Leskovec, J.: Human wayfinding in information networks. In: Proceedings of the 21st International Conference on World Wide Web (WWW 2012), pp. 619–628. ACM, New York (2012)
7. Jain, V., Varma, M.: Learning to re-rank: query-dependent image re-ranking using click data. In: Proceedings of the 20th International Conference on World Wide Web (WWW 2011), pp. 277–286. ACM, New York (2011)
8. Kleinberg, J.: The small-world phenomenon: an algorithmic perspective. In: Proceedings of the Thirty-Second Annual ACM Symposium on Theory of Computing (STOC 2000), pp. 163–170. ACM, New York (2000)

Resorting Relevance Evidences to Cumulative Citation Recommendation for Knowledge Base Acceleration

Jingang Wang[1], Lejian Liao[1(✉)], Dandan Song[1], Lerong Ma[1,2],
Chin-Yew Lin[3], and Yong Rui[3]

[1] School of Computer Science, Beijing Institute of Technology, Beijing 100081, China
{bitwjg,liaolj,sdd}@bit.edu.cn
[2] College of Mathematics and Computer Science,
Yan'an University, Shaanxi 716000, China
malerong_bit@bit.edu.cn
[3] Knowledge Mining Group, Microsoft Research, Beijing 100080, China
{cyl,yongrui}@microsoft.com

Abstract. Most knowledge bases (KBs) can hardly be kept up-to-date due to time-consuming manual maintenance. Cumulative Citation Recommendation (CCR) is a task to address this problem, whose objective is to filter relevant documents from a chronological stream corpus and then recommend them as candidate citations with certain relevance estimation to target entities in KBs. The challenge of CCR is how to accurately category the candidate documents into different relevance levels, since the boundaries between them are vague under the current definitions. To figure out the boundaries more precisely, we explore three types of relevance evidences including entities' profiles, existing citations in KBs, and temporal signals, to supplement the definitions of relevance levels. Under the guidance of the refined definitions, we incorporate these evidences into classification and learning to rank approaches and evaluate their performance on TREC-KBA-2013 dataset. The experimental results show that all these approaches outperform the corresponding baselines. Our analysis also reveals various significances of these evidences in estimating relevance levels.

Keywords: Cumulative citation recommendation · Knowledge base acceleration · Information filtering

1 Introduction

Knowledge Bases (KBs), such as Wikipedia, have shown great power in many applications such as question answering, entity linking and entity retrieval. However, most KBs are maintained by human editors, which are hard to keep up-to-date because of the wide gap between limit number of editors and millions of

---

This work was partially done when the first author was visiting Microsoft Research.

© Springer International Publishing Switzerland 2015
J. Li and Y. Sun (Eds.): WAIM 2015, LNCS 9098, pp. 169–180, 2015.
DOI: 10.1007/978-3-319-21042-1_14

entities. As reported in [8], the median time lag between the emerging date of a cited article and the date of the citation created in Wikipedia is almost one year. Moreover, some less popular entities cannot attract enough attentions from editors, making the maintenance more challenging. These burdens can be released if relevant documents are automatically detected as soon as they emerge and then recommended to the editors with certain relevance estimation. This task is studied as Cumulative Citation Recommendation (CCR) in Text REtrieval Conference (TREC) Knowledge Base Acceleration (KBA)[1] Track. CCR is defined as filtering a temporally-ordered stream corpus for documents related to a predefined set of entities in KBs and estimating the relevance levels of entity-document pairs. In this paper, we follow the 4-point scale relevance settings of TREC-KBA-2013, as illustrated in Table 1. We consider two scenarios in CCR: (i) **vital + useful**: detecting relevant (*vital* and *useful*) documents, and (ii) **vital only**: detecting *vital* documents only.

There are two challenges in CCR. (1) It is too time-consuming to process all the documents in the stream corpus. To guarantee the coverage and diversity of the recommendation, a real-life CCR system need include as many documents as possible to filter relevant ones for target entities. (2) Besides, according to current definitions, it is difficult to figure out the explicit differences between the four relevance levels, especially for *vital* and *useful*. Moreover, in the ground truth of TREC-KBA-2013 dataset, there also exist many inconsistent annotations due to the vague boundaries between relevance levels.

Table 1. Heuristic definitions of relevance levels by TREC-KBA-2013

Vital	timely info about the entity's current state, actions, or situation. This would motivate a change to an already up-to-date knowledge base article.
Useful	possibly citable but not timely, e.g. background bio, secondary source.
Neutral	Informative but not citable, e.g. tertiary source like Wikipedia article itself.
Garbage	No information about the target entity could be learned from the document, e.g. spam.

In this paper, we focus on the aforementioned two challenges. For the former one, we implement an entity-centric query expansion method to filter the corpus and discard irrelevant documents as many as possible beforehand. For the latter one, given an entity E, we resort extra evidences to enrich the definitions of relevance levels, including the profile page of E, existing citations for E in KB and some temporal signals. Entities' profile pages are helpful in entity disambiguation since they contain rich information about the target entities. Moreover, most KB entities already possess several citations which are cited into KB because of their relatedness to the target entities. In our opinion, these two evidences are helpful to differentiate relevant and irrelevant documents. However, it is not enough to detect *vital* documents accurately merely with these two evidences

---

[1] http://trec-kba.org/

because vital documents are required to contribute some novel knowledge about the target entity. Hence we introduce temporal signals to support the detection of *vital* documents. For an KB entity, the occurrences of its relevant documents, especially *vital* documents, are not distributed in the stream corpus evenly along a timeline. When some significant event related to it happens, an entity is spotlighted in a short time period during which its *vital* documents emerge heavily and exceed the average frequency of occurrence. Therefore, the difficulty of detecting *vital* documents can be reduced as along as we detect these "bursty" time periods. Accordingly, we propose the following assumptions as supplements to the definitions of relevance levels.

1. The relevant (i.e., *vital* and *useful*) documents are more related to the profile page and existing citations than irrelevant ones.
2. The relevant documents occurring in the "bursty" time periods where the entity is spotlighted are more likely to be detected as *vital*.

Under the guidance of the refined definitions, we then undertake a comparison study on classification and learning to rank approaches to CCR.

To the best of our knowledge, this is the first work endeavoring to improve the definitions of relevance levels of CCR. Under the refined definitions, all experimental approaches outperform their own baselines. Besides, the classification approach performs best and achieve the state-of-the-art performance on TREC-KBA-2013 dataset. The proposed assumptions are proved to be reasonable and effective. In addition, the experimental results show that our filtering method is effective, making the whole procedure more efficient and practical.

2   Related Work

Entity Linking. Entity linking describes the task of linking a textural name of an entity to a knowledge base entry. In [5], cosine similarity was utilized to rank candidate entities based on the relatedness of the context of an entity mention to a Wikipedia article. [6] proposed a large-scale named entity disambiguation approach through maximizing the agreement between the contextual information from Wikipedia and the context of an entity mention. [14] implemented a system named WikiFy! to identify the important concepts in a document and automatically linked these concepts to the corresponding Wikipedia pages. Similarly, [11] parsed the Wikipedia to extract a concept graph, measuring the similarity by means of the distance of co-occurring terms to candidate concepts. Besides identifying entity mentions in documents, CCR is required to evaluate the relevance levels between the document and the mentioned entity.

Knowledge Base Acceleration. TREC KBA has hosted CCR track since 2012. In the past tracks, there were three mainstream approaches submitted by the participants: query expansion [13,17], classification, such as SVM [12] and Random Forest [2–4], and learning to rank approaches [1,17]. In [10], a graph-based entity filtering method is implemented through calculating the similarity of word co-occurring graphs between an entity's profile and a document. Transfer learning is employed to transfer the keyword importance learned from training entities to query entities in [18]. All these work focus on utilizing the training data adequately to improve the performance of a CCR system, yet the key contribution of our work is the improvement of the definitions of relevance levels of CCR. With the help of the refined definitions, we can differentiate the relevant and irrelevant documents more accurately.

3   Problem Statement and Dataset

Given an entity E from a KB (e.g., Wikipedia or Twitter) and a document D, our goal is to generate a confidence score $r(E, D) \in (0, 1000]$ for each document-entity pair, representing the citation-worthiness of D to E. The higher $r(E, D)$ is, the more likely D can be considered as a citation for E.

The TREC-KBA-2013 dataset is utilized as our experimental dataset, which consists of a stream corpus and a predefined target entity set. The stream corpus[2], is a time-ordered document collection and contains nearly 1 billion documents published between Oct. 2011 and Feb. 2013. The target entity set is composed of 141 entities: 121 Wikipedia entities and 20 Twitter entities. These entities consist of 98 persons, 24 facilities and 19 organizations. The documents from Oct. 2011 to Feb. 2012 are annotated as training data and the remainder from Mar. 2012 to Feb. 2013 as testing data. Each document-entity pair is annotated as one of the four relevance levels.

4   Filtering

In previous study [8], most relevant documents mention the target entity explicitly. A naïve filtering strategy is to retain the documents mentioning a target entity name at least once and discard those without mentioning any target entity. Nevertheless, this strategy has two disadvantages. (1) It misses lots of relevant documents for entities with aliases, because entities can be referred by different surface forms in documents. (2) It cannot differentiate two entities with the same name. For example, when we filter the stream corpus for *"Michael Jordan"*, a machine learning scientist, the filtering results also include documents referring to the basketball player *"Michael Jordan"*.

An entity-centric query expansion method is proposed to address the above problems. We first expand as many reliable surface forms as possible for each target entity. Then we extract related entities as expansion terms. Finally, we expand the naïve filtering query (i.e., entity name) with the surface forms and these related entities.

---

[2] http://trec-kba.org/kba-stream-corpus-2013.shtml

Surface Form Expansion. For a Wikipedia entity, we treat the redirect[3] names as its surface forms. For a Twitter entity, we extract its display name as its surface form. Take *@AlexJoHamilton* as an example, we extract its display name *Alexandra Hamilton* as a surface form.

Query Expansion. We first construct a document collection for each target entity E, denoted as $C(E)$. $C(E)$ is composed of documents including the profile page of E, existing citations for E in KB and the documents labeled as *vital* and *useful* for E in training data. Then we perform named entity recognition for each document $d \in C(E)$ and obtain a related entity set $R(E)$ for E. The related entities in $R(E)$ are expansion terms we leveraged to improve the naïve filtering query. The relevance score between a document d referring to E is estimated as:

$$rel(d, E) = w(E, e) \cdot \sum_{e \in R(E)} occ(d, e) \tag{1}$$

where $occ(d, e)$ is the number of occurrences of related entity e in d. $w(E, e)$ serves as the prior weight of e to E. Here we use e's IDF (inverse document frequency) in $C(E)$ to estimate its prior importance to E.

$$w(E, e) = log\frac{|C(E)|}{|\{j : e \in d_j\}|} \tag{2}$$

where $|C(E)|$ is the number of documents in $C(E)$, and $|\{j : e \in d_j\}|$ is the number of documents containing related entity e. Through setting a relevance threshold, the documents with low relevance scores are removed from the stream corpus, and the remaining documents compose a candidate relevant document collection $RC(E)$. In the following work, we perform all experiments on $RC(E)$ instead of the entire stream corpus for the sake of efficiency.

5   Relevance Estimation

5.1   Relevance Evidences

Profile. Each KB entity possesses a profile page either in Wikipedia or on Twitter. The profile page can be leveraged to differentiate relevant and irrelevant documents for an entity. We make an assumption that relevant documents are more similar with profile page in comparison to irrelevant ones. We verify the assumption in various approaches. In query expansion, we extract named entities from the profile page as expansion terms for a target entity. In supervised approaches, we develop some similarity features. In Wikipedia, entities' profile pages are usually organized as different sections. Each section introduces a specific aspect of the target entity. Relevant documents for a target entity are possibly highly related with a few of these sections rather than all of them. Therefore, we calculate similarities between the documents and different sections respectively instead of the whole profile page. These features are listed in the second block of Table 2, denoted as profile features.

---

[3] http://en.wikipedia.org/wiki/Wikipedia:Redirect

Citations. There usually exists a list of supplementary citations for an entity in Wikipedia. These existing citations are extremely valuable in identifying relevant documents. In the refined definitions of relevance levels, we assume that relevant documents are more similar with existing citations than irrelevant ones. We also verify it in various approaches. In query expansion, named entities are extracted from the existing citations as expansion terms. In supervised approaches, the similarity between each citation and the document is utilized as training features. These features are listed in the third block of Table 2, denoted as citation features.

Temporal Signals. CCR is to filter relevant documents from a temporally-ordered stream corpus and entities are evolving with the passage of time. However, semantic features cannot capture the dynamic characteristics of entities in the stream corpus. Temporal signals have been considered to make up for this deficiency [2,4,17].

The view statistics[4] of an entity's Wikipedia page is adopted as a useful signal to capture if something important to the entity happens during a given time period. Nevertheless, Twitter entities do not have off-the-shelf daily view statistics as Wikipedia entities. Alternatively, we employ Google Trends[5] statistics, which shows how often a particular search term is entered relative to the total search-volume. Based on these statistics, we can capture temporal signals for each entity by detecting its bursy activities.

For an entity E, we have a daily view/search statistics sequence $\mathbf{v} = (v_1, v_2, \cdots, v_n)$. We detect bursts of E based on $\mathbf{v}$ with a tailored moving average (MA) method [16]. More concretely, for each item v_i in $\mathbf{v}$,

1. Calculate moving average sequence of length w as

$$MA_w(i) = \frac{v_i + v_{i-1} + \cdots + v_{i-w+1}}{w}$$

2. Calculate cutoff $c(i)$ based on previous MA sequence $Pre_{MA} = (MA_w(1), \cdots, MA_w(i))$ as

$$c(i) = mean(Pre_{MA}) + \beta \cdot std(Pre_{MA}).$$

3. Detect bursty day sequence $\mathbf{d}$, where $\mathbf{d} = \{i | MA_w(i) > c(i)\}$.
4. Calculate daily bursty weights

$$\mathbf{w} = \{w_i | \frac{MA_w(i)}{c(i)}, i \in \mathbf{d}\}$$

The moving average length can be varied to detect long-term or short-term bursts. We set the moving average length as 7 days (i.e., $w = 7$). The cutoff value is empirically set as 2 times the standard deviation of the MA (i.e., $\beta = 2$).

---

[4] https://dumps.wikimedia.org/other/pagecounts-raw/
[5] http://www.google.com/trends/

Moreover, we compact the consecutive days in **d** into bursty periods. The bursty weight for each period is calculated as the average weight of all the bursts in this period.

Given a document D and an entity E, we define a bursty value $b(D, E)$ to represent their temporal relation. Let t be the timestamp of D. If t is in one of E's bursty periods, denoted as $[t_{start}, t_{end}]$, then $b(D, E)$ is calculated by Equation 3. Otherwise, $b(D, E)$ would be set as 0.

$$b(D, E) = (1 - \frac{t - t_{start}}{t_{end} - t_{start}}) \cdot bw_{(t_{start}, t_{end})}(E), \quad t \in [t_{start}, t_{end}] \tag{3}$$

where $1 - \frac{t - t_{start}}{t_{end} - t_{start}}$ is a decaying coefficient reflecting the intuition that the documents appear at the beginning of a burst are more informative than those appear at the end.

5.2 Approaches

Query Expansion. We consider query expansion as one baseline, since each document obtains a relevance score (i.e., $rel(D, E)$) after the filtering step as described in Section 4. The relevance score can be converted to the confidence score conveniently.

Classification. CCR is usually considered as a binary classification task which categories documents into different relevance levels. In our internal test, random forests outperforms other classifiers, such as SVM and logistic regression. So we employ random forests classifier in our experiments. Two classifiers are built with different features, denoted as **Class** and **Class+** respectively. **Class** is built with the basic features in Table 2, while **Class+** is built with the whole features in Table 2.

Learning to Rank. If we consider the different relevance levels as an ordered sequence, i.e., $vital > useful > neutral > garbage$, CCR becomes a learning to rank (LTR) task. Two ranking models are built, denoted as **Rank** and **Rank+** respectively. **Rank** is a LTR model built with the basic features in Table 2, while **Rank+** is a LTR model built with the whole features in Table 2.

6 Experimental Evaluation

6.1 Evaluation Metrics

A CCR system is fed with documents in a chronological order and outputs a confidence score in the range of $(0, 1000]$ for each document-entity pair. There are two metrics to evaluate the performance: $\max(F(avg(P), avg(R)))$ and $\max(SU)$. Scaled Utility (SU) is a metric introduced in filtering track to evaluate the ability of a system to separate relevant and irrelevant documents in a stream [15].

Table 2. Feature set employed in supervised approaches

Feature	Description
Basic Features	
$N(E_{rel})$	# of entity E's related entities found in its profile page
$N(D, E)$	# of occurrences of E in document D
$N(D, E_{rel})$	# of occurrence of the related entities in D
$FPOS(D, E)$	first occurrence position of E in D
$FPOS_n(D, E)$	$FPOS(D, E)$ normalized by document length
$LPOS(D, E)$	last occurrence position of E in D
$LPOS_n(D, E)$	$LPOS(D, E)$ normalized by document length
$Spread(D, E)$	$LPOS(D, E) - FPOS(D, E)$
$Spread_n(D, E)$	$Spread(D, E)$ normalized by document length
Profile Features	
$Sim_{cos}(D, S_i(E))$	cosine similarity between D and the i_{th} section of E's profile
$Sim_{jac}(D, S_i(E))$	jaccard similarity between D and the i_{th} section of E's profile
Citation Features	
$Sim_{cos}(D, C_i)$	cosine similarity between D and the i_{th} citation of E
$Sim_{cos}(D, C_i)$	jaccard similarity between D and the i_{th} citation of E
Temporal Feature	
$burst\_value(D, E)$	$burst\_value$ calculated by Equation 3

A cutoff value is varied from 0 to 1000 with some step size and the documents with the scores above the cutoff are treated as positive instances. Correspondingly, the documents with the scores below the cutoff are negative instances.

The primary metric $\max(F(avg(P), avg(R)))$ is calculated as follows: given a cutoff c and an entity E_i, $P_i(c)$ and $R_i(c)$ are calculated respectively, then macro-average them in all entities, $avg(P) = \frac{\sum_{i=1}^{N} P_i(c)}{N}$, $avg(R) = \frac{\sum_{i=1}^{N} R_i(c)}{N}$, where N represents the quantities of entities in the target entity set. Therefore, F is actually a function of the relevance cutoff c, and we select the maximum F to evaluate the overall performance of a CCR system. In a similar manner, $\max(SU)$ is calculated as an auxiliary metric.

6.2 Filtering Evaluation

This section evaluates the filtering performance of our entity-centric QE method in comparison to the official baseline of TREC-KBA-2013 track, which is a baseline that annotators manually generate a list of reliable queries of each entity for filtering. We calculate the best recall of different methods through setting the cutoff as 0, in which case all remaining documents after filtering are considered as positive instances. Meanwhile, corresponding precision, $F1$ and SU measures are taken into account. The results are reported in Table 3.

From the perspective of SU, our entity-centric QE method achieves the best filtering results in both scenarios. Moreover, in *vital only* scenario, the QE method outperforms the baseline on all metrics. In *vital + useful* scenario, although recall of our QE method is not best, it achieves the overall best $F1$

and *SU*. This reveals that our filtering method can comprise between precision and recall to reach an overall optimal filtering performance.

Table 3. Filtering results when setting cutoff value as zero to maximize recall

Method	Vital				Vital + Useful			
	P	R	F1	SU	P	R	F1	SU
Official Baseline	.166	.705	.268	.139	.503	**.855**	.634	.523
QE	**.175**	**.721**	**.281**	**.146**	**.520**	.850	**.645**	**.544**

6.3   Relevance Estimation Evaluation

Results and Discussion. All the results of our approaches are reported in Table 4. For reference, the official baseline and two top-ranked approaches in TREC-KBA-2013 track are also included. The official baseline assigns a *"vital"* rating to each document that matches a surface form of an entity and estimates a confidence score based on the length of the observed name [9]. The UMass approach derives a sequential dependent retrieval model which scores documents by frequency of unigrams, bigrams and windowed bigrams of the target entity name [7]. The UDel approach is a query expansion approach, which tunes the weights of the expansion terms with training data [13]. Figure 1 illustrates the

Table 4. Results of all experimental approaches. All the measures are reported by the KBA official scorer with cutoff-step-size=10. Best scores are typeset boldface.

Run	Vital Only		Vital + Useful	
	$\max(F(avg(P), avg(R)))$	$\max(SU)$	$\max(F(avg(P), avg(R)))$	$\max(SU)$
QE	.281	.170	.645	.544
UniClass	.291	.219	.644	.544
UniClass+	**.300**	.222	**.660**	**.568**
UniRank	.285	**.260**	.644	.544
UniRank+	.290	.253	.651	.560
OfficialBaseline	.267	.174	.637	.531
UMass [7]	.273	.247	.610	.496
UDel [13]	.267	.158	.611	.515

macro-averaged recall and precision measures of the approaches listed in Table 4. The parallel curves are contour lines of $\max(F(avg(P), avg(R)))$ measure. The approaches in upper right perform better than the lower left ones.

As shown in Table 4, in *vital + useful* scenario, both UDEL and UMASSthe approaches cannot perform as well as the official baseline. In *vital only*, UMass approach beats the official baseline less than 1% on overall $\max(F)$. Nevertheless, all our approaches outperform the official baseline notably. Moreover, our approaches outperform the others on separate measures for Wikipedia and Twitter entities respectively.

Generally, supervised approaches have more potential than unsupervised approaches in both scenarios. Even the classification and LTR approaches merely

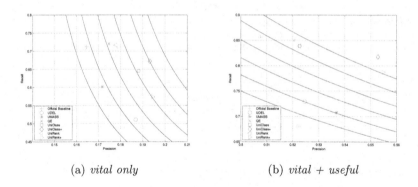

(a) *vital only*   (b) *vital + useful*

Fig. 1. Macro-averaged recall versus precision with curves of constant F1

utilizing the basic feature set, i.e., **UniClass** and **UniRank**, achieve comparative performance with unsupervised approaches. After being augmented with the
supplemental features, supervised approaches achieve more promising results.

As illustrated in 1(a) and 1(b), though our QE approach can achieve high
recall among all approaches, the precision is not satisfactory. This may be
resulted from our equally weighting strategy for expansion terms. UDel approach
weights expansion terms with the help of training data and achieves higher precision than our QE approach in both scenarios. We believe our query expansion
approach can be improved by introducing better weighting strategies.

UniClass+ outperforms **UniClass** on all metrics prominently. Since the
only difference is whether to include 3 extra types of features developed from
relevance evidences, the new features are indeed helpful in CCR. Furthermore,
according to Figure 1, these features can improve both precision and recall of
classification approaches. Similar to classification approaches, in both scenarios,
UniRank+, achieves better results than its baseline **UniRank**.

6.4 Feature Analysis

In this section, we concentrate on **UniClass+** and explore the impacts of the
proposed features in two scenarios. We perform a fine-grained feature analysis
with the help of information gain (IG). Table 5 demonstrates the IG values of
the proposed features in different scenarios. The *maximum*, *mean* and *median*
IGs achieved by basic features are also reported for reference (i.e., *max*, *mean*,
min in Table 5).

Generally, the three types of proposed features perform better than the basic
features in both scenarios. As we have stated, CCR is not just a text classification
task, so basic text features are not so significant as the developed feavotures according to the refined definitions. The temporal feature, i.e., *burst_value* works best in
vital only scenario, verifying our assumption that *vital* documents are more likely
to appear in "bursty" periods of an entity. The temporal feature also works consid-
erably well in *vital + useful* scenario, revealing that *useful* documents distribute

depending on *vital* documents to a certain extent. This phenomenon inspires us to explore more precise boundary between *vital* and *useful* in future.

Table 5. Information gain values of different features

Feature	Information Gain	
	vital only	vital+useful
$Burst\_Value(D)$	**0.130**	**0.287**
$avg(Sim_{cos}(D, S_i(E)))$	0.069	0.145
$avg(Sim_{jac}(D, S_i(E)))$	0.058	0.150
$avg(Sim_{cos}(D, C_i))$	0.028	0.083
$avg(Sim_{jac}(D, C_i))$	0.052	0.108
max	0.121	0.175
$mean$	0.037	0.053
$median$	0.010	0.039

7   Conclusion

The objective of CCR is filtering vitally relevant documents from a huge time-ordered stream corpus and then separating them into different relevance levels. The key challenge is how to classify candidate documents into different relevance levels accurately due to the unclear definitions of them. Apart from the labeled data, we proposed two assumptions based on three relevance evidences to enrich the definitions. The relevance evidences include the profiles of KB entities, existing citations in KB and temporal signals. Under the guidance of the refined definitions, we investigated three mainstream CCR approaches in two scenarios. The experimental results validated the effectiveness of our assumptions. In addition, we developed an entity-centric query expansion method to filter the volume stream corpus and reduce the amount of documents into a considerable size.

Acknowledgments. The authors would like to thank the anonymous reviewers for their helpful comments. This work is funded by the National Program on Key Basic Research Project (973 Program, Grant No. 2013CB329600), National Natural Science Foundation of China (NSFC, Grant Nos. 61472040 and 60873237), and Beijing Higher Education Young Elite Teacher Project (Grant No. YETP1198).

References

1. Balog, K., Ramampiaro, H.: Cumulative citation recommendation: classification vs. ranking. In: SIGIR 2013, pp. 941–944. ACM (2013)
2. Balog, K., Ramampiaro, H., Takhirov, N., Nørvåg, K.: Multi-step classification approaches to cumulative citation recommendation. In: OAIR 2013, pp. 121–128 (2013)
3. Berendsen, R., Meij, E., Odijk, D., Rijke, M.d., Weerkamp, W.: The university of amsterdam at trec 2012. In: TREC (2012)
4. Bonnefoy, L., Bouvier, V., Bellot, P.: A weakly-supervised detection of entity central documents in a stream. In: SIGIR 2013, pp. 769–772 (2013)

5. Bunescu, R., Pasca, M.: Using encyclopedic knowledge for named entity disambiguation. In: EACL 2006, pp. 9–16 (2006)
6. Cucerzan, S.: Large-scale named entity disambiguation based on Wikipedia data. In: EMNLP-CoNLL. pp. 708–716. Association for Computational Linguistics, Prague, June 2007
7. Dietz, L., Dalton, J., Balog, K.: Umass at trec 2013 knowledge base acceleration track. In: TREC (2013)
8. Frank, J.R., Kleiman-Weiner, M., Roberts, D.A., Niu, F., Zhang, C., Re, C., Soboroff, I.: Building an Entity-Centric Stream Filtering Test Collection for TREC 2012. In: TREC 2012 (2012)
9. Frank, J., Bauer, S.J., Kleiman-Weiner, M., Roberts, D.A., Triouraneni, N., Zhang, C., Rè, C.: Evaluating stream filtering for entity profile updates for trec 2013. In: TREC (2013)
10. Gross, O., Doucet, A., Toivonen, H.: Named entity filtering based on concept association graphs. CICLing (2013)
11. Han, X., Zhao, J.: Named entity disambiguation by leveraging wikipedia semantic knowledge. In: CIKM 2009, pp. 215–224 (2009)
12. Kjersten, B., McNamee, P.: The hltcoe approach to the trec 2012 kba track. In: TREC 2012 (2012)
13. Liu, X., Darko, J., Fang, H.: A related entity based approach for knowledge base acceleration. In: TREC (2013)
14. Mihalcea, R., Csomai, A.: Wikify!: linking documents to encyclopedic knowledge. In: CIKM, pp. 233–242. ACM (2007)
15. Robertson, S., Soboroff, I.: The trec 2002 filtering track report. In: TREC (2002)
16. Vlachos, M., Meek, C., Vagena, Z., Gunopulos, D.: Identifying similarities, periodicities and bursts for online search queries. In: SIGMOD, pp. 131–142. ACM (2004)
17. Wang, J., Song, D., Lin, C.Y., Liao, L.: Bit and msra at trec kba ccr track 2013. In: TREC (2013)
18. Zhou, M., Chang, K.C.C.: Entity-centric document filtering: boosting feature mapping through meta-features. In: CIKM, pp. 119–128. ACM (2013)

Relevance Search on Signed Heterogeneous Information Network Based on Meta-path Factorization

Min Zhu[1], Tianchen Zhu[1], Zhaohui Peng[1(✉)], Guang Yang[1], Yang Xu[1],
Senzhang Wang[2], Xiangwei Wang[3], and Xiaoguang Hong[1]

[1] School of Computer Science and Technology, Shandong University, Jinan, China
zm13181@163.com, ztc1319@sina.cn, {pzh,hxg}@sdu.edu.cn,
loggyt@yeah.net, zzmylq@gmail.com
[2] School of Computer Science and Engineering, Beihang University, Beijing, China
szwang@buaa.edu.cn
[3] State Grid Shandong Electric Power Company, Jinan, China
shandongwangxw@163.com

Abstract. Relevance search is a primitive operation in heterogeneous information networks, where the task is to measure the relatedness of objects with different types. Due to the semantics implied by network links, conventional research on relevance search is often based on meta-path in heterogeneous information networks. However, existing approaches mainly focus on studying non-signed information networks, without considering the polarity of the links in the network. In reality, there are many signed heterogeneous networks that the links can be either positive (such as trust, preference, friendship, etc.) or negative (such as distrust, dislike, opposition, etc.). It is challenging to utilize the semantic information of the two kinds of links in meta-paths and integrate them in a unified way to measure relevance.

In this paper, a relevance search measure called SignSim is proposed, which can measure the relatedness of objects in signed heterogeneous information networks based on signed meta-path factorization. SignSim firstly defines the atomic meta-paths and gives the computing paradigm of similarity between objects with the same type based on atomic meta-paths, with collaborative filtering using positive and negative user preferences. Then, on basis of the combination of different atomic meta-paths, SignSim can measure the relatedness between objects with different types based on multi-length signed meta-paths. Experimental results on real-world dataset verify the effectiveness of our proposed approach.

Keywords: Relevance search · Signed heterogeneous information network · Meta-path factorization

1 Introduction

Heterogeneous information networks are logical networks involving multiple typed objects and multiple typed links denoting different relationships, such as bibliographic networks, social media networks, and the knowledge network encoded in Wikipedia [1, 19]. In many heterogeneous information networks, the links could have positive or

© Springer International Publishing Switzerland 2015
J. Li and Y. Sun (Eds.): WAIM 2015, LNCS 9098, pp. 181–192, 2015.
DOI: 10.1007/978-3-319-21042-1_15

negative polarity denoting the positive or negative views and opinions of people. For instance, in Epinions network, consumers rate the products expressing viewpoints of likes or dislikes; in Slashdot Zoo, users can tag others as friends or foes. Such activities constitute the meaningful signed heterogeneous information networks, where links can be positive ("like", "trust") or negative ("dislike", "distrust").

In recent years, relevance or similarity search on heterogeneous information networks has attracted remarkable research interest [6, 11, 12, 13, 14]. By considering different linkage paths in the network, one could derive various relevance semantics, therefore, conventional work on relevance search usually take advantage of the various meta-paths which may contain different relevance semantics between objects. A meta-path is a path consisting of a sequence of relationships defined between different types of objects, namely the structural paths at the meta level. However, conventional research mainly focuses on non-signed information network, without considering the polarity of the links in the network.

The relevance could be computed simply based on meta-path in non-signed information networks. However, it is challenging to define the semantic information of meta-paths with negative links in a signed heterogeneous information network, especially when several different negative links exist simultaneously in one path, making the semantic more ambiguous. Directly utilizing previous meta-path-based methods to relevance search in signed heterogeneous networks might get undesirable or even totally opposite results. Therefore, it is challenging to model the meta-paths with both negative and positive links for relevant search in signed heterogeneous information networks.

In this paper, we propose a novel relevance search approach called SignSim to measure the relatedness of objects with different types in signed heterogeneous information network based on signed meta-path factorization. SignSim firstly defines the atomic meta-paths, based on which the computing paradigm of the similarity between objects with the same type, with collaborative filtering using positive and negative user preferences. Then, on basis of the combination of different atomic meta-paths, SignSim can measure the relatedness between objects with different types based on multi-length signed meta-paths.

The main contributions of this paper can be summarized as follow. (1)For the first time, to the best of our knowledge, we investigate the problem of relevance search in signed heterogeneous information networks. (2)A novel signed meta-path factorization based approach named SignSim for relevance search in signed heterogeneous information networks is proposed. To measure the relatedness between objects of different types, SignSim can effectively capture the positive and negative link semantics along the meta-paths. (3)Experiments on real-world dataset demonstrate the effectiveness of SignSim, by comparison with state-of-the-art approaches on non-signed heterogeneous information network.

The rest of the paper is organized as follows. We introduce the concepts and definitions in Section 2. The details of SignSim are given in Section 3. In section 4, extensive experiments on real data are performed to evaluate the effectiveness of our method. We discuss related work in Sections 5, and conclude the study in Section 6.

2   Problem Statement

In this section, we introduce some concepts and notations to help us state the studied problem.

Definition 1. (Signed Heterogeneous Information Network): A signed heterogeneous information network is an undirected graph G =(V, E) with an object type mapping function τ: V→A and link type mapping function φ: E→R and |A|>1, |R|>1. Each object $v \in V$ belongs to a particular object type $\tau(v) \in A$ and each link $e \in E$ belongs to a particular relation $\varphi(e) \in R$. For each link $e \in E$, it has polarity $s(e) \in \{Positive, Negative\}$.

As an example, a toy IMDB network is given in Fig.1. It is a typical signed heterogeneous information network, containing five types of objects: users (U), movies (M), genres (G), directors (D), and actors (A).

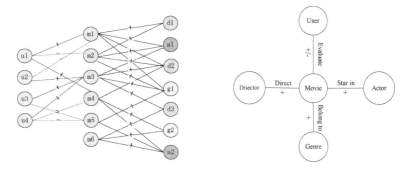

Fig. 1. A signed IMDB network   **Fig. 2.** The signed network schema of IMDB

Definition 2. (Signed Network Schema): The signed network schema, denoted as $T_G=(A, R)$, is a meta-level template for a signed heterogeneous network G=(V, E) with the object type mapping τ: V→A and the link type mapping φ: E→R. In network schema, the nodes are the types of objects, and the edges are relationships between types, guiding the exploration of the semantics in the network.

For the IMDB network defined in Fig.1, the network schema is shown in Fig.2. The links between users and movies denote the rating or rated-by relationships with positive or negative polarities, while the links between movies and other types of entities only have the positive polarity.

Definition 3. (Signed Meta-path): The signed meta-path is a directed path defined on the signed network schema $T_G = (A, R)$ of a graph G. A meta-path can be formally denoted as $\mathcal{P} = A_0 \xrightarrow{R_1} A_1 \xrightarrow{R_2} \ldots \xrightarrow{R_L} A_L$, which is a meta-level description of a path instance between two objects with type A_0 and type A_1 respectively, usually abbreviated as $R_1R_2 \cdots R_L$ or $A_0A_1 \cdots A_L$.

Definition 4. (Atomic Meta-path): Atomic meta-path is a minimum part of a signed meta-path that can be used to compute the similarity between objects of the same type.

Definition 5. (Meta-path Factorization): Meta-path factorization splits a signed meta-path into multiple atomic meta-paths, and then the relatedness between objects from different types can be computed based on the signed meta-paths.

For instance, U (users)–>M (movies)–>U (users)–>M (movies) is one of the meta-path of the network schema shown in Fig.2. It denotes that two users have rated the same movie, and the second user also rated another movie. To get the relatedness between users and movies based on meta-path UMUM, SignSim uses meta-path factorization and splits the meta-path into UMU and UM. UMU is an atomic meta-path, based on which the similarity between users can be measured. UM is not an atomic meta-path, but based on UM the relevance between users and movies can be computed directly. Then SignSim combines the computing results and gets the final relevance.

Problem Statement. (Relevance Search on Signed Heterogeneous Network): In a signed heterogeneous information network G=(V, E), for a given node s ∈ V and target type T, how to find the nodes in V of type T which are most relevant to s.

3   SignSim: A Relevance Measure

In this section, we elaborate a signed meta-path-based relevance framework and a novel relevance measure under this framework, SignSim. Taking Fig.1 for example, SignSim calculates the relatedness between users and movies based on multiple signed meta-paths.

3.1   Framework

Due to the signed meta-path has positive or negative polarity, we cannot directly use the meta-path-based relevance measure to compute the relevance between objects from different types in the signed heterogeneous information network. Therefore, SignSim first splits the signed meta-path, getting computable atomic meta-paths, and then combines the computing results.

SignSim mainly has two points: (1) It splits the meta-path into multiple atomic meta-paths, and computes the similarity between objects of the same type based on the atomic meta-paths. (2) It determines the possibility space of the target nodes, and computes the relevance between objects of different types based on the combination of the similarity computed by various atomic meta-paths.

The signed meta-paths can be obtained by traversing on the network schema using BFS (breadth first search) [7]. Then the atomic meta-paths can be got by meta-path factorization. In meta-path factorization, we always find the first match to the atomic meta-path, and then repeat the above procedure on the remaining meta-path. We separate out the mismatch path and call them **redundant meta-path**. For example, from the IMDB network schema shown in Fig.2, we can get the meta-path $\mathcal{P} = U \xrightarrow{R_1} M \xrightarrow{R_2} A \xrightarrow{R_3} M \xrightarrow{R_4} A \xrightarrow{R_5} M$. It can be decomposed to redundant meta-path $\mathcal{P}_1 = U \xrightarrow{R_1} M$, atomic meta-path $\mathcal{P}_2 = M \xrightarrow{R_2} A \xrightarrow{R_3} M$ and $\mathcal{P}_3 = M \xrightarrow{R_4} A \xrightarrow{R_5} M$. The algorithm of meta-path factorization is shown in Algorithm 1.

Algorithm 1. Meta-path Factorization

Input: $\mathcal{P}$: the meta-path; **m**: the number of nodes of the meta-path

Output : $\mathcal{P}_i$: the atomic meta-path

1 j ← 1, k ← 2, k' ← 1
2 **while** k<=m **do**
3   **for** i=k'; i<k; i++ **do**
4   **if** the i-th node and the k-th node of meta-path are the same **then**
5   k' ← k
6   output path between two nodes as the atomic meta-path
7   **end if**
8   **end for**
9   k ← k+1
10 **end while**
11 output all remaining paths as the redundant meta-path

The main idea of SignSim can be expressed by following three formulas.

$$\mathcal{P} = \mathcal{P}_1\mathcal{P}_2...\mathcal{P}_i...\mathcal{P}_n \tag{1}$$

Formula (1) represents the decomposition of meta-path $\mathcal{P}$, and $\mathcal{P}_i$ $(1 <= i <= n)$ is the atomic meta-path or redundant meta-path.

$$U_i = \bigcup_{o_{i-1}\in U_{i-1}} s(o_{i-1}, o_i \mid \mathcal{P}_{i-1}) \ and \ U_1 = o_1 \tag{2}$$

In formula (2), U_i is the possible target node space of the node o_i, $s(o_{i-1}, o_i \mid \mathcal{P}_{i-1})$ is the set of the object o_i positively related to object o_{i-1} based on meta-path $\mathcal{P}_{i-1}$.

$$signsim(o_1, o_n \mid \mathcal{P}) = \sum_{o_2 \in U_2}\sum_{o_3 \in U_3}...\sum_{o_{n-1}\in U_{n-1}} sim(o_1, o_2 \mid \mathcal{P}_1)...sim(o_{n-2}, o_{n-1} \mid \mathcal{P}_{n-2})sim(o_{n-1}, o_n \mid \mathcal{P}_{n-1}) \tag{3}$$

In formula (3), $sim(o_{i-1}, o_i \mid \mathcal{P}_{i-1})$ is the relatedness between objects o_{i-1} and o_i based on meta-path $\mathcal{P}_{i-1}$.

3.2   Atomic Meta-path-Based Similarity Measure

In this section, taking IMDB network as an example, we present how to construct atomic meta-paths and measure the similarity between objects of the same type. Based on Algorithm 1, we find out three kinds of atomic meta-paths.

Atomic Meta-path: UMU. The atomic meta-path UMU describes the two users seeing the same movie. It contains two edges and both of them are signed. The similarity between the two users is measured based on this atomic meta-path.

We construct an $n*m$ adjacency matrix $W_{UM} = \left[v_{ij}\right]_{n\times m}$ between users and movies in the network, denoting the degree of user's preference to the movie. The m is the number of movies and n is the number of users. We denote the row of the matrix with vector u^i, in which $i(1<=i<=n)$ is the row number. The similarity between the two users u_i and u_j is measured by cosine similarity based on vector space model. That is:

$$v_{ij} = \begin{cases} 1 & user\ u_i\ enjoy\ movie\ m_j \\ -1 & user\ u_i\ doesn't\ enjoy\ movie\ m_j \\ 0 & user\ u_i\ never\ see\ movie\ m_j \end{cases}$$

$$sim\ (u_i, u_j \mid UMU\ ) = \cos\ (u^i, u^j) = \frac{u^i * u^j}{|u^i| * |u^j|} \tag{4}$$

If $sim\ (u_i, u_j \mid UMU) \geq \zeta$, the user u_i is considered as similar to the user u_j. ζ is a threshold parameter, which will be discussed in the experiments.

Atomic Meta-path: MUM. The atomic meta-path MUM describes the two movies seen by the same user. It contains two edges and both of them are signed. The similarity between two movies is measured based on the atomic meta-path.

We can measure the similarity between two movies based on the same adjacency matrix as presented above. We denote the row of matrix with vector m^i. The similarity measure between movies m_i and m_j is:

$$sim\ (m_i, m_j \mid MUM\ ) = \cos\ (m^i, m^j) = \frac{m^i * m^j}{|m^i| * |m^j|} \tag{5}$$

Atomic Meta-path: MPM. The atomic meta-path MPM describes the two movies having common attributes (genres, directors, or actors). Here **P** represents **G** (genres), **D** (directors), and **A** (actors). It contains two edges and neither of them is signed. It is used to represent the relationship between movies and their various properties. The similarity between two movies can be measured by this atomic meta-path, just like formula (4) and (5).

3.3 Signed Meta-path-based Relevance Measure

For non-atomic meta-paths, we measure the relevance between objects by signed meta-path factorization. Different paths represent different semantics, thus path selection is essential to measure the relevance between objects from different types. The length of the meta-path may also affect the performance, and the previous works have shown that the accuracy will decrease as the length increases [6, 14]. In this section, we only study the meta-paths whose length is less than five.

Redundant Meta-path: UM. The redundant meta-path UM, containing a signed edge, describes the movie seen by the user, and the relevance between user u and movie m can be measured based on it. The relevance based on redundant meta-path can be computed directly without meta-path factorization. That is:

$$sim\ (u_i, m_j \mid UM\ ) = \frac{v_{ij}}{|D_s(\ u_i |UM\ )| * |D_s(\ m_j |UM\ )|}$$

$$v_{ij} = \begin{cases} 1 & user\ u_i\ enjoy\ movie\ m_j \\ -1 & user\ u_i\ doesn't\ enjoy\ movie\ m_j \\ 0 & user\ u_i\ never\ see\ movie\ m_j \end{cases}$$

$$D_s(\ u_i\ |UM\ ) = \begin{cases} D_+(u_i |UM\ ) & v_{ij} = 1 \\ D_-(u_i |UM\ ) & v_{ij} = -1 \end{cases} \tag{6}$$

where $D_+(u_i|UM)$ is the neighbor set of node u_i based on positive edges of UM, and $D_-(u_i|UM)$ is the neighbor set of u_i based on negative edges of UM.

Meta-path: UMUM. The meta-path UMUM contains three edges and all of them are signed. For measuring the relevance between users and movies based on UMUM, we split it into one atomic meta-path UMU and one redundant meta-path UM, and compute the similarity or relevance respectively, then combine the computing results. Intuitively, the factorization means we find out the most similar users to the source user, and then we collect a set of favorite movies of these most similar users to the source user. The procedure can be formally depicted as follows:

$$U_2 = s(u, u_2 | UMU)$$
$$signsim(u, m | UMUM) = \sum_{u_2 \in U_2} sim(u, u_2 | UMU) sim(u_2, m | UM) \qquad (7)$$

where $s(u, u_2 | UMU)$ is the set of users that u is similar to, and $sim(u_2, m | UM)$ is the relatedness between u_2 and m based on meta-path UM.

Meta-path: UMUMUM. This meta-path contains five edges and all of them are signed. For measuring the relevance between users and movies, we split the meta-path UMUMUM into UMU, UMU, and UM. Similarly, we first find out a set U_2 of most similar users with the source user based on meta-path UMU. Then, we find out a set U_3 of most similar users with users of set U_2 based on meta-path UMU. Finally, we collect a set of favorite movies of these users of set U_3. The procedure can be formally depicted as follows:

$$U_2 = s(u, u_2 | UMU)$$
$$U_3 = \bigcup_{u_2 \in U_2} s(u_2, u_3 | UMU)$$
$$signsim(u, m | UMUMUM) = \sum_{u_2 \in U_2} \sum_{u_3 \in U_3} sim(u, u_2 | UMU) sim(u_2, u_3 | UMU) sim(u_3, m | UM) \qquad (8)$$

Meta-path: UMPM. This atomic meta-path contains three edges and one of them is a signed edge. For measuring the relevance between users and movies, we split this meta-path into one redundant meta-path and one atomic meta-path, which are UM and MPM.

Meta-path: UMPMPM. This meta-path contains five edges and one of them is signed. For measuring the relevance between users and movies, we could split it into one redundant meta-path and two atomic meta-paths, which are UM, MPM and MPM.

Meta-path: UMPMUM. This meta-path contains five edges and of which three are signed. For measuring the relevance between user and movie, we could split this meta-path into three meta-paths, which are UM, MPM, and MUM.

Meta-path: UMUMPM. This meta-path contains five edges and of which three are signed. For measuring the relevance between user and movie, we could split it into three meta-paths, which are UMU, UM, and MPM.

4   Experiments

4.1   Experimental Setup

Dataset. In this section, we evaluate the effectiveness of the proposed SignSim approach with comparison to existing relevance search algorithms on the real dataset hetrec2011-movielens [18]. The dataset contains 2,113 users, 10,197 movies, 20 movie genres, 4,060 directors, and 11,019 actors. We extract the first two actors according to the actor ranking of movies. On average, there are 2.04 genres, 2 actors and 1 director per movie. By calculating the average ranking of movies given by users, a signed heterogeneous information network can be built on this dataset which contains five types of objects: user, movie, director, actor and genre.

Evaluation Metric. To evaluate our approach, we first sort all the rating records in the dataset according to the timestamp. For each user, the latest 30% movies he or she has rated are selected as testing data and the old ones are selected as training data. When conducting relevance search, we generate a list of K movies named Ru for each user u. If the testing movie appears in the result list, we call it a hit. The mean hit of movie rating can be calculated as follows:

$$MR_u = \frac{\sum_{m \in R_u} I(m \in T_u) r(u, m)}{\sum_{m \in R_u} I(m \in T_u)}$$

where I($\cdot$) is an indicator function, R_u is a set of top-K movies recommended to user u, T_u is the set of testing movies of user u, $r(u,m)$ is the rating on movie m by user u. We use MR_u to evaluate the relevance search results. We select the following approaches as baselines:

MatrixCal: MatrixCal is the naïve algorithm to calculate relevance between users and movies. User's preference matrix was calculated to represent the preference to the movie features (such as directors, actors, genres). The feature matrix of movies was also obtained. Then the preference matrix and feature matrix were multiplied together.
HeteSim: HeteSim[14] can measure the relatedness of objects with the same or different types in heterogeneous networks. The relatedness of object pairs is defined based on the search path that connects two objects through following a sequence of node types.

4.2   Experimental Results

Case Study. In the following experiments, we set parameters ζ=0.1 and K=5 that ζ is the similarity threshold of objects in the same type and K is a parameter to control the number of movies recommended, and single out a user with ID 7815 as the source user.

Table 1 shows the top K search results of SignSim with varied meta-paths. We compare MR_u of SignSim with two baselines, MatrixCal and HeteSim, and show the results in Table 2. In both tables, the bold items mean the searched movies are hit, and the figures in brackets denote the score rated by the user.

As we can see in Table 1, the search results based on short meta-paths are much better than those based on long meta-paths, so the length of meta-path is critical to measure the relevance of objects. Table 2 shows the results with different algorithms. The results given by SignSim are better than the two baselines.

Table 1. The results of different signed meta-paths in SignSim

UMUM	UMPM	UMPMUM	UMUMUM
The Usual Suspects (5.0)	**Sicko (4.5)**	Red Dust	The Shawshank Redemption
The Shawshank Redemption	Heavenly Creatures	Ruby Cairo	**The Usual Suspects (5.0)**
American Beauty (4.5)	King Kong	The Bad and the Beautiful	**American Beauty (4.5)**
Shichinin no samurai (3.5)	JLG/JLG -autoportrait de décembre	Arabian Nights	**Shichinin no samurai (3.5)**
Trainspotting	The Endurance: Shackleton's Legendary Antarctic Expedition	Small Faces	Monty Python and the Holy Grail

Table 2. The results of different approaches

SignSim	MatrixCal	HeteSim
The Usual Suspects (5.0)	This Is Spinal Tap	**Sicko (4.5)**
The Shawshank Redemption	Finding Neverland	Star Wars: Episode I - The Phantom Menace
American Beauty (4.5)	Casanova	Super Size Me
Shichinin no samurai (3.5)	Recount	Unprecedented: The 2000 Presidential Election
Trainspotting	The Bourne Supremacy	Believers

Quantitative Comparison. In this section, we quantitatively compare the proposed SignSim with two baselines. Fig.3 shows the results of different algorithms. One can see that the average rating for all users calculated by SignSim is higher than that by HeteSim and MatrixCal.

As mentioned above, the length of meta-path can significantly affect the relevance search performance. In SignSim, the accuracy of the relevance decreases as the meta-path length increases. Fig.4 shows the effect of different length of meta-paths.

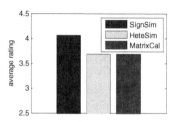

Fig. 3. The results of different algorithms

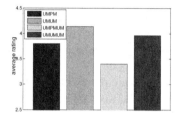

Fig. 4. The results of various meta-paths

Parameter Analysis. In this section, we will study the effect of parameter K and threshold ζ. K is used to control the amount of movies that we recommend to users and ζ is used to control the similarity between the same types of objects.

Fig.5 shows the results with various K values. One can see from Fig.5 that the performance shows a decrease trend with the increase of parameter K, which means a larger K can hurt the accuracy.

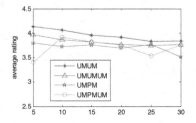

Fig. 5. The effect of parameter K

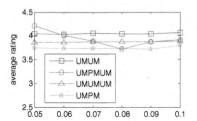

Fig. 6. The effect of parameter ζ

Fig.6 shows the performance of SignSim with the varied threshold parameters. One can see from Fig.6 that the performance of SignSim increases as the ζ increases for the meta-paths of UMUM, UMPM and UMUMUM, which means a higher similarity between the same type of objects helps to improve the final search accuracy, although the growth trend is not significant. As for UMPMUM before $\zeta=0.08$, the performance decrease as the ζ increases and after $\zeta=0.08$, the performance increase slowly. On the whole, SignSim can get a fairly better performance when $\zeta=0.06$.

As to meta-path UMUMPM and UMPMPM, due to the space limit, we do not describe them in detail here. Their results are similar to those of other meta-paths.

5   Related Work

Relevance search on information network has gained wide attentions from researchers in link prediction [1, 7, 8, 9, 10], clustering [3], similarity query [6], text mining [2], etc. The most related work to relevance search is similarity search. Similarity calculation is used to measure the degree of similarity between two nodes, commonly used in data mining and natural language processing. Research on similarity search has made many significant achievements in recent years. Without considering the polarity of the edge in heterogeneous information network, i.e. links are all positive, similarity search broadly divided into two types: feature-based approach and link-based approach.

Feature-based measurement method is based on the characteristics of the object. Vector space model (VSM) is the most widely used similarity calculation model, in which objects are represented as vectors. However, feature-based approaches do not take the links between objects into account, so they cannot apply to network data.

Link-based approach is based on topological similarity of the object. SimRank [4] is a general algorithm determined only by the similarity of structural context, where two objects are similar if their neighbors are similar. Due to its relatively high computational complexity, most of the follow-up studies focus on how to improve the efficiency of the algorithm [17]. SCAN [2] is a structure clustering algorithm, which calculates the similarity between two objects through comparing their neighbor sets, while [6] proposed a meta-path-based similarity measure PathSim, which considers that objects are not only share similar visibility in the network but also strongly connected with each other. However, this approach just considers the objects with the same type. Different from the above study, HeteSim [14] is able to measure the correlation between two objects of different types based on arbitrary search path in heterogeneous information network. However, all of these methods don't consider the link polarity, i.e., their studies are built on a fundamental assumption that all edges are positive.

There are some works [5, 13, 15, 16] discussing the trust prediction task in social networks which have both positive and negative links. The majority of those studies focus on homogeneous networks which have only one type of objects and one type of links. However, many networks in the real world, such as Epinions, Slashdot Zoo, and Wikipedia, are both signed and heterogeneous composed of multiple types of objects and multiple types of links. Therefore it is necessary to study new methods to address knowledge discovery in signed heterogeneous network.

6   Conclusion and Future Work

In the paper, we study the relevance search problem in signed heterogeneous information networks. A novel relevance search approach called SignSim is proposed. Based on signed meta-path factorization, SignSim is able to find the most relevant objects with various types to a given object in the signed heterogeneous information networks. SignSim first splits the signed meta-path into multiple atomic meta-paths. By utilizing the atomic meta-paths, SignSim next proposes to measure the similarity between objects of the same type. Finally, SignSim integrates various similarity results to measure the relatedness between objects from different types based on non-atomic signed meta-paths. Experimental results on a real-world dataset verify the effectiveness of the approach. In this paper, we only study the relevance based on individual signed meta-paths. Our future work will consider the combination of different signed meta-paths to get global relevance.

Acknowledgments. This work was supported by the National Natural Science Foundation of China (Grant No.61303005, No.61170052), the Natural Science Foundation of Shandong Province of China (Grant No.ZR2013FQ009), and the Technology Project of State Grid (Grant No.2012GWK515).

References

1. Sun,Y., Han, J., Yan, X., Yu, P.: Mining knowledge from interconnected data: a heterogeneous information network analysis approach. In: Proceedings of the VLDB Endowment, pp. 2022–2023. ACM Press, New York (2012)

2. Xu, X., Yuruk, N., Feng, Z., Schweiger, T.A.J.: SCAN: a structural clustering algorithm for networks. In: Proceedings of the 13th ACM SIGKDD International Conference on Knowledge Discovery and Data Mining, pp. 824–833. ACM Press, New York (2007)
3. Lao, N., Cohen, W.: Fast query execution for retrieval models based on path constrained random walks. In: Proceedings of the 16th ACM SIGKDD International Conference on Knowledge Discovery and Data Mining, pp. 881–888. ACM Press, New York (2010)
4. Jeh, G., Widom, J.: Simrank: a measure of structural-context similarity. In: Proceedings of the Eighth ACM SIGKDD International Conference on Knowledge Discovery and Data Mining, pp. 538–543. ACM Press, New York (2002)
5. Ye, J., Cheng, H., Zhu, Z., Chen, M.: Predicting positive and negative links in signed social networks by transfer learning. In: WWW, Switzerland (2013)
6. Sun, Y., Han, J., Yan, X., Yu, P., Wu, T.: Pathsim: meta path-based Top-k similarity search in heterogeneous information networks. In: PVLDB, pp. 992–1003 (2011)
7. Sun, Y., Han, J., Aggarwal, C., Chawla, N.V.: When will it happen? : relationship prediction in heterogeneous information networks. In: Proceedings of the Fifth ACM International Conference on Web Search and Data Mining, pp. 663–672. ACM Press, New York (2012)
8. Sun, Y., Barber, R., Gupta, M., Aggarwal, C.C., Han, J.: Co-author relationship prediction in heterogeneous information networks. In: Proceedings of the 2011 International Conference on Advances in Social Networks Analysis and Mining, pp. 121–128. IEEE Computer Society Press, Washington (2011)
9. Sun, Y., Norick, B., Han, J., Yan, X., Yu, P.S., Yu, X.: Integrating meta-path selection with user-guided object clustering in heterogeneous information networks. In: Proceedings of the 18th ACM SIGKDD International Conference on Knowledge Discovery and Data Mining, pp. 1348–1356. ACM Press, New York (2012)
10. Tong, H., Faloutsos, C., Pan, J.Y.: Fast random walk with restart and its applications. In: Proceedings of the Sixth International Conference on Data Mining, pp. 613–622. IEEE Computer Society Press, Washington (2006)
11. Liben-Nowell, D., Kleinberg, J.: The link-prediction problem for social networks. J. Am. Soc. Inf. Sci. Technol., 1019–1031 (2007). John Wiley & Sons, Inc. Press
12. Lichtenwalter, R.N., Lussier, J.T., Chawla, N.V.: New perspectives and methods in link prediction. In: Proceedings of the 16th ACM SIGKDD International Conference on Knowledge Discovery and Data Mining, pp. 243–252. ACM Press, New York (2010)
13. Leskovec, J., Huttenlocher, D., Kleinberg, J.: Predicting positive and negative links in online social networks. In: Proceedings of the 19th International Conference on World Wide Web, pp. 641–650. ACM Press, New York (2010)
14. Shi, C., Kong, X., Huang, Y., et al.: HeteSim: a general framework for relevance measure in heterogeneous networks. In: IEEE Transactions on Knowledge and Data Engineering, pp. 2479–2492. IEEE Computer Society, Washington (2014)
15. Symeonidis, P., Tiakas, E., Manolopoulos, Y.: Transitive node similarity for link prediction in social networks with positive and negative links. In: Proceedings of the Fourth ACM Conference on Recommender Systems, pp.183–190. ACM Press, New York (2010)
16. DuBois, T., Golbeck, J., Srinivasan, A.: Predicting trust and distrust in social networks. In: Agent and Multi-Agent Systems: Technologies and Applications, pp. 122–131. Springer Berlin Heidelberg Press, Heidelberg (2011)
17. Sun, L., Cheng, R., Li, X., Cheung, D.W., Han, J.: On Link-based similarity join. In: Proceedings of 37th International Conference on Very Large Data Bases (2011)
18. hetrec2011-movielens. http://grouplens.org/datasets/hetrec-2011/
19. Wang, S.Z., Yan, Z., Hu, X., Yu, P.S., Li, Z.J.: Burst time prediction in cascades. In: Proceedings of 29th AAAI Conference on Artificial Intelligence (2015)

Improving the Effectiveness of Keyword Search in Databases Using Query Logs

Jing Zhou, Yang Liu$^{(\boxtimes)}$, and Ziqiang Yu

School of Computer Science & Technology, Shandong University, Jinan, China
{zhoujing1990,zqy800}@gmail.com, yliu@sdu.edu.cn

Abstract. Using query logs to enhance user experience has been extensively studied in the Web IR literature. However, in the area of keyword search on structured data (relational databases in particular), most existing work has focused on improving search result quality through designing better scoring functions, without giving explicit consideration to query logs. Our work presented in this paper taps into the wealth of information contained in query logs, and aims to enhance the search effectiveness by explicitly taking into account the log information when ranking the query results. To concretize our discussion, we focus on schema-graph-based approaches to keyword search (using the seminal work DISCOVER as an example), which usually proceed in two stages, candidate network (CN) generation and CN evaluation. We propose a query-log-aware ranking strategy that uses the frequent patterns mined from query logs to help rank the CNs generated during the first stage. Given the frequent patterns, we show how to compute the maximal score of a CN using a dynamic programming algorithm. We prove that the problem of finding the maximal score is NP-hard. User studies on a real dataset validate the effectiveness of the proposed ranking strategy.

1 Introduction

The success of keyword queries as a common way of Web search and exploration has spurred much interest in the research community in supporting effective and efficient keyword search in relational databases. It allows information retrieval (IR) from the databases by simply giving a set of keywords, without requiring users to know either query languages (such as SQL) or the database schema. A large body of literature has appeared in this area, which can be broadly classified into two categories: the schema-graph-based approach (e.g., DISCOVER [1], and SPARK [2]) and the data-graph-based approach (e.g., BANKS [3], and Blinks [4]).

Our work presented in this paper is along the direction of schema-graph-based approaches. For a given query consisting of one or more keywords, the schema-graph-based approach first locates the relations that contain the keywords, and for each such relation, generates a tuple set that consist of only the tuples matching a given keyword. It then generates candidate networks (CNs) where each CN corresponds to a join expression of the tuple sets. The final step is to evaluate all the generated CNs to produce join networks of tuples,

© Springer International Publishing Switzerland 2015
J. Li and Y. Sun (Eds.): WAIM 2015, LNCS 9098, pp. 193–206, 2015.
DOI: 10.1007/978-3-319-21042-1_16

which are presented as answers to the users. Recent work along this direction has attempted to improve the effectiveness of the search through better ways of ranking the final results. For example, some authors (e.g., [2,5,6]) address the effectiveness issue by leveraging the relevance-ranking strategies which have been proved effective over text data.

Despite the recent advances in keyword search over databases, very few work explicitly incorporate query feedback into the ranking of query results. For example, in DISCOVER, the CN generation step simply outputs all the generated CNs in ascending order by the number of joins. User preferences are not explicitly considered during the whole process. This is also true for other existing keyword search methods. In contrast, a user is more likely to find the answer he/she is interested in if his/her preference (captured through search history, etc.) is taken into account when ranking the results.

As an example, consider the following scenario. Company XYZ is a wholesale supplier with geographically distributed warehouses, each of which serves several sales districts. A database D is used to manage the information of the company's products and customers. D consists of seven tables with the following schema: $warehouse(warehouseID, \ldots)$, $district(districtID, warehouseID, \ldots)$, $customer(customerID, districtID, warehouseID, \ldots)$, $item(itemID, \ldots)$, $stock(itemID, warehouseID, \ldots)$, $order(orderID, districtID, warehouseID, customerID, \ldots)$, $orderline(orderID, number, itemID, \ldots)$. We assume that the warehouses are named $W_1, W_2, \ldots, W_m$ and items $I_1, I_2, \ldots, I_n$. Intuitively, users of D from different departments of the company would want different information from the database even when they issue the same query. For example, when an employee from the sales department issues a query "W_i, I_j", she is likely to prefer retrieving information regarding the sales of I_j in warehouse W_i. Therefore, search results corresponding to the join $item \bowtie orderline \bowtie order \bowtie warehouse$ should be promoted towards the top of the ranked list of results. This preference can be naturally reflected in the query log through past queries issued by her or her colleagues in the sales department. In contrast, an employee from the distribution department, who often checks stock and distributes goods from warehouses to stores, may prefer the stock information of item I_j in the warehouse W_i ($item \bowtie stock \bowtie warehouse$) for the same search. Again, this preference can be reflected in the log of past queries.

In this paper, we introduce a new ranking strategy to adapt the ranking of query results to user preferences. Query logs, which record the queries along with the results chosen by users for each query, are the source of user feedback. We first mine the frequent patterns in the query log. For a given query, we then score all CNs obtained by a standard CN-generating algorithm, such as that from DISCOVER, by a new scoring function that combines the score based on the user query log and the score on the CN size through normalization and weighting. The re-ranked CNs can better reflect user preferences. The above scoring process involves a NP-hard problem, which is proved. To solve the problem, we present a dynamic programming algorithm. The experiments we conducted on the DBLP dataset demonstrate the effectiveness of our strategy.

Our main contributions can be summarized as follows.

- We propose a novel ranking strategy to re-rank the *CN*s utilizing frequent patterns mined from user query logs.
- We prove the hardness result on the scoring problem, and provide an optimal dynamic programming algorithm.
- Extensive experiments and user studies are conducted to evaluate the proposed ranking strategy, and confirm its effectiveness.

The rest of the paper is arranged as follows. Section 2 provides an overview of related work. Section 3 defines some related basic concepts. Section 4 presents our ranking strategy based on query log and the NP-hard problem. Section 5 reports the experimental results. We conclude this paper in Section 6.

2   Related Work

Keyword search in relational databases has been an active area of research [1, 3, 5, 7–10]. Existing approaches can be categorized into data graph based and schema graph based. The schema graph based approach [1, 5, 7] executes the querying process by two steps: *CN* generation and *CN* evaluation. We take DISCOVER as an example. The schema of the relational database is modeled as a directed graph for which the node represents the corresponding relation and the edge indicates the key-foreign key constraint between two relations. To generate all *CN*s, a set of join expressions are constructed by breadth-first traversing the tuple set graph expanded from the schema graph. In the evaluation step, a execution plan is generated to evaluate all the generated *CN*s.

To improve the query efficiency, DISCOVER has given several rules such as pruning condition to avoid generating unnecessary tuple trees and designed a greedy algorithm to produce a near-optimal execution plan. More studies [1, 5, 11] have been done to further improve the query efficiency, and much work has also been done to improve effectiveness such as [2, 6]. Liu [6] identifies the formalization of four new factors (tuple tree size formalization, etc.) to improve the ranking formula in [5]. However, existing work on the effectiveness issue primarily focuses on returning results with basic semantics, while few have considered adapting user preferences.

In the IR community, relevance feedback has proven effective in interactive IR. However, the methods of relevance feedback in IR cannot be directly applied to the context of keyword search in databases becasue in the IR and Web search context, the entities are existing documents, Web pages, etc., where as in the database context, the entities are join networks of tuples, etc., which are dynamically assembled based on the query. Moreover, in the database context, user preferences are often reflected in the structure of the result (e.g., which *CN* this result is based on), which is a non-existent issue in IR and Web search.

User feedback has also been used in keyword search-based data integration to help learn how to correctly integrate data. In the literature of keyword search in databases, some previous works have also considered user feedback. Gao and

Paper

Pid	Title	Cid
P1	A hidden **Markov** model information retrieval system	C1
P2	Topics over time: a non **Markov** continuous-time model of topical trends	C2
P3	An HMM acoustic model incorporating various additional knowl-edge sources	C3
P4	ILDA: interdependent LDA model for learning latent aspects and their ratings from online product reviews	C4

Author

Aid	Name
A1	Miller
A2	Xuerui Wang
A3	**Markov**
A4	Moghuddam

Conference

Cid	Name
C1	22nd International ACM SIGIR Conference
C2	12th ACM SIGKDD international conference
C3	8th Conference of the International Speech Communication Association
C4	34th International ACM SIGIR Conference

Write

Aid	Pid
A1	P1
A2	P2
A3	P3
A4	P4

PaperCitation

Pid	CitedPid
P2	P4

Fig. 1. DBLP database sample

Yu [12] employ query logs for keyword search, but the purpose is different from ours: the query logs are used to help improve the effectiveness of keyword query cleaning. Peng et al. [13] aim at better reformulating a user's initial query to retrieve more relevant query results in relational databases by applying user feedback. However, this work is still based on the vector space model and applies the IR-Style ranking for query reformulation without considering the information (i.e., frequent patterns) in the query logs.

3   Preliminaries

We first introduce some terms and notations used throughout this paper, and formulate the problem that this paper focuses on.

We consider a relational database with n relations $R_1, \ldots, R_n$. Each relation R_i has m_i attributes $a_1^i, \ldots, a_{m_i}^i$.

Definition 1. (Labeled Directed Graph). *Given a relational database D, we define the schema graph of D as a Labeled Directed Graph (LDG) $G = (V, E)$. Each node $v \in V$ represents the corresponding relation in D, and each edge $e \equiv v_i \to v_j$ ($v_i, v_j \in V$, $e \in E$) corresponds to a primary-key-foreign-key relationship between the relations represented by v_i and v_j. We assign unique ids (i.e., label) to all nodes and edges respectively.*

Fig. 1 depicts a sample of five tables from the DBLP biography database [14]. The tables *Paper* and *Author* contain information on papers and researchers respectively; table *Conference* contains conference information. Table *PaperCitation* stores the citation relationships between papers; and table *Write* records the $m : n$ relationships between authors and papers. The LDG of the sample DBLP database from Figure 1 is shown in Figure 2.

Given a query $Q = \{k_1, \ldots, k_m\}$, where k_i is a keyword, we can obtain a set of basic tuple sets $\overline{R}_i^{k_j}$ ($i = 1, \ldots, n$, and $j = 1, \ldots, m$). The basic tuple set $\overline{R}_i^{k_j}$ consists of all tuples of relation R_i that contain the keyword k_j. Then the basic

Fig. 2. the LDG of DBLP database (Nodes and edges are labeled with number respectively)

Fig. 3. Labeled free tree form of a *CN*

tuple sets are processed to produce tuple sets R_j^K for the non-empty subset K of Q. R_i^K, a *non-empty tuple set* that contains the tuples of R_i that contain all keywords of K and no other keywords, is defined as $R_i^K = \{t | t \in R_i \wedge \forall k \in K, t \text{ contains } k \wedge \forall k \in Q - K, t \text{ does not contain } k\}$. For example, $Paper^{Markov}$ is the set {P1,P2}, and $Paper^{LDA}$ is {P4}. The database relations that appear in the schema graph is *free tuple set* denoted as $R^{\{\}}$ which means that the relation R does not have tuples that contain a keyword. The non-empty tuple sets combine with the schema graph of the database by adding corresponding edges to form the tuple set graph G^{TS}.

Definition 2. (Candidate Network). *A candidate network (CN) is a join network of tuple sets formed by traversing G^{TS} in a breadth-first mode.*

A *CN* is a portion of G^{TS} and can be considered as labeled free tree which belongs to the family of *free trees*-the connected, acyclic and undirected graphs. Figure 3 illustrates the labeled free tree of one *CN* for the query "*Markov,LDA*", which represents the join network $Paper^{Markov} \bowtie PaperCitation \bowtie Paper^{LDA}$.

Definition 3. (Query Log). *A query log L is a set of entries. Each user has his own query log. For a specific user, each entry in his/her user log records the candidate network (in the form of labeled free tree) that the user chose to visit when all the candidate networks were presented to him/her in answering a given query. In short, L contains the chosen results of a user for one or more queries.*

Definition 4. (Mining Frequent Patterns from Query Logs). *For a specific user (or a group of similar users) u, his/her entries in the query log form a set L_u where each recorded CN $\mathbf{c} \in L_u$ is a labeled free tree. For a given pattern p (a free tree), we say that p occurs in a logged CN $\mathbf{c}$ or $\mathbf{c}$ supports p if p is isomorphic to a subtree of $\mathbf{c}$. The support of a pattern p is the number of CNs in L_u that supports p. A pattern p is said to be* frequent *if its support, $sup(p)$, is no less than a predefined minimum support* **minsup**. *The problem of mining frequent patterns from query logs is to compute, for a given user u and a log L, the set $\mathcal{P}(L) = \{p | sup(p) \geq minsup\}$.*

The frequent patterns in the query log can be obtained using the algorithms *FreeTreeMiner* described in [15]. *FreeTreeMiner*, which applies the bottom up *Apriori* method [16], first computes all frequent subtrees with 2, 3 and 4 vertices using brute-force method, based on which larger candidate frequent subtrees can be generated by *Apriori*. Each candidate would be checked if it is really frequent. Iteratively executing the steps above, all frequent subtrees would be generated.

Given a keyword query Q over relational database D, the first stage of the schema graph based approach produces the set of *CN*s. These generated *CN*s are simply ranked according to their sizes. For example, DISCOVER adopts the following formula for scoring a *CN* **c**:

$$Score_{SIZE}(\mathbf{c}) = 1/Size(\mathbf{c}) \tag{1}$$

By Equation (1), smaller *CN*s are ranked before larger ones, and ties are broken arbitrarily. In the second stage, the final results are obtained by evaluating the *CN*s. Users get query results ranked by the sizes of their corresponding *CN*s, which could be very different from the users' real needs. As an example, consider the following case. Suppose that a user issues a query *"Markov, LDA"*. The *CN*s (i) $Paper^{Markov} \bowtie Conference \bowtie Paper^{LDA}$ and (ii) $Paper^{Markov} \bowtie Write \bowtie Author \bowtie Write \bowtie Paper^{LDA}$ are included in the results of the *CN* generation step. According to Equation (1), *CN* (i) is ranked higher than (ii) as it has less joins and hence a smaller size. But if the query log contains entries related to this user's past queries, we can take them into consideration when ranking the *CN*s. For example, we consider an extreme case where there is no pattern $Paper \bowtie Conference$ and the support of the pattern $Author \bowtie Write \bowtie Paper$ is very large. Intuitively, *CN* (ii) should be ranked higher than (i) as the preference of the user can be clearly inferred from his/her search history.

The problem we study in this paper, is how to adapt to user preferences through query logs. This can be reduced to the problem of ranking the candidate networks using information from query logs.

4   Ranking with Query Logs

In this section, we discuss how to take user feedback information into consideration when ranking the *CN*s. Before presenting the proposed ranking strategy, let us first take a look at a straightforward way of using feedback information for ranking.

4.1   A First Attempt

A simple method to incorporate the query log information is to assign, for each user, a degree of preference (e.g., $p \in [0, 1]$) for each table in the database based on frequency of that table appearing in the log. The score of a generated *CN* for a given query can be computed by a linear combination of the preference degree of each table involved and the size of the *CN*. However, this does not work in some cases as the score may be dominated by a minority (sometimes even one) of the tables in the *CN*. For example, for the query *"Markov, LDA"*, the candidate network $Paper^{Markov} \bowtie Write \bowtie Author \bowtie Write \bowtie Paper^{LDA}$

may be ranked higher than $Paper^{Markov} \bowtie PaperCitation \bowtie Paper^{LDA}$ if the degree of preference for *Author* is very high for that user. However, although the user has a strong preference for *Author*, it is very likely that for this particular query the user would prefer a pattern in which one paper cites another. In this example, the high preference degree of a single table *Author* has dominated the scoring of the *CN*.

Another reason why this may not work is that in some cases, the join of frequent tables may not be frequent. For example, it is possible that two tables, say *Paper* and *Author*, are both of high frequency, but the join $Paper \bowtie Write \bowtie Author$ may be rare in this log. In this case, any *CN* with this join as a component should not be ranked high despite the high frequency of *Paper* and *Author*. Intuitively, instead of considering the frequency of single tables, we should focus more on the frequency of those "join structures". This leads us to develop the methods described in the sequel.

4.2 Ranking Functions

We seek to augment the scoring function in Equation (1) using the frequent patterns mined from query logs. Let $\mathcal{P}(L_u)$ (or $\mathcal{P}(L)$ when there is no ambiguity) denote the set of frequent patterns mined from log L for a given user (or group of users) u. For a *CN* **c**, we define $FS(\mathbf{c})$ to be the set of frequent subtrees from $\mathcal{P}(L)$ such that each subtree is also part of **c**, i.e., $FS(\mathbf{c}) = \{p|p$ is a subtree of $\mathbf{c} \wedge p \in \mathcal{P}(L)\}$. The set of edges in **c** not covered by $FS(\mathbf{c})$, together with their corresponding vertices, constitute another set denoted by $NFS(\mathbf{c})$. Naturally, $NFC(\mathbf{c})$ and $FS(\mathbf{c})$ do not have any overlapping edges.

We define a *proper partition* $P_{\mathbf{c}}$ of a *CN* **c** to be a complete non-overlapping cover of **c** by a combination of elements from $FS(\mathbf{c})$ and $NSF(\mathbf{c})$ such that

- There is no overlap between any pair of elements; and
- The union of the edges in all of the elements in the combination is equal to the set of edges in **c**.

Obviously, each edge of **c** is contained in exactly one element of the combination. We can assign a score to proper partition $P_{\mathbf{c}}$ as follows.

$$score_{PAR}(P_{\mathbf{c}}) = \sum_{FS_i \in CFS} score(FS_i), \qquad (2)$$

$$score(FS_i) = (Size(FS_i)/Size(\mathbf{c}))^t \cdot N(sup(FS_i)) \qquad (3)$$

where *CFS* denotes the set of frequent subtrees used in the partition and $sup(FS_i)$ is the support of $FS_i \in CFS$. The configurable parameter t $(t > 0)$ is used to control the degree of preference for larger frequent structures, and $N(\cdot)$ is a normalization function to be described next. Notice that elements in $NFS(\mathbf{c})$ do not contribute the scoring of the partition.

In Equation (3), normalization is applied to the support. The support of a frequent pattern mined from the query log ranges from *minsup* to an unknown large number. If the value of $sup(FS_i)$ is too big, the score of the *CN* that contains the subtree FS_i will become unreasonably large. Therefore, normalization must be done to limit the influence of the support value. In particular, when the support is extremely large, its effect should be dampened even more. Based on the above consideration, we use a sigmoid function as a starting point for normalizing the support values, which takes the form of $sigmoid(x) = 1/(1+e^{-\alpha x})$, where the weight parameter α controls the linearity of the curve. In our case, the range of support is $[minsup, +\infty)$. As $minsup$ is greater than zero, the range of the sigmoid function is $(0.5, 1)$. However, the range of Equation (1) is $[0,1]$. So, we have to scale the range of the sigmoid function to the range of $(0,1)$ by the transformation $N(x) = 2 \cdot (sigmoid(x) - 0.5)$.

Note that there may exist many proper partitions for a candidate network. Let $\mathcal{P}_\mathbf{c}$ denote the set of all proper partitions of $\mathbf{c}$, each of which has a corresponding score computed by Equation (2). The largest such score is used as the *query log score*, as indicated in Equation (4). The rule comes from the intuition that we always want to get a partition in which the frequent subtrees have both larger support and larger size. But in fact, smaller patterns in $\mathcal{P}(L)$ have larger supports. Thus, a trade-off is needed. So, a candidate network is assigned the largest combination score as the log score. N is the set of all combination scores.

$$Score_{LOG}(\mathbf{c}) = max\{score_{PAR}(P_\mathbf{c}) | P_\mathbf{c} \in \mathcal{P}_\mathbf{c}\} \tag{4}$$

Finally, we combine the original size-based score and the query log score by weighting as in Equation (5), where λ is the weight of the original score and controls the relative importance between the two parts. If $\lambda = 1$, then the new scoring function is the same as the one used in DISCOVER.

$$Score(\mathbf{c}) = \lambda \cdot Score_{SIZE}(\mathbf{c}) + (1 - \lambda) \cdot Score_{LOG}(\mathbf{c}) \tag{5}$$

To illustrate the advantage of this new scoring function, consider the example we give in Section 3. For the query {*"Markov, LDA"*}, two example *CNs* (i) $Paper^{Markov} \bowtie Conference \bowtie Paper^{LDA}$ and (ii) $Paper^{Markov} \bowtie Write \bowtie Author \bowtie Write \bowtie Paper^{LDA}$ are assigned with an score respectively by our ranking functions. As $P(L)$ has no the pattern $Paper \bowtie Conference$ while the pattern $Author \bowtie Write \bowtie Paper$ has a large support, $S(i) < S(ii)$. Apparently, by incorporating the query log, the generated candidate networks can be ordered emerging user preferences.

Algorithm 1 summarizes the procedure to score and rank all generated candidate networks of a given query using the new scoring function. Since there can exist a large number of proper partitions for a given *CN*, the most complex and time consuming part of this procedure is to compute the largest score of a $\mathbf{c}$ over all such proper partitions.

4.3   Complexity Result

We show that to find the largest score corresponding to the best *proper partition* of a candidate network $\mathbf{c}$ based on the sets $FS(\mathbf{c})$ and $NFS(\mathbf{c})$ is a NP-hard problem.

We first introduce the notations that will be used. Let Ψ be the set of all edges in the *CN* $\mathbf{c}$; $\mathbb{S} = FS(\mathbf{c}) \cup NFS(\mathbf{c})$. We can represent a subtree with the corresponding subset of Ψ. Each element $S \in \mathbb{S}$ has a score $Score(S)$.

Algorithm 1. Scoring generated candidate networks

Require:
 The set of generated candidate networks, $\mathcal{C}$, with size $\leq T_{max}$;
 The set of frequent patterns for the given user (or group of users) mined from query log L, $\mathcal{P}(L)$;
Ensure:
 The list of CNs in descending order by score, l;
1: Initialize list l;
2: **for** each $\mathbf{c} \in \mathcal{C}$ **do**
3: Find the set of frequent patterns in $\mathcal{P}(L)$ that are subtrees of $\mathbf{c}$, i.e., $FS(\mathbf{c})$;
4: Compute, out of all proper partitions $P_{\mathbf{c}}$ of $\mathbf{c}$, the largest $Score_{LOG}(\mathbf{c})$;
5: Compute $\mathbf{c}$'s score using Equation 5;
6: Insert $\mathbf{c}$ into the sorted list l (in descending order of score);
7: **end for**
8: **return** l;

Definition 5. (Best Proper Partition Problem). *With Ψ and $\mathbb{S}$ as input, the best proper partition problem is to find a set $\mathcal{S}^* \subseteq \mathbb{S}$ such that each element of Ψ appears in only one element of $\mathcal{S}^*$ and $\mathcal{S}^*$ maximizes $Score(\mathcal{S}^*) = \sum_{S \in \mathcal{S}^*} score(S)$.*

Our problem can be considered an optimization problem formulated as follows.

- *Instance:* Given a set of elements Ψ, and a set of subsets of Ψ, $\mathbb{S} = \{S_1, S_2, \ldots, S_n\}$, where $S_i \subseteq \Psi$ and has a score w_i.
- *Question:* Find a set $\mathcal{S}^* \subseteq \mathbb{S}$, which ensures that each element of Ψ appears in one and only one element of $\mathcal{S}^*$ and $\sum_{S_i \in \mathcal{S}^*} w_i$ is maximized.

Correspondingly, the decision version of the *best proper partition problem*, the *best proper partition decision problem*, can be formulated as follows.

- *Instance:* Given a set of elements Ψ, and a set of subsets of Ψ, $\mathbb{S} = \{S_1, S_2, \ldots, S_n\}$, where $S_i \subseteq \Psi$ and has a score w_i, and a constant B.
- *Question:* Is there a set $\mathcal{S}^* \subseteq \mathbb{S}$ such that each element of Ψ just in one and only one element of $\mathcal{S}^*$ and $\sum_{S_i \in \mathcal{S}^*} w_i \geqslant B$?

Theorem 1. *The best proper partition problem is NP-hard.*

It is sufficient to prove best proper partition decision problem is NP-Complete. We can apply the *restriction* technique which shows the NP-Completeness of an NP problem by stating that a special case of the problem is NP-Complete. By limiting $B = min\{w_i | S_i \subseteq \mathcal{S}^*\}$, the decision problem can be restricted to the *exact cover problem*, a problem known to be NP-Complete. Then, the decision problem is proved to be NP-Complete. Hence, the *best proper partition problem* is NP-hard.

4.4   A Dynamic Programming Solution

We show that the *best proper partition problem* can be solved by dynamic programming. Assuming that the elements in $\mathbb{S}$ are numbered, we define a set of indicator variables x_i for a given set $\mathcal{S} \subseteq \mathbb{S}$ such that $x_i = 1$ if the i-th element in $\mathbb{S}$ appears in $\mathcal{S}$, and $x_i = 0$ otherwise. Then an indicator vector $(x_1, x_2, \ldots x_{|\mathbb{S}|})$ can be formed.

Our problem can be considered as maximizing the following function with respect to $\mathcal{S}$

$$F(\mathcal{S}) = \sum_{i=1}^{|\mathbb{S}|} x_i \cdot score(S_i)$$

subject to the constraints: (i) $S_i \cap S_j = \varnothing$, if $x_i = 1, x_j = 1, 1 \leq i,j \leq |\mathbb{S}|$, (ii) $\bigcup_{i=1}^{|\mathbb{S}|} x_i S_i = \Psi$ and (iii) $x_i \in \{0, 1\}$, where $score(S_i)$ is calculated by Equation (3). The constraint (i) ensures that for a partition no pairs of subtrees share the same edges, and (ii) makes sure that a partition can cover all the edges in Ψ.

The exists a recursive structure that allows us to use dynamic program to solve the optimization problem.

Theorem 2. *Define $F^*(\mathcal{S})$ as the maximum score for the CN c given the set $\mathcal{S} \subseteq \mathbb{S}$. Then the optimal solution is given by $F^*(\mathcal{S}) = \max_i\{F^*(\mathcal{S} - S_i) + Score(S_i)\}$.*

Based on Theorem 2, we propose a dynamic programming algorithm to compute the optimal partition and its corresponding score. Algorithm 2 performs the dynamic programming and outputs the optimal combination score. The time complexity of the dynamic programming algorithm is $O(2^{|\mathbb{S}|})$.

Algorithm 2. Dynamic Programming Algorithm

Require:
    Ψ; $\mathbb{S}$; CN
Ensure:
    Max: the optimal log score of the candidate network
 1: compute $score(S_i), S_i \in \mathbb{S}$;
 2: $Max = F(\mathbb{S})$;
 3: **return** Max;

 4: **Procedure** $F(\mathcal{S})$
 5: **if** $\mathcal{S} = \varnothing$ **then**
 6:     **return** 0;
 7: **else**
 8:     $Max = \max_{S_i \in \mathcal{S}}\{F(\mathcal{S} - S_i) + score(S_i)\}$;
 9: **end if**
10: **End Procedure**

---

5 Experiments

We conduct experiments to evaluate the proposed dynamic programming algorithm (*Dynamic*) and compare it with existing approaches that do not consider user feedback.

5.1 Dataset and Settings

Due to the lack of publicly available databases with query logs, we use the DBLP database[1] in our experiments and build our own query log through a controlled user study. The DBMS used is MySQL with default configurations. We build indexes for all primary keys and foreign keys. Full-text indexes are built for all

---

[1] http://dblp.uni-trier.de/xml/

textual attributes. The experiments are conducted on a workstation with two 2.33GHz Intel Core2 Duo processors and 2GB of main memory.

The query set comes from a user study. Ten graduate students from different research areas participate in our experiment as query initiators. They formulated 60 "meaningful" queries consisting of varying number of keywords related to their research areas. For each participant, we applied 3-fold cross-validation on his queries. The queries were randomly divided into three sets. In each trial, two folds were used as the training set to generate the query log and the other was testing set. For each query in the training set, with the predefined parameters (such as T_{max}), all of the generated CNs, the number of which may range from tens to hundreds , were presented to the corresponding participant in sequence, in ascending order of size; the participant was asked to choose "yes" or "no" for each CN according to whether that CN meets his/her requirement. The "yes" CNs were recorded in the query log. Up to here, each participant had his own query log in each trail. We set $minsup$=10.By the settings above, each participant has more than 200 frequent subtrees on average and the corresponding supports range from 10 to about 300. For each query in testing set, all the generated CNs were ranked by our methods and presented to the participant. The participant assessed the result quality using a six-point scale ranging from 0 to 5 (5="perfect" and 0="bad").

For the algorithm *Dynamic*, we conduct experiments to evaluate the performance of them by DBLP database. The parameters that involve in the experiments are illustrates in Table 1 with explanation. And Table 2 shows some sample queries from a user. A user is required to issue queries that are reasonable. Generally, a query contains at least two keywords and no more than four.

Table 1. Parameter under investigation

λ	weight of the original score
t	the preference for larger structure in a CN
T_{max}	the maximum allowed CN size
α	the Sigmoid function parameter
K	top-K results

Table 2. Query example

Q_1:bender, p2p	Q_6:Hardware, luk, wayne
Q_2:sigmod, xiaofang	Q_7:Ishikawa, P2P, Yoshiharu
Q_3:fagin, middleware	Q_8:hongjiang, Multimedia, zhang
Q_4:Owens, VLSI	Q_9:vldb, xiaofang
Q_5:p2p, Steinmetz	Q_{10}:intersection, nikos

5.2 Effectiveness

To measure the effectiveness, we adopt four metrics, namely, *Normalized Discounted Cumulative Gain* (NDCG), *Precision at K* (P@K), *11-point Precision/Recall* and *F-measure*. Each is described in detail as follows.

- **NDCG at K:** For a given query q in the testing set, the ranked candidate networks are assessed manually to compute NDCG@K. The motivation of using NDCG@K is to pay more attention to the top-ranked results.

- **Precision at K:** P@K shows the fraction of the candidate networks ranked in top K results that are preferred by the user. In our settings, we define that a candidate network assessed with 3 or larger is preferred. The position of preferred candidate networks within top K is unconcerned. As the most intuitive metric, P@K measures the overall user satisfaction with the top K results.
- **11-point Precision/Recall:** For a query result, this metric reports the precision that is measured at the 11 recall levels of 0.0, 0.1, 0.2, ..., 1.0. In our experiments, 11-pt Precision/Recall is the average result for all the testing queries.
- **F-measure:** Given K, we can compute the precision and the recall at K. Here, the candidate networks with 3 or larger points are preferred. F-measure is the harmonic mean of precision and recall.

Effect of Parameter t, α and λ of *Dynamic*. Set $T_{max} = 7$ and K=10. Table 3 shows the experiment results. Each line in Table 3 indicates that when fix the value of t and then vary α and λ, the best setting of α and λ and the corresponding result are presented (we do not show other settings here which only generate worse results). And we can conclude that when t (preference to larger frequent subtrees) is 4, NDCG@10 reaches the best value 0.890 with $\alpha = 0.01$ and $\lambda = 0.1$.

Table 3. NDCG@10 for varying t, α and λ

t	2	3	4	5	6
α	0.07	0.09	0.01	0.025	0.001
λ	0.7	0.5	0.1	0.1	0.3
NDCG@10	0.836	0.880	**0.890**	0.876	0.873

Effectiveness Comparison for DISCOVER, *Dynamic*. We use DIS-COVER as a baseline here as its scoring function directly corresponds to the "original score" part of the proposed new strategy. Fig. 4 shows that incorporating query log results in significant improvements over DISCOVER. In Fig. 4(a) ($T_{max} = 7$), when K increases, *Dynamic* always outperforms DISCOVER by a considerable margin. This can also be seen in Fig. 4(b) which shows the effect of K on Precision. Now, we vary T_{max} as Fig. 4(c) (K=10). The gap between our proposed algorithms and DISCOVER remains similar; meanwhile, each of them progresses with a downward trend when T_{max} increases.

Overall Effectiveness Comparison for DISCOVER, *Dynamic*. Figure 5(a) ($T_{max} = 7$) shows the 11-points precision/recall graph for DISCOVER, *Dynamic*, in which the precision goes down with recall growing. In the global perspective, *Dynamic* behaves well with points (0.1,0.988), ..., (0.9,0.230). DIS-COVER takes the worst performance. Meanwhile, as an auxiliary, Figure 5(b) ($T_{max} = 7$) presents the F-measure value by varying K. The three preserve some differences as in the previous case.

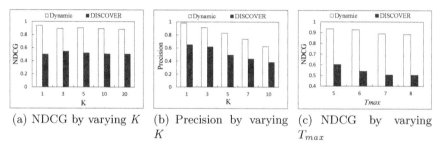

(a) NDCG by varying K (b) Precision by varying (c) NDCG by varying
                           K T_{max}

Fig. 4. Effectiveness comparison for DISCOVER, *Dynamic*

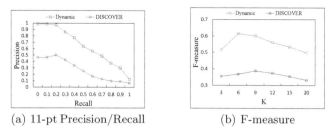

(a) 11-pt Precision/Recall   (b) F-measure

Fig. 5. Overall effectiveness comparison for DISCOVER, *Dynamic*

6   Conclusion

Existing work on keyword search in databases has considered the problem of improving the search effectiveness extensively. However, few work has explicitly taken user preferences into consideration when ranking the query results. In this paper, by introducing user feedback to the problem of ranking candidate networks, we have proposed a new ranking strategy to adapt to user preferences. As this new ranking strategy involves a NP-hard problem, we provide the Dynamic Programming algorithm. We have evaluated the proposed strategy by the DBLP dataset through a user study, which verifies the effectiveness of our strategy.

Acknowledgments. This work was supported in part by the 973 Program (2015CB352502), the National Natural Science Foundation of China Grant (61272092), the Shandong Provincial Natural Science Foundation Grant (ZR2012FZ004), the Science and Technology Development Program of Shandong Province (2014GGE27178), the Taishan Scholars Program, and NSERC Discovery Grants. The authors would like to thank the anonymous reviewers, whose valuable comments helped improve this paper.

References

1. Hristidis, V., Papakonstantinou, Y.: DISCOVER: keyword search in relational databases. In: VLDB (2002)
2. Luo, Y., Lin, X., Wang, W., Zhou, X.: SPARK: top-k keyword query in relational databases. In: SIGMOD, pp. 115–126 (2007)

3. Hulgeri, A., Nakhe, C.: Keyword searching and browsing in databases using banks. In: ICDE (2002)
4. He, H., Wang, H., Yang, J., Yu, P.S.: Blinks: ranked keyword searches on graphs. In: SIGMOD, pp. 305–316 (2007)
5. Hristidis, V., Gravano, L., Papakonstantinou, Y.: Efficient IR-style keyword search over relational databases. In: VLDB, pp. 850–861 (2003)
6. Liu, F., Yu, C., Meng, W., Chowdhury, A.: Effective keyword search in relational databases. In: SIGMOD, pp. 563–574 (2006)
7. Agrawal, S., Chaudhuri, S., Das, G.: DBXplorer: a system for keyword-based search over relational databases. In: ICDE (2002)
8. Kacholia, V., Pandit, S., Chakrabarti, S., Sudarshan, S., Desai, R., Karambelkar, H.: Bidirectional expansion for keyword search on graph databases. In: VLDB, pp. 505–516 (2005)
9. Yu, X., Shi, H.: CI-Rank: ranking keyword search results based on collective importance. In: ICDE (2012)
10. Ganti, V., He, Y., Xin, D.: Keyword++: A framework to improve keyword search over entity databases. VLDB 3(1–2), 711–722 (2010)
11. Markowetz, A., Yang, Y., Papadias, D.: Keyword search on relational data streams. In: SIGMOD (2007)
12. Gao, L., Yu, X., Liu, Y.: Keyword query cleaning with query logs. In: Wang, H., Li, S., Oyama, S., Hu, X., Qian, T. (eds.) WAIM 2011. LNCS, vol. 6897, pp. 31–42. Springer, Heidelberg (2011)
13. Peng, Z., Zhang, J., Wang, S., Wang, C.: Bring user feedback into keyword search over databases. In: Proc. of the 3rd Workshop on Electronic Government Technology and Application, pp. 210–214 (2009)
14. Zeng, Z., Bao, Z., Ling, T.W., Lee, M.L.: iSearch: an interpretation based framework for keyword search in relational databases. In: KEYS, pp. 3–10 (2012)
15. Chi, Y., Yang, Y., Muntz, R.: Indexing and mining frequent subtrees. In: ICDE (2003)
16. Agrawal, R., Srikant, R.: Fast algorithms for mining association rules in large databases. In: VLDB, pp. 487–499 (1994)

cluTM: Content and Link Integrated Topic Model on Heterogeneous Information Networks

Qian Wang[1], Zhaohui Peng[1]([✉]), Senzhang Wang[2], Philip S. Yu[3,4],
Qingzhong Li[1], and Xiaoguang Hong[1]

[1] School of Computer Science and Technology, Shandong University, Jinan, China
wangqian8636@gmail.com, {pzh,lqz,hxg}@sdu.edu.cn
[2] School of Computer Science and Engineering, Beihang University, Beijing, China
szwang@buaa.edu.cn
[3] Department of Computer Science, University of Illinois at Chicago, Chicago, USA
psyu@uic.edu
[4] Institute for Data Science, Tsinghua University, Beijing, China

Abstract. Topic model is extensively studied to automatically discover the main themes that pervade a large and unstructured collection of documents. Traditional topic models assume the documents are independent and there are no correlations among them. However, in many real scenarios, a document may be interconnected with other documents and objects, and thus form a text related heterogeneous network, such as the DBLP bibliographic network. It is challenging for traditional topic models to capture the link information associated to diverse types of objects in such a network. To this end, we propose a unified Topic Model cluTM by incorporating both the document content and various links in the text related heterogeneous network. cluTM combines the textual documents and the link structures by the proposed joint matrix factorization on both the text matrix and link matrices. Joint matrix factorization can derive a common latent semantic space shared by multi-typed objects. With the multi-typed objects represented by the common latent features, the semantic information can be therefore largely enhanced simultaneously. Experimental results on DBLP datasets demonstrate the effectiveness of cluTM in both topic mining and multiple objects clustering in text related heterogeneous networks by comparing against state-of-the-art baselines.

1 Introduction

As a powerful way to discover the hidden semantics of the document collection, topic models are extensively studied and successfully applied to many text mining tasks, such as information retrieval [1] and document clustering [2]. Traditional topic models assume that the documents are independent and do not consider the correlation among them. With the flourish of Web application, textual documents such as papers, blogs and product reviews, are not only getting richer, but also interconnecting with other objects like users in various ways; and

© Springer International Publishing Switzerland 2015
J. Li and Y. Sun (Eds.): WAIM 2015, LNCS 9098, pp. 207–218, 2015.
DOI: 10.1007/978-3-319-21042-1_17

therefore form text related heterogeneous information networks [3]. Take the bibliographic data shown in Figure 1 as an example. There are three types of objects, papers, authors, and venues in such a heterogeneous network. These objects form two types of relationships: the *author-write-paper* relationship between authors and papers, and the *venue-publish-paper* relationship between venues and papers. It is challenging for traditional topic models to capture the rich information, especially the link information in such a text related heterogeneous information network.

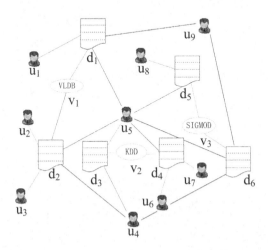

Fig. 1. The bibliographic heterogeneous network with three types of objects: papers, authors, venues and two types of links: *author-write-paper*, *venue-publish-paper*

Traditional topic models, such as latent semantic analysis (LSA) [4], probabilistic latent semantic analysis (PLSA) [5], and latent dirichlet allocation (LDA) [6] focus on purely utilizing the textual information to discover topics with the assumption that the documents are *i.i.d.* (independent and identical distributed). With the explosive of interconnected textual contents with rich link information, the assumption may not hold and traditional models become less effective. Although some attempts, such as LaplacianPLSI [7], NetPLSA [8], and iTopicmodel [9] have been conducted to combine topic modeling with link information in a homogeneous network, how to integrate various types of links associated to different types of objects into a unified topic model is still less studied.

Although the links among documents as well as other types of objects might be helpful for analyzing text, it is non-trivial to handle the rich heterogeneous information in a unified framework. First, it is challenging to model the semantic information of links such as the *author-write-paper* and *venue-publish-paper* relationships. Different from text, link structure is a totally different type of information and can not be easily added to traditional topic models in a straightforward manner. Second, there are usually several types of different objects in a heterogeneous information network. Different types of objects may have their own inherent information and should be treated differently. How to use the different types of objects and integrate them in a unified way with the textual information also makes the studied problem challenging.

In this paper, we propose a unified topic model named cluTM by incorporating the heterogeneous link information into topic modeling. cluTM learns a latent semantic space by jointly factorizing the document-phrase matrix and the link matrices with latent semantic analysis. The basic idea is that the textual documents and link information in the heterogeneous information networks have similar latent semantic features. For example, in the bibliographic data, a paper contains several topics. Likewise, the researchers and venues also have their preferred research topics associated to related papers. The inherent connections between contents and links can be therefore constructed by assuming that the text matrix and link matrices share the same latent semantic features. With such an assumption, all the objects in the heterogeneous information network are projected into a unified latent semantic space based on the common latent semantic features. In the unified latent semantic space, each object is represented as a vector. Topics of documents and clusters of other types of objects can be easily obtained by calculating the similarity of the vectors.

We summarize the main contributions of this paper as follows:

- study the novel problem of topic mining and multi-objects clustering simultaneously in a text related heterogeneous information network;
- propose a unified topic model to seamlessly integrate the content of textual documents and links by joint matrix factorization;
- extensive experiments on DBLP dataset show the effectiveness of the proposed model on topic modeling and object clustering by comparing against traditional topic models.

The rest of this paper is organized as follows. Section 2 introduces some basic concepts. We elaborate cluTM in Section 3. Section 4 presents the extensive experiment results. We discuss the related work in section 5 and finally conclude our work in section 6.

2   Preliminaries

In this section, we formally introduce several related concepts and notations to help us state the problem.

Definition 1. Information Network [3]. *Given a set of objects from K types $\mathcal{X} = \{X_k\}_{k=1}^{K}$, where X_k is a set of objects belonging to the k_{th} type, a graph $G =< V, E >$ is called an information network on objects $\mathcal{X}$, if $V = \mathcal{X}$, and E is a binary relation on V. Specifically, we call such an information network* **heterogeneous information network** *when $K \geq 2$.*

Definition 2. Text Information Network. *An information network $G =< V, E >$ with K types of objects is called a text information network if there* *exists at least one type of text object in the network, i.e. $\exists X_k \in \mathcal{X}$ that the type of X_k is text. Specifically, we call a text information network **heterogeneous text information network** when $K \geq 2$.*

DBLP Bibliographic Network Example. We use the DBLP bibliographic network as an example to illustrate the heterogeneous text information network. As shown in Figure 1, there are three types of objects, i.e., authors A, venues VE and papers D, and two types of links among papers, authors, and venues. The type of paper object is text. The bibliographic network can be denoted as $G = (D \cup A \cup VE, E)$, where E is a set of edges that describe the relationships between papers $D = \{d_1, ..., d_n\}$, authors $A = \{a_1, ..., a_l\}$ as well as venues $VE = \{ve_1, ..., ve_o\}$.

In our model, each topic can be represented as a set of meaningful frequent phrases [10], definited as follows.

Definition 3. *Meaningful Frequent Phrases Meaningful frequent phrases are defined as the phrases that capture the main themes of the document collection. Meaningful frequent phrases lay a foundation for the readability of the discovered topics. They can be represented as $MFP = \{mfp_1, mfp_2, ..., mfp_M\}$, where mfp_m denoting the m_{th} meaningful frequent phrase.*

3   cluTM: Incorporating Text and Links in a Unified Framework

We will first revisit the classic LSA model that is widely used to discover topics of document by matrix factorization. Motivated by LSA model, we next introduce how to conduct the matrices factorization on the *document-author* matrix and *document-venue* matrix. Finally, we elaborate how to combine the content and link information by joint matrix factorization with a assumption that these matrices share the same latent semantic space.

3.1   LSA on Document-Phrase Matrix

We use the classic LSA model to discover the latent topics of documents. The key idea of LSA model is to project documents as well as terms into a relatively low dimensional vector space, namely the latent semantic space, and produce a set of topics associated with documents [4].

In our model, documents are represented as a bag of meaningful frequent phrases. Consider the analysis of document-phrase matrix $M_{D-MFP} \in R^{n \times m}$, and it is a sparse matrix whose rows represent documents, and columns represent phrases, where n is the number of documents and m is the number of meaningful frequent phrases. Singular vector decomposition [12] is performed on matrix M_{D-MFP} as follows:

$$M_{D-MFP} = U_{D-MFP}\Sigma_{D-MFP}V_{D-MFP}^T \tag{1}$$

where U_{D-MFP} and V_{D-MFP} are orthogonal singular matrices U_{D-MFP}^T $U_{D-MFP} = V_{D-MFP}^T V_{D-MFP} = I$ (I is the identity matrix) and Σ_{D-MFP} is a diagonal matrix containing the singular values of M_{D-MFP}.

Given an integer k ($k << rank(M_{D-MFP})$), LSA only remains the first k singular vectors, $U_{D-MFP} \in R^{n \times k}$, $V_{D-MFP} \in R^{m \times k}$, and sets all but the largest k singular values to zero, $\Sigma_{D-MFP} \in R^{k \times k}$. The matrix $Y = U_{D-MFP} \Sigma_{D-MFP}$ ($Y \in R^{n \times k}$) defines a new representation of documents that each column corresponds to a topic and each row is a k-dimensional vector representing the weights of a document in the k topics. Therefore, the LSA approximation of M_{D-MFP} can be obtained by $Y V_{D-MFP}^T$.

This can be transformed into an optimization problem that aims to approximate matrix M_{D-MFP} with $Y V_{D-MFP}^T$ as follows,

$$min \, ||M_{D-MFP} - Y V_{D-MFP}^T||_F^2 + \gamma_1 ||V_{D-MFP}||_F^2 \qquad (2)$$

where $|| \cdot ||_F$ is the Frobenius norm, γ_1 is the parameter, $\gamma_1 ||V_{D-MFP}||_F^2$ is a regularization term to improve the robustness. The $i-th$ row vector of Y can be considered as the latent semantic feature vector of document d_i.

3.2   Link Matrices Factorization

Taking the bibliographic network in Figure 1 as an example again, the relationships between papers and authors as well as papers and venues can be represented by link matrices $M_{D-A} \in R^{n \times l}$ and $M_{D-VE} \in R^{n \times o}$, respectively. l is the number of authors and o is the number of venues. In LSA model, a document contains several topics with each topic associated with a set of frequently used terms. Likewise, an author also has several preferred research topics with each research topic associated with a set of related papers. If we consider the authors and papers as documents and words respectively, we can use the similar idea to LSA to analyze the latent semantic of the $author - paper$ link. Motivated by above idea, the link matrices M_{D-A} can also be factorized by SVD as follows,

$$M_{D-A} = U_{D-A} \Sigma_{D-A} V_{D-A}^T \qquad (3)$$

where U_{D-A} and V_{D-A} are orthogonal matrices $U_{D-A}^T U_{D-A} = V_{D-A}^T V_{D-A} = I$ and the diagonal matrix Σ_{D-A} contains the singular values of M_{D-A}.

Likewise, each venue prefers to accept papers of some particular research topics. The latent topics of a venue preferring can be obtained by factorizing the document-venue matrix using SVD as follows,

$$M_{D-VE} = U_{D-VE} \Sigma_{D-VE} V_{D-VE}^T \qquad (4)$$

where U_{D-VE} and V_{D-VE} are orthogonal matrices $U_{D-VE}^T U_{D-VE} = V_{D-VE}^T V_{D-VE} = I$ and the diagonal matrix Σ_{D-VE} contains the singular values of M_{D-VE}.

Similar to LSA model, we also only keep the first k singular vectors and set the other singular values to zero. For the matrix M_{D-A} and M_{D-VE}, we use the matrix $Y_{D-A} = U_{D-A}\Sigma_{D-A}$ and $Y_{D-VE} = U_{D-VE}\Sigma_{D-VE}$ to represent the document respectively. Thus the matrices M_{D-A} and M_{D-VE} can be represented as follows

$$M_{D-A} \approx Y_{D-A}V_{D-A}^T \qquad (5)$$

$$M_{D-VE} \approx Y_{D-VE}V_{D-VE}^T \qquad (6)$$

where V_{D-A} is a $l \times k$ matrix, V_{D-VE} is a $o \times k$ matrix, and Y_{D-A}, Y_{D-VE} are the latent semantic feature matrices. Each column of Y_{D-A} and Y_{D-VE} represents a topic and each row is $k-$dimensional vector representing the weights of a document in the k topics. Therefore, Y_{D-A} and Y_{D-VE} are very similar to the matrix Y. In our model, to combine the content of textual documents and link information in the heterogeneous text information network, we assume that they share the latent semantic feature Y, i.e. $Y_{D-A} = Y_{D-VE} = Y$.

3.3   Combing Content and Link by Joint Matrix Factorization

Based on the assumption discussed above, the document-phrase matrix M_{D-MFP} and link matrices M_{D-A}, M_{D-VE} are connected by the latent semantic feature Y, that is, the latent feature for content is tied to the latent feature for links. Our model aims to find a latent semantic feature Y that best explains the semantic captured by M_{D-MFP} and M_{D-A}, M_{D-VE} simultaneously. Furthermore, different types of objects and links reflect distinctive semantics of a heterogeneous text information network, so they should be treated differently. To achieve these goals, we propose a joint matrix factorization framework to fuse them into such an unified optimization problem,

$$\begin{aligned}
&minJ(Y, V_{D-MFP}, V_{D-A}, V_{D-VE}) \\
&=min\{\lambda(||M_{D-MFP} - YV_{D-MFP}^T||_F^2 + \gamma_1||V_{D-MFP}||_F^2) \\
&\quad + \alpha(||M_{D-A} - YV_{D-A}^T||_F^2 + \gamma_2||V_{D-A}||_F^2) \\
&\quad + \beta(||M_{D-VE} - YV_{D-VE}^T||_F^2 + \gamma_3||V_{D-VE}||_F^2)\}
\end{aligned} \qquad (7)$$

where λ, α and β ($\lambda > 0$, $\alpha > 0$, $\beta > 0$) are parameters to balance the relative importance of document-phrase matrix M_{D-MFP} and link matrices M_{D-A}, M_{D-VE}. We set a constraint $\lambda + \alpha + \beta = 1$. γ_1, γ_2 and γ_3 are regularization parameters that improve the robustness. V_{D-MFP}, V_{D-A} and V_{D-VE} are $m \times k$, $l \times k$, and $o \times k$ matrix respectively. Y is a $n \times k$ matrix. Note that if $\alpha = 0$, $\beta = 0$, thus $\lambda = 1$, the unified topic model boils down to the LSA model on document-phrase matrix.

The optimization problem aims to simultaneously approximate M_{D-MFP}, M_{D-A}, M_{D-VE} by YV_{D-MFP}^T, YV_{D-A}^T, YV_{D-VE}^T respectively, a product of two low-dimensional matrices with regularizations. The joint optimization illustrated in Eq.7 can be solved by using the standard Conjugate Gradient (CG) method. The gradients for the object function J are computed as follows:

$$\frac{\partial J}{\partial V_{D-MFP}} = \lambda(V_{D-MFP}Y^TY - M_{D-MFP}^TY) + \lambda\gamma_1 V_{D-MFP} \qquad (8)$$

$$\frac{\partial J}{\partial V_{D-A}} = \alpha(V_{D-A}Y^TY - M_{D-A}^TY) + \alpha\gamma_2 V_{D-A} \tag{9}$$

$$\frac{\partial J}{\partial V_{D-VE}} = \beta(V_{D-VE}Y^TY - M_{D-VE}^TY) + \beta\gamma_3 V_{D-VE} \tag{10}$$

$$\begin{aligned}\frac{\partial J}{\partial Y} =& \lambda(YV_{D-MFP}^TV_{D-MFP} - M_{D-MFP}V_{D-MFP}) \\ &+ \alpha(YV_{D-A}^TV_{D-A} - M_{D-A}V_{D-A}) \\ &+ \beta(YV_{D-VE}^TV_{D-VE} - M_{D-VE}V_{D-VE})\end{aligned} \tag{11}$$

The new optimal latent semantic feature Y is to capture both the document-phrase matrix M_{D-MFP} and the link matrices M_{D-A}, M_{D-VE} in the heterogeneous text information network.

A unified latent semantic space can be constructed based on the obtained optimal latent semantic feature Y. All the objects in the heterogeneous information network are projected into the unified latent semantic space in which each paper, meaningful frequent phrase, author and venue is represented by a k-dimensional vector. According to the similarity calculation of vectors, we can get the topics. Analogously, the author clusters and venue clusters also can be obtained by similarity calculation.

4   Evaluations

In this section, we evaluate cluTM on the real dataset. First, we introduce the experiment setup, including the dataset and evaluation metric. Then we show the experimental results from the following three aspects: case study, parameters analysis, and quantitive comparison with baselines.

4.1   Dataset and Metric

We evaluate cluTM on the Digital Bibliography and Library Project (DBLP) dataset. In our experiments, we select papers from DBLP of four research areas, i.e. database (DB), data mining (DM), information retrieval (IR) and artificial intelligence (AI). The selected dataset contains 1200 papers, 1576 authors and 8 conferences. We extract 1660 meaningful frequent phrases from these papers. The heterogeneous text information network of this dataset contains three types of objects: papers, authors and venues, and two types of links: *paper-author* link and *paper-venue* link. There are 3139 *paper-author* links and 1200 *paper-venue* links in total. Link matrices M_{D-A}, M_{D-VE} are constructed from the heterogeneous text information network, and the element value in matrix M_{D-MFP} is obtained by using the $tf - idf$ weight of the phrases. As we select the papers from four research areas, we set the number of topics k to be 4.

For a quantitative evaluation, we use F1-measure as the metric. In our experiments, there are four topic clusters. For each topic cluster, we calculate the Precision and Recall with regard to each given category. Specifically, for the obtained

Table 1. The topic representation generated by cluTM

Topic 1	Topic 2	Topic 3	Topic 4
database systems	data mining	information retrieval	artificial intelligence
database management	data analysis	language models	machine learning
relational databases	data clustering	web search	knowledge-based algorithms
data integration	classification algorithms	learn to rank	knowledge based systems
distributed databases	knowledge discovery	search engine	expert systems
query processing	mining problems	keyword search	pattern recognition
distributed computing	rule learning	document retrieval	knowledge engineering
query optimization	pattern matching	semantic search	user interface

cluster label j and the true cluster label i, the precision can be calculated by $Precision(i,j) = \frac{n_{ij}}{n_j}$, and the recall can be calculated by $Recall(i,j) = \frac{n_{ij}}{n_i}$, where n_{ij} is the number of members of category i in cluster j, n_i is the number of members in the given category i, and n_j is the number of members in cluster j. Based on precision and recall, the F1-measure of cluster j and i can be calculated by

$$F1(i,j) = \frac{2 \times Precision(i,j) \times Recall(i,j)}{Precision(i,j) + Recall(i,j)}. \tag{12}$$

The F1-measure of the whole clustering results is defined as a weighted sum over all the categories as follows: $F1 = \sum_i \frac{n_i}{n} max_j F1(i,j)$.

4.2 Experimental Results

We first analyze the topic modeling results with case studies. Then we discuss the effect of parameters on performance. Finally, experiments are conducted to compare the performance of object clustering with different models.

Table 2. Topics discovered by PLSA and TMBP-Regu

Topics discovered by PLSA				Topics discovered by TMBP-Regu			
Topic 1	Topic 2	Topic 3	Topic 4	Topic 1	Topic 2	Topic 3	Topic 4
data	data	information	problem	data	database	information	learning
database	mining	retrieval	algorithm	database	mining	web	based
systems	learning	web	paper	query	algorithm	retrieval	knowledge
query	based	based	reasoning	databases	clustering	search	model
system	clustering	learning	logic	systems	classification	based	problem
databases	classification	knowledge	based	queries	based	text	reasoning
management	algorithm	text	time	system	algorithms	language	system
distributed	image	search	algorithm	processing	rules	user	logic

Topic Analysis with Case Study. In our model, we set parameters $\lambda = 0.6$, $\alpha = 0.3$, and $\beta = 0.1$ due to the better performance based on our empirical experiment results. The topic modeling results are shown in Table 1. Each discovered topic is represented as a set of meaningful frequent phrases.

PLSA [5] and TMBP-Regu [11] are selected as baselines. The most representative terms generated by PLSA and TMBP-Regu on the DBLP dataset are shown in Table 2. Compared with the results in Table 2, the results shown in Table 1 is easier to understand the meanings of the four topics by meaningful frequent phrases, i.e., "database systems", "data mining", "information retrieval", and "artificial intelligence". cluTM and TMBP-Regu achieve better performance than PLSA by considering the heterogeneous text information network.

For the first three topics discovered by PLSA and TMBP-Regu, although different algorithms select different terms, all these terms can reveal the topics to some extent. For Topic 4, the topics such as "artificial intelligence", derived from cluTM is obviously better than the terms "problem, algorithm, paper" derived by PLSA and "learning, based, knowledge" derived by TMBP-Regu. Therefore, from the view of readability of the topics, cluTM is better than PLSA and TMBP-Regu by representing the topics by a set of meaningful frequent phrases.

Parameter Analysis. In our model, there are three essential parameters, λ, α and β in joint matrix factorization. In this section, we study the effect of these parameters on the performance of the proposed cluTM.

Table 3. The effect of parameters on paper, author, and venue

F1-measure		α (Paper)									
		0	0.1	0.2	0.3	0.4	0.5	0.6	0.7	0.8	0.9	1
λ	0	0.1213	0.1235	0.1268	0.1306	0.1339	0.1373	0.1410	0.1452	0.1436	0.1367	0.1305
	0.2	0.2555	0.2561	0.2627	0.2692	0.2747	0.2783	0.2906	0.2869	0.2724	-	-
	0.4	0.5212	0.5263	0.5408	0.5621	0.5881	0.5840	0.5516	-	-	-	-
	0.6	0.7098	0.7417	0.7544	**0.7857**	0.7336	-	-	-	-	-	-
	0.8	0.5406	0.5538	0.5511	-	-	-	-	-	-	-	-

F1-measure		α (Author)									
		0	0.1	0.2	0.3	0.4	0.5	0.6	0.7	0.8	0.9	1
λ	0	0.2317	0.2365	0.2472	0.2521	0.2598	0.2636	0.2684	0.2739	0.2691	0.2619	0.2508
	0.2	0.4256	0.4310	0.4539	0.4623	0.4807	0.4885	0.4982	0.4914	0.4749	-	-
	0.4	0.5815	0.5872	0.6138	0.6294	0.6581	0.6563	0.6226	-	-	-	-
	0.6	0.7501	0.7863	0.7949	**0.8332**	0.8058	-	-	-	-	-	-
	0.8	0.6274	0.6500	0.6388	-	-	-	-	-	-	-	-

F1-measure		α (Venue)									
		0	0.1	0.2	0.3	0.4	0.5	0.6	0.7	0.8	0.9	1
λ	0	0.3122	0.3243	0.3294	0.3361	0.3459	0.3536	0.3623	0.3789	0.3685	0.3544	0.3206
	0.2	0.5209	0.5262	0.5337	0.5462	0.5629	0.5813	0.5896	0.5735	0.5572	-	-
	0.4	0.6047	0.6231	0.6475	0.6602	0.6828	0.6893	0.6579	-	-	-	-
	0.6	0.7936	0.8315	0.8520	**0.8668**	0.8476	-	-	-	-	-	-
	0.8	0.6418	0.6596	0.6302	-	-	-	-	-	-	-	-

As mentioned in section 3.3, these parameters are used to balance the relative importance of document-phrase matrix M_{D-MFP}, link matrices M_{D-A} and

M_{D-VE}. When $\alpha = 0$, $\beta = 0$, the joint regularization framework boils down to the LSA model. Since $\lambda + \alpha + \beta = 1$, we vary λ from 0 to 1 by step 0.2 and α from 0 to 1 by step 0.1 respectively. Tables 3 report the results with the varied parameter values.

Tables 3 show that the best performance is obtained with $\lambda = 0.6$, $\alpha = 0.3$, thus $\beta = 0.1$. When $\lambda < 1$, the joint regularization framework takes into account both textual documents and links in the heterogeneous text information network. We observe that the performance is improved over the LSA model ($\lambda = 1$) when incorporating link information. One can also observe that the *document−author* matrix M_{D-A} is more important than the *document − venue* matrix M_{D-VE} in the joint regularization framework by the different values of α, β. Note that with the decrease of λ, the performance becomes worse and even worse than the standard LSA. This is mainly because cluTM relies more on the topic consistency between the content of textual documents and links while ignores the intrinsic topic of the textual documents. Due to the superior performance, we empirically set $\lambda = 0.6$, $\alpha = 0.3$, $\beta = 0.1$ in the following experiments.

Clustering Performance Comparison of Objects. We apply cluTM on the task of object clustering. The discovered topics can also be regarded as clusters. We can obtain the clustering results of other objects similarly.

The proposed cluTM is compared with the following two state-of-the-art baselines: latent semantic analysis (LSA), and LSA-PTM [10]. Table 4 reports the clustering performance comparison on different methods.

Table 4. Clustering performance comparasion

Metric	F1-measure			
Object	Paper	Author	Venue	Average
LSA	0.5359	0.6040	0.6990	0.6130
LSA-PTM	0.7535	0.7861	0.8136	0.7844
cluTM	0.7857	0.8332	0.8668	0.8286

For the DBLP data, cluTM and LSA-PTM cluster all types of objects in different groups by considering both the textual documents and the link information. As one can see, both cluTM and LSA-PTM achieve better performance than LSA. This shows that integrating the heterogeneous network structures into topic modeling does help us better cluster the objects. Meanwhile, compared with LSA-PTM, cluTM is consistently better on all the three types of objects. This is mainly because LSA-PTM combines the textual content and heterogeneous network structures as two independent stages, while cluTM combines the textual documents and the heterogeneous network structures into a joint regularization framework such that they can mutually enhance each other.

5   Related Work

Topic modeling is an unsupervised approach to automatically discover the latent semantic of document collections. It has attracted a lot of attention in multiple types of text mining tasks, such as information retrieval [1], geographical topic discovery [13], topic level information diffusion modeling in social media [18].

Many topic models, such as latent semantic analysis (LSA) [4], probabilistic latent semantic analysis (PLSA) [5] and Latent Dirichlet Allocation (LDA) [6] have been successfully applied or extended to many data analysis problems, including document clustering and classification [7,14], author-topic modeling [15,16]. However, most of these models merely consider the textual documents while ignore the network structures. Several proposed topic models, such as LaplacianPLSI [7], NetPLSA [8] and iTopicmodel [9] have combined topic modeling and network structures, but they only emphasize on the homogeneous networks, such as document network and co-authorship network. Recent study [10] integrates the heterogeneous network structures into topic modeling, however, it combines the textual documents and the heterogeneous network structures as two independent stages. Our model combines the textual documents and heterogeneous network structures into a joint regularization framework in which the textual content analysis and heterogeneous network analysis can mutually enhance each other. Experimental results prove the effectiveness of our model.

Link analysis has been a hot topic for a few years since the advent of Pagerank and HITS. Many techniques have been proposed to analyze the heterogeneous networks. For example, [17] proposed a Co-HITS algorithm for bipartite graph analysis. Graph-based methods have been widely and successfully applied in data mining and information retrieval, such as text classification [14], and document re-ranking [19]. However, most of existing work treats different objects uniformly. Our work is different from them, as we focus on heterogeneous information networks and propose a joint regularization framework, in which different types of objects are treated in a different way.

6   Conclusion

In this paper, we proposed a unified topic model cluTM to effectively discover topics of documents and cluster objects of various types simultaneously on heterogeneous text information networks. cluTM first conducted latent semantic analysis on the content of textual documents and factorized the link matrices of objects by SVD separately; then fused all the matrices into a single, compact feature representation by joint matrix factorization to find the common latent feature. By projecting all the objects in the heterogeneous text information networks into the unified latent semantic space, topics of documents and clusters of other objects could. be finally obtained by calculating their similarity. We evaluated cluTM on DBLP bibliographic dataset against several state-of-the-art baselines. Experimental results showed the effectiveness of cluTM.

Acknowledgments. This work was supported in part by US NSF (Grant No.CNS-1115234), NSFC (Grant No.61170052, No.61100167), and the Natural Science Foundation of Shandong Province of China (Grant No.ZR2013FQ009).

References

1. Wei, X., Croft, W.B.: LDA-based document models for ad-hoc retrieval. In: SIGIR, pp. 178–185 (2006)
2. Xie, P.T., Xing, E.P.: Integrating document clustering and topic modeling. In: UAI, pp. 694–703 (2013)
3. Sun, Y.Z., Han, J.W., Yan, X., Yu, P.S.: Mining knowledge from interconnected data: a heterogeneous information network analysis approach. In: VLDB Endowment, pp. 2022–2023 (2012)
4. Deerwester, S.C., Dumais, S.T., Landauer, T.K., Furnas, G.W., Harshman, R.A.: Indexing by latent semantic analysis. Journal of the American Society for Information Science and Technology, 391–407 (1990)
5. Hofmann, T.: Probabilistic latent semantic indexing. In: SIGIR, pp. 50–57 (1999)
6. Blei, D.M., Ng, A.Y., Jordan, M.I.: Latent dirichlet allocation. Journal of Machine Learning Research, 993–1022 (2003). ACM Press, New York
7. Cai, D., Mei, Q., Han, J., Zhai, C.: Modeling hidden topics on document manifold. In: CIKM, pp. 911–920 (2008)
8. Mei, Q., Cai, D., Zhang, D., Zhai, C.: Topic modeling with network regularization. In: WWW, pp. 101–110 (2008)
9. Sun, Y.Z., Han, J.W., Gao, J., Yu, Y.T.: Itopicmodel: information network-integrated topic modeling. In: ICDM, pp. 493–502 (2009)
10. Wang, Q., Peng, Z., Jiang, F., Li, Q.: LSA-PTM: a propagation-based topic model using latent semantic analysis on heterogeneous information networks. In: Wang, J., Xiong, H., Ishikawa, Y., Xu, J., Zhou, J. (eds.) WAIM 2013. LNCS, vol. 7923, pp. 13–24. Springer, Heidelberg (2013)
11. Deng, H., Han, J., Zhao, B., Yu, Y.: Probabilistic topic models with biased propagation on heterogeneous information networks. In: KDD, pp. 1271–1279 (2011)
12. Golub, G.H., Reinsch, C.: Singular value decomposition and least squares solutions. Numerische Mathematic, 403–420 (1970). Springer, Heidelberg
13. Yin, Z., Cao, L., Han, J., Zhai, C.: Geographical topic discovery and comparison. In: WWW, pp. 247–256 (2011)
14. Zhu, S., Yu, K., Chi, Y., Gong, Y.: Combining content and link for classification using matrix factorization. In: SIGIR, pp. 487–494 (2007)
15. Steyvers, M., Smyth, P., Rosen-Zvi, M., Griffiths, T.L.: Probabilistic author-topic models for information discovery. In: KDD, pp. 306–315 (2004)
16. Tang, J., Zhang, R.M., Zhang, J.: A topic modeling approach and its integration into the random walk framework for academic search. In: ICDM, pp. 1055–1060 (2008)
17. Deng, H., Lyu, M.R., King, I.: A generalized Co-HITS algorithm and its application to bipartite graphs. In: KDD, pp. 239–248 (2009)
18. Wang, S.Z., Hu, X., Yu, P.S., Li, Z.J.: MMRate: inferring multi-aspect diffusion networks with multi-pattern cascades. In: KDD, pp. 1246–1255 (2014)
19. Deng, H., Lyu, M.R., King, I.: Effective latent space graph-based re-ranking model with global consistency. In: WSDM, pp. 212–221 (2009)

Recommender Systems

Learning to Recommend with User Generated Content

Yueshen Xu[1](✉), Zhiyuan Chen[2], Jianwei Yin[1], Zizheng Wu[1], and Taojun Yao[3] .

[1] School of Computer Science and Technology, Zhejiang University, Hangzhou, China
{xyshzjucs,zjuyjw,zizhengwu}@zju.edu.cn
[2] Department of Computer Science, University of Illinois at Chicago, Chicago, USA
czyuanacm@gmail.com
[3] Alipay.com Co. Ltd, Hangzhou, China
taojun.ytj@alibaba-inc.com

Abstract. In the era of Web 2.0, user generated content (UGC), such as social tag and user review, widely exists on the Internet. However, in recommender systems, most of existing related works only study single kind of UGC in each paper, and different types of UGC are utilized in different ways. This paper proposes a unified way to use different types of UGC to improve the prediction accuracy for recommendation. We build two novel collaborative filtering models based on Matrix Factorization (MF), which are oriented to user features learning and item features learning respectively. In the user side, we construct a novel regularization term which employs UGC to better understand a user's interest. In the item side, we also construct a novel regularization term to better infer an item's characteristic. We conduct comprehensive experiments on three real-world datasets, which verify that our models significantly improve the prediction accuracy of missing ratings in recommender systems.

1 Introduction and Related Work

Recommender systems have been an indispensable component in e-commerce sites, which help users select their favorite products from a large number of candidates. There are mainly three kinds of recommendation algorithms, i.e., CF (Collaborative Filtering)-based, content-based and hybrid approaches. Among all CF-based algorithms, models based on MF (Matrix Factorization) have been verified to achieve satisfactory accuracy in rating prediction, and therefore widely studied in academia and industry [9]. Meanwhile, content-based recommendation algorithms still play an important role in recommender systems [15,18].

User generated content (denoted by **UGC** in this paper) refers to the various kinds of content created by users on web sites [14], including social tag, user review, question answer, blog, tweet, etc. User generated content widely exists in e-commerce sites and social networking sites, and has been employed in recommendation problems, such as tag-aware recommendation and review-based recommendation [10,13,23]. For example, Liang et al. [10] proposed a weighted tag-based top-N recommendation algorithm. Each item and each user had a profile consisting of tags, which were used to calculate the similarity between each

© Springer International Publishing Switzerland 2015
J. Li and Y. Sun (Eds.): WAIM 2015, LNCS 9098, pp. 221–232, 2015.
DOI: 10.1007/978-3-319-21042-1_18

pair of items and each pair of users. Although UGC has been verified to be helpful in existing works, there are still several problems stated below:

1. Not every web site allows users to tag products. For example, Ebay[1] and Epinion[2] do not allow consumers to tag products.

2. Since the item-tag space is highly sparse, it is difficult to get accurate results when we conduct similarity computation between items or tags.

3. The problems of synonym and polysemy are ubiquitous, since words that are used in UGC are highly personalized and 60% of words are personally proprietary [3].

4. The existing researches only focus on one kind of UGC, and utilize different kinds of UGC in very different ways. This makes it hard to find common features that are shared among various types of UGC (such as tags and reviews).

To solve the above problems, we study various types of UGC in recommender systems. Since user reviews and social tags are the two common types of UGC, in this paper, we mainly focus on reviews and tags. Because user reviews almost exist in all e-commerce sites and social networking sites, our work can be used in extensive web sites. We utilize topic modeling technique to transform highly spare word space into low topic space. In the topic space, we compute similarity for each two items and each two users. Also, topic modeling techniques can find the semantic relations among words. Besides, we integrate both types of UGC into recommendation algorithms in a unified manner.

There have been many works that utilize the social or trust relationship beyond the single user-item rating matrix to help infer users' preferences [11,12,22]. These methods assume that a user's preference tends to be influenced by his friends or trusted people. Indeed, social recommendation makes a great progress and improves the prediction accuracy. However, in many dominating e-commerce sites, like Amazon[3], Ebay[4], Newegg[5] and Jingdong[6], there are no social relationships so that these sites can hardly benefit from the social recommendation techniques. In such a case, we still have to infer a user's interest according to his consumption records. Note that a user's interest is usually related to some certain topics. For example, if John is a fan of *Harry Potter*, he may not only like the related movies, but also the related books, DVDs, gadgets and clothes. Additionally, he may also mark tags and write reviews on the related items. Although these products are in different categories, they have common features that can be inferred by UGC. However, most of existing works just conduct collaborative filtering based on users' feedbacks on each individual product separately. In contrast, in this paper, we study a user's interest distribution under different topics based on clustering to better understand his preference.

---

[1] http://www.ebay.com/

[2] http://www.epinions.com/

[3] http://www.amazon.com/

[4] http://www.ebay.com/

[5] http://www.newegg.com/

[6] http://www.jd.com/

In recent years, several researchers propose to use item descriptions, which are edited by website editors, to help learn item features [6,16,21]. Their recommendation algorithms are based on the collaborative topic regression model (CTR) [21], which employs LDA (Latent Dirichlet Allocation) [4] to learn the intrinsic features of items. However, there are two problems in these works:

1. Item descriptions are static and usually fail to distinguish two products that are under the same category. This is because a large part of words used in item descriptions overlap with each other. For example, the descriptions of laptops that are produced by *Samsung* and *Lenovo* are similar, since both of them consist of similar words that are used to depict product features, such as *memory*, *cpu* and *price*.

2. It is hard to infer a user's preference through item descriptions, since they are independent to users.

In contrast, UGC can emphasize items' characteristics. For example, in Last.fm[7], among all tags that the song *My Heart Will Go On* receives, *love*, *pop*, *ballad* and *soundtrack* are frequently used, which can well describe this song's main topics. Also, UGC can reflect a user's preference well. For example, if Bill wrote a lot of reviews containing words like *superhero*, *technology* and *fiction*, it can be inferred that Bill may like movies such as *Iron Man* and *Batman*, since these movies have related themes with *fiction* and *superhero*.

The main contributions of this paper are summarized as follows:

1. It studies the function of UGC in learning users' interests and learning items' characteristics.
2. It proposes a user-oriented collaborative filtering model and an item-oriented collaborative filtering model. It utilizes different types of UGC in a unified way in recommender systems.
3. It conducts sufficient experiments on three real-world datasets, which attest the effectiveness of proposed models.

2 Matrix Factorization with User Generated Content in User Side

As expounded in Sect. 1, a user's interest is specific to certain topics. For example, if Tom is a fan of Jackie Chan, he may not only watch his movies, but also buy relevant posters, clothes and books, even though these items are in different categories. In this paper, we employ the user' consumption records on different item groups to infer a user's interest. In the rest of the paper, the word *item* refers to all kinds of things that are provided to consume on the Internet, such as electronic products, restaurants, songs and movies.

2.1 Matrix Factorization Model

In this paper, we employ Matrix Factorization (denoted by **MF** in this paper) as the basic prediction model for its effectiveness and popularity [9]. As a typical

---

[7] http://www.last.fm/

latent factor model, MF can factorize the high dimensional space into two low latent dimensional spaces. Concretely, let $R \in \mathbb{R}^{M \times N}$ represent the rating matrix, which is usually extremely sparse. M, N are the numbers of users and items separately. Through MF-based technique, R can be estimated by the multiplication of latent user feature matrix U ($U \in \mathbb{R}^{M \times D}$) and latent item feature matrix V ($V \in \mathbb{R}^{N \times D}$), where D is the number of latent features. D is far smaller than M and N. Each entry R_{ij} ($1 \leq i \leq M, 1 \leq j \leq N$) can be estimated as the inner product of the two corresponding column vectors of U and V as

$$R_{ij} \approx U_i^T V_j, \tag{1}$$

where U_i is the ith column vector of U, and V_j is the jth column vector of V. To minimize the squared estimation error, the objective function of MF is constructed as follows:

$$\min_{U,V} \mathcal{E} = \frac{1}{2} \sum_{i=1}^{M} \sum_{j=1}^{N} I_{ij} (R_{ij} - U_i^T V_j)^2 + \frac{\lambda_U}{2} \|U\|_F^2 + \frac{\lambda_V}{2} \|V\|_F^2, \tag{2}$$

where I_{ij} is an indicator function. If user i gave a rating to item j before, I_{ij} equals 1. Otherwise, I_{ij} equals 0. $\| \cdot \|_F^2$ denotes the *Frobenius* norm, and λ_U, λ_V are both small constants. Gradient descent algorithm can be used to achieve the local optima of U and V, and thus R can be filled completely by estimating those missing values according to Eq. 1.

2.2   Topic Analysis for Items Through LDA

In this subsection, we aim to find the shared topics among items through topic modeling technique.

UGC records a user's personal impression and understanding on an item, which also reflects this item's characteristics. In UGC, a word that has been used frequently and uniquely is usually corresponding to one of the item's significant features. Therefore, an item's specialty can be emphasized by its highly frequent words that are written in UGC. For example, in Douban[8], among all tags that the movie *Titanic* receives, *love*, *romantic*, *disaster* and *USA* are frequently used, which can well describe this movie's main topics. We combine the words used in UGC together to compose the whole words set W as follows:

$$W = \{w_t \mid w_t \text{ is a word extracted from } UGC\} \ (1 \leq t \leq T)$$

where T is the number of all unique words. Then, each item j will own a word vector $\boldsymbol{W}(j) = (n(j)_{w_1}, n(j)_{w_2}, \ldots, n(j)_{w_t}, \ldots, n(j)_{w_T})$, in which $n(j)_{w_t}$ ($1 \leq t \leq T$) is the times of word w_t that item j has received. Specially, if word w_t has never been applied to item j, $n(j)_{w_t}$ in $\boldsymbol{W}(j)$ will equal 0. Thus, we can construct a weighted item-word co-occurrence matrix (a toy example is shown in Fig. 1), in which each row is $\boldsymbol{W}(j)$ corresponding to each item j.

---

[8] http://www.douban.com/

	fantasy	love	...	James Cameron
Titanic	0	50	...	3
X-Men	30	0	...	0
⋮	⋮	⋮	⋮	⋮
Avatar	40	10	...	2

Fig. 1. A toy example of item-word co-occurrence matrix

Topic modeling refers to a series of techniques used for latent semantic analysis in text mining and information retrieval [2]. In this paper, we chose Latent Dirichlet Allocation (LDA) [4] as the topic modeling technique for its popularity. In LDA, a document is assumed to be constructed in the generative process. In a corpus, for each document j, a topic k is chosen according to the document-topic multi-nominal distribution $\boldsymbol{\Theta}$, where $\boldsymbol{\Theta} = \{\boldsymbol{\theta}_j\}_{j=1}^N$. $\boldsymbol{\theta}_j = (\theta_{j1}, \theta_{j2}, \ldots, \theta_{jK})$ denotes the topic distribution under the document j, and $\theta_{jk} = p(k|j)$ $(1 \leq k \leq K)$ is the probability that topic k is chosen for document j. K and N are the numbers of topics and documents respectively. Then, a word w is sampled from the assigned topic k according to the topic-word multi-nominal distribution $\boldsymbol{\Phi}$, where $\boldsymbol{\Phi} = \{\boldsymbol{\phi}_k\}_{k=1}^K$. Therefore, the probability distribution of topics over a document describes the hidden aspects of this document. As for each item, words extracted from UGC compose its text descriptions. Thus, in this paper, each item is regarded as a document with a collection of words $\boldsymbol{W}(j)$, and the item-word co-occurrence matrix is the corpus. With collapsed Gibbs sampling [8], we estimate the posterior parameters in LDA, i.e., $\boldsymbol{\theta}_j$ $(1 \leq j \leq N)$ and $\boldsymbol{\phi}_k$ $(1 \leq k \leq K)$.

2.3   User Interest Distribution

After topic analysis through LDA, to measure the topic similarity between two items j and h, we calculate the cosine similarity [1] between $\boldsymbol{\theta}_j$ and $\boldsymbol{\theta}_h$ as

$$Sim(j, h) = cos(\boldsymbol{\theta}_j, \boldsymbol{\theta}_h) = \frac{\boldsymbol{\theta}_j \cdot \boldsymbol{\theta}_h}{\|\boldsymbol{\theta}_j\| \times \|\boldsymbol{\theta}_h\|}, \tag{3}$$

where $\boldsymbol{\theta}_h$ is the topic distribution vector of item h.

To find items that share similar topics, items are clustered into Q groups based on their similarities of topic distribution vectors (see Eq. 3). We use K-Means as the clustering algorithm for its easy implementation and popularity. For each user, we count the number of his historical consumption records on each cluster C_q $(1 \leq q \leq Q)$, by summing the times of his feedbacks on all items that belong to cluster C_q. Thus, each user i will be associated with a vector that represents his consumption distribution on Q clusters. In this paper, we call this vector as user i's *interest vector* $\boldsymbol{I}_i$:

$$\boldsymbol{I}_i = (n_i(C_1), n_i(C_2), \ldots, n_i(C_q), \ldots, n_i(C_Q)) \ (1 \leq q \leq Q)$$

where $n_i(C_q)$ represents the total times of user i's feedbacks on cluster C_q. Therefore, we can construct a user-interest distribution matrix, with a toy example shown in Fig. 2. In this paper, we use both ratings record and interest vector to infer a users' preference instead of only ratings. It is because a user's interest is relatively stable, and is not easily affected by one or two unpleasant consumption experiences. For example, that once Tom gave a low rating to a T-shirt printed with a portrait of Jachie Chan does not mean that he will not be a fan of Jachie Chan anymore.

	C_1	C_2	...	C_Q
Tom	0	14	...	0
John	11	0	...	6
⋮	⋮	⋮	⋮	⋮
Bill	7	0	...	2

Fig. 2. A toy example of user-interest distribution matrix

Based on users' interest vectors, the interest similarity of user i and l is calculated with cosine similarity as:

$$Sim(i,l) = cos(\boldsymbol{I}_i, \boldsymbol{I}_l) = \frac{\boldsymbol{I}_i \cdot \boldsymbol{I}_l}{\|\boldsymbol{I}_i\| \times \|\boldsymbol{I}_l\|}, \tag{4}$$

where $\boldsymbol{I}_l$ is user l's interest vector. For user i, to find the neighbors who share similar interests, his neighborhood $L(i)$ is constructed by selecting the S most similar users with him, denoted as $L(i) = \{l \mid Sim(i,l) \text{ is in the top } S \text{ similarity list}\}$. Therefore, the weight of each neighbor l can be calculated as

$$e_{il} = \frac{Sim(i,l)}{\sum_{l' \in L(i)} Sim(i,l')}, \tag{5}$$

where $e_{il} \in [0,1]$.

2.4 Matrix Factorization with User Topic Regularization

Since a user's interest is positively correlated to his latent feature vector, the feature vectors of users having similar interests also tend to be similar. Therefore, user i's feature vector U_i should be similar to those of his neighbors'. To model such a kind of similarity relation, the average feature vector of his neighbors is calculated as

$$\bar{U}_l = \sum_{l \in L(i)} e_{il} U_l.$$

For each user, we try to minimize the difference between each pair of U_i and $\bar{U}_l$ as

$$\min \sum_{i=1}^{M} \|U_i - \sum_{l \in L(i)} e_{il} U_l\|_F^2. \tag{6}$$

We call Eq. 6 *user topic regularization (UTR)*, and integrate it into the MF model to build the UTR-based MF model (**UTR-MF**) as

$$\min_{U,V} \mathcal{E}_u = \frac{1}{2} \sum_{i=1}^{M} \sum_{j=1}^{N} I_{ij} (R_{ij} - U_i^T V_j)^2 + \frac{\lambda_U}{2} \|U\|_F^2 + \frac{\lambda_V}{2} \|V\|_F^2$$
$$+ \frac{\alpha}{2} \sum_{i=1}^{M} \|U_i - \sum_{l \in L(i)} e_{il} U_l\|_F^2, \tag{7}$$

where α is a regulatory factor to regulate the effect of UTR. Gradient descent algorithm can also be used to achieve the local optima of U and V.

3   Matrix Factorization with User Generated Content in Item Side

Since an item's specialty can be reflected by UGC, it can be inferred that two items that share similar UGC tend to share similar latent features. For example, the DVDs and books, which are affiliated to the movie *The Lord of the Rings*, receive many similar words in their reviews in Amazon. Naturally, they also share many similar latent features. After topic analysis presented in Sect. 2.2, for each item j, those items that are highly similar to it (see Eq. 3) compose j's neighborhood $H(j)$: $H(j) = \{h \mid Sim(j,h)$ *is in the top S similarity list*$\}$, where S is the size of $H(j)$. Similar to Eq. 5, each neighbor h's weight is

$$w_{jh} = \frac{Sim(j,h)}{\sum_{h' \in H(j)} Sim(j,h')}, \tag{8}$$

where $w_{jh} \in [0,1]$. Since item j's topic distribution is similar to those of its neighbors' in $H(j)$, it can be inferred that their latent feature vectors should also be similar to each other. This is because a large part of latent factors reflect the item's characteristics on different topics. To utilize this similarity relation, the average feature vector of the whole neighborhood is computed as

$$\bar{V}_h = \sum_{h \in H(j)} w_{jh} V_h,$$

Similar to the user topic regularization, we construct the *item topic regularization (ITR)* to minimize the difference among items' feature vectors as

$$\min \sum_{j=1}^{N} \|V_j - \sum_{h \in H(j)} w_{jh} V_h\|_F^2. \tag{9}$$

The item topic regularization (ITR) is integrated into the MF model to build
the ITR-based MF model (**ITR-MF**) as

$$\min_{U,V} \ \mathcal{E}_v = \frac{1}{2} \sum_{i=1}^{M} \sum_{j=1}^{N} I_{ij}(R_{ij} - U_i^T V_j)^2 + \frac{\lambda_U}{2} \|U\|_F^2 + \frac{\lambda_V}{2} \|V\|_F^2$$
$$+ \frac{\alpha}{2} \sum_{j=1}^{N} \|V_j - \sum_{h \in H(j)} w_{jh} V_h\|_F^2, \tag{10}$$

where α is a regulatory factor, which is the same with the notation α in Eq. 7.
The partial derivatives of Eq. 10 over U_i and V_j are computed as follows:

$$\frac{\partial \mathcal{E}_v}{\partial U_i} = \sum_{j=1}^{N} I_{ij}(R_{ij} - U_i^T V_j)(-V_j) + \lambda_U U_i,$$

$$\frac{\partial \mathcal{E}_v}{\partial V_j} = \sum_{i=1}^{M} I_{ij}(R_{ij} - U_i^T V_j)(-U_i) + \lambda_V V_j + \alpha(V_j - \sum_{h \in H(j)} w_{jh} V_h) \tag{11}$$
$$+ \alpha \sum_{g \in G(j)} (V_g - \sum_{h' \in H(g)} w_{gh'} V_{h'}) \times (-w_{gj}),$$

where $G(j)$ contains those items whose neighborhoods include item j. Finally,
gradient descent algorithm can be employed to achieve local optima of U and V
based on the two partial derivatives.

4   Experiments and Evaluation

In this section, we conduct experiments on real-world datasets to evaluate our
proposed models' performance.

4.1   DataSet and Evaluation Metrics

We use three real-word datasets in the following experiments, which are *Movie-lens* dataset, *Last.fm* dataset and *Yelp* dataset. *Movielens* dataset and *Last.fm*
dataset are published by GroupLens research group[9] in the workshop *HetRec'11*
[5]. *Movielens* dataset contains 2113 users, 10197 movies, 13222 tags and ratings.
Last.fm dataset contains 1892 users, 17632 artists, 11946 tags and listening times
of users to artists. We map listening times into the range 1~5 as ratings. *Yelp*
dataset is published by *Yelp*[10], which contains 23152 users, 11537 restaurants,
reviews and ratings.

For *Movielens* dataset and *Yelp* dataset, we divide the whole dataset into the
training set and testing set according to each user's consumption timestamps.

---

[9] http://www.grouplens.org
[10] https://www.kaggle.com/c/yelp-recsys-2013

For example, in the training set of 80% density, if a user has given 10 ratings in time order, the first 8 old ratings will be allocated to the training set, and the last 2 new ratings will be the testing data. Since there is no timestamp in *Last.fm* dataset, we randomly select a certain percentage of the whole data as the training set and the remaining data will be the testing set. In this paper, we test the prediction accuracy on training sets with four different data densities, which are 60%, 70%, 80% and 90% respectively.

We use *Root Mean Squared Error (RMSE)* and *Mean Absolute Error (MAE)* as the evaluation metrics, which are shown below:

$$RMSE = \sqrt{\frac{1}{|T_S|} \sum_{R_{ij} \in T_S} (R_{ij} - \hat{R}_{ij})^2}, \quad MAE = \frac{1}{|T_S|} \sum_{R_{ij} \in T_S} |R_{ij} - \hat{R}_{ij}|,$$

where T_S and $|T_S|$ represent the testing set and its size. R_{ij}, $\hat{R}_{ij}$ are the real rating and corresponding estimated rating in T_S respectively.

4.2 Performance Comparison and Parameter Setting

Five well-known models are chosen as baselines to compare with our models:

1. **UserCF**: This is the well-known user-based collaborative filtering algorithm. The missing ratings are predicted with the historical records of each user's similar neighbors [17].
2. **ItemCF**: This is the well-known item-based collaborative filtering algorithm, and each missing value is predicted with the historical records of each item's similar neighbors [20].
3. **PMF** (Probabilistic MF model): This model provides the probabilistic interpretation of MF model. The rating matrix and latent feature matrices are all assumed to follow Gaussian distribution [19].
4. **TF-IDF MF**: This method first constructs the item content matrix, in which each entry represents the TF-IDF (term frequency-inverse document frequency) weight of each attribute. Then, the rating matrix and the content matrix are factorized together [7]. The original work also utilizes the user profile information. But since in this paper we do not introduce user profiles, we eliminate the term that factorizes the user profile matrix. We call this model *TF-IDF MF* model.
5. **CTR**: It is the collaborative topic regression model (CTR) proposed in [21].

The default parameter setting is as follows: the number of topics (K), the number of clusters (Q), the number of latent features (D) and the regulatory factor (α) are all set to 10. The neighborhood sizes (S) of items and users are set to 100. λ_U, λ_V are set to 0.1 in all experiments.

4.3 Performance Comparison in Social Tag Case

This subsection presents the model results when UGC is the social tag. The experiments are conducted on *Movielens* dataset and *Last.fm* dataset.

Table 1. Accuracy Comparison (Smaller RMSE and MAE mean better performance)

Approach	Training Set Density (TD) — *Movielens*							
	TD=60%		TD=70%		TD=80%		TD=90%	
	RMSE	MAE	RMSE	MAE	RMSE	MAE	RMSE	MAE
UserCF	0.8405	0.6290	0.8303	0.6213	0.8270	0.6194	0.8240	0.6170
ItemCF	0.8885	0.6421	0.8722	0.6309	0.8689	0.6281	0.8697	0.6245
PMF	0.8312	0.6259	0.8225	0.6191	0.8146	0.6133	0.8139	0.6122
TF-IDF MF	0.8303	0.6250	0.8213	0.6180	0.8123	0.6109	0.8118	0.6099
CTR	0.8211	0.6184	0.8109	0.6103	0.8030	0.6042	0.8011	0.6019
UTR-MF	**0.8175**	**0.6141**	**0.8085**	**0.6075**	**0.7999**	**0.6007**	**0.7974**	**0.5985**
ITR-MF	**0.8086**	**0.6077**	**0.7982**	**0.5996**	**0.7893**	**0.5929**	**0.7867**	**0.5904**

Table 2. Accuracy Comparison (Smaller RMSE and MAE mean better performance)

Approach	Training Set Density (TD) — *Last.fm*							
	TD=60%		TD=70%		TD=80%		TD=90%	
	RMSE	MAE	RMSE	MAE	RMSE	MAE	RMSE	MAE
PMF	0.5125	0.3941	0.5091	0.3913	0.5006	0.3854	0.4891	0.3777
TF-IDF MF	0.5121	0.3936	0.5028	0.3867	0.4881	0.3767	0.4725	0.3649
CTR	0.4815	0.3702	0.4698	0.3608	0.4619	0.3533	0.4494	0.3437
UTR-MF	**0.4771**	**0.3645**	**0.4670**	**0.3560**	**0.4573**	**0.3482**	**0.4426**	**0.3397**
ITR-MF	**0.4444**	**0.3278**	**0.4334**	**0.3202**	**0.4244**	**0.3174**	**0.4082**	**0.3014**

Tables 1 and 2 show that for both datasets, our proposed models UTR-MF and ITR-MF outperform other baseline models in all cases of training set densities. For example, in *Last.fm* dataset, on average over four training set densities, ITR-MF achieves 14.98% improvement than PMF, and 8.18% improvement than CTR for RMSE. Such an improvement demonstrates that user topic regularization (see Eq. 6) and item topic regularization (see Eq. 9) are both effective. In detail, it means that those users who have similar interest distributions indeed tend to have similar rating behaviors, and such a similar relation can be reflected by user topic regularization effectively. Also, it indicates that the items which have similar topic distributions indeed tend to be given similar ratings. Additionally, the prediction accuracy of ITR-MF is higher than that of UTR-MF in all cases. It means that item topic regularization can make more contributions to improving MF model's performance than user topic regularization. It also indicates that users' interests are harder to infer than items' features. Compared with all baseline models, the improvements achieved by UTR-MF and ITR-MF are both significant according to the paired t-tests ($p < 0.0001$).

4.4 Performance Comparison in User Review Case

This subsection compares the prediction accuracy of models when UGC is user review. The parameter setting is the same with that in Sect. 4.3, and the experiment is conducted on *Yelp* dataset.

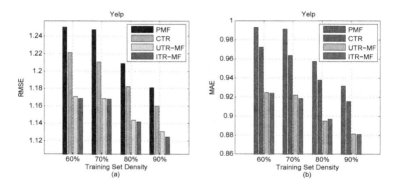

Fig. 3. Precision Comparison in User Review Case

It is shown in Fig. 3 that UTR-MF and ITR-MF get higher prediction accuracy than other baseline models in all training set densities. For example, in the case that the training set density is 60%, ITR-MF gains 6.58% improvement than PMF, and 4.33% improvement than CTR for RMSE. It is verified that for reviews, user topic regularization can also help infer users' interests, and item topic regularization can also help infer items' characteristics. Since user reviews exist in almost all e-commerce sites and social networking sites, it indicates that our models have wide applicability. The improvements achieved by UTR-MF and ITR-MF are also both significant according to the paired t-tests ($p < 0.0001$). Besides, it can be seen that along with the increasing of training set density, the prediction errors decline. This is because more training data can provide more historical records to learn user and item latent features more accurately.

5   Conclusion

This paper proposes two MF-based collaborative filtering models, which utilize UGC to solve the problem of rating prediction in recommender systems. Sufficient experiments on real-world datasets verify the effectiveness of our models. First, this paper demonstrates that UGC, such as tags and reviews, can be integrated into the MF model in a unified way to significantly improve the prediction accuracy. Second, this paper verifies that users' interests and items' features can indeed be reflected by different types of UGC. Third, this paper proposes two novel regularization terms, which can model the similarity between each pair of users and each pair of items effectively. These contributions are instructive for the utilization of other kinds of UGC to build more accurate recommender systems. One example is that we plan to employ tweets to infer users' interests in social networking sites.

Acknowledgments. This paper is supported by National Natural Science Foundation (No.61272129), National High-Tech Research Program (No.2013AA01A213), New-Century Excellent Talents Program (No.NCET-12-0491), Zhejiang Provincial Natural

Science Foundation (LR13F020002), Science and Technology Program of Zhejiang (No.2012C01037-1) and China Scholarship Council.

References

1. Adomavicius, G., Tuzhilin, A.: Toward the next generation of recommender systems: A survey of the state-of-the-art and possible extensions. IEEE TKDE **17**(6), 734–749 (2005)
2. Aggarwal, C.C., Zhai, C.: Mining Text Data. Springer, New York (2012)
3. Bischoff, K., Firan, C.S., Nejdl, W., Paiu, R.: Can all tags be used for search? In: ACM CIKM, pp. 193–202 (2008)
4. Blei, D.M., Ng, A.Y., Jordan, M.I.: Latent dirichlet allocation. JMLR **3**, 993–1022 (2003)
5. Cantador, I., Brusilovsky, P., Kuflik, T.: HetRec workshop. ACM RecSys, New York (2011)
6. Chen, C., Zheng, X., Wang, Y., Hong, F. and Lin, Z.: Context-aware collaborative topic regression with social matrix factorization for recommender systems. In: AAAI, pp. 9–15 (2014)
7. Fang, Y., Si, L.: Matrix co-factorization for recommendation with rich side information and implicit feedback. In: HetRec (workshop of RecSys), pp. 65–69 (2011)
8. Griffiths, T.L., Steyvers, M.: Finding Scientific Topics. In: PNAS (2004)
9. Koren, Y., Bell, R., Volinsky, C.: Matrix factorization techniques for recommender systems. Computer **42**(8), 30–37 (2009)
10. Liang, H., Xu, Y., Li, Y., Nayak, R., Tao, X.: Connecting users and items with weighted tags for personalized item recommendations. In: Hypertext, pp. 51–60 (2010)
11. Liu, X., Aberer, K.: SoCo: a social network aided context-aware recommender system. In: WWW, pp. 781–802 (2013)
12. Ma, H., Zhou, D., Liu, C., Lyu, M.R., King, I.: Recommender systems with social regularization. In: ACM WSDM, pp. 287–296 (2011)
13. McAuley, J.J., Leskovec, J.: Hidden factors and hidden topics: understanding rating dimensions with review text. In: ACM RecSys, pp. 165–172 (2013)
14. Moens, M.-F., Li, J., Chua, T.-S.: Mining user generated content. In: Chapman and Hall/CRC (2014)
15. Pandora. Music genome project. http://www.pandora.com/about/mgp
16. Purushotham, S., Liu, Y.: Collaborative topic regression with social matrix factorization for recommendation systems. In: IEEE ICML, pp. 759–766 (2012)
17. Resnick, P., Iacovou, N., Suchak, M., Bergstrom, P., Riedl, J.: Grouplens: an open architecture for collaborative filtering of netnews. In: CSCW, pp. 175–186 (1994)
18. Rovi. Recommendations api version 2.0. http://prod-doc.rovicorp.com/mashery/index.php/Recommendations
19. Salakhutdinov, R., Mnih, A.: Probabilistic matrix factorization. In: NIPS
20. Sarwar, B., Karypis, G., Konstan, J., Reidl, J.: Item-based collaborative filtering recommendation algorithm. In: WWW, pp. 285–295 (2001)
21. Wang, C., Blei, D.M.: Collaborative topic modeling for recommending scientific articles. In: ACM SIGKDD, pp. 448–456 (2011)
22. Yang, X., Steck, H., Liu, Y.: Circle-based recommendation in online social networks. In: ACM SIGKDD, pp. 1267–1275 (2012)
23. Zhang, Y., Lai, G., Zhang, M., Zhang, Y., Liu, Y., Ma, S.: Explicit factor models for explainable recommendation based on phrase-level sentiment analysis. In: ACM SIGIR, pp. 83–92 (2014)

Combining Positive and Negative Feedbacks with Factored Similarity Matrix for Recommender Systems

Mengshuang Wang, Jun Ma$^{(\boxtimes)}$, Shanshan Huang, and Peizhe Cheng

School of Computer Science and Technology, Shandong University, Jinan, China
{frost_sdu,huangshanshansdu,aquariuscpz}@163.com, majun@sdu.edu.cn

Abstract. Traditional collaborative filtering algorithms like ItemKNN, cannot capture the relationships between items that are not co-rated by at least one user. To cope with this problem, the item-based factor models are put forward to utilize low dimensional space to learn implicit relationships between items. However, these models consider all user's rated items equally as positive examples, which is unreasonable and fails to interpret the actual preferences of users. To tackle the aforementioned problems, in this paper, we propose a novel item-based latent factor model, which can consider user's positive and negative feedbacks while learning item-item correlations. In particular, for each user, we divide his rated items into two different parts, i.e., positive examples and negative examples, depending on whether the rating of the item is above the average rating of the user or not. In our model, we assume that the predicted rating of an item should be boosted if the item is similar to most of the positive examples. On the contrary, the predicted rating should be diminished if the item is similar to most of the negative examples. The item-item similarity is approximated by an inner product of two low-dimensional item latent factor matrices which are learned using a structural equation modeling approach. Comprehensive experiments on two benchmark datasets indicate that our method has significant improvements as compared with existing approaches in both rating prediction and top-N recommendation.

Keywords: Recommender systems · Collaborative filtering · Similarity matrix

1 Introduction

With the development of Internet and Web 2.0, users have to face the information overload problem when searching for what they are interested in from the massive data. Recommender systems can benefit users by presenting personalized recommendations based on their preferences and the available information. Furthermore, a successful recommender system facilitates the interaction of customers in online communities and promotes the development of e-commerce. In

© Springer International Publishing Switzerland 2015
J. Li and Y. Sun (Eds.): WAIM 2015, LNCS 9098, pp. 233–246, 2015.
DOI: 10.1007/978-3-319-21042-1_19

recent decades, many recommendation algorithms have been proposed in different application scenarios [1–3]. Among the existing methods, collaborative filtering (CF) [4] approaches utilize users' implicit or explicit feedbacks to give customized recommendations without domain knowledge, while content-based [5,6] approaches need to use the profiles of items and users. CF methods have become one of the most widely used recommendation approaches.

Typically, CF methods can be classified into two main categories. The first, referred to as neighborhood-based methods [7–10], are centered on computing the similarities between users/items. And subsequently missing ratings are predicted based on these similarities. However, with the increasing number of users and items, the complexity is too high to make online recommendations. The second, referred to as model-based methods [11–13], employ a machine learning algorithm to build a model which transforms features of both items and users to the same latent factor space.

In real world scenarios, users typically provide feedbacks to only a few items out of thousands of items. So the user-item rating matrix is quite sparse and the available few feedbacks are particularly significant. Therefore, it is not easy for traditional methods like ItemKNN to find the transitive relations between items which have not been co-rated by at least one user. Meanwhile, the model-based methods like NSVD alleviate this problem by projecting the matrix onto a low dimensional space. However, there is an inherent limitation that they fail to capture the accurate inferences of the users by treating all rated items equally without discrimination. As a result, they ignore the differences among items and weaken the impact of items that have unique characteristics, which may lead to poor performance.

In general, items can be classified into two main categories for a certain user, namely the favourite ones and the disliked ones. Correspondingly in recommendation system, we can divide all the rated items into two different parts for one user, i.e., positive examples and negative examples, depending on whether the rating value is above the average rating of the user or not. By distinguishing the positive examples from the negative examples, the rare feedbacks can be fully utilized and the user's interest can be excavated more accurately. The positive feedbacks of a user are the examples of items which this user will prefer to. On the contrary, the negative feedbacks are the representative items which this user will dislike. In other words, for a target user, the predicted rating of an item should be boosted if it is similar to most of the user's positive feedbacks and diminished if it is similar to most of the user's negative feedbacks.

Based on the above analysis, in this paper, we propose a novel approach that combines both the positive and the negative feedbacks to learn latent item-item similarity matrix. This method is a new variant of item-oriented CF along with matrix factorization techniques. In our approach, for a target user, the predicted rating of an unrated item is computed based on the learned similarities of this item with other positive examples and negative examples. To evaluate the performance of our proposed method, we conduct extensive experiments on MovieLens and EachMovie datasets. The experimental results shows that it's a most convenient and effective way to infer user preferences by utilizing classified feedbacks. It also confirms that our method is superior to the state-of-the-art methods not only in ratings predictions but also in top-N recommendation.

The prime contributions of this paper are summarized as follows:

(i) We propose a modified method that considers both the positive and the negative examples to learn latent item-item similarities. In this way, it can more accurately capture the user's preference and make the favorite items rank in the top especially on a small scale.

(ii) We integrate some concepts in neighborhood-based methods into a latent factor model to excavate the transitive relations between items. The item-item similarity matrix can be learned as an inner production of two low dimensional item latent factor matrices.

(iii) We perform comparative experiments on different datasets in order to verify the effect of classified examples. Then, we investigate and analyze the impact of some parameters referred to similarity agreement and dimension.

The remainder of this paper is organized as follows. In section 2, we introduce the baseline estimate method. Section 3 provides a detailed description for the modeling, learning and prediction process of the proposed model. We give a brief review of the datasets and evaluation metrics in section 4. The experimental results are reported in section 5. Finally, we conclude the paper and present some directions for future work in section 6.

2   Baseline Estimate

In recommender systems, some users tend to give higher ratings and some items receive higher ratings than others. As a result, typical CF data exhibit enlarges the effects of those users and items. In order to adjust the data by accounting for these effects, it is customary to encapsulate them in the baseline estimate function [8] as follows:

$$b_{ui} = \mu + b_u + b_i \tag{1}$$

The unknown rating is estimated by b_{ui} that accounts for the user and item biases. The overall average rating is denoted by μ. The parameters b_u and b_i are the user and item biases, respectively, which indicate the observed deviations of user u and item i from the average. For example, if we want to estimate the baseline for a movie's rating like *Gone with Wind* by user Smith, we get the value of μ (the average rating over all movies) as 3.3 stars at first. Because *Gone with Wind* is a better movie, its rating is higher than the average with 0.8 stars. Regarding to personal preference, Smith prefers to rate 0.4 stars lower than the average as a critical user. Above all, we will estimate the baseline for *Gone with Wind* by Smith 3.7 stars through calculating 3.3 - 0.4 + 0.8 in terms of Eqution 1. Hence, the parameters of this model are estimated to solve the least squares problem:

$$\min_{b_*} \sum_R (r_{ui} - \mu - b_u - b_i)^2 + \lambda(\sum_u b_u^2 + \sum_i b_i^2) \tag{2}$$

where R is the set of observed ratings and r_{ui} is the ground truth value. The first term aims to find the best fit values (e.g., b_u, b_i) for the given ratings. The latter is the regularization term that can avoid overfitting by penalizing the magnitudes of these parameters.

3   Combining Positive and Negative Feedbacks with Factored Similarity Matrix for Recommender Systems

In this section, we propose a new item-based latent factor model which can learn latent item correlations by considering users' positive feedbacks and negative feedbacks. For convenience, we named the proposed method PNSM (Combining Positive and Negative Feedbacks with Factored Similarity Matrix for Recommender Systems).

3.1   Combining Positive Feedbacks with Factored Similarity Matrix (PNSM$_1$ for short)

In reality, the user-item rating matrix is very sparse, since users usually provide feedbacks to only a handful of items out of thousands or millions of items. Traditional methods based on similarity rely on so few neighborhood relations that cannot capture the dependencies between items especially which have not been co-rated by at least one user. To overcome this problem, methods based on matrix factorization like NSVD thereby implicitly learn better relationships between items. However, these methods treat all rated items as the positive examples equally. But in fact, for one user, her rated items can be sorted into the favorite items and the dislike ones (corresponding to the positive and the negative examples respectively) depending on whether the rating value is above the average or not. Naturally the feedbacks (the positive feedbacks and the negative feedbacks) of these two kind of items should have a different effect on rating prediction.

To further explore the different influence between feedbacks, first we only take the positive feedbacks into consideration in modeling process and then introduce how it can benefit recommender systems. The positive feedbacks can facilitate the user's preference, therefore for an item the predictive rating can be boosted if it is similar to most of the positive examples (items that ratings are higher than the average of the user). The estimated value for a given user u on item i is computed as:

$$\hat{r}_{ui} = b_{ui} + |R(u)^+|^{-\alpha} \sum_{j \in R(u)^+} q_i^T p_j \tag{3}$$

Importantly, the b_{ui} is derived as explained in Sec. 2. $R(u)$ is the set of items rated by user u. $R(u)^+$ is the set of the positive examples whose ratings are higher than the average rating of the user u and the number of items in this set is denoted by $|R(u)^+|$. q_i, p_j are both the learned item latent factors and their inner product can be regarded as the similarity to the item whose rating is being estimated. Specially, p_j is the positive feedback learned from the positive item space matrix where item's rating is higher than the average rating of the user. The parameter α is a user specified parameter between 0 and 1, and it controls the number of neighborhood items that are similar to the positive examples. Afterwards, the term $|R(u)^+|^{-\alpha}$ can represent the degree of agreement to their similarity weight.

In the structural equation model, the recommendation score for a user u on an item i is calculated as an aggregation of the positive examples with the corresponding product of p_j latent vectors from $R(u)^+$ and q_i latent vector. This method is named PNSM$_1$ for short. The essential difference between PNSM$_1$ and NSVD is that a flexible coefficient α is proposed to compromise the influence of their similarity, as the domain is focus on the positive examples rather than mixed records.

In PNSM$_1$, the various parameters are learned by minimizing the following regularized optimization problem:

$$\min_{b_*, p_j, q_i} (r_{ui} - \hat{r}_{ui})^2 + \lambda(||b_u||^2 + ||b_i||^2 + ||q_i||^2 + \sum_{j \in R(u)+} ||p_j||^2) \qquad (4)$$

where $\hat{r}_{ui}$ is the estimated value for user u and item i (as in Equation 3). The regularization terms are used to avoid overfitting and λ is the regularization weight for latent factor vectors, user bias vector and item bias vector.

3.2 Combining Positive and Negative Feedbacks with Factored Similarity Matrix (PNSM$_2$ for short)

In PNSM$_1$, we only integrate the positive feedbacks in rating prediction. However, as referred to before, the negative feedbacks also make a difference in recommendation process. In other words, the item can receive a lower score if it's most similar to the negative examples (items that the user rated lowly). In general terms, the positive feedbacks can facilitate the prediction while the negative feedbacks can weaken it on the contrary. There is an assumption that for a certain user if the item is similar to most of the positive/negative examples, the user appreciates/dislikes this item as well. Hence to take full advantages of both feedbacks and discriminate them, we exploit a more comprehensive approach based on PNSM$_1$. Such method can explain the recommendations in terms of all items previously rated by users and we abbreviate it to PNSM$_2$. To better provide accurate evaluation, the prediction of the rating that user u gives to item i can be inferred as:

$$\hat{r}_{ui} = b_{ui} + |R(u)^+|^{-\alpha} \sum_{j \in R(u)+} q_i^T p_j - |R(u)^-|^{-\beta} \sum_{k \in R(u)-} q_i^T y_k \qquad (5)$$

where y_k is the negative feedback learned from item space matrix where item's rating is lower than the average rating of the user, while p_j represents the positive feedback. Contrary to $|R(u)^+|$, $R(u)^-$ is the set of the negative examples whose ratings are lower than the average rating of user u and the number of items in this set is denoted by $|R(u)^-|$. In addition, parameter β is similar to α and both of them are user specified parameters between 0 and 1. Their values are dependent on the properties of datasets and the best performing values can be determined empirically.

In a sense, Equation 5 provides a two tier model for prediction. The first tier (i.e., $b_{ui} = \mu + b_u + b_i$) generally describes attributes of users and items, without taking account of any involved interactions. The rest tier illustrates the interaction among item profiles and can excavate implicit transitive relations. It can also be split up into two parts in detail.

Formally, the former part, $|R(u)^+|^{-\alpha} \sum_{j \in R(u)^+} q_i^T p_j$, represent the promotion that the positive examples provide. More specifically, given an item i and taking any positive example j, their inner product ($q_i^T p_j$) can be regarded as the similarity weight to boost the prediction. Therefore, the item will receive a high score by adding the aggregation of positive similarity weights to the formula. Namely, we can make the favorite items rank as top as possible and recommend the most likely items to users. In comparison, the latter part, $|R(u)^-|^{-\beta} \sum_{k \in R(u)^-} q_i^T y_k$, are the impaired feedbacks provided by the negative examples. Concretely, as a negative similarity weight, the inner product of a certain item i and a negative example j can diminish the prediction. In this case, we subtract the cumulative values of these similarity weights to weaken the influence of negative items in the model. Besides, we use $|R(u)^+|^{-\alpha}$ and $|R(u)^-|^{-\beta}$ to control the degree of agreement to the positive similarity weights and the negative similarity weights respectively. This is also one of the important differences between PNSM and NSVD.

To better understand, the former part are the cumulative similarities between item i and all of the positive examples in $R(u)^+$ when $\alpha = 0$. Under the circumstances, item i can be rated high, even though only one positive example is similar to it. Considering another case in which $\alpha = 1$, these part amount to the average similarities between i and the positive examples in $R(u)^+$. Then item i can receive a high score when almost all of the items in $R(u)^+$ are similar to it. As far as we know, these are two extreme cases with different settings. In principle, it is usually that the right choice will be somewhere in between. So do the parameter β and the latter part.

Model parameters are estimated at a pre-processing stage. In general, the values of involved parameters are determined by minimizing the associated regularized squared error function:

$$\min_{b_u, b_i, p_j, q_i, y_k} (r_{ui} - \hat{r}_{ui})^2 + \lambda(b_u^2 + b_i^2 + ||q_i||^2 + \sum_{j \in R(u)^+} ||p_j||^2 + \sum_{k \in R(u)^-} ||y_k||^2) \quad (6)$$

The residual sum of squares, $(r_{ui} - \hat{r}_{ui})^2$, measures how well the linear model fits the training data. The regularization terms are used to avoid overfitting and λ is the regularization weight for biases and several latent factor vectors. The optimization problem can be solved by employing a stochastic gradient descent method. The derivatives of the parameters are as follows:

$$\frac{\partial f}{\partial b_u} = -e_{ui} + \lambda b_u$$

$$\frac{\partial f}{\partial b_i} = -e_{ui} + \lambda b_i$$

$$\frac{\partial f}{\partial q_i} = -e_{ui}(s - t) + \lambda q_i$$

$$\frac{\partial f}{\partial p_j} = -e_{ui} \cdot s + \lambda p_j, \quad \textbf{for all } j \in R(u)^+$$

$$\frac{\partial f}{\partial y_k} = e_{ui} \cdot t + \lambda y_k, \quad \textbf{for all } k \in R(u)^-$$

where $e_{ui} \overset{def}{=} r_{ui} - \hat{r}_{ui}$, $s = |R(u)^+|^{-\alpha} \sum_{j \in R(u)^+} p_j$ and $t = |R(u)^-|^{-\beta} \sum_{k \in R(u)^-} y_k$. In each iteration, we modify the parameters by moving in the opposite direction of the gradient.

In Algorithm 1, we present a concrete process for learning the parameters. As we can see, latent factor vectors are all initialized with small random values that are non-negative. We would repeat the iteration until the number of iterations has reached a predefined threshold or the error on the validation set tends to convergence.

4   Experimental Settings

4.1   Datasets

In this research, we evaluated the performance of the recommender algorithms on two different real datasets, including MovieLens[1], EachMovie[2]. ML100K is a subset of data obtained from the MovieLens research project. It is an available movie rating dataset and all collecting ratings are in the scale of 1-to-5 star. It contains 100000 ratings by 943 users on 1682 different items. Each user had rated at least 20 movies. For further research, we chose a subset of users who rated more than 100 items from EachMovie dataset and it contains 1196106 ratings by 7471 users on 1619 items. For rating prediction, we perform 5-fold cross validation to measure MAE and RMSE evaluators in our experiments. In each fold,, we use 80% of data as the training set and the remaining 20% as the test set on ML100K dataset. For ranking accuracy, we exactly select 10 rating records per user to fill the test set when evaluate the performance of methods on NDCG values.

4.2   Evaluation Metrics

Originally, we have followed a common practice to evaluate rating prediction and the first evaluation are frequently-used based on error metrics, namely Mean

---

[1] http://www.grouplens.org/node/12
[2] http://www.cs.cmu.edu/lebanon/IR-lab/data.html

Algorithm 1. PNSM-ModelLearning

1:    **input:** $\gamma \leftarrow$ learnrate
2:            $\lambda \leftarrow$ regularization weight
3:    **output:** b_u, b_i, q_i, p_j, y_k
4:    *iteration* $\leftarrow 0$
5:    Initial p_j, q_i, y_k all with small random values
      in $(0, 1)$, keeping them non-negative
6:
7: **while** *iteration* $<$ *maxIter* or error on validation set decreases **do**
8:    **for** each $r_{ui} \in R$ **do**
9:        $b_u \leftarrow b_u - \gamma \frac{\partial f}{\partial b_u}$
10:       $b_i \leftarrow b_i - \gamma \frac{\partial f}{\partial b_i}$
11:       $q_i \leftarrow q_i - \gamma \frac{\partial f}{\partial q_i}$
12:
13:       **for** each $j \in R(u)^+$ **do**
14:           $p_j \leftarrow p_j - \gamma \frac{\partial f}{\partial p_j}$
15:       **end for**
16:
17:       **for** each $k \in R(u)^-$ **do**
18:           $y_k \leftarrow y_k - \gamma \frac{\partial f}{\partial y_k}$
19:       **end for**
20:
21:    **end for**
22:    *iteration* $\leftarrow$ *iteration* $+ 1$
23: **end while**

Absolute Error (MAE) and Root Mean Square Error (RMSE), which are defined as:

$$MAE = \frac{\sum_{u,i} |r_{ui} - \hat{r}_{ui}|}{N} \tag{7}$$

$$RMSE = \sqrt{\frac{\sum_{u,i}(r_{ui} - \hat{r}_{ui})^2}{N}} \tag{8}$$

where $\hat{r}_{ui}$ denotes the predicted rating user u gives to item i and N is the number of tested ratings. Since above-mentioned measures focus on the error between ground truth and predicted rating, they are not a natural fit for evaluating the top-N recommendation task. Rather, we employ an alternative prediction evaluator named Normalized Discounted Cumulative Gain (NDCG) [14] to directly measure top-N performance.

It is notable that the measure significantly sharpens the difference between the approaches over what a traditional accuracy measure could show. The NDCG metric is evaluated over the top-n items on the ranked item list and defined as:

$$NDCG(T, k) = \frac{1}{|T|} \sum_{u \in T} Z_u \sum_{p=1}^{k} \frac{2^{R(u,p)} - 1}{log(1 + p)} \tag{9}$$

where T is the set of users in testing dataset and $R(u,p)$ is the p-th item rating returned by the recommender on the ranked list for user u. Z_u is a normalization factor calculated so that the NDCG of the optimal ranking has a value of 1. The lower ranked items in predicted list that should be higher in ground truth can be penalized with the discounting factor $log(1+p)$. So the NDCG is very sensitive to the ratings of the highest ranked items and it is highly beneficial to measure the ranking quality in recommender systems where the relevance of items at the top positions are far more significant than those at low.

4.3 Comparison Algorithms

We compare the performance of our models against the achieved by baseline estimate[8], ItemBased[15], RSVD (Regularized SVD [16]), NSVD [7] and FISM (Factored Item Similarity Methods [17]). We also compare the performance of two proposed models PNSM$_1$ and PNSM$_2$. We present the best performing model with the explored parameter space for each method.

5 Results

In this section, we present the quality of the recommender algorithms mentioned before. The experimental evaluation is made up of two parts. At first, we discuss the comparison results with other competing methods on all the datasets. Then, we study the effect of various parameters of PNSM on the recommendation performance.

5.1 Comparison with Other Approaches

Here shows the overall results of comparison partners for the recommendation task on datasets. For each method, we explored the following parameter space to benefit the best performing model. Empirical analysis, the learning rate decreased at a rate of 0.97 after one iteration. For itembased CF, we used cosine similarity to measure item-item correlations. We set the dimension of latent feature to be 160 in all SVD-based methods. For PNSM$_1$ and PNSM$_2$, we set the regulation parameter $\lambda = 0.01$, $\alpha = 0.5$ and $\beta = 0.6$. In addition, as to evaluate NDCG, we have zoomed in on N in the range [1...10].

Figure 1(a) presents the MAE on all methods while Figure 1(b) shows the RMSE values on Movielens dataset. From the result, firstly, we can see that ItemBased method outperforms the baseline with great ascension, which means neighborhood similarity contributes to rating prediction. Secondly, all of the referred SVD-based models make a significant improvement compared to baseline, which indicates that the decomposition of matrix is necessary to capture the underlying interactions between items. Obviously, considering all these together, our models perform better than the other methods with the positive and negative feedbacks. As can be seen, PNSM$_1$ is the second best method both on MAE and RMSE. We believe that the positive feedbacks can facilitate predictions.

We also observe that $PNSM_2$ yields best performance in most cases. It has a 4.77 percent improvement on MAE and 4.16 percent improvement on RMSE compared with the baseline on average. This verifies that the negative feedbacks and the positive feedbacks supplement each other and help to find more accurate inferences of users.

Table 1 and Table 2 show the experimental results of NDCG values on ML100K and EachMovie datasets respectively. In this experiment, we take the different top-n numbers with a step size of 2. For each algorithm, we report the NDCG values with the different top-N numbers denoted by NDCG@n, which takes the mean of the NDCG value over the multiple permutations of top-n. For each column, we have highlighted the performance achieved by the best method. The values shown in the bottom row are the performance improvements achieved by $PNSM_2$ over the baseline.

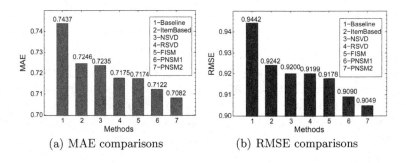

(a) MAE comparisons   (b) RMSE comparisons

Fig. 1. Error matrices comparisons on Movielens

Table 1. NDCG comparisons on ML100K (Bold typeset indicates the best performance. ** indicates statistical significance at $p < 0.001$ and * indicates statistical significance at $p < 0.01$ compared to the second best.)

Methods	NDCG@2	NDCG@4	NDCG@6	NDCG@8	NDCG@10
Baseline	0.710231	0.755715	0.802629	0.846698	0.891105
NSVD	0.731076	0.763308	0.806784	0.852213	0.896343
RSVD	0.753483	0.788281	0.823899	0.865538	0.906164
FISM	0.746136	0.775814	0.816049	0.859586	0.901284
$PNSM_1$	0.763501**	0.789676*	0.826981*	0.868643*	0.908464*
$PNSM_2$	**0.767875****	**0.794709****	**0.831230****	**0.868996***	**0.909579***
	8.12 %	5.16 %	3.56 %	2.63 %	2.07 %

From the results, we observe that $PNSM_1$ and $PNSM_2$ yield better performance under most of the evaluation conditions. It verifies the assumption that recommendation can be improved if we consider the positive and negative feedbacks separately. In particular, $PNSM_1$ provides a better top-N recommendation since the positive feedbacks can help stand out the most relative items. Moreover, because the negative feedbacks help to filter out irrelevant items and

Table 2. NDCG comparisons on EachMovie (Bold typeset indicates the best performance. ** indicates statistical significance at $p < 0.001$ and * indicates statistical significance at $p < 0.01$ compared to the second best.)

Methods	NDCG@2	NDCG@4	NDCG@6	NDCG@8	NDCG@10
Baseline	0.717188	0.764475	0.809590	0.851897	0.886023
NSVD	0.727989	0.769205	0.818331	0.859360	0.890469
RSVD	0.791263	0.825774	0.861321	0.890892	0.916047
FISM	0.772024	0.811722	0.848733	0.881317	0.908155
$PNSM_1$	0.792878*	0.825963*	0.864258*	0.894252*	0.917713*
$PNSM_2$	**0.795960****	**0.827317****	**0.864581***	**0.895560***	**0.918901***
	10.98 %	8.22 %	6.79 %	4.77 %	3.71 %

denoise data, $PNSM_2$ outperforms the other methods in most cases compared to $PNSM_1$. It manifests that not only the positive but also the negative feedbacks play an indispensable role in top-N recommendation. Viewed from the row both in Table 1 and Table 2, all the methods perform better along with the increase of the top-n number. This is because that it can be easier to give a sufficient ranking with more items so that the preferences of users can be estimated more accurately. We also observe that the percentage of improvement based on NDCG@n values of $PNSM_2$ over baseline is inversely proportional to top-n number. That is, it can be noted $PNSM_2$ can better recommend top popular items even with a small n number. In addition, compared to ML100K, we can find that our methods achieve fairly good performance on EachMovie dataset which is much denser. It confirms that the more sufficient the positive and negative feedbacks there are, the better recommendation the method provides.

In summary, the model either $PNSM_1$ or $PNSM_2$ can significantly outperform other comparison partners on MAE, RMSE and NDCG values, which verifies the assumption that it is highly effective to separate the positive feedbacks from the negative ones both in rating prediction and top-N recommendation.

5.2   Impact of the Dimension

As stated previously, we utilized an inner product of two low dimensional latent factor matrices to learn item-item similarities. When the dimension varies, the performance of latent factor models is different with keeping other parameters in constant. We conduct experiments on ML100K dataset with the dimension from 20 to 160 and plot the results of MAE in Figure 2. As can be seen from the figure, our recommendation performance becomes better as the value of dimension increases. When the dimension is 160, we have the best MAE measure.

5.3   Impact of Neighborhood Agreements α, β

In this study, one of the main advantages is that our method incorporates the positive and the negative feedbacks to provide recommendations. Accordingly, another two important parameters α and β respectively control the positive and negative neighborhood agreement between the items rated by users. Figure 3 presents how α and β affect the performance of $PNSM_2$ on ML100K dataset. In

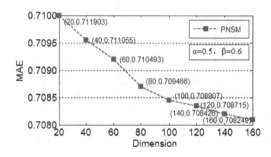 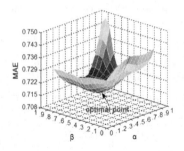

Fig. 2. Impact of Dimension on Movielens   **Fig. 3.** Impact of α and β

the experiments, different values of α and β are set from 0 to 1 with a step size of 0.1 as the dimension is 100. We vary them to better understand the roles they played in the optimal recommendation and compare the impact of neighborhood agreement on the performance.

In the figure, they have similar trends as α and β increase. When α and β are small, they have little effect on the performance because both the positive and the negative implicit correlations of items are ignored. When they increase to 1, these implicit feedbacks overwhelm the rating information and cause the descendence of performance. We can also see that no matter how β varies, the performance tends to be steadily poor when α is close to zero. It is the same to the case in which β is small. This result is consist with our expectation because the positive and negative feedbacks supplement each other while only taking either aspect into account have little effect. In addition, as α/β is around 1, there is a quadratic curve in the figure and MAE value is minimum on the bottom. Specially when α is 0.5 and β is 0.6, we have the best performance. That is, on average, for both the positive examples and negative examples, a substantial number of neighborhood items need to have a high similarity value. This is confirmed that it's necessary to integrate two kind of feedbacks and give them a separation to make recommendations.

6   Conclusion

In this paper, we proposed an extended method that combines both the positive and negative feedbacks of users while learning latent item-item correlations. It's a new variant of item-oriented collaborative filtering algorithm based on matrix factorization techniques. To exploit effective features in recommendation process, we separated the positive examples from the negative ones and treated them differently to capture accurate preference of users. Further, we employed some parameters to control the agreement degree of the neighborhood items, and it's helpful to filter out useless information and denoise data. To evaluate the performance of our proposed method, we conducted a comprehensive set of experiments on ML100K and EachMovie datasets in terms of MAE and NDCG.

The method combining both the positive and negative feedbacks had a statistical significant improvement than other comparison partners, which implies that the classified feedbacks contribute more to recommendation performance. The experimental results showed that our method is significantly effective in both rating prediction and top-N recommendation.

In the future, we will focus on enriching the proposed model by incorporating more relevant factors, such as social trusts, etc.

Acknowledgments. This work was supported by Natural Science Foundation of China (61272240, 71402083, 6110315).

References

1. Kantor, P.B., Rokach, L., Ricci, F., Shapira, B.: Recommender systems handbook. Springer (2011)
2. Ren, Z., Liang, S., Meij, E., de Rijke, M.: Personalized time-aware tweets summarization. In: Proceedings of the 36th International ACM SIGIR Conference on Research and Development in Information Retrieval, pp. 513–522. ACM (2013)
3. Ren, Z., Peetz, M.-H., Liang, S., van Dolen, W., de Rijke, M.: Hierarchical multi-label classification of social text streams. In: Proceedings of the 37th International ACM SIGIR Conference on Research Development in Information Retrieval, pp. 213–222. ACM (2014)
4. Goldberg, D., Nichols, D., Oki, B.M., Terry, D.: Using collaborative filtering to weave an information tapestry. Communications of the ACM **35**(12), 61–70 (1992)
5. Mooney, R.J., Roy, L.: Content-based book recommending using learning for text categorization. In: Proceedings of the 5th ACM Conference on Digital Libraries, pp. 195–204. ACM (2000)
6. Pazzani, M.J., Billsus, D.: Content-based recommendation systems. In: Brusilovsky, P., Kobsa, A., Nejdl, W. (eds.) Adaptive Web 2007. LNCS, vol. 4321, pp. 325–341. Springer, Heidelberg (2007)
7. Paterek, A.: Improving regularized singular value decomposition for collaborative filtering. In: Proceedings of KDD cup and workshop, vol. 2007, pp. 5–8 (2007)
8. Koren, Y.: Factorization meets the neighborhood: a multifaceted collaborative filtering model. In: Proceedings of the 14th ACM SIGKDD International Conference on Knowledge Discovery and Data Mining, pp. 426–434. ACM (2008)
9. Rendle, S., Freudenthaler, C., Gantner, Z., Schmidt-Thieme, L.: Bpr: bayesian personalized ranking from implicit feedback. In: Proceedings of the 25th Conference on Uncertainty in Artificial Intelligence, pp. 452–461. AUAI Press (2009)
10. Ning, X., Karypis, G.: Slim: sparse linear methods for top-n recommender systems. In: Proceedings of the 11th International Conference on Data Mining, pp. 497–506. IEEE (2011)
11. Mnih, A., Salakhutdinov, R.: Probabilistic matrix factorization. In: Advances in Neural Information Processing Systems, pp. 1257–1264 (2007)
12. Koren, Y., Bell, R., Volinsky, C.: Matrix factorization techniques for recommender systems. Computer **42**(8), 30–37 (2009)

13. Rennie, J.D.M., Srebro, N.: Fast maximum margin matrix factorization for collaborative prediction. In: Proceedings of the 22nd International Conference on Machine Learning, pp. 713–719. ACM (2005)
14. Järvelin, K., Kekäläinen, J.: Cumulated gain-based evaluation of ir techniques. ACM Transactions on Information Systems (TOIS) **20**(4), 422–446 (2002)
15. Sarwar, B., Karypis, G., Konstan, J., Riedl, J.: Item-based collaborative filtering recommendation algorithms. In: Proceedings of the 10th International Conference on World Wide Web, pp. 285–295. ACM (2001)
16. Funk, S.: Netflix update: Try this at home, 2006 (2011). http://sifter.org/~simon/journal/20061211.html
17. Kabbur, S., Ning, X., Karypis, G.: Fism: factored item similarity models for top-n recommender systems. In: Proceedings of the 19th ACM SIGKDD International Conference on Knowledge Discovery and Data Mining, pp. 659–667. ACM (2013)

Review Comment Analysis for Predicting Ratings

Rong Zhang[1,2], Yifan Gao[1,2], Wenzhe Yu[1,2], Pingfu Chao[1,2], Xiaoyan Yang[3], Ming Gao[1,2(✉)], and Aoying Zhou[1,2]

[1] Institute for Data Science and Engineering, East China Normal University, Shanghai, China
{rzhang,mgao,ayzhou}@sei.ecnu.edu.cn, {yfgao,wyu,pfchao}@ecnu.edu.cn
[2] Shanghai Key Laboratory of Trustworthy Computing, East China Normal University, Shanghai, China
[3] Advanced Digital Sciences Center, Illinois at Singapore Pte. Ltd., Singapore, Singapore
xiaoyan.yang@adsc.com.sg

Abstract. Rating prediction is a common task in recommendation systems that aims to predict a rating representing the opinion from a user to an item. In this paper, we propose a comment-based collaborative filtering (CCF) approach that captures correlations between hidden aspects in review comments and numeric ratings. The idea is motivated by the observation that the opinion of a user against an item is represented by different aspects discussed in review comments. In our approach, we first explores topic modeling to discover hidden aspects from review comments. Profiles are then created for users and items separately based on the discovered aspects. In the testing stage, we estimate the aspects of comments based on the profiles of users and items because the comments are not available when testing. Lastly, we build final systems by utilizing the profiles and traditional collaborative filtering methods. We evaluate the proposed approach on a real data set. The experimental results show that our prediction systems outperform several strong baseline systems.

1 Introduction

Recommendation system is among the most popular, simple and useful tasks in web applications. It aims to recommend some items for users by analyzing behaviors of users. The approaches on this topic can be grouped into two types [17]: Content-based recommendations and collaborative filtering (CF). The content-based approaches build a profile for each user or item to capture its properties [1]. For example, a restaurant profile could include the menus it serves, its location, and so on. Then the profiles are used to predict whether a user likes an item. Compared with the content-based approaches, the traditional collaborative filtering approach makes prediction based on the ratings expressed by similar users, e.g. the scores for items previously viewed or purchased.

With the prevalence of the Web, customer review comments, which are usually associated with numeric ratings, have become commonly available on many

© Springer International Publishing Switzerland 2015
J. Li and Y. Sun (Eds.): WAIM 2015, LNCS 9098, pp. 247–259, 2015.
DOI: 10.1007/978-3-319-21042-1_20

e-commerce sites. For example, users give reviews containing comments and ratings to products they bought or services they received on *Yelp*[1] and *Dianping*[2]. These review comments contain abundant information about the opinions and preferences of users, which could be valuable to recommendation systems. However, in the traditional approaches the review comments are often ignored. More specifically, the collaborative filtering approaches just consider rating scores while the content-based approaches utilize the predefined description of users and items.

In this paper, we propose a new approach, named Comment-based Collaborative Filtering (CCF), by utilizing the review comments to predict rating scores. Compared with the traditional collaborative filtering approaches, our approach makes prediction based on the comments given by similar users rather than ratings. In our approach, the relationships between users and interdependencies among products are connected via analyzing the review comments. We first apply a topic model to analyze each comment to obtain hidden aspects/topics which represent the opinions or preferences of users. Then, profiles of users and items are represented by the hidden topics. Finally, we build prediction systems with different classification models to predict rating scores based on the profiles. The systems are further enhanced by combining the traditional collaborative filtering model. The main issue of comment-based prediction is how to automatically generate simulated comments because the comments are not available when testing. We tackle with this issue by utilizing the profiles of users and items.

To investigate the effect of the CCF approach, we evaluate our systems on a real data set, which is created on Dianping.com, the largest website for restaurant reviews in China. The experimental results show that the CCF approach yields better performance than strong baselines. Our main contributions are: First, we combine a probabilistic topic model in review comments with classifiers in rating prediction; Second, in our approach, the profiles of items are generated by analyzing the review comments written by the users who share similar interests; Third, our model can deal with multiple tasks to predict any kinds of service rating that has ratings and review comments.

2 Related Work

Existing collaborative-filtering techniques [10,11,18] typically rely on analyzing past user rating behavior to make predictions on new items for a user, in which text reviews are not utilized. On the other hand, there have been ample work [4,9,14,15,19,20] on review analysis from aspect discovery [4,9,20], sentiment analysis [19] to opinion mining [14,15], etc. However, none of them studies the linkage between ratings and text reviews for making recommendations. [7,16] studies review rating prediction by combining text analysis techniques.

---

[1] http://www.yelp.com
[2] http://www.dianping.com

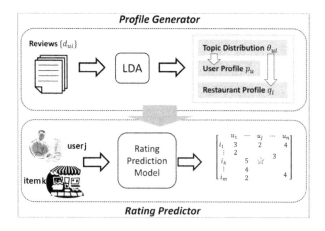

Fig. 1. System Flowchart

[16] models reviews using bag-of-opinions[3] representation, which is more expressive than unigram and n-gram representations. A linear model is then learned based on opinion roots, modifiers and negation words for review rating prediction. [7] models ratings as a function of pre-defined aspects and their polarities discovered from reviews. However, these two models are designed for predicting rating of a given review, where user-item relationship is not captured. Therefore they cannot be directly applied to a recommender system. Besides, the opinions learned by [16] tend to be conclusive comments that convey little information about different aspects of a particular item.

The work most related to ours is [13], in which HFT (Hidden Factors as Topics) is proposed to combine ratings with review text for product recommendation. HFT aligns hidden factors in ratings with hidden topics in reviews to create user/item profiles, which are then fit into SVD [10], a matrix factorization model to make rating predictions. The problem with HFT is that each time, review text is associated with one of the two dimensions, i.e., either from items' perspective (by grouping reviews by items) or from users' perspective (by grouping reviews by users), which means the hidden topics discovered only reflect the hidden factors of ratings from one dimension. The profiles of the other dimension are forced to be aligned to the same hidden factor space. We overcome this issue by proposing a novel model that considers both dimensions in discovering the common hidden factor space from review text.

Table 1. Table of Notations

Symbol	Description
K	# of latent topics, i.e., the dimension of the latent factor space
r_{ui}	rating of item i by user u
$\hat{r}_{ui}$	predicted rating of item i by user u
d_{ui}	review of item i by user u
q_i	profile of item i, $q_i \in [0,1]^K$
p_u	profile of user u, $p_u \in [0,1]^K$
θ_{ui}	topic distribution vector for d_{ui}, $\theta_{ui} \in [0,1]^K$

3   CCF Model

The flowchart of our approach to predict ratings is presented in Figure 1. The input to our system is a review corpus and a rating matrix. LDA model is used to uncover hidden aspects in review text and generates user/item profiles with a K-dimensional topic. Combing profiles with rating matrix, we can predict rating of an item, to which a user never given a review with regression models. As shown in Figure 1, Our model consists of two main components: *profile generator* and *rating predictor*.

Profile generator accepts item reviews $\{d_{ui}\}$ as input and outputs user/item profiles: p_u for user u and q_i for item i.

Rating predictor accepts a pair of user u and item i as input and outputs the rating user u would give to item i.

The two functional components are connected where the output of the profile generator, including topic distributions θ_{ui}, user profile p_u and item profile q_i, are fed into the rating predictor for training regression models and making predictions.

Based on the profiles, we also design a browsing tool to enable efficient access to *representative reviews* of items.

3.1   Topic Analysis

Similar to probabilistic latent semantic analysis (pLSA), Latent Dirichlet Allocation(LDA) is a generative probabilistic model which regards each document d is composed of a K-dimensional topic distribution θ [2]. Meanwhile, words in document d state each of topic k with probability ϕ. It's easily seen that LDA reduces dimensions of the vector space with referencing topic dimension between documents and words. In this work, LDA is used for analyzing aspects hidden in user reviews.

As shown in Figure 2, a document is a sequence of N words and a corpus is a collection of M documents. α and β are two parameters for drawing θ and

---

[3] An opinion here contains a subjectivity clue with positive or negative polarity.

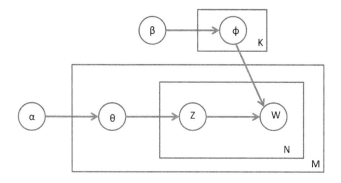

Fig. 2. Graphical model representation of LDA

ϕ separately from the prior probability of the Dirichlet distributions. Here, let z denote topic. From decuments to topics and from topics to words, topic distribution θ and word distribution ϕ follow multinomial distributions. Therefore, processing of the corpus and selecting of parameters are crucial in the application of LDA to get an ideal result. We will discuss the selecting of parameters in part of experiments.

3.2   Profile Generator

Similar to [10], we map users and items to a common latent space $\mathcal{S}$. In this work, we try to discover $\mathcal{S}$ from the review text. The intuition is that reviews, though in the format of unstructured free text, contain information about user preferences and opinions on different aspects of items. These aspects 'hidden' in the review text may well reflect the latent factors that affect the user ratings. To find these hidden aspects and construct the latent space, we apply standard Latent Dirichlet Allocation (LDA) to reviews, as LDA or topic modeling is well recognized as an effective tool for text analysis, extracting prevalent, meaningful and potentially overlapping topics from pure texts. As our experiments show, the topics discovered by LDA are well interpretable and have good correspondence with human-labeled common aspects in restaurant reviews.

Let d_{ui} denote the review of item i by user u. Different from HFT, we treat each d_{ui} as one document. We apply LDA on the review corpus $\{d_{ui}\}$ and discover K topics represented it. Let θ_{ui} denote the topic distribution of d_{ui} generated by LDA. Define D_u as the set of reviews written by user u, and D_i as the set of reviews written for item i. Each user u (or item i) is associated with a profile p_u (or q_i), which is a vector from $\mathcal{S}$. In our system, $\mathcal{S} = [0,1]^K$. For a given user u, we define her profile p_u as follows:

$$p'_{uj} = \frac{\sum_i \theta_{uij}}{|D_u|}$$

$$p_{uj} = \frac{p'_{uj}}{\sum_j p'_{uj}}, j \in [1, K] \tag{1}$$

where $p_u = (p_{u1}, p_{u2}, \cdots, p_{uK})$, p_{uj} is the distribution on the jth topic for user u, and θ_{uij} is the distribution on the jth topic for review d_{ui}. Similarly, we define profile q_i for item i as:

$$q'_{ij} = \frac{\sum_u \theta_{uij}}{|D_i|}$$

$$q_{ij} = \frac{q'_{ij}}{\sum_j q'_{ij}}, j \in [1, K] \tag{2}$$

In summary, profile p_u/q_i is the normalized average topic distribution over all reviews of a given user u/item i. The notations are summarized in Table 1.

3.3   Representation of Samples

Given a pair of user u and item i, we want to predict the rating $\hat{r}_{ui}$ the user u would give to item i. Recommendations are then made based on $\hat{r}_{ui}$ of items that u have not rated/visited.

To predict ratings, we rely on the intuition that hidden topics discovered from review text define the latent factors that affect the ratings. Prediction models (described in Section 3.4) learn the relationship between ratings r_{ui} and topic distributions θ_{ui} of d_{ui} on training data and predict the ratings on test data.

Training Samples. As described above, each sample d_{ui} in the training data is represented as a feature vector which has K dimensions (a topic distribution generated by LDA),

$$d_{ui} = [\theta_{ui1}, ..., \theta_{uij}, ..., \theta_{uiK}] \tag{3}$$

The prediction model is trained on the feature vectors of the training data.

Test Samples. When testing, the review comment d_{ui} is not available since user u has not rated item i. Thus, we generate $\hat{d}_{ui}$ based on the profiles p_u and q_i. Each dimensional value of $\hat{d}_{ui}$ is estimated as,

$$\theta'_{uij} = p_{uj} q_{ij}$$

$$\hat{\theta}_{uij} = \frac{\theta'_{uij}}{\sum_j \theta'_{uij}}, j \in [1, K] \tag{4}$$

Then, $\hat{d}_{ui}$ is represented as,

$$\hat{d}_{ui} = [\hat{\theta}_{ui1}, ..., \hat{\theta}_{uij}, ..., \hat{\theta}_{uiK}] \tag{5}$$

$\hat{d}_{ui}$ is then fed into the prediction model to predict $\hat{r}_{ui}$.

3.4   Prediction Models

In our systems, we use different models to learn the relationships between the discovered topics and ratings.

Linear Regression (LR). is a standard regression analysis model used extensively in practical applications. Supposing characteristics and the results are linear, parameters are easier to fit and studied rigorously [5]. A rating is predicted with multiple linear regression by the following function:

$$\hat{r}_{ui} = W^T d_{ui} + \varepsilon_{ui} \tag{6}$$

where $W = (W_1, ..., W_K)$, W_j is the weight of the jth topic, and ε_{ui} is an error variable.

Gradient Boosted Regression Trees(GBRT). Gradient Boosted Regression Trees[6] (GBRT) is a machine learning technique for regression problems, which produces a prediction model in the form of an ensemble of decision trees. Similar to other boosting methods, GBRT combines weak learners into a single strong learner. The target is to learn a model F that predicts values, minimizing the mean squared error to the true values on development sets in each iteration,

- Initialize model with a constant value $F_0(d_{ui})$
- For $m = 1, ..., M$:
- $F_m(d_{ui}) = F_{m-1}(d_{ui}) + \gamma_m h_m(d_{ui})$, $\gamma_m = \arg\min_\gamma \sum_{(u,i)} L(r_{ui}, F_{m-1}(d_{ui}) + \gamma h_m(d_{ui}))$

where, $h_m(d)$ is a decision tree and $L(r, \bullet)$ is a loss function.

Random Forest (RF). Random forest[3] is an ensemble learning method for classification that construct a multitude of decision trees at training time and make decisions by combining the outputs of individual trees. The training algorithm for random forest applies the general idea of bagging to tree learners. The algorithm repeatedly selects a random subset with replacement of the features and fits trees to the training examples,

- For $b = 1, ..., B$:
- 1. Sample, with replacement of the features for training examples from the training data; call X_b, Y_b.
- 2. Train a decision tree f_b on X_b, Y_b.

When testing, predictions for a new example $\hat{d}_{ui}$ can be made by averaging the predictions from all the individual trees on $\hat{d}_{ui}$,

$$\hat{r}_{ui} = \frac{1}{B} \sum_{b=1}^{B} f_b(\hat{d}_{ui}) \tag{7}$$

3.5   Enhanced Systems

So far, we have described the CCF systems based on different machine learning techniques. In this section, we enhance the CCF systems by combining a traditional collaborative filtering approach.

There are many collaborative filtering methods that have been studied recently. Among them, Bias From Mean (BFM)[8] is a very efficient method with low computational cost. In BFM, the rating is predicted by,

$$\beta_{ui} = \overline{r_u} + \frac{1}{n} \sum_{v \in Z_i} (r_{vi} - \overline{r_v}) \tag{8}$$

where, Z_i refers to all the users except u who rated item i and $\overline{r_u}$ is the average value of ratings user u gives.

The prediction models take β_{ui} as an additional feature. For example, the new prediction function for the LR model is,

$$r_{ui} = W^T d_{ui} + W_\beta \beta_{ui} + \varepsilon_{ui} \tag{9}$$

where W_β is the weight for the new feature. All the weights are re-trained on the training data.

4   Experiments

In this section, we evaluate our systems on a real data set.

4.1   Dataset

We crawl the review data from *Dianping* which is the biggest restaurant review site in China to evaluate the performance of our system. The reviews contain user IDs, restaurant IDs, numeric ratings ($[1, 5]$), and comments. We filter out the reviews that do not have comments. Finally, our dataset consists of 3.62M reviews written by 638.6K users for 48.7K restaurants. The detailed statistics of the data sets are summarized in Table 2, where "Original" refers to the data we crawl from the site and "Filtered" refers to the data after filtering. In Figure 3, we report the log-log plot of the distributions in terms of the number of users or restaurants with various amount of reviews. The two distributions on users and restaurants follow the power law. We randomly split the whole data set into two sets: 80% as training data and 20% test data. We also randomly select 10% of training data as development data to tune the parameters of the systems.

4.2   Evaluation Metric

We report the mean absolute error (MAE) of the rating predictor, i.e.,

$$MAE = \frac{1}{N} \sum_{u,i \in T} |\hat{r}_{ui} - r_{ui}|$$

where T is the test set, N is the total number of predicted ratings, $\hat{r}_{ui}$ and r_{ui} are the predicted rating and the user assigned rating scores for user u and item i, respectively.

Table 2. Statistics on the Dianping data

	#ofUsers	#ofRestaurants	#ofComments
Original	703.4K	51.5K	4.41M
Filtered	638.6K	48.7K	3.62M

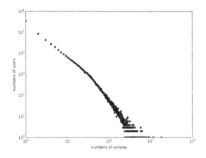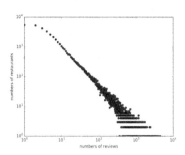

Fig. 3. The log-log plot on the number of comments and the number of users/restaurants

4.3 Systems

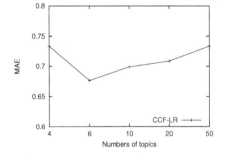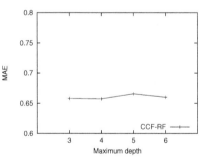

Fig. 4. Different numbers of topics for LDA

Fig. 5. Different values of maximum depth for RF

In the experiments, we compare our systems with other approaches that are listed as follows:

- **BFM**: BFM is a collaborative filtering (CF) system based on Bias From Mean described in [8].
- **SlopeOne**: SlopeOne[12] is an item-based CF algorithm which is famous for its simplicity and efficiency. Essentially, Slope One is a simple form of regression model.

- **HFT**: As mentioned in the related work, HFT[13] combines rating dimensions with latent review topics for product recommendation and it is the work most similar to ours. The implementation of HFT is available online[4].
- **CCF-LR** and **CCF-LR+**: CCF-LR refers to the CCF system based on Linear Regression (LR) model and CCF-LR+ refers to the enhanced system that include the CCF-LR system and Bias From Mean.
- **CCF-GBRT** and **CCF-GBRT+**: CCF-GBRT refers to the CCF system based on (GBRT) model and CCF-GBRT+ refers to the enhanced system that include the CCF-GBRT system and Bias From Mean.
- **CCF-RF** and **CCF-RF+**: CCF-RF refers to the CCF system based on Random Forest (RF) model and CCF-RF+ refers to the enhanced system that include the CCF-RF system and Bias From Mean.

4.4 Topic Analysis

We use GibbsLDA++[5] to perform topic analysis and set the hyper-parameters $\alpha = 0.2$, and $\beta = 0.1$, and the number of iterations is set to 1000. As for the number of topics, we run the CCF-LR system with different values. The experiment results are shown in Figure 4. From the figure, we find that the system achieves the best score when the number of topics is 6. In the following sections, we set the number of topics as 6 for LDA in the experiments.

4.5 Tree Depth for Random Forest

Here we investigate the effect of the maximum depth of the tree which is a very important parameter for Random Forest. We run the CCF-RF system with different values for the maximum depth. The results are shown in Figure 5. The figure shows that the CCF-RF system provides the best score when the maximum depth is 4. In the following sections, we set the maximum depth as 6 for the RF model in the experiments.

4.6 Main Results

Table 3 shows the main results of predicting ratings on the test data. From the table, we find that BFM performs the best among the baselines. Compared with HFT which also utilizes comments for prediction, our CCF systems achieve better scores. The results also show that the enhanced systems can perform better than the single systems. Among all the systems, CCF-RF+ obtains the highest MAE score. This fact indicates that the information the CCF model learns is different from the information that the traditional CF models learn.

---

[4] http://www.cseweb.ucsd.edu/~jmcauley/
[5] http://gibbslda.sourceforge.net/

Table 3. Main results of predicting ratings

Single System	MAE	Enhanced System	MAE
BFM	0.6066		
SlopeOne	0.6885		
HFT	0.6803		
CCF-LR	0.6765	CCF-LR+	0.6004
CCF-GBRT	0.6606	CCF-GBRT+	0.5970
CCF-RF	0.6575	CCF-RF+	0.5938

Table 4. Topic words generated by LDA

Topic1	Topic2	Topic3	Topic4	Topic5	Topic6
好	好	好	菜	服务员	好
good	good	good	dish	waiter	good
味道	味道	味道	味道	好	不错
taste	taste	taste	taste	good	very good
甜	不错	小	不错	菜	环境
sweet	very good	small	very good	dish	environments
蛋糕	好吃	面	好	店	店
cake	delicious	noodle	good	restaurant	restaurant
不错	饭	不错	鱼	团	菜
very good	rice	very good	fish	group	dish
小	小	汤	辣	差	味道
small	small	soup	spicy	bad	taste
奶茶	大	店	好吃	态度	感觉
tea with milk	big	restaurant	delicious	manner	feeling
面包	牛肉	好吃	香	东西	朋友
bread	beef	delicious	good smell	goods	friends
好吃	鱼	鸡	虾	钱	价格
delicious	fish	chicken	shrimp	money	price
茶	锅	大	大	少	小
tea	hot pots	big	big	few	small

4.7   Further Analysis

In the main experiments, our comment-based systems have shown their efficiency in predicting ratings. Here, we investigate the topic words generated by LDA to know how the systems work. The TOP10 topic words (of 6 topics) are shown in Table 4. From the table, we can infer the topics from the words although there are some overlapped words among the topics. The corresponding topics of Topic[1-6] are: Sweets and drinks, Meat and hot pots, Noodles, Vegetable dishes, Services, and Environments. We also find that there are some words that are not related to the topics. The prediction systems still have room to improve further. We leave it in future work.

5 Conclusion

In this paper, we present a comment-based collaborative filtering approach that novelly explores the connections between ratings and review comments. In particular, we utilize the hidden topics discovered in review comments to model users and items in a latent space. Our prediction systems build on several learning models to union predicted ratings with latent topic distributions. The systems are evaluated on a real data set from the biggest restaurant review site in China. The experiments demonstrate the effectiveness of our proposed method in rating prediction over state-of-art approaches.

Acknowledgments. This work is partially supported by National Science Foundation of China (Grant No.61232002, 61402177 and 61332006), and Program for Innovative Research Team in Yunnan University under grant No. XT412011. Xiaoyan Yang is supported by Human-Centered Cyber-physical Systems (HCCS) programme by A*STAR in Singapore.

References

1. Blanco-Fernández, Y., Pazos-Arias, J.J., Gil-Solla, A., Ramos-Cabrer, M., López-Nores, M., García-Duque, J., Fernández-Vilas, A., Díaz-Redondo, R.P., Bermejo-Muñoz, J.: A flexible semantic inference methodology to reason about user preferences in knowledge-based recommender systems. Knowledge-Based Systems **21**(4), 305–320 (2008)
2. Blei, D.M., Ng, A.Y., Jordan, M.I.: Latent dirichlet allocation. The Journal of machine Learning research **3**, 993–1022 (2003)
3. Breiman, L.: Random forests. Machine learning **45**(1), 5–32 (2001)
4. Brody, S., Elhadad, N.: An unsupervised aspect-sentiment model for online reviews. In: Human Language Technologies: The 2010 Annual Conference of the North American Chapter of the Association for Computational Linguistics, pp. 804–812. Association for Computational Linguistics (2010)
5. Freedman, D.: Statistical models: theory and practice. Cambridge University Press (2009)
6. Friedman, J.H.: Stochastic gradient boosting. Computational Statistics & Data Analysis **38**(4), 367–378 (2002)
7. Ganu, G., Elhadad, N., Marian, A.: Beyond the stars: improving rating predictions using review text content. In: WebDB (2009)
8. Herlocker, J.L., Konstan, J.A., Borchers, A., Riedl, J.: An algorithmic framework for performing collaborative filtering. In: Proceedings of the 22nd Annual International ACM SIGIR Conference on Research and Development in Information Retrieval, pp. 230–237. ACM (1999)
9. Jo, Y., Oh, A.H.: Aspect and sentiment unification model for online review analysis. In: Proceedings of the Fourth ACM International Conference on Web Search and Data Mining, pp. 815–824. ACM (2011)
10. Koren, Y., Bell, R.: Advances in collaborative filtering. In: Recommender Systems Handbook, pp. 145–186. Springer (2011)
11. Koren, Y., Bell, R., Volinsky, C.: Matrix factorization techniques for recommender systems. Computer **42**(8), 30–37 (2009)

12. Lemire, D., Maclachlan, A.: Slope one predictors for online rating-based collaborative filtering (2005)
13. McAuley, J., Leskovec, J.: Hidden factors and hidden topics: understanding rating dimensions with review text. In: Proceedings of the 7th ACM Conference on Recommender Systems, pp. 165–172. ACM (2013)
14. Pang, B., Lee, L.: Opinion mining and sentiment analysis. Foundations and trends in information retrieval **2**(1–2), 1–135 (2008)
15. Popescu, A.-M., Etzioni, O.: Extracting product features and opinions from reviews. In: Natural language Processing and Text Mining, pp. 9–28. Springer (2007)
16. Qu, L., Ifrim, G., Weikum, G.: The bag-of-opinions method for review rating prediction from sparse text patterns. In: Proceedings of the 23rd International Conference on Computational Linguistics, pp. 913–921. Association for Computational Linguistics (2010)
17. Rajaraman, A., Ullman, J.D.: Mining of massive datasets, Chapter 9. Cambridge University Press (2012)
18. Sarwar, B., Karypis, G., Konstan, J., Riedl, J.: Item-based collaborative filtering recommendation algorithms. In: Proceedings of the 10th International Conference on World Wide Web, pp. 285–295. ACM (2001)
19. Titov, I., McDonald, R.: A joint model of text and aspect ratings for sentiment summarization. Urbana **51**, 61801 (2008)
20. Titov, I., McDonald, R.: Modeling online reviews with multi-grain topic models. In: Proceedings of the 17th International Conference on World Wide Web, pp. 111–120. ACM (2008)

Adaptive Temporal Model for IPTV Recommendation

Yan Yang, Qinmin Hu$^{(\boxtimes)}$, Liang He, Minjie Ni, and Zhijin Wang

Shanghai Key Laboratory of Multidimensional Information Processing,
Department of Computer Science and Technology,
East China Normal University Shanghai, Shanghai 200241, China
{yanyang,qmhu,lhe}@cs.ecnu.edu.cn, mjni@ica.stc.sh.cn, zhijin@ecnu.cn

Abstract. How to help the IPTV service provider make the program recommendation to their clients is the problem we propose to solve in this paper. Here we offer an adaptive temporal model to identify multiple members under a shared IPTV account. The time intervals are first detected and defined in each account. Then, the preference similarity is calculated among the intervals to extract the members. After that, we evaluate our model on the industrial data sets by a famous IPTV provider. The experimental results show that our proposed model is promising and outperform the state-of-the-art algorithms with low computational complexity and versatility without user feedback. Furthermore, the proposed model has been officially adopted by the IPTV provider and applied in their IPTV systems with excellent user satisfaction in 2013.

1 Introduction and Motivation

With the explosive growth of multimedia resources, customized content/program recommendation is becoming very important and necessary for the Internet Protocol Television (IPTV) users. Usually for the IPTV clients, a family shares an IPTV account. This situation makes the members hard to retrieve their programs directly and the IPTV system is as well difficult to recommend the programs to the members. Therefore, it is a critical challenge for the IPTV system to identify the members and make the accurate personalized recommendation.

We collect our data from a famous IPTV service provider. Table 1 presents the sample logs under an IPTV account "84AA70", in which we observe that (1) genres of programs are vary under the same account during a day; (2) the similar genres are often demanded at the same period. Then we imagine this scenario that there could be three family members: kids, mom and dad. The kids love to play cartoon after dinner, the mother watches costume drama in the afternoon and the father prefers the action and gangster movies at night. In the traditional IPTV recommender systems, frequent or recent genres are often recommended to the account, regardless of the time difference and member preference. At the above scenario, it is definitely failed when cartoon has been recommended to the dad.

© Springer International Publishing Switzerland 2015
J. Li and Y. Sun (Eds.): WAIM 2015, LNCS 9098, pp. 260–271, 2015.
DOI: 10.1007/978-3-319-21042-1_21

Table 1. Sample Data: account logs

Account ID	Date	Program ID	Genre	Start	End
84AA70	3/1	1002200032	Cartoon	19:59	20:21
84AA70	3/3	1002030011	Cartoon	20:42	21:04
84AA70	3/13	1004190009	Cartoon	21:47	22:10
84AA70	3/8	1011090004	Action	22:32	23:23
84AA70	3/12	1011050002	Action	23:23	23:33
84AA70	3/14	1102230005	Action	1:55	3:19
84AA70	3/12	0910180041	Costume	19:37	20:24

Therefore, we make the assumptions as follows: (1)an IPTV account is shared by one or more members in the family; (2)members have diverse interest genres; (3)there is only a member watching the TV at a time interval.

In this paper, we propose an adaptive temporal model to distinguish the potential members for personalized recommendation, which makes the recommendation to a specific member individually, instead of a shared account. The temporal information in the logs is fully used to dynamically define time intervals fitting to each family. We first compare the genre types through computing the similarity of the genre preference. Then, we apply a graph to identify the latent members, where the nodes are the time intervals and the arcs stand for the similarity of the genres. After that, we recommend the personalized genres to the identified members. Our experimental results show the proposed model achieves the improvements as highly as 200% over the baselines. Here the baselines are adopted as the standard user collaborative filtering (CF) algorithm [5,6] and item CF algorithm [9]. Currently,The adaptive temporal model has been well integrated in the IPTV system on the real-time data by the data provider.

The rest of the paper is organized as follows. In Section 2, we present the overview of the AT-IPTV system, followed by the proposed adaptive temporal model in Section 3, including the noise definition, the time interval detection and the member identification. We describes the experimental setup in Section 4, such as the data collection and the evaluation methods. Section 5 shows the experimental results, and Section 6 presents the analysis and discussions. The related work is presented in Section 7. Finally, we draw our conclusions and future work in Section 8.

2   Overview of the AT-IPTV System

There is a general situation in our data that a family shares an IPTV account and members have program preferences. Figure 1 presents the picture of the system. Here we focus on the genre/program recommendation to each member instead of the whole shared account so that our proposed approach always deal with data under an account.

The input of the system is the account log. The output is the genre/program recommendations to each member. Therefore, there is a cycle that the feedback will be collected as the log, once the recommendations have been proposed.

The most important step is the adaptive temporal model which is designed to identify the members in a shared account. First of all, the non-overlapped time intervals are observed and defined. We then identify members through computing the program similarities among the time intervals. Third, the log of each member is extracted correspondingly. After that, we apply the recommendation system to make the specific recommendations to each member, where the user CF and the item CF algorithms are adopted.

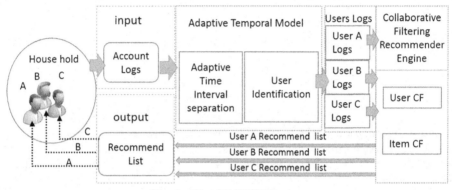

Fig. 1. The AT-IPTV System

3 Adaptive Temporal Model

Throughout this paper, we formally define the terms as (1) the account set $A = (a_1, a_2, \ldots, a_n)$; (2) the interval set $T_a = (t_1, \ldots, t_k)$, for each account a; (3) the program set $P_a = (p_1, \ldots, p_j)$, where $p_j \in T_a$; (4) the program genre set $G = (g_1, g_2, \ldots, g_m)$, (5) the member set $U_a = (u_1, u_2, \ldots, u_l)$ and (6) the original consuming time as $X = (x_1, x_2, \ldots, x_q)$.

3.1 Noise Definition

There are noises in the data if the watching period is very short. In order to get rid of the noises, we do statistics to measure how to set the period as the countable one.

Then, we put all the consuming time $X = (x_1, x_2, \ldots, x_q)$ together, regardless of the influences of the accounts and the genre types. The expectation of the consuming time is around 1758 seconds and the variance is around 257547. Here we put the expectation equation and the variance equation as

$$E(X) = \bar{X} = \sum_{i=1}^{q} x_i \tag{1}$$

$$D(X) = S^2 = \frac{1}{q-1} \sum_{i=1}^{q} (x_i - \bar{x})^2 \tag{2}$$

In the experiments, if we take the mean value 1758 seconds as the criteria to extract the data, we find that almost 41.3% data have been cleaned, which makes the data very sparse.

There is a very interesting observation that the data can be fit to a normal distribution with respect to the sample mean and the sample variance. Then, we decide to adopt the definition of the confidence interval. The confidence level α we set is 0.10 after we test it from 0.05 to 0.15. Finally, the confidence interval is around $[577, 2939]$. Hence, we take 577 seconds and define all the consuming periods less than 577 seconds as the noises mathematically.

3.2   Adaptive Time Interval Detection

In order to split the consuming time conveniently, we use 600 seconds (10 minutes) which is approximate to 577 seconds to filter the noises.

Figure 2 presents a typical program consuming distribution under a shared account in 30 days. Here the noises have been removed. The x axis denotes the 24 hours a day. The y axis stands for the frequency that this account watches the TV as the same time in 30 days.

It is easy to see that most of the members watch the TV at noon, in the evening and at night throughout the midnight respectively. Intuitively, we initialize the whole day, twenty-four hours, as 143 slips. The length of each slip is 10 minutes.

Figure 2 demonstrates that the active period whose frequency is larger than 1 is our research target. Then, we define the active periods with the start point A_i and the end point B_i. The slips in $[A_i, B_i$ are detected as the time intervals with the length of 10 minutes.

Based on the time interval detection and the watching history, we next step examine the relationships among the time interval and the program types. Then, we can identify the members from a shared account.

3.3   Member Identification

Focusing on the genre types and the active consuming periods, we get the following conclusions: (1) there are obvious different types of programs consumed every day; (2) similar programs are selected during the same time periods each day.

We assume that a specific member prefers one or several genre types of programs. In order to identify multiple members, we are motivated to consider the similarity of the consumed programs among the time intervals. Then, the

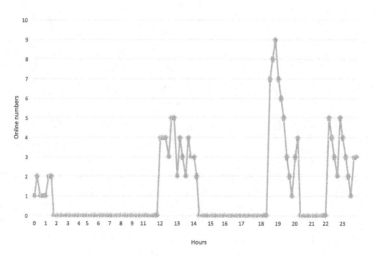

Fig. 2. Distribution of Time Consumption

more similar of two time intervals, the more possible the two intervals belong to a same member.

Under a shared account a, we get its corresponding time interval set as $T_a = (t_1, \ldots, t_i, \ldots, t_k)$, the program set as $P_a = (p_1, \ldots, p_i, \ldots, p_w)$ and the genre set as $G_a = (g_1, g_2, \ldots, g_m)$. Then, we set a rating matrix to represent the relationships among the genre set and the time interval set under the account a as

$$R_a = G_a \times T_a = (r_{g_i}^{t_j})_{m \times k} \tag{3}$$

where $r_{g_i}^{t_j}$ stands for the ratio of the program genre type g_i when they are consumed at the time interval t_j.

We define $r_{g_i}^{t_j}$ as

$$r_{g_i}^{t_j} = \frac{\sum_{p_z \epsilon g_i} x_{p_z}^{t_j}}{\sum_{g_i \epsilon G_a} \sum_{p_z \epsilon g_i} x_{p_z}^{t_j}} \tag{4}$$

where the genre type $g_i \epsilon G_a$, program $p_z \epsilon g_i$ and $x_{p_z}^{t_j}$ denote the length of the time interval in which the member watches the program p_z.

After that, we define the similarity set $S = (s^1, \ldots, s^q)$ with $q = C_{k^2}$ among the time interval set $T_a = (t_1, \ldots, t_i, \ldots, t_k)$, based on the rating matrix R_a as

$$S_a = (s^1, s^2, \ldots, s^q), q = C_k^2$$

$$= \begin{pmatrix} < r^{t_1}, r^{t_2} > & \cdots & \cdots & < r^{t_1}, r^{t_k} > \\ & \cdots & \cdots & \\ & < r^{t_i}, r^{t_{i+1}} > \cdots & < r^{t_i}, r^{t_k} > \\ & & & < r^{k-1}, r^{tl_k} > \end{pmatrix} \tag{5}$$

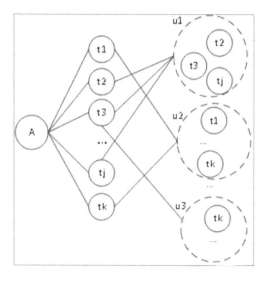

Fig. 3. Time Intervals for Multiple Members

The Cosine similarity formula is applied in the following equation to compute the similarities among the time interval set T_a.

$$s^i = Similarity(r^{t_j}, r^{t_k}) = \frac{R^{t_j} \cdot R^{t_k}}{||R^{t_j}|| \cdot ||R^{t_k}||}$$
$$= \frac{\sum_{g \epsilon G_a}(r_g^{t_j} \cdot r_g^{t_k})}{\sqrt{\sum_{g \epsilon G_a}(r_g^{t_j})^2} \cdot \sqrt{\sum_{g \epsilon G_a}(r_g^{t_k})^2}} \tag{6}$$

where $R^{t_j} = \{r_{g_1}^{t_j}, r_{g_2}^{t_j}, \dots, r_{g_m}^{t_j}\}$ and $R_a = \{R^{t_1}, R^{t_2}, \dots, R^{t_k}\}$.

Next, we identify the members in a graph shown as Figure 4, where the nodes in the graph are the time intervals T_a and the edges are the similarities S_a.

Since we compute all the similarities between every two time intervals t_j and t_k, there is an edge for each two nodes. However, our purpose is to distinguish the members in this graph. Hence, we define a parameter β as a threshold to normalize the similarity values, which decides whether two nodes should be connected. For example, in Figure 4, there are six intervals for an account. We set β as 0.7 (this number can be locally optimized in the experiments), then six nodes are divided into three groups. The similarity values less than β are normalized as 0. Finally, there are three members identified from this account.

4 Experimental Setup

4.1 Data Set

We collect our data from an industrial IPTV provider. To de-identify the private information of an account, the data are randomly selected and only partial of

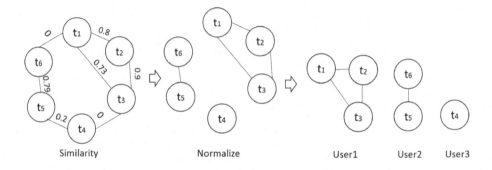

Fig. 4. Graphical Member Identification

the logs from March 1, 2011 and March 31, 2011. As table 1 shown, the data contains the programs, the corresponding genres, the start time and the end time under every account.

As we define the noises in Section 3.1 and the time interval detection in Section 3.2, we clean the data first at the pre-processing stage. We then have 10, 145 accounts with the corresponding 770, 169 records. The sample data will be available to download only for the research purpose.

As to the training data set and the testing data set, we make 75% of the data which are 611, 016 records as the training data, 159, 153 records as the testing data.

4.2   Evaluation Methods

The precision and recall scores are adopted to evaluate whether our proposed adaptive temporal model can substantially make efficient recommendation. The equations are shown as follows.

$$Precision@N = \frac{\sum_{u \in A} |List(a, N) \cap P(a)|}{\sum_{a \in A} |P(a)|} \tag{7}$$

$$Recall@N = \sum_{u \in A} |List(a, N) \cap P(a)| \sum_{u \in A} |List(a, N)| \tag{8}$$

where $List(a, N)$ is the recommending list to account a with the top N recommendations, $P(a)$ is the programs of account a. $List(a, N) \cap P(a)$ is the co-occurrences rating in the training dataset and the test dataset of account a.

5   Experimental Results

User collaborative filtering (UCF) and item collaborative filtering (ICF) algorithms are most frequently in the recommendation systems. Therefore, we adopt

Table 2. Precision of Genre/Video Recommendation of Users

TopN	UCF		AT-UCF		ICF		AT-ICF	
	Genre	Video	Genre	Video	Genre	Video	Genre	Video
1	0.258	0.022	0.496	0.041	0.083	0.031	0.104	0.089
2	0.217	0.016	0.415	0.033	0.065	0.027	0.096	0.067
3	0.181	0.017	0.345	0.041	0.067	0.03	0.106	0.077
4	0.173	0.016	0.316	0.041	0.061	0.024	0.094	0.061
5	0.157	0.017	0.287	0.038	0.061	0.022	0.092	0.055

Table 3. Recall of Genre/Video Recommendation of Members

TopN	UCF		AT-UCF		ICF		AT-ICF	
	Genre	Video	Genre	Video	Genre	Video	Genre	Video
1	0.125	0.01	0.13	0.041	0.04	0.015	0.027	0.023
2	0.183	0.014	0.188	0.033	0.063	0.026	0.05	0.035
3	0.21	0.022	0.215	0.041	0.097	0.044	0.083	0.06
4	0.251	0.029	0.242	0.041	0.117	0.046	0.099	0.064
5	0.262	0.036	0.252	0.038	0.147	0.052	0.12	0.072

Table 4. Precision of Genre/Video Recommendation of a Shared Account

TopN	UCF		AT-UCF		ICF		AT-ICF	
	Genre	Video	Genre	Video	Genre	Video	Genre	Video
1	0.38	0.052	0.89	0.103	0.09	0.08	0.22	0.16
2	0.315	0.042	0.629	0.072	0.125	0.05	0.18	0.135
3	0.28	0.035	0.5	0.076	0.13	0.067	0.183	0.147
4	0.274	0.031	0.426	0.073	0.145	0.077	0.177	0.117
5	0.248	0.029	0.371	0.075	0.144	0.07	0.168	0.102

Table 5. Recall of Genre/Video Recommendation of a Shared Account

TopN	UCF		AT-UCF		ICF		AT-ICF	
	Genre	Video	Genre	Video	Genre	Video	Genre	Video
1	0.162	0.009	0.094	0.019	0.038	0.015	0.094	0.019
2	0.269	0.015	0.154	0.026	0.107	0.019	0.154	0.026
3	0.359	0.019	0.235	0.041	0.167	0.037	0.235	0.041
4	0.462	0.022	0.303	0.052	0.248	0.058	0.303	0.052
5	0.509	0.026	0.359	0.067	0.308	0.065	0.359	0.067

these two methods as the baselines and output the top N recommended genres and programs for evaluation.

Table 2 and 3 present the personalized recommendations with the members identification. Considering the current traditional recommendations to a shared account, we also generate recommendations to the whole account. Table 4 and 5 show the results which are merged based on the personalized recommendations.

Note that all results contain the genre type and the program type. AT-UCF and AT-ICF are the proposed model on UCF and ICF. Top N lists for different methods are evaluated, where $N = \{1, 2, 3, 4, 5\}$. Note that we set the local optimization value as $\beta = 0.79$ after tuning.

6   Analysis and Discussions

Here we further analysis and discuss the results in Table 2, 3, 4 and 5. First, we compare the difference of the recommendation lists under the genre type and the program type. We then discuss the influence of top N recommendations. After that, the influence of the proposed adaptive temporal model is analyzed.

6.1   Investigation of UCF and ICF

The experimental results on the general UCF and ICF algorithms show that the UCF works better than ICF if we do the genre recommendation. Correspondingly, the ICF outperforms UCF if we do the video recommendation. Then, we get the pairs of (user, genre) and (item, video) as the strong alliances. We will continue to confirm the similar conclusions in the proposed adaptive approach.

6.2   Influence of Adaptive Temporal Model

From Table 2 and 4, we can see that the proposed adaptive temporal model AT_UCF and AT_ICF outperforms the general UCF and ICF, at both precision and recall levels. Specifically, when $N = 1$ which means that we only make a recommendation, our proposed model on UCF is precisely 200% better than the baseline. Also, for different Ns, the proposed method is promising on all the results.

Similarly, if we carefully examine the genre level and the video level respectively, we can draw the same conclusions that AT_UCF is good at the genre recommendation and AT_ICF at video. The same strong pairs of (user, genre) and (item, video) have been proved.

Therefore, we suggest genre recommendation if the IPTV system focuses on the user experience. Otherwise, we suggestion program recommendation if the system prefers recommending a program precisely or for the advertisement purpose. We can also see that this conclusion is consistent at both the account level and the identified user level.

6.3   Influence of Top N

We discuss the top 5 recommendations on genre and program separately. For the genre recommendation, it's very interesting that the top 1 genre achieves the best user performance and the improvements become small when we put more recommendations to the list. For the program recommendation, the item lists obviously dominate the performance. On the contrary to the genre recommendation, the larger the top N list in the program recommendation, the better performance ICF and AT-ICF achieve. The two conclusions confirm that the IPTV users focus on the genre types, instead of single program. This results one of our motivations at the introduction part.

7   Related Work

Compared to search engines, recommendation systems provide another way to push information to users. In the IPTV domain, the recommender systems are designed to automatically provide useful videos/programs or genres to users. To the best of our knowledge, only a few previous work has been done on user identification within a shared account. Most of them were based on the contextual information.

Masanobu Abe et al. [1] presented a contextual method which used a private unified remote controller to get users' history, preferences and other contextual information. We collect the similar information from the account logs which are provided by the service provider.

G. Adomavicius presented a multi-dimensional context-aware recommendation model at [2],[3]. The authors claimed to take user contextual information into consideration and proposed context information adapting collaborative filtering and the content based filtering. Said et al. [8] applied time contextual information to split profile into several sub-profile, then directly integrated context into model at the contextual modeling stage. In our proposal, we consider the relationships between the time interval and the genre information together and then identity the members.

Amy Zhang et al. [11] developed a composite account model based on unions of linear subspaces, then adopted subspace clustering algorithms for the identification task. Yang et al. [10] proposed a web session clustering method based on hyper-graph to find characteristics of user preferences in different periods.

Kim et. al. [7] claimed that they proposed an efficient collaborative recommendation method for IPTV services, which solved the scalability and sparsity problems which the conventional algorithms suffer from in the IPTV environment characterized by the large numbers of users and contents. The users were grouped by similar preferences and the group profile information was utilized for recommending contents in a more specialized manner to the target user.

[4] proposed to provide diversified recommendations. The authors investigated an approach to obtain more diversified recommendations using an aggregation method based on various similarity measures. Their work was evaluated using three experiments: the two first ones were lab experiments and show

that aggregation of various similarity measures improved accuracy and diversity. The last experiment involved real users to evaluate the aggregation method as proposed. The results showed that the proposed method allowed the balance between accuracy and diversity of recommendations.

8   Conclusions and Future Work

The contribution of this paper is four-fold. First, we propose an adaptive temporal model to identify multiple users within a shared account, which helps the system make specific recommendation to each member. Second, we dynamically split the consuming time of every day as time intervals, and then extract the users by computing the preference similarity among the intervals. Third, our proposed model is developed as an independent module such that it can be generally applied in any IPTV system. Fourth, our experiments on the data show that our model is superiority and promising.

In the future, we will focus on detecting the drift of user interests in the IPTV systems. This is also our ongoing work.

Acknowledgments. This research is funded by the Shanghai Science and Technology Commission Foundation (No. 14511107000) and the National Key Technology R&D Program (No.2012BAH93F02). We also thank all the anonymous reviewers for their suggestions on this work.

References

1. Abe, M., Morinishi, Y., Maeda, A., Aoki, M., Inagaki, H.: A life log collector integrated with a remote-controller for enabling user centric services. IEEE Transactions on Consumer Electronics **55**(1), 295–302 (2009)
2. Adomavicius, G., Sankaranarayanan, R., Sen, S., Tuzhilin, A.: Incorporating contextual information in recommender systems using a multidimensional approach. ACM Transactions on Information Systems (TOIS) **23**(1), 103–145 (2005)
3. Adomavicius, G., Tuzhilin, A.: Context-aware recommender systems. In: Recommender Systems Handbook, pp. 217–253. Springer (2011)
4. Candillier, L., Chevalier, M., Dudognon, D., Mothe, J.: Multiple similarities for diversity in recommender systems. International Journal On Advances in Intelligent Systems **5**(3 and 4), 234–246 (2012)
5. Chen, Q., Yang, Y., Hu, Q., He, L.: Locating query-oriented experts in microblog search. In: Proceedings of Workshop on Semantic Matching in Information Retrieval Co-located with the 37th International ACM SIGIR Conference on Research and Development in Information Retrieval, SMIR@SIGIR 2014, Queensland, Australia, July 11, 2014, pp. 16–23 (2014)
6. Herlocker, J.L., Konstan, J.A., Terveen, L.G., Riedl, J.T.: Evaluating collaborative filtering recommender systems. ACM Trans. Inf. Syst. **22**(1), 5–53 (2004)
7. Kim, M.-W., Song, W.-M., Song, S.-Y., Kim, E.-J.: Efficient collaborative recommendation with users clustered for IPTV services. In: Lee, G., Howard, D., Ślęzak, D., Hong, Y.S. (eds.) ICHIT 2012. CCIS, vol. 310, pp. 409–416. Springer, Heidelberg (2012)

8. Said, A., De Luca, E.W., Albayrak, S.: Inferring contextual user profiles-improving recommender performance. In: Proceedings of the 3rd RecSys Workshop on Context-Aware Recommender Systems (2011)
9. Sarwar, B., Karypis, G., Konstan, J., Riedl, J.: Item-based collaborative filtering recommendation algorithms. In: Proceedings of the 10th International Conference on World Wide Web, WWW 2001, pp. 285–295 (2001)
10. Yang, M.-H., Gu, Z.-M.: Mining user's behavioral patterns based on hypergraph clustering. Journal-Guangxi Normal University Natural Science Edition **24**(4), 163 (2006)
11. Zhang, A., Fawaz, N., Ioannidis, S., Montanari, A.: Guess who rated this movie: Identifying users through subspace clustering (2012). arXiv preprint arXiv:1208.1544

RPCV: Recommend Potential Customers to Vendors in Location-Based Social Network

Yuanliu Liu[1], Pengpeng Zhao[1(✉)], Victor S. Sheng[2], Zhixu Li[1],
An Liu[1], Jian Wu[1], and Zhiming Cui[1]

[1] School of Computer Science and Technology, Soochow University, Suzhou, China
sudalyl@gmail.com, {ppzhao,zhixuli,anliu,jianwu,szzmcui}@suda.edu.cn
[2] Computer Science Department, University of Central Arkansas, Conway, USA
ssheng@uca.edu

Abstract. Location-based social network has received much attention recently. It provides rich information of social and spatial context for researchers to study users' behaviors from different aspects. A number of recent efforts focus on recommending locations, users, activities, and social medias for users. Unlike previous works, we intend to make recommendations for vendors, assisting vendors in finding potential customers in location-based social network. We propose a framework to recommend potential customers to vendors (called RPCV) in location-based social network effectively and efficiently. To find the best set of customers, RPCV takes both spatial relations and user preference into consideration. A reverse spatial-preference kRanks algorithm, which effectively combines spatial relations with user preference, is also proposed. Our experimental results on real datasets from Foursquare and Brightkite show that our framework has higher performance than other state-of-the-art approaches.

Keywords: Location based social network · Recommendation

1 Introduction

With the rapid development of location-acquisition and mobile communication technologies, a number of location-based social network services have emerged in recent years (e.g., Foursquare, BrightKite, Gowalla and so on), which result in a new concept of online social media named location-based social network (LBSN). LBSN allows users to share their locations and find out local points of interest, which bridges the gap between the physical and digital worlds. Foursquare, as a representative LBSN, has achieved one billion users' check-ins within two years since its foundation. The average amount of check-ins made by users per day was about three million in Foursquare, and this number is still growing. Such rapid growth of LBSN has brought the convenience to collect user data, which consists of the spatial, social and rating aspects. These real user data enables a deeper understanding of users' preferences and behaviors.

© Springer International Publishing Switzerland 2015
J. Li and Y. Sun (Eds.): WAIM 2015, LNCS 9098, pp. 272–284, 2015.
DOI: 10.1007/978-3-319-21042-1_22

In recent years, a numerous researchers make contributions to study on user-centered recommendations in LBSN, such as user recommendations [1], activity recommendations [2] and social media recommendations [3]. However, increasing demands encourage researchers to study on vendor-centered recommendations, which can help vendors grow their business by recommending potential customers. Besides the user preference, vendor-centered recommendations need to consider more users' real-time requirements and spatial location relations than user-centered recommendations. For example, a user makes a check-in at a restaurant after he has a meal with his friends. There is a great possibility that they will do some entertainment activities after the meal. This is a good time point for recommendation service providers to push useful information which meats the user's real-time needs. As we know, users' demands may be changed over time, and outdated information is worthless for them. Therefore if the service provider can catch users' real-time requirements at a right time point, it will improve their recommendation quality, and can help vendors to do marketing better. It is important to consider spatial relations while recommending customers to local vendors. Figure 1 gives an example to illustrate the importance of spatial locations. Whereby, A and B are two customers, V1 and V2 are two vendors, A is the top-1 customer of V1 and V2. If we want to recommend one customer to V1 and V2, traditional works that use similar approaches like top-k query will recommend customer A to both of them. In reality, customer A treats V2 as his top-1 vendor. There is a great possibility that he will choose V2 instead of V1. As we know, recommending service is not free, and vendors should pay for each recommendation. Obviously, top-k query is not a good solution, since it does not consider the spatial relations among vendors. Therefore, to save vendors' cost and improve the quality of recommendations, it is necessary to make better use of spatial relations among vendors. In conclusion, to recommend potential customers to vendors should fulfill two essential requirements: (1) Users' requirements and (2) Vendors' profits. The former one requires to find users' true preferences, which guarantees the relevance of a vendor to an individual user. The latter one encourages the service providers to maximize the profit of a vendor when the vendor has a fixed budget.

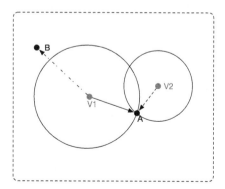

Fig. 1. An example of recommending scenario

In this paper, we propose a framework that can Recommend Potential Customers to Vendors in LBSN, named RPCV. It aims at selecting relevant potential customers for vendors in real-time, and considers both of the two essential requirements. Instead of considering vendors separately, we propose a reverse spatial-preference kRanks algorithm, which considers the influence of other vendors on customers. What's more, this algorithm effectively combines spatial relations with user preference and can catch users' real-time requirements. By this algorithm, we can find k customers whose ranks for a given vendor are highest among all customers. The contributions of this paper are summarized below:

- We propose a recommendation framework that can recommend potential customers to vendors and work in a real-time environment.
- We propose a reverse spatial-preference kRanks algorithm, which effectively combines spatial relations with user preference, and can also catch users' real-time requirements.
- We evaluate our work in two real datasets from Foursquare and BrightKite. Experiments show that our strategy has a higher precision than others.

The rest of this paper is organized as follows. We first review related work in Section 2. Then, Section 3 presents the system framework of our RPCV. Sections 4 and 5 introduce our novel user preference model and reverse spatial-preference kRanks approach. Section 6 analyzes experimental results. Finally, we conclude this paper in Section 7.

2   Related Work

Recommendations in LBSN. Due to the rapid development of location-based services, real traces of user locations and activities have been collected and used in several researches [4–6]. The location dimension bridges the gap between the virtual online social network and physical world, giving rise to new challenges and opportunities in recommending systems. Besides location, the temporal, spatial, and social aspects are also available in LBSN. Therefore, more and more researchers try to make efforts to mine users' mobile behaviors in LBSN. In [7], Huiji et al. proposed to model temporal effects of human mobile behavior in LBSN, and strong temporal cyclic patterns have been observed in user movement with their correlated spatial and social effects. In [8,9], the authors observed strong heterogeneity across users with different characteristic geographic scales of interactions across ties. While Ye et al. [10] investigated the geographical influence [11] and social influence [12] for location recommendation, they discovered that user preference plays a more important role in contributing to the recommendation than social and geographical influences do.

Ranking Query. In general, the ranking query returns top k tuples with maximal (minimal) ranking scores by employing a user-defined scoring function. A general linear model is one of the most widely adopted models among existing scoring function. However, the top-k query under linear model is mainly focused on the perspective of users, which aims at finding a set of products to match their preferences. However, vendors would prefer to know who would be interested in them, which are totally opposite. In [13], reverse top-k query is used to return an unordered result set of users, and these users are all treating the given vendors as one of their top-k vendors. Nevertheless, it can not work in our research, since it can not guarantee 100% coverage of any given query.

3   System Overview

In this section, we illustrate the framework architecture of RPCV shown in Figure 2. In order to avoid missing users' real-time requirements, we designed it in a burst mode and triggered by users' check-in behaviors. When the start event is triggered, our RPCV will start from its reverse spatial-preference kRanks module. It begins to recommend a specific number of potential customers to preliminary defined vendors. As presented in the figure, the reverse spatial-preference kRanks module requests user preference from a user preference module. The main task of user preference module is to compute user preference with our user preference model. Our user preference model handles this task by leveraging social, historical, rating and spatial data. The details of the two modules will be explained in Section 4 and 5. Finally, the framework outputs a series of potential users for preliminary defined vendors.

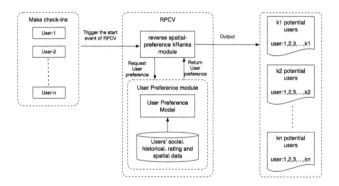

Fig. 2. System Framework

4   User Preference Model

In reality, users prefer more relevant information with respect to their preferences [14]. Therefore, in this section, we introduce a novel model to build user preference, which is an indispensable part of our reverse spatial-preference kRanks approach. The preference of user u_i to a vendor v_j can be described by different profiles. In this paper, we use a two-dimension vector $w_{i,j}$ to represent this preference. The first dimension of this vector is the degree of a user preferring the vendor. This preference can be computed as a preference score $S_{u_i}^p(v_j)$ by leveraging users' social relations and historical behaviors. The second dimension is the sensibility of the spatial distance, which is computed by a spatial score. The spatial score can be measured by different metrics. A common approach for the spatial score computation is based on the distance between the check-in locations and users' home. Then, we can find out the user's sensibility of the spatial distance, denoted as $S_{u_i}^{sd}(v_j)$. Finally, the vector $w_{i,j} = <S_{u_i}^p(v_j), S_{u_i}^{sd}(v_j)>$ can be used to represent the preference of the user u_i for the vendor v_j. And we will normalize $S_{u_i}^p(v_j)$ and $S_{u_i}^{sd}(v_j)$ to make their summation to be 1.

4.1   Analyzing Social and Historical Behaviors

Previous works find users' friends may influence their behaviors [8,15]. To inspect our guess, we capture about 2M users and 1M venues across US from foursquare and compared the number of common check-ins between two friends and two strangers. As shown in Table 1(a), on average, a pair of friends share about 13.527 check-ins, while a pair of strangers share only 4.831 check-ins. This indicates that friends have strong influence to a user's check-in behaviors.

Table 1. Social And Historical Behaviors

(a) Average number of check-ins between two users

Type	number
Between friends	13.527
Between strangers	4.831

(b) Correspondences between language model and LBSN model

Language Model	LBSN Model
Word	Check-in location
Phrase	Daily check-in sequence
Sentence	Weekly check-in sequence
Paragraph	Monthly check-in sequence
Document	Individual check-ins
Corpus	Check-in collection

When mining user preference, users' historical behaviors are important factors that cannot be ignored. While capturing a user's historical behaviors, two important properties must be considered. The first one is that a user tends to go a few places many times and many places a few times. This means the history approximately follows a power-law distribution. Another property is that the historical behaviors have short-term effect. This means the previous check-ins have different strengths to the latest check-in. As shown in Table 1(b), there are many common features between language processing and LBSN mining. Besides, the check-in data and the document have similar structure. Therefore, we prefer to use a state-of-the-art language processing tech called hierarchical Pitman-Yor process (HPY) [16] to capture users' historical behaviors.In addition, the HPY is a hierarchical extension of Pitman-Yor process [17].

Since social and historical behaviors are useful for mining user preference, we leverage them to model user preference in this paper. We will combine the two behaviors together to evaluate user preference, using Equation $S_{u_i}^p(v_j) = \eta S_{u_i}^f(v_j) + (1-\eta) P_{u_i}^H(v_j)$. In the equation, $S_{u_i}^f(v_j)$ represents the social score, and $P_{u_i}^H(v_j)$ represents the historical score. The parameter $\eta \in [0,1]$ in the equation is used to adjust the relative importance of the parts.

4.2 Social Score Computation

The opinions data about how much a user u_i likes a vendor v_j can be extracted from similarities, which can be inferred from user's ratings and check-in location histories in LBSN. In this paper, we use a simple but effective computation approach by leveraging users' ratings. Table 2(a) shows an example about the closeness between users while Table 2(b) shows the normalized rating score $s_{u_i}(v_j)$ (with a value between zero and one) of a vendor v_j given by a user u_i. As we know, different friends have different social effects on a user. So in this paper, we use a weight $cw_{i,k}$ to represent the closeness u_k towards u_i, where u_k is a friend of the user u_i. We use $f(u_i)$ to represent the friends of the user u_i. The weight can be evaluated from many metrics, such as the number of common friends, the number of common check-ins and so on. We constrain the total weight of the closenesses to be one. Thus, the friends' recommending score of the user u_i on a vendor v_j can be computed by the equation as follows.

$$S_{u_i}^f(v_j) = \sum_{k=1}^{|f(u)|} cw_{i,k} \cdot s_{u_k}(v_j) \tag{1}$$

Table 2. An example of social links and rating scores

(a) Social Links		(b) Rating Score				
User	Friends and Closeness		v_1	v_2	v_3	v_4	v_5
u_1	$< u_2, 0.25 >, < u_3, 0.75 >$	u_1	0.2	0.8	0.6		0.4
u_2	$< u_1, 0.40 >, < u_3, 0.60 >$	u_2		0.4	0.8	0.4	
u_3	$< u_1, 0.80 >, < u_2, 0.20 >$	u_3	0.8		0.8	0.4	0.2

4.3 Hierarchical Pitman-Yor Process

As mentioned above, the historical score is generated by HPY and the HPY. In this subsection, we will explain how HPY works. The HPY process assumes that the earlier a word is mentioned, the less importance the word has. It is an n-gram model that captures the short-term effect without losing the power-law property in distribution. In HPY, G_u denotes the probability of the next check-in while given a history context u:

$$G_u \sim PY(d_{|u|}, \gamma_{|u|}, G_{\pi(u)}) \tag{2}$$

In (2), $d_{|u|} \in [0, 1)$ is a discount parameter to control the power-law property, $\gamma_{|u|}$ is the strength parameter, $\pi(u)$ is the suffix of u and $G_{\pi(u)}$ is probability of next check-in in the history context u. And the PY is a Pitman-Yor process. All of these parameters are under context u which consisting of all but the earliest check-ins. $G_{\pi(u)}$ is then computed with parameters $d_{|\pi(u)|}$, $\gamma_{\pi(u)}$ and $G_{\pi(\pi(u))}$. This process is repeated until we get the empty historical context $\emptyset$:

$$G_\emptyset \sim PY(d_0, \gamma_0, G_0) \tag{3}$$

The m locations can be represented by the location space L, and $m = |L|$. The base distribution G_0 is a uniform distribution providing by a prior probability. It satisfies $G_0(l) = 1/m$, where $G_0(l)$ is the probability of a location $l \in L$ being check-in.

$$
\begin{aligned}
&P_u^{HPY}(c_{n+1} = l | c_1, c_2, \ldots, c_n) \\
&= \frac{N_{ul} - t_{ul} d_{|u|}}{\gamma_{|u|} + n_u} + \frac{\gamma_{|u|} + t_u d_{|u|}}{\gamma_{|u|} + n_u} G_{\pi(u)}(c_{n+1} = l | c_1, c_2, \ldots, c_n)
\end{aligned} \tag{4}
$$

Where N_{ul} is the number of check-ins at l following the historical context u and $n_u = \sum_l N_{ul} \cdot t_u = \sum_l t_{ul}$ is the sum of all t_{ul}, which is a latent variable satisfying:

$$
\begin{cases}
t_{ul} = 0 & if\ N_{ul} = 0 \\
0 \leq t_{ul} \leq N_{ul} & if\ N_{ul} > 0
\end{cases} \tag{5}
$$

5   Reverse Spatial-Preference kRanks

In this section, we will explain our reverse spatial-preference kRanks algorithm. It effectively combines spatial relations with user preference. Its tree-based pruning approach improves its efficiency and makes it applicable in a real-time environment. It guarantees that we do not lose the chance to catch users' real-time requirements.

5.1   Basic Concepts

In our work, we want to know who will be interested in a vendor. This is totally opposite to top-k query. Reverse top-k query seems to meet our requirements, but it cannot ensure 100% coverage for any given vendor and the size of the result set sometimes is larger than k. However, it is necessary to recommend a specific number of customers to vendors. Therefore, we propose a reverse spatial-preference kRanks algorithm, which can effectively combine spatial relations with user preference and return a ranked result set of k customers for any given vendors. For any given vendor, we use an attribute vector $v = <r, z>$ to represent it, where r is the average ratings given by all users and z is the z-value calculated by its location. The z-value here is generated by the well-known Z-order approach [18]. With the help of Z-order approach, we can estimate the distance between a user and a vendor through the difference value of their z-value quickly. What's more, the Z-order makes it possible to take the spatial location information into the vendor's attribute vector, which can help avoid rebuilding the R-tree of our vendors. The details of R-tree will be explained later. In this paper, a score function is defined as $f(w,v) = \sum_{i=1}^{d} w^{(i)} \cdot v^{(i)}$, i.e., the inner production of the user's preference vector w and the vendors attribute vector v. For example, if a user's preference vector for the vendor $v = < 4.2, 65 >$ is $w = < 0.3, 0.7 >$, and the z-value of the user is 23, then the value of the score function is $f(w,v) = (0.3 * 4.2 + 0.7 * (65 - 23))$.

Definition 1 *(rank(u,q)). Given a vendor point vector set D, a specific user u, and a query point vector q, the rank of q for u is $rank(u,q) = |S|$, where $|S|$ is the cardinality of S, and S is a subset of D. And $\forall p_i \in S$, there must have $f(u.w, p_i) < f(u.w, q)$; For $\forall p_j \in (D - S)$, we have $f(u.w, p_j) \geq f(u.w, q)$.*

Definition 2 *(reverse spatial-preference kRanks query). Given a vendor point vector set D, a user set U, a positive integer k (the size of the result set), and a query point vector q, reverse spatial-preference kRanks query returns a set SU, $SU \subset U, |SU| = k$ and $\forall u_i \in SU, \forall u_j \in (U - SU)$, $rank\,(u_i.w, q) \leq rank\,(u_j.w, q)$*

For any given user u in U and any given point p in D, we compute $rank(u.w, q)$ one by one, and keep k weights with the smallest $rank(u.w, q)$ for query point q. The complexity is $O(m \cdot n)$ where m is the number of customers and n is the number of vendors. Note that this is used as a baseline to compare with our tree-based pruning approach.

5.2 Tree-Based pruning approach

The baseline method above evaluates every customer and vendor pair. As we mentioned that its time complexity is $O(m \cdot n)$. Fortunately, we can use an R-tree to avoid some parts of computations, i.e., building an R-tree to index all vendors by their attribute vectors. In the R-tree, we use r denote a minimal bounding rectangle (MBR), and then use $r.L$ and $r.U$ to denote the bottom left and top right points of r respectively.

Definition 3 *(Dominance). We say a point p_1 dominates a point p_2 (denoted as $p_1 \prec p_2$), if and only if the following two conditions hold: $(1)\forall l, p_1^l \leq p_2^l$, $(2)\exists j, p_1^j < p_2^j$*

According to the definition 3, we have $f(u.w, p_1) < f(u.w, p_2)$ when $p_1 \prec p_2$, since $\forall i, w^{(i)} \geq 0$. Therefore, there are two facts we can obtain. Given a query point q, a specific user u_i, and an MBR r in R-tree, if $f(u_i.w, q) < f(u_i.w, r.L)$, then $\forall p \in r$, we have $f(u_i.w, q) < f(u_i.w, p)$. And if $f(u_i.w, r.U) < f(u_i.w, q)$, $\forall p \in r$, we have $f(u_i.w, p) < f(u_i.w, q)$.

Algorithm 1 illustrates the main steps of our tree-based pruning approach (TPA). We index all vendors by an R-tree R offline. During processing, it computes ranking score by traversing R-tree. According to the two facts, the nodes satisfying $f(u_i.w, q) < f(u_i.w, r.L)$ or $f(u_i.w, r.U) < f(u_i.w, q)$ can be pruned

---

Algorithm 1. Tree-based pruning approach (R, U, q, k)

Input: an R-tree R; user set U; query vendor vector q; number of return customers k;
Output: an array contains k most valuable customers

1: Q and Q' represent two queues
2: A denote an array to record the rank of each user
3: $minRank \leftarrow$ the number of objects in R
4: **for** each u_i in U **do**
5: 　　Empty Q and Q'
6: 　　$enqueue(Q, R.root)$
7: 　　**while** $r = dequeue(Q) \neq \emptyset$ **do**
8: 　　　　**if** $f(u_i.w, r.U) < f(u_i.w, q)$ **then**
9: 　　　　　　$A[u_i] \leftarrow A[u_i] + |r|$
10: 　　　　**else if** $f(u_i.w, q) < f(u_i.w, r.L)$ **then**
11: 　　　　　　continue
12: 　　　　**else if** r *is a leaf node* **then**
13: 　　　　　　$enqueue(Q', r)$
14: 　　　　**else**
15: 　　　　　　$\forall r'$ (r' is r's child), $enqueue(Q, r')$
16: 　　　　**end if**
17: 　　**end while**
18: 　　**if** $A[u_i] \leq minRank$ **then**
19: 　　　　**for** r *in the queue* Q' **do**
20: 　　　　　　**for** *point* p *in* r **do**
21: 　　　　　　　　**if** $f(u_i.w, q) < f(u_i.w, p)$ **then**
22: 　　　　　　　　　　$A[u_i] \leftarrow A[u_i] + 1$
23: 　　　　　　　　**end if**
24: 　　　　　　**end for**
25: 　　　　**end for**
26: 　　**end if**
27: 　　$minRank \leftarrow$ the kth minimal value in A
28: **end for**
29: **return** k *smallest entries in* A

---

safely. Otherwise, if the node is not a leaf node, it will be appended to the queue Q for further processing. The queue Q' keeps the tracks of all un-pruned leaf nodes. And the global variable $minRank$ is used to describe the kth maximal entry in array A. That is, at least k customers treat q as one of his $top-minRnak$ favorite objects till now. Finally, the algorithm returns k smallest entries in A. TPA prunes computations from the perspective of data points and reduces time complexity to $O(m_r \cdot n)$, where $m_r = |R|, n = |D|$.

6 Experiments

6.1 Experimental Settings

In this subsection, we introduce the experimental settings. All codes are written in JAVA, and run on a single machine with Intel CPU/2.8GHZ and 8GB memory.

Table 3. Information of two datasets

	Foursquare	Brightkite
Amount of users	18,107	26,915
Amount of check-ins	2,073,740	4,666,732
Amount of links	231,148	261,982
Amount of unique locations	43,063	751,176

Data Set. Our experiments are based on two real datasets from Foursquare and Brightkite. Both of them allow users to make check-ins at physical locations through mobile phones and let their online friends know where they are. All the information about the two datasets is listed in Table 3.

Evaluation Metrics. In this paper, we use the average precision as our evaluation metric. We compare our approach with top-k query and reverse top-k query. Actually, this is a strict evaluation measurement, as a user may still like a vendor even if he did not visit it. In addition, there is a probability that the user could forget to make a check-in when he visits the vendor. Therefore, our framework is actually more effective than the experimental results shown later. Besides, we compare the time consumption of our tree-based pruning approach with a baseline method.

6.2 Experiment Results

Reverse Spatial-preference KRanks Effectiveness. Reverse spatial-preference kRanks (short for RSPKR) is the core of our RPCV. It is responsible to deal with the input query and output potential customers for a querying vendor. Thus, we first evaluated the effectiveness of our reverse spatial-preference kRanks approach. We use the real check-ins of users recommended to the querying vendor as the ground truth, and compare RSPKR with top-k query (i.e. return top-k preferred customers, denoted as TopK) and reverse top-k query (i.e. return a set of unordered customers who treat the given vendor as their top-k vendors, denoted as RTopK), in terms of their precisions. The performance of three approaches on the two datasets are shown in Figure 3(a) and 3(b) respectively. The X-axis of both figures represents the number of recommending customers, and the Y-axis of both figures represents the precision value.

As shown in Figure 3(a) and 3(b), our RSPKR is more effective than the other two approaches. The TopK is the worst, since it does not consider the opinion of a user towards other vendors. RTopK is better than TopK, because it chooses the customers who treat the querying vendor as their top-k vendors. However, the size of its return customers may be larger or smaller than k. For some unpopular vendors, RTopK even returns none customers. As mentioned in the previous section, the results of RTopK are unordered. Thus, while the size of the results is larger than k, we should choose the top k customers for the querying vendor, but we do not know the rank of these customers. These reasons

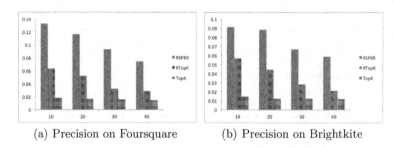

(a) Precision on Foursquare   (b) Precision on Brightkite

Fig. 3. Effectiveness Comparison

cause that RTopK is less effective than our RSPKR. The experimental results show that it is helpful to consider spatial relations among vendors.

Efficiency Comparison. Our reverse spatial-preference kRanks uses tree-based pruning approach to accelerate the query processing. Therefore, we compare the tree-based pruning approach (TPA) with baseline method mention before, in terms of different numbers of customers and different numbers of vendors respectively. The comparisons of time consumption are shown in Figure 4(a) and 4(b). In both figures, the unit of their Y-axis is seconds. Note that Figure 4(a) shows the performance changes while the number of customers increases from 100K to 500K and the number of vendors is 20K. Figure 4(b) shows the performance changes under the number of vendors increases from 10K to 50K and the number of customers is 400K.

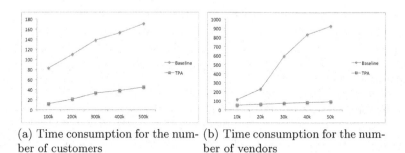

(a) Time consumption for the num-   (b) Time consumption for the number of customers   ber of vendors

Fig. 4. Time Consumption

Figure 4(a) and 4(b) show that the processing time of both methods increases with both the increment of the number customers and vendors. From this two figure, we can also see that our tree-based pruning approach is far more efficient than the baseline method under the two different scenarios. Comparing these two figures, we can see that the computation time grows significantly quicker with the increment of the number of vendors, comparing with the time under the increment of the number of customers. This is because the computation time of each customer grows when the number of vendors increases.

7 Conclusions

In this paper, we present RPCV, a recommending framework that aims at recommending potential customers to vendors in a real-time environment. With the help of our reverse spatial-preference kRanks approach, we effectively combines spatial relations with user preference. The experimental studies on two real datasets demonstrate that our framework outperforms existing strategies. An interesting direction for future work is to consider temporal and/or semantic features.

Acknowledgments. This work was partially supported by Chinese NSFC project (61170020, 61402311, 61440053), and the US National Science Foundation (IIS-1115417).

References

1. Levandoski, J.J., Sarwat, M., Eldawy, A., Mokbel, M.F.: Lars: a location-aware recommender system. In: 2012 IEEE 28th International Conference on Data Engineering (ICDE), pp. 450–461. IEEE (2012)
2. Zheng, V.W., Zheng, Y., Xie, X., Yang, Q.: Collaborative location and activity recommendations with gps history data. In: Proceedings of the 19th International Conference on World Wide Web, pp. 1029–1038. ACM (2010)
3. Bouidghaghen, O., Tamine, L., Boughanem, M.: Personalizing mobile web search for location sensitive queries. In: 2011 12th IEEE International Conference on Mobile Data Management (MDM), vol. 1, pp. 110–118. IEEE (2011)
4. Ashbrook, D., Starner, T.: Using gps to learn significant locations and predict movement across multiple users. Personal and Ubiquitous Computing **7**(5), 275–286 (2003)
5. Liao, L., Patterson, D.J., Fox, D., Kautz, H.: Learning and inferring transportation routines. Artificial Intelligence **171**(5), 311–331 (2007)
6. Lin, J., Xiang, G., Hong, J.I., Sadeh, N.: Modeling people's place naming preferences in location sharing. In: Proceedings of the 12th ACM International Conference on Ubiquitous Computing, pp. 75–84. ACM (2010)
7. Gao, H., Tang, J., Hu, X., Liu, H.: Modeling temporal effects of human mobile behavior on location-based social networks. In: Proceedings of the 22nd ACM International Conference on Conference on Information & Knowledge Management, pp. 1673–1678. ACM (2013)
8. Scellato, S., Mascolo, C., Musolesi, M., Latora, V.: Distance matters: geo-social metrics for online social networks. In: Proceedings of the 3rd Conference on Online Social Networks, pp. 8–8 (2010)
9. Scellato, S., Noulas, A., Lambiotte, R., Mascolo, C.: Socio-spatial properties of online location-based social networks. ICWSM **11**, 329–336 (2011)
10. Ye, M., Yin, P., Lee, W.C.: Location recommendation for location-based social networks. In: Proceedings of the 18th SIGSPATIAL International Conference on Advances in Geographic Information Systems, pp. 458–461. ACM (2010)

11. Ye, M., Yin, P., Lee, W.C., Lee, D.L.: Exploiting geographical influence for collaborative point-of-interest recommendation. In: Proceedings of the 34th International ACM SIGIR Conference on Research and Development in Information Retrieval, pp. 325–334. ACM (2011)

12. Ye, M., Liu, X., Lee, W.C.: Exploring social influence for recommendation: a generative model approach. In: Proceedings of the 35th International ACM SIGIR Conference on Research and Development in Information Retrieval, pp. 671–680. ACM (2012)

13. Vlachou, A., Doulkeridis, C., Kotidis, Y., Norvag, K.: Reverse top-k queries. In: 2010 IEEE 26th International Conference on Data Engineering (ICDE), pp. 365–376. IEEE (2010)

14. Yuan, S.T., Tsao, Y.W.: A recommendation mechanism for contextualized mobile advertising. Expert Systems with Applications **24**(4), 399–414 (2003)

15. Humphreys, L.: Mobile social networks and social practice: A case study of dodgeball. Journal of Computer-Mediated Communication **13**(1), 341–360 (2007)

16. Teh, Y.W.: A bayesian interpretation of interpolated kneser-ney (2006)

17. Pitman, J., Yor, M.: The two-parameter poisson-dirichlet distribution derived from a stable subordinator. The Annals of Probability, 855–900 (1997)

18. Van Dam, A., Feiner, S.K.: Computer graphics: principles and practice. Pearson Education (2014)

Mining Dependencies Considering Time Lag in Spatio-Temporal Traffic Data

Xiabing Zhou, Haikun Hong, Xingxing Xing, Wenhao Huang,
Kaigui Bian$^{(\boxtimes)}$, and Kunqing Xie

Key Laboratory of Machine Perception, Ministry of Education,
Peking University, Beijing 100871, China
bkg@pku.edu.cn

Abstract. Learning dependency structure is meaningful to characterize causal or statistical relationships. Traditional dependencies learning algorithms only use the same time stamp data of variables. However, in many real-world applications, such as traffic system and climate, time lag is a key feature of hidden temporal dependencies, and plays an essential role in interpreting the cause of discovered temporal dependencies. In this paper, we propose a method for mining dependencies by considering the time lag. The proposed approach is based on a decomposition of the coefficients into products of two-level hierarchical coefficients, where one represents feature-level and the other represents time-level. Specially, we capture the prior information of time lag in spatio-temporal traffic data. We construct a probabilistic formulation by applying some probabilistic priors to these hierarchical coefficients, and devise an expectation-maximization (EM) algorithm to learn the model parameters. We evaluate our model on both synthetic and real-world highway traffic datasets. Experimental results show the effectiveness of our method.

Keywords: Dependency · Time lag · Highway traffic analysis

1 Introduction

Mining dependencies is meaningful to characterize causal or statistical relationships that exist among variables of interest and quantify them. The problem of mining dependencies between variables in complex systems, such as economics, biological systems, traffic systems, climate change, etc., is important and fundamental. Given these multiple variables, the goal is to use available variables to make precise prediction of future events and trends. In addition to this primary goal, an important task is to identify dependencies between these variables wherein, data from one variable significantly help in marking predictions about another variable. For example, economists want to know whether burning natural gas is a causal factor for the global warming, so they need to mine whether the global warming depends on burning nature gas.

Graphical modeling techniques which use Bayesian networks and other causal networks have been considered as a viable option for modeling dependencies in

© Springer International Publishing Switzerland 2015
J. Li and Y. Sun (Eds.): WAIM 2015, LNCS 9098, pp. 285–296, 2015.
DOI: 10.1007/978-3-319-21042-1_23

the past [1–3]. Statistical tests such as specific hypothesis tests have also been designed to identify causality between the various temporal variables [4]. However, most of them either do not consider the time lag, or only use a predefined value. In traffic temporal data, time-lagged relationships are crucial towards understanding the linkages and influence of the change between relative entrance ramps and exit ramps. These relationships are lagged in time because vehicles from entrance ramps do not affect vehicle flow of exit ramps at the same time but only at a later time. One such important time-lagged pattern in traffic is relative entrance ramps learning for exit ramps. Fig. 1 shows the Origin-Destination(OD) matrix[5–7][1] of a highway traffic network, where the rows and the columns denote the entrance and exit ramps in the highway, respectively, and the values of the matrix represent the vehicle counts rushed from the entrances to the exits. The entries with brighter color denote larger vehicle counts and darker ones represent small vehicle counts. We say that there exits dependency between an entrance ramp and an exit ramp when the corresponding entry of OD matrix is bright.

Owing to its dependencies to the traffic system, its understanding has the potential to aid forecasts of vehicle flow at exit ramps.

One important model for mining causal dependencies considering time lag is Granger causality [9], which is a widely accepted notion of causality in econometrics. In particular, it says that time series A causes B, if the current observations in A and B together, predict the future observations in B significantly more accurately, than the predictions obtained by using just the current observations in B. Recently, there has been a surge of methods that combine this notion of causality with regression algorithms [10–12]. However, they either do not emphasize the concept and important of time lag, or make all history data and current data together to predict, which leads to learning more irrelevant attributions.

Fig. 1. OD matrix of highway traffic network of a province in China. The rows represent the entrance stations, and then columns represent the exit stations. If the traffic flow from the entrance station to exit station is non-zero, the pixel is white.

In this paper, we proposed a method to cope with mining dependencies of spatio-temporal traffic data. This method is based on the idea that combine the notion of causality with regression algorithm, called Two-Level Hierarchies with time Lag lasso (TLHL). TLHL decomposes the regression coefficients into a product between a feature-level component and a time-level component. Such a decomposition is very natural from the theory, namely, a specific regression coefficient is equal to zero if either of its two components is zero; Furthermore, the feature-level control the dependencies of prediction variables and responding variables; the time-level component represents the choice of time lag. Specifically, the TLHL model places Gaussian and Cauchy distributions for the component

---

[1] The OD matrix can provide important spatial correlation information for the traffic behaviors and is a widely used tool in the traffic analysis [8].

coefficients as priors to control the model complexity. With the Gaussian likelihood, we devise an efficient expectation maximization (EM) algorithm [13] to learn the model parameters. Moreover, we evaluate our model on both synthetic and real-world traffic data, and the conducted results show that the TLHL model is very effective in mining dependencies due to considering the time lag.

The remainder of this paper is organized as follows. In Section 2, we briefly review background and preliminaries. Section 3 presents the proposed method. Experimental studies are reported in Section 4. We conclude this paper and present future directions in Section 5.

2   Background and Preliminaries

2.1   Background

Estimating the dependency structure in traffic networks plays an important role. Most of the existing works on dependency structure discovery do not consider the time lag. For example, Meinshausen etc. [14] learn the dependency structure by using lasso; Yuan etc. [10] use group lasso to cope with this problem; Han etc. [8] detect dependency relationships in traffic system by Gaussian graphical model. However, time lag is an important factor for discovering the traffic pattern, especially in spatio-temporal traffic data.

In traffic system, it takes time for a vehicle from an entrance ramp to an exit ramp, thus, we need to consider this time when discovering the dependencies between entrance ramps and exit ramps. Traditional methods either use the same time stamp data of entrance ramps and exit ramps [8], or use neighbor history data of current time stamp immediately [15,16]. [17,18] plug different values of time lag in the model, which set time lag as a constant. Granger graphical model has a parameter called the maximum lag. The maximum lag for a set of time series signifies the number of time units one must look into the past to make accurate predictions of current and future events. However, all the past time stamp data in this method are treated uniformly. The disadvantage is that it does not consider time lag accurately, and always learns more irrelevant attributions.

2.2   Problem Definition

Given d random variables $\boldsymbol{X} = \{x_1, \cdots, x_d\}$, and each variable x_i has n observations, $x_i = (x_i^1, \cdots, x_i^n)^T$. Let $\boldsymbol{Y} = \{y^i, \cdots, y^n\}$ be the response variable. We aim to learn the dependency structures between prediction variables and response variable.

In traffic analysis, $\boldsymbol{X}$ denotes the vehicle counts collected from the entrance ramps, and $\boldsymbol{Y}$ is the vehicle counts of the exit ramps. The tasks are predicting the vehicle counts for the exit ramps, since the traffic experts are eager to know how many vehicles will pass through some important exit ramps. We aim to encode the dependency structures between entrance ramps and exit ramps. The vehicles from relative entrance ramps of exit ramp are useful for predicting the vehicle flow of this exit ramp, thus, the challenge is how to learn the relative entrance ramps of exit ramp more accurately. To solve this, we make full use of time lag, which is a prior information in traffic data. Time lag is caused by travel time, and the mean travel time can be obtained by history statistic. However, different vehicles might have different travel time, the mean travel time is not used as a fixed time lag because it has fluctuation. Thus, we learn the distribution of time lag to choose the relative past time data of entrance ramps for exit ramps more flexibly and accurately.

3   Proposed Method

In this section, we introduce the proposed TLHL model. TLHL model is based on the idea that combines the notion of causality with regression algorithm. The traditional regression coefficient is decomposed into products of two-level hierarchical coefficients. They represents feature-level and time-level respectively. Specially, we propose a probabilistic framework for TLHL model, and devise an EM algorithm to infer the model parameters.

3.1   The TLHL Model

First we propose a model for the component coefficients introduced previously. Most of the lasso-based algorithms solve the following optimization problem [10,14,19,20]:

$$\min_{\beta} \sum_{i=1}^{n} L(y^i, \beta^T x^i + b) + \lambda R(\beta), \tag{1}$$

where $\beta = (\beta_1, \cdots, \beta_i)$ is the coefficient vector, $L(\cdot, \cdot)$ is the loss function, and $R(\cdot)$ is a regularizer that encodes different sparse pattern of β.

In this paper, we propose a hierarchical model where each coefficient in β is decomposed into products of multiple hierarchical coefficients. In order to learn the dependencies between variables and time lag of each variables simultaneously, we consider the two-level hierarchies,

$$\beta_j = \alpha_j \sum_{l=1}^{L} \gamma_{jl}$$

where β_j is the jth element in vector β, α_j represents the feature-level coefficient, and γ_{jl} denotes the time-level coefficient of lth time lag with respect to the jth feature.

In order to improve the efficiency of dependencies and time lag learning, we use some prior knowledge about time lag. For example, in traffic data, some travel time can be obtained from history data, or computed by road length and vehicle speed. However, different vehicles have different speeds, which leads to the fluctuation of time lag. Let $K(i) = exp(-\frac{(i-\mu_j)^2}{\delta_j^2})$ be the prior information restriction about time lag at the ith time stamp for the jth feature, where the μ_j is mean time lag for vehicles from jth entrance ramp to the target exit ramp, and it is a known information. Thus, we have:

$$\beta_j = \alpha_j \sum_{l=1}^{L} K(l)\gamma_{jl} \tag{2}$$

This two-level hierarchies with time lag lasso objection is established as follow:

$$\min_{\beta} \sum_{i=1}^{n}(y^i - \sum_{j=1}^{d} \alpha_j \sum_{l=1}^{L} exp(-\frac{(l-\mu_j)^2}{\delta_j^2})\gamma_{jl}x_{jl}^i - b)^2 + \lambda_1\|\alpha\| + \lambda_2\|\gamma\|, \tag{3}$$

where $\alpha = (\alpha_1, \cdots, \alpha_d)$ and $\gamma = (\gamma_{11}, \cdots, \gamma_{1L}, \cdots, \gamma_{dL})$.

3.2 A Probabilistic Framework for TLHL Model

In this section, we give the probabilistic interpretation for introducing our probabilistic model. For a regression problem, we use normal distribution to define the likelihood for y^i:

$$y^i \sim \mathcal{N}(\sum_{j=1}^{d} \sum_{l=1}^{L} \beta_{jl}x_{jl}^i + b, \sigma), \tag{4}$$

where $\mathcal{N}(\mu, s)$ denotes a normal distribution with mean μ and variance s^2. Then we need to specify the prior over the parameter β_j in β. Since β_j is represented by α_j and γ_{jl}, instead we define priors over the component coefficients. The component coefficients corresponding to the feature-level and time-level are placed in two probabilistic priors. We assume time-level coefficient follows a norm distribution:

$$\gamma_{jl} \sim \mathcal{N}(0, \theta_{jl}^2). \tag{5}$$

For feature-level coefficient, a Cauchy prior is placed:

$$\alpha_j \sim \mathcal{C}(0, \phi_j), \tag{6}$$

where $\mathcal{C}(a, b)$ denotes the Cauchy distribution [21] with the probability density function defined as:

$$p(x; a, b) = \frac{1}{\pi b[(\frac{x-a}{b})^2 + 1]},$$

where a and b represent the location and scale parameters respectively. The Cauchy prior is widely used for feature learning [22,23]. Moreover, in order to obtain sparse hyperparameter θ_{jl}, we place the Jeffreys prior [24] over the θ_{jl}:

$$p(\theta_{jl}) \propto \frac{1}{\theta_{jl}} \tag{7}$$

Eqs.(4)-(7) define the probabilistic model of TLHL. In the next, we discuss how to learn the model parameters.

3.3   Parameter Inference

In the EM algorithm, $\boldsymbol{\theta} = \{\theta_{jl}\}_{j \in \mathbb{N}_d, l \in \mathbb{N}_L}$, where $\mathbb{N}_d$ is the index set $\{1, \cdots, d\}$, are treated as hidden variables, and the model parameters are denoted by $\boldsymbol{\Theta} = \{\boldsymbol{\delta}, b, \sigma, \boldsymbol{\phi}\}$ where $\boldsymbol{\delta} = \{\delta_j\}_{j \in \mathbb{N}_d}$, and $\boldsymbol{\phi} = \{\phi_j\}_{j \in \mathbb{N}_d}$. In the following, we give the details in the EM algorithm.

E-step: we construct the Q-function as

$$Q(\boldsymbol{\Theta}|\boldsymbol{\Theta}^t) = \mathbb{E}[\ln p(\boldsymbol{\Theta}|y, \boldsymbol{\theta})] = \int p(\boldsymbol{\theta}|y, \boldsymbol{\Theta}^t) \ln p(\boldsymbol{\Theta}|y, \boldsymbol{\theta}) d\theta.$$

where $\boldsymbol{\Theta}^t$ denotes the estimate of $\boldsymbol{\Theta}$ in the tth iteration. It is easy to get

$$\ln p(\boldsymbol{\Theta}|\boldsymbol{y}, \boldsymbol{\theta}) \propto \ln p(\boldsymbol{y}|\boldsymbol{\delta}, b, \sigma, \boldsymbol{\gamma}, \boldsymbol{\alpha}) + \ln p(\boldsymbol{\alpha}|\boldsymbol{\phi}) + \ln p(\boldsymbol{\gamma}|\boldsymbol{\theta})$$

$$\propto -\sum_{i=1}^{n} \frac{1}{2\sigma^2}(y^i - \sum_{j=1}^{d} \alpha_j \sum_{l=1}^{L} exp(-\frac{(l-\mu_j)^2}{\delta_j^2})\gamma_{jl} x_{jl}^i - b)^2 - n\ln\sigma$$

$$-\sum_{j=1}^{d} \ln p(\frac{\alpha_j^2}{\phi_j^2} + 1) - \sum_{j=1}^{d}\sum_{l=1}^{L} \frac{\gamma_{jl}^2}{2\theta_{jl}^2} - \sum_{j=1}^{d} \ln \phi_j$$

and $p(\theta_{jl}|\boldsymbol{y}, \boldsymbol{\Theta}^t) \propto p(\theta_{jl})p(\gamma_{jl}^t|\theta_{jl})$. Then we compute the following expectation:

$$\mathbb{E}[\frac{1}{2\theta_{jl}}|\boldsymbol{y}, \boldsymbol{\Theta}^t] = \frac{\int_0^\infty \frac{1}{2\theta^2} p(\theta_{jl})p(\gamma_{jl}^t|\theta_{jl})d\theta_{jl}}{\int_0^\infty p(\theta_{jl})p(\gamma_{jl}^t|\theta_{jl})d\theta_{jl}} = \frac{1}{[2\gamma_{jl}^t]^2}.$$

Let $\nu_{jl} = \frac{1}{[2\gamma_{jl}^t]^2}$ and we can finally get

$$Q(\boldsymbol{\Theta}|\boldsymbol{\Theta}^t) = -\sum_{i=1}^{n} \frac{1}{2\sigma^2}(y^i - \sum_{j=1}^{d} \alpha_j \sum_{l=1}^{L} exp(-\frac{(l-\mu_j)^2}{\delta_j^2})\gamma_{jl} x_{jl}^i - b_i)^2 - n\ln\sigma$$

$$-\sum_{j=1}^{d} \ln p(\frac{\alpha_j^2}{\phi_j^2} + 1) - \sum_{j=1}^{d}\sum_{l=1}^{L} \nu_{jl}\gamma_{jl}^2 - \sum_{j=1}^{d} \ln \phi_j.$$

M-step: We maximize $Q(\boldsymbol{\Theta}|\boldsymbol{\Theta}^t)$ to update the estimates of $\boldsymbol{\alpha}, \boldsymbol{\gamma}, b, \sigma$ and $\boldsymbol{\delta}$.

(1) For the estimation of $\boldsymbol{\alpha}$, we have to solve the following optimization problem:

$$\min J = \sum_{i=1}^{n} \frac{1}{2\sigma^2}(y^i - \sum_{j=1}^{d} \alpha_j \sum_{l=1}^{L} exp(-\frac{(l-\mu_j)^2}{\delta_j^2})\gamma_{jl} x_{jl}^i - b)^2$$

$$+ \sum_{j=1}^{d} \ln p(\frac{\alpha_j^2}{\phi_j^2} + 1) + \sum_{j=1}^{d} \ln \phi_j.$$

$$(8)$$

By setting the derivatives of J with respect to ϕ to zero, we only consider $\min_{\phi} \sum_{j=1}^{d} \ln p(\frac{\alpha_j^2}{\phi_j^2} + 1) + \sum_{j=1}^{d} \ln \phi_j$, and then get

$$\phi_j = |\alpha_j|. \tag{9}$$

By plugging the solution in Eq.(9), we simplify Eq.(8) as

$$\min \bar{J} = \sum_{i=1}^{n} \frac{1}{2\sigma^2} (\hat{y}^i - \sum_{j=1}^{d} \alpha_j \sum_{l=1}^{L} \hat{x}_{jl}^i)^2 + \sum_{j=1}^{d} \ln |\alpha_j|, \tag{10}$$

where $\hat{y}^i = y^i - b$ and $\hat{x}_{jl}^i = K(l)\gamma_{jl}x_{jl}^i$. Problem (10) is non-convex since the second term of the objective function is non-convex. To solve this problem, we use the majorization-minimization (MM) algorithm [25] to solve this problem. For numerical stability, we slightly modify Eq.(10) by replacing the second term with $\sum_{j=1}^{d} \ln(|\alpha_j| + \eta)$, where η is a tuning parameter. We denote the solution obtained in the t'th iteration in the MM algorithm as $\alpha_j^{t'}$. Thus, in the $(t'+1)$th iteration, we only need to solve a weighted ℓ_1 minimization problem [26,27]:

$$\sum_{i=1}^{n} \bar{\sigma}(\hat{y}^i - \sum_{j=1}^{d} \alpha_j \sum_{l=1}^{L} \hat{x}_{jl}^i)^2 + \lambda \sum_{j=1}^{d} \frac{|\alpha_j|}{|\alpha_j^{t'}| + \eta}, \tag{11}$$

where $\bar{\sigma} = \frac{1}{2\sigma^2}$, which can be solved by some mature solvers, such as LASSO-style solvers.

(2) For the estimation of γ, we need to solve:

$$\min \tilde{J} = \sum_{i=1}^{n} \bar{\sigma}(\hat{y}^i - \gamma_l \tilde{x}_l^i)^2 + \sum_{j=1}^{d} \sum_{l=1}^{L} \nu_{jl} \gamma_{jl}^2, \tag{12}$$

where $\gamma_l = (\gamma_{1l}, \cdots, \gamma_{dl})$, $\tilde{x}_l^i = (\alpha_1 x_{1l}^i, \cdots, \alpha_d x_{dl}^i)^T$. We use a gradient method such as conjugate gradient to optimize Eq.(12). The subgradient with respect to γ_l is

$$\frac{\partial \tilde{J}}{\partial \gamma_l} = 2\bar{\sigma}(\tilde{X}_l \tilde{X}_l^T \gamma_l - \tilde{X}_l \hat{Y}) + 2D, \tag{13}$$

where $\tilde{X}_l = (\tilde{x}_l^i, \cdots, \tilde{x}_l^n)$, $\hat{Y} = (\hat{y}^1, \cdots, \hat{y}^n)^T$, and $D = (\nu_{1l}\gamma_{1l}, \cdots, \nu_{dl}\gamma_{dl})^T$.

(3) For the estimation of b, σ and δ, we solve:

$$\min \hat{J} = \sum_{i=1}^{n} \frac{1}{2\sigma^2} (y^i - \sum_{j=1}^{d} \sum_{l=1}^{L} exp(-\frac{(l - \mu_j)^2}{\delta_j^2})\bar{x}_{jl}^i - b)^2 + n \ln \sigma, \tag{14}$$

where $\bar{x}_{jl}^i = \alpha_j^{t+1}\gamma_{jl}^{t+1}x_{jl}^i$. We set the derivatives of Eq.(14) with respect to each of them to zero and get

$$b^{t+1} = \frac{1}{n} \sum_{i=1}^{n} [y^i - \sum_{j=1}^{d} \sum_{l=1}^{L} exp(-\frac{(l - \mu_j)^2}{\delta_j^2})\bar{x}_{jl}^i], \tag{15}$$

$$\sigma^{t+1} = \sqrt{\frac{1}{n}\sum_{i=1}^{n}[y^i - \sum_{j=1}^{d}\sum_{l=1}^{L}exp(-\frac{(l-\mu_j)^2}{\delta_j^2})\bar{x}_{jl}^i - b^{t+1}]^2}, \qquad (16)$$

For the estimation of δ_j, we also use a gradient method. The gradient can be calculated as

$$\frac{\partial \hat{J}}{\partial \delta_j} = \sum_{i=1}^{n}\frac{1}{(\sigma^{t+1})^2}\{[y^i - b^{t+1} - \sum_{j=1}^{d}\sum_{l=1}^{L}exp(-\frac{(l-\mu_j)^2}{\delta_j^2})\bar{x}_{jl}^i]$$

$$\cdot \sum_{l=1}^{L}exp(-\frac{(l-\mu_j)^2}{\delta_j^2}\bar{x}_{jl}^i) \cdot \frac{2(l-\mu_j)^2}{\delta_j^3}\}. \qquad (17)$$

4   Experiments

In this section, we evaluate our proposed method on both synthetic dataset and real-world traffic dataset. The compared methods include: lasso without considering time lag [19], lasso with fixed time lag and Granger Lasso [28].

4.1   Synthetic Data

Setting: We first evaluate the effectiveness of our method on synthetic data. We simulate a regression problem with $d = 20$ features (prediction observations). The corresponding observations are generated from $y = X\beta + \varepsilon, \varepsilon \sim \mathcal{N}(0,\sigma^2)$. We vary n from 40 to 100, and set $\sigma = 0.1$. The true coefficient of feature-level is set as

$$\beta^* = (2\ 2\ 0\ 0\ 0\ 2\ 2\ 0\ 0\ 0\ 0\ 2\ 2\ 2\ 2\ 0\ 0\ 0\ 0\ 0)^T.$$

In order to simulate the corresponding observations generated from prediction observations of different time lag, we randomly generate mean time lag $\mu_j | j \in \mathbb{N}_d$ of each feature and standard deviation δ_j. The maximum lag is set as $L = 10$. For each feature, only 5 out of coefficients of time-lag-level $\gamma_{jl} | l \in \mathbb{N}_L, j \in \mathbb{N}_d$ have values from $[0, 1]$. Finally, we use $exp(-\frac{(l-\mu_j)^2}{\delta_j^2}) \cdot \gamma_{jl}$ as a coefficient of the jth feature at the lth time. We set $X \sim \mathcal{N}(0, \mathbf{S}_{\mathbf{P}\times\mathbf{P}})$ with $s_{ii} = 1, \forall i$ and $s_{ij} = 0.5$ for $i \neq j$. We generate n samples for training as well as 20 samples for testing.

We use the mean squared error (MSE) $(\frac{1}{n}\sum_{t=1}^{n}\|Y_i - X_i\beta\|_2^2)$ and mean absolute percentage error (MAPE) $(\frac{1}{n}\sum_{i=1}^{n}|\frac{y_i - x_i\beta}{y_i}|)$ to evaluate the prediction error and use the F1-score to evaluate the performance in terms of feature selection, where the F1-score is computed as follows:

$$Pre = \frac{|\{i|\beta_i \neq 0, \beta_i^* \neq 0\}|}{|\{i|\beta_i \neq 0\}|}, \quad Rec = \frac{|\{i|\beta_i \neq 0, \beta_i^* \neq 0\}|}{|\{i|\beta_i^* \neq 0\}|}, \quad F1 = \frac{2 \cdot Pre \cdot Rec}{Pre + Rec}.$$

Results: We compare the methods when the sample size n is varying. Fig.3 shows the values of MSE and MAPE. The methods with considering time lag outperform traditional lasso algorithm. Specially, TLHL performs best. Due to the

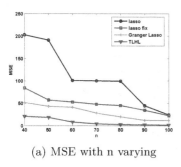

(a) MSE with n varying

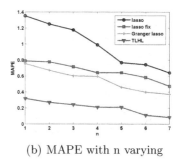

(b) MAPE with n varying

Fig. 3. Results on Synthetic datasets

method of Lasso with fixed time lag only consider one value of lag, the lag fluctuating cannot be captured. Granger Lasso considers all the history prediction observations of lag equally, and it focuses on making best prediction, thus, many unrelated dependencies are learned, and might appear over fitting. Granger Lasso has a better prediction accuracy on training samples than that on test

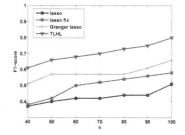

Fig. 2. F1-score with n varying

samples. Fig.2 gives the F1-score, which reflects the accuracy of dependency structures. We can see that the result of Granger Lasso is worse than TLHL, because the value of recall is small. Our method, TLHL, shows the best effect no matter on prediction accuracy or on dependencies learning.

4.2   Highway Traffic Data

Description and Setting: We evaluate our methods on real-life traffic data. The features in this traffic dataset are observations collected from sensors located on ramps in a highway traffic network. Each observation is the vehicle count during 15 minutes interval.

There are total 236 traffic stations, which correspond to 236 ramps, i.e., $d = 236$. The corresponding observations are collected at time interval 10:00-10:15 AM from 2010/8/1 to 2010/10/31 (n=92). All prediction observations are collected from 236 entrance ramps at 5:00-10:00 AM (L=20). The mean time lag of each entrance ramp is average travel time from this entrance ramp to exit ramp. The data in this experiment are all normalized.

Because there is no ground truth for dependency structure in real traffic data, F1-score cannot be measured. Nevertheless, the dependencies detected is the most important for traffic prediction. We use MSE to evaluate the accuracy. Specifically, the dependency structure information can help domain experts to predict vehicle flow with relative prediction algorithm, thus, we combine local

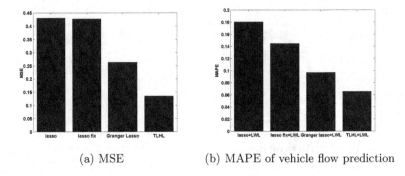

(a) MSE   (b) MAPE of vehicle flow prediction

Fig. 4. Results on Highway traffic datasets

weighted learning (LWL) [29] with our vehicle flow information of entrance ramps, which is learned by dependency structure learning algorithm. The value of MAPE is the prediction results.

Results: Fig.4(a) shows the MSE of all methods. As can be seen, our proposed TLHL is consistently among the best. Time lag is the key feature of hidden in traffic analysis. It takes time for a vehicle from one entrance ramp to exit ramp. It is not accurate for only using the same time stamp data of entrance ramp and exit ramp, because some of vehicles from that entrance ramp do not arrive at the same time stamp. That is why the algorithm without considering time lag does not show the satisfied results. Similarly, different vehicle might has different time lag, using fixed time lag also does not obtain the accurate results.

The dependency structures obtained by dependency learning methods are essentially important for the analysis of traffic systems, such as vehicle flow prediction, anomaly detection. Domain experts can obtain the information of upper stations (entrance ramps) from the accurate dependencies, and predict the state of lower stations (exit ramps). Fig.4(b) shows the results of exit ramp flow prediction algorithm with dependency learning methods. We revise LWL prediction algorithm by monitoring the vehicle flow of entrance ramps. Due to our TLHL method can obtain the accurate dependency structure of entrance ramps and exit ramps, we can get precise information. The TLHL outperforms all other methods in terms of MAPE.

5   Conclusion

In this paper, we propose a two-level hierarchies with time lag lasso to cope with dependency structure learning. We decompose the traditional regression coefficients into products of two-level hierarchical coefficients, where each one represents the different levels of information. Specifically, in order to learn the time-level structure more accurately, we use the prior time lag information. We develop a probabilistic model to interpret how to construct the regularization form for model parameters. For experimental studies, we demonstrate the effectiveness of our method on both synthetic data and real-life highway traffic data. The results show that our TLHL model can achieve significant improvements in both the datasets compared with other methods.

For future work, it is interesting to extend the static dependency structure learning to deal with time-varying observations. We can follow the evolvement of the dependencies and time lag in a network.

Acknowledgments. Research was supported by the National Natural Science Foundation of China under Grant No. 61473006.

References

1. Moneta, A., Spirtes, P.: Graphical models for the identification of causal structures in multivariate time series models. In: JCIS (2006)
2. Friedman, N., Nachman, I., Per, D.: Learning bayesian network structure from massive datasets: the "sparse candidate" algorithm. In: Proceedings of the Fifteenth Conference on Uncertainty in Artificial Intelligence, pp. 206–215 (1999)
3. Silva, R., Scheines, R., Glymour, C., Spirtes, P.: Learning the structure of linear latent variable models. The Journal of Machine Learning Research **7**, 191–246 (2006)
4. Spirtes, P., Glymour, C.N., Scheines, R.: Causation, prediction, and search, vol. 81. MIT press (2000)
5. Lee, S., Heydecker, B., Kim, Y.H., Shon, E.-Y.: Dynamic od estimation using three phase traffic flow theory. Journal of Advanced Transportation **45**(2), 143–158 (2011)
6. Barceló, J., Montero, L., Bullejos, M., Serch, O., Carmona, C.: A kalman filter approach for the estimation of time dependent od matrices exploiting bluetooth traffic data collection. In: Transportation Research Board 91st Annual Meeting, number 12–3843 (2012)
7. Zhao-cheng, H., Zhi, Y.: Dynamic od estimation model of urban network [j]. Journal of Traffic and Transportation Engineering **5**(2), 94–98 (2005)
8. Han, L., Song, G., Cong, G., Xie, K.: Overlapping decomposition for causal graphical modeling. In: Proceedings of the 18th ACM SIGKDD International Conference on Knowledge Discovery and Data Mining, KDD 2012, pp. 114–122 (2012)
9. Granger, C.W.J.: Testing for causality: a personal viewpoint. Journal of Economic Dynamics and control **2**, 329–352 (1980)
10. Yuan, M., Lin, Y.: Model selection and estimation in regression with grouped variables. Journal of the Royal Statistical Society. Series B (Statistical Methodology) **68**(1) (2006)
11. Li, F., Zhang, N.R.: Bayesian variable selection in structured high-dimensional covariate spaces with applications in genomics. Journal of the American Statistical Association **105**(491), 1202–1214 (2010)
12. Bahadori, M.T., Liu, Y.: Granger causality analysis in irregular time series. In: SDM, pp. 660–671 (2012)
13. Dempster, A.P., Laird, N.M., Rubin, D.B.: Maximum likelihood from incomplete data via the em algorithm. Journal of the Royal Statistical Society, Series B **39**(1), 1–38 (1977)

14. Meinshausen, N., Bühlmann, P.: High-dimensional graphs and variable selection with the lasso. The Annals of Statistics, 1436–1462 (2006)
15. Gao, Y., Sun, S.: Multi-link traffic flow forecasting using neural networks. In: 2010 Sixth International Conference on Natural Computation (ICNC), vol. 1, pp. 398–401 (2010)
16. Sun, S., Huang, R., Gao, Y.: Network-scale traffic modeling and forecasting with graphical lasso and neural networks. Journal of Transportation Engineering 138(11), 1358–1367 (2012)
17. Kawale, J., Liess, S., Kumar, V., Lall, U., Ganguly, A.: Mining time-lagged relationships in spatio-temporal climate data. In: 2012 Conference on Intelligent Data Understanding (CIDU), pp. 130–135 (2012)
18. Yang, S.: On feature selection for traffic congestion prediction. Transportation Research Part C: Emerging Technologies 26, 160–169 (2013)
19. Tibshirani, R.: Regression shrinkage and selection via the lasso. Journal of the Royal Statistical Society. Series B (Methodological) 58, 267–288 (1996)
20. Zhao, P., Rocha, G., Yu, B.: Grouped and hierarchical model selection through composite absolute penalties. Department of Statistics, UC Berkeley. Tech. Rep, 703 (2006)
21. Arnold, B.C., Brockett, P.L.: On distributions whose component ratios are cauchy. The American Statistician 46(1), 25–26 (1992)
22. Carvalho, C.M., Polson, N.G., Scott, J.G.: Handling sparsity via the horseshoe. In: International Conference on Artificial Intelligence and Statistics, pp. 73–80 (2009)
23. Hernández-Lobato, D., Hernández-Lobato, J.M., Dupont, P.: Generalized spike-and-slab priors for bayesian group feature selection using expectation propagation. The Journal of Machine Learning Research 14(1), 1891–1945 (2013)
24. Qi, Y.A., Minka, T.P., Picard, R.W., Ghahramani, Z.: Predictive automatic relevance determination by expectation propagation. In: Proceedings of the Twenty-first International Conference on Machine Learning, ICML 2004, pp. 85–92 (2004)
25. Lange, K., Hunter, D.R., Yang, I.: Optimization transfer using surrogate objective functions. Journal of Computational and Graphical Statistics 9(1), 1–20 (2000)
26. Candes, E.J., Wakin, M.B., Boyd, S.P.: Enhancing sparsity by reweighted ℓ_1 minimization. Journal of Fourier analysis and applications 14(5–6), 877–905 (2008)
27. Wipf, D., Nagarajan, S.: Iterative reweighted ℓ_1 and ℓ_2 methods for finding sparse solutions. IEEE Journal of Selected Topics in Signal Processing 4(2), 317–329 (2010)
28. Arnold, A., Liu, Y., Abe, N.: Temporal causal modeling with graphical granger methods. In: Proceedings of the 13th ACM SIGKDD International Conference on Knowledge Discovery and Data Mining, KDD 2007, pp. 66–75 (2007)
29. Meng, S., Lei, H., Kunqing, X., Guojie, S., Xiujun, M., Guanhua, C.: An adaptive traffic flow prediction mechanism based on locally weighted learning. Acta Scientiarum Naturalium Universitatis Pekinensis 46(1), 64–68 (2010)

Location Semantics Protection Based on Bayesian Inference

Zhengang Wu[1,2](✉), Zhong Chen[1], Jiawei Zhu[1], Huiping Sun[1], and Zhi Guan[1]

[1] Institute of Software, School of EECS, MoE Key Lab of High Confidence Software Technologies (PKU), MoE Key Lab of Network and Software Security Assurance (PKU), Peking University (PKU), Beijing, China
wuzhengang@pku.edu.cn, {chen,sunhp}@ss.pku.edu.cn, zhujw.happy@163.com, guanzhi1980@gmail.com
[2] China Academy of Information and Communications Technology, Beijing, China

Abstract. In mobile Internet, popular Location-Based Services (LBSs) recommend Point-of-Interest (POI) data according to physical positions of smartphone users. However, untrusted LBS providers can violate location privacy by analyzing user requests semantically. Therefore, this paper aims at protecting user privacy in location-based applications by evaluating disclosure risks on sensitive location semantics. First, we introduce a novel method to model location semantics for user privacy using Bayesian inference and demonstrate details of computing the semantic privacy metric. Next, we design a cloaking region construction algorithm against the leakage of sensitive location semantics. Finally, a series of experiments evaluate this solution's performance to show its availability.

Keywords: Location privacy protection · Location semantics · Bayesian inference · Spatial cloaking

1 Introduction

Location Based Services (LBSs), as the representative of context-aware services, can recommend accurate and timely information according to user locations. The wide application of LBSs (such as Check-ins, Navigation, Maps and Mobile Social Networks) is benefit from the widespread availability of wireless networks and smart devices with built-in positioning modules. However, LBS gets involved in the problematic concern about location privacy because of its operating mechanism. Generally, in popular LBS-based applications and systems, real-time user locations from the LBS clients (e.g. some specific APPs installed in smartphones) as the vital contextual information need to be reported to the corresponding LBS providers in the on-demand manner. As a result, massive user locations are readily collected by potential adversaries via some untrusted servers and connection channels in mobile Internet.

Following privacy protection of relational databases, existing techniques of location privacy preservation have aimed at constructing the cloaking region

© Springer International Publishing Switzerland 2015
J. Li and Y. Sun (Eds.): WAIM 2015, LNCS 9098, pp. 297–308, 2015.
DOI: 10.1007/978-3-319-21042-1_24

under general privacy metrics such as k-anonymity and l-diversity to generalize exact user locations into custom extended spatial regions. Although effectively achieving a limited guarantee for location privacy, these techniques are vulnerable to Location Semantics Attack [1]. Intuitively, for a target user of LBS, the entire or major part of a k-anonymity cloaking region may be annotated with a similar sensitive semantic label such as Cancer Treatment Hospitals, and therefore adversaries can breach his privacy by learning his poor health status with a high probability.

Contributions. This solution involves three-fold contributions. First, the proposed approach LSRG models the process which extracts sensitive semantics form user requests and measures the degree of the semantics leakage. Second, this paper introduces a spatial cloaking method for preserving sensitive semantics on user locations. Third, its performance is demonstrated experimentally under different configurations by our adjusting crucial parameters.

Outline. The rest of the article is organized as follow. The 2nd section describes background. Section 3 and Section 4 shows two major parts of this work, extracting sensitive location semantics and constructing the cloaking region respectively. And Section 5 evaluates the performance of this solution through experiments. Section 6 reviews related works and the last section makes a summary.

2   Structure and Motivation

System Description. Following the popular three-tire architecture [2] for location privacy protection, our solution runs on this middle server in Figure 1. An LBS provider holds massive Point-of-Interest (POI) records which are meaningful location points over real maps. This middle server as a Trusted-Third-Party (TTP) is deployed between mobile clients and LBS servers to protect location privacy. First, for a user, this middle server extends exact locations into cloaking regions where all POI results are ready for forwarding. Second, this middle server refines and dispatchs POIs to corresponding users.

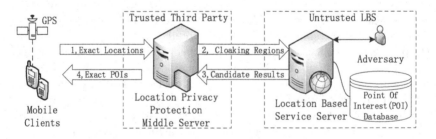

Fig. 1. System Model

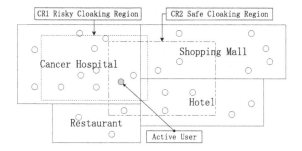

Fig. 2. Location Semantics Attack

Problem Setting and Motivation. Intuitively, the adversary analyzes published locations by mining semantic information for a target user. For example, a user Alice reports her current location coordinate *loc* to an untrusted LBS provider Malice in real activities. Next, Malice learns that a cancer hospital is located in the location point *loc* after querying public POI databases and map services such as Baidu Maps, Tencent Maps and Google Maps. Finally, Malice learns that Alice's health is poor with a high probability since some LBS requests are linked with meaningful labels.

The crux of Location Semantics Attack is that the adversary holds the public background knowledge on POI databases as same as users. In Figure 2, the cloaking region $CR1$ discloses that the active user is probably a cancer patient since all requests in $CR1$ are from cancer hospitals and the poor health status is one of his sensitive attributes. By contrast, these requests of $CR2$ are dispersed into various semantic regions such as hospitals, malls, restaurants and hotels and thus it is safer if the distribution of the adversary's guessing is uniform over these regions without additional information.

A recent work [1] has aimed at the semantic safety. The adversary may learn sensitive semantics on various locations. The majority of existing cloaking methods fail to capture the semantic risk. A cloaking region can still leak some risky semantics information in spite of satisfying the k-anonymity rule, since the major or entire part of the cloaking region which holds k users in a snapshot of LBS requests may be mapped into a risky semantic label such as infectious hospitals.

3   Evaluating Location Privacy

Generally, in client-side of LBS, when visiting LBS, a user submits a location request (U, L, T) where the users identifer U, the raw location L and the timestamp T. e.g. A request is (Alice,(116.42284,39.908063),'12:00'). In server-side of LBS, a POI entry is defined as a tuple (L, S, D). The raw location L is a pair (lng, lat) refers to the longitude lng and the latitude lat. The semantic label S refers to a meaningful brief name on this raw location L. The detail content D is a readable text to describe this raw location. e.g. a POI is ((116.42284,39.908063), 'bank', 'The Bank of China'). Thus, a location request discloses that this user may execute a personal activity about this semantic information 'bank'.

We model the causal relationship among raw locations (i.e. location coordinates), semantic labels (i.e. the meaning name of the raw location in real maps) and privacy risks (i.e. possible privacy disclosure events on special semantic labels) and naturally measure the belief of privacy risks using the probability of the privacy disclosure events on any raw locations and regions (the section/set of raw locations).

3.1   Modeling Location Semantics

As shown in Figure 3, this graphical model Location Semantics Risk Graph (in short LSRG) describes the privacy risk belief of a location request when the adversary eavesdrops this request after knowing semantic information of raw locations.

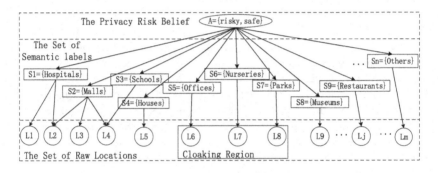

Fig. 3. Location Semantics Risk Graph

Definition 1. *Location Semantics Risk Graph is a three-tier directed acyclic graph $G = (V, E)$. The node collection V falls into three mutually exclusive subsets, the set of raw locations V_l, the set of semantic labels V_s and the privacy risk belief A. A directed edge $e = <a, b> \in E$ refers to the dependency belief between its start point a and its end point b that is a conditional probability $P(b|a) > 0$.*

To simplify this model properly, we adopt an assumption that events of locations are independent of one another and so those of semantic labels are. i.e. There are no edges which connect two peer nodes of locations or semantic labels. Clearly, connections between locations and semantic labels have a many-to-many relationship, referring to edges from the location node set V_l to the semantics node set V_s. This is consistent with real-world experiences. A building on a location may be comprehensive with offices and shopping centers. Similarly, hospitals may be dispersed into different regions in a real city.

3.2   Inferring Privacy Risks

The binary class variable A that is the root node of LSRG denotes the probability event that $A = A_t$ if the adversary learns privacy information of users via location semantics inference and otherwise $A = A_f$. In brief, the event A_t means a risky request and A_f refers to a safe request form the perspective of user privacy.

The evidence variable O_{loc} is the observed location information (e.g. a region) which the middle server submits into an untrusted LBS server for forwarding the user request. O_{loc} actually is a cloaking region in the generalization-based location privacy protection schemes. O_{loc} is a subset of location nodes V_l. i.e. $O_{loc} \subseteq V_l$. Without cloaking, O_{loc} holds only one location that is the user's current location. But after cloaking locations, O_{loc} becomes a continuous spatial region which includes the current location.

Therefore, the posterior probability $P(A_t|O_{loc})$ refers to the conditional probability for the privacy disclosure event A_t of a request on the published location information O_{loc}. Naturally, $P(A_t|O_{loc})$ can be used to measure the privacy risk degree. The privacy risk $P(A_t|O_{loc})$ can be calculated by the Bayesian rule as follow.

$$PrivacyRisk \overset{def}{=} P(A_t|O_{loc}) = \frac{P(O_{loc}|A_t)P(A_t)}{P(O_{loc}|A_t)P(A_t) + P(O_{loc}|A_f)P(A_f)} \quad (1)$$

3.3   Estimating Parameters

Computing the posterior belief needs to obtain three prior beliefs $P(O_{loc}|A_t)$, $P(O_{loc}|A_f)$ and $P(A_t)$ which are estimated by given samples and the Maximum Likelihood Estimation (MLE). For all cloaked location-based requests, each published location information $O_{loc} \subseteq V_l$ can be decomposed into a series of basic locations, relying on specific methods of clustering or partitioning spatial data for original location coordinates. Basic locations in Figure 3 are the m-order collection of leaf nodes, $V_l = \{L_1, \cdots, L_m\}$. Therefore, for $A \in \{A_t, A_f\}$, $P(O_{loc}|A) = \sum_{l \in O_{loc}} P(l|A)$ where $P(l|A)$ is the condition probability for a basic location $l \in V_l$. Naturally, repetitive computation steps can be reduced using precalculated beliefs $P(l|A)$ of all basic locations.

The Prior Belief $P(O_{loc}|A_t)$. Without loss of generality, the adversary observes a spatial region O_{loc} which refers to a set of locations as the evidence. Specifically, throughout a middleware for location privacy protection, the cloaking region R_{ca} is actually the observed region. i.e. $O_{loc} = R_{ca}$. This belief can be calculated as follow.

$$P(O_{loc}|A_t) = \sum_{l \in O_{loc}} \sum_{s \in pa(l)} [P(l|s)P(s|A_t)] \quad (2)$$

For an untrusted LBS, the adversary's ability involves two-fold factors. The first factor is the location semantics knowledge. The adversary can access public

POI databases and thus obtains corresponding semantics information on locations. Next, the semantics risk knowledge is the other factor. The adversary's intention relies on semantics labels for learning sensitive attributes of a target user and thus different location semantics implies different risk levels for location privacy. Based on the intuitive understanding, we can estimate this belief $P(O_{loc}|A_t) = \sum_{l \in O_{loc}} P(l|A_t)$ under the LSRG model, after knowing these two factors which express as two condition probabilities $P(s|A_t)$ and $P(l|s)$ respectively where $s \in pa(l) \subseteq V_s$ and $l \in V_l$. For simplicity, $pa(l)$ denotes the set of parent nodes of the node l in LSRG.

The location semantics knowledge can be computed using $\widehat{P}(l|s) = \frac{F(l,s)}{F(s)}$ where the function $F(x)$ is the metric of the event x. We assume that for a semantics label s the adversary's attack is the spatial uniform distribution over the region of this semantics s and so the metric function $F(x)$ should be the area of the region meeting the event (l, s) or (s). i.e. $P(l|s) = \frac{Area(l,s)}{Area(s)}$. However, computation of exact areas of massive irregular regions over a real map will generally consume intensive resources since popular online map services fail to provide related data directly. As a practical alternate, we can employ the number of POI entries in the region meeting specific semantics conditions. i.e. $P(l|s) = \frac{count(l,s)}{count(s)}$. Given the POI database which the untrusted LBS holds, the function $count(s)$ counts up the number of POI entries whose semantics label is s and the function $count(l, s)$ refers to the number of POI entries whose semantics label is s and meanwhile whose location coordinates fall in the spatial cell annotated by l. For convenience, we use the pyramid structure [2] based on Quad-Tree to index POI entries in the 4^n grid and in fact the location semantics knowledge reflects the inherent feature of POI databases over real maps.

The semantics risk knowledge can be estimated using the frequency of risky events which are annotated by the semantics label $s \in V_s$ over all risky events. i.e. $\widehat{P}(s|A_t) = \frac{count(s,A_t)}{count(A_t)}$. Given a sample dataset of risky events, we can make a statistic analysis on the frequency of risky events grouped by semantic categories such as 'hospitals', 'offices' and so on. $count(s, A_t)$ adds up the number of risky events with the semantics label s and $count(A_t)$ is the total number of all risky events. e.g. 50 risky events on 'hospitals' exist in 100 risky events and thus we can learn the belief $P(s = hospitals|A_t) = 0.5$ on the semantics information 'hospatials' based on this sample. Note that, all events of a sample are classified into defined catalogs (semantics labels). i.e. Each event relates to only one label, and for all semantics labels $\sum_{s \in V_s} P(s|A_t) = 1$.

The Prior Belief $P(O_{loc}|A_f)$. The probability of safe requests on the observed region O_{loc} denotes this prior belief $P(O_{loc}|A_f)$. By collecting requests via a safe LBS, this MLE is obtained by Equation 3. Since O_{loc} is a set of basic locations on this partitioned maps, the probability of each basic location $P(l|A_f) = \frac{count(l,A_f)}{count(A_f)}$ can be calculated from the safe request sample. $count(l, A_f)$ is the number of safe requests on this basic location l and $count(A_f)$ is the total number of all requests on the safe sample dataset.

$$\widehat{P}(O_{loc}|A_f) = \frac{count(O_{loc}, A_f)}{count(A_f)} = \sum_{l \in O_{loc}} \frac{count(l, A_f)}{count(A_f)} \qquad (3)$$

The Prior Belief $P(A_t)$. Intuitively, the prior belief $P(A_t)$ implies the trust status of the entire LBS system including related network connections. Given an event sample of accessing an LBS, the MLE of $P(A_t)$ can express as the frequency of past request events in Equation 4 where the class variable A_t refers to events of risky requests. By the sample, $count(A_t)$ is the number of violated request events where user sensitive information is disclosed and $count(A)$ denotes the number of all events on both risky and safe requests simply. Note that, $P(A_t) + P(A_f) = 1$.

$$\widehat{P}(A_t) = \frac{count(A_t)}{count(A)} \qquad (4)$$

4 Cloaking Published Locations

This section describes a cloaking region construction method to protect location semantics. Based on the aforementioned privacy risk evaluation method, we design Algorithm 1 which can recursively construct a (k, l, t)-Secure Cloaking Region (for short, (k, l, t)-SCR) to meet three privacy requirements. First, k-anonymity[2,3] means that the cloaking region holds k different users at least. Second, l-diversity[2,4] means that the cloaking region covers l different locations (or spatial cells) at least. Third, t-safety ensures that the semantics safety of the cloaking region is larger than a threshold t. This can be defined as follow.

Definition 2. *A cloaking region O_{loc} meets t-safety if and only if its semantics safety $P(A_f|O_{loc}) = 1 - P(A_t|O_{loc}) \geq t$.*

Definition 3. *(k, l, t)-Secure Cloaking Region is a cloaking region which satisfies k-anonymity, l-diversity and t-safety.*

In the pyramid structure[2], a location point falls into a rectangular region linked with a node of the Quad-Tree. Each non-root node has only one parent node. Importantly, each non-leaf node has four child nodes like a cross and thus the non-root node has the only vertical or horizontal neighbor node in the four quadrants of the cross. For convenience, two notations $VNode$ and $HNode$ refer to the vertical neighbor and the horizontal one of $Node$ respectively.

The bottom-up Algorithm 1 can recursively create a continuous region from leaf to root along the Quad-Tree by gradually merging neighbors and check whether these candidate regions satisfy the pre-defined privacy profile. First, for k-anonymity, the region's request amount defines the anonymity degree. Here, $Node.N$ is the request amount in the region referred by $Node$. Second, for l-diversity, the region's area denoted by $Area(Node)$ measures the diversity degree. We employ the number of cells in the region $Node$ to count $Area(Node)$ since all cells occupy the same area as the basic unit of the Quad-Tree partitioned

---

Algorithm 1. SCR($k, l, t, Node$)

---

1 **if** $Area(Node) \geq MaxArea$ **then**   /* Restrict oversize. */
2 $\quad$ **return** $CR \leftarrow \varnothing$;   /* Cloaking fails. */

3 $Risk \leftarrow P(A_t|O_{loc} = \{Node\})$; $Safety(Node) = 1 - Risk$;
4 **if** $Node.N \geq k \wedge Area(Node) \geq l \wedge Safty(Node) \geq t$ **then**
5 $\quad$ **return** $CR \leftarrow \{Node\}$;
6 **else**
7 $\quad$ $(VNode, HNode) \leftarrow GetNeighbors(Node)$;
8 $\quad$ $VN \leftarrow VNode.N + Node.N$; $HN \leftarrow HNode.N + Node.N$;
9 $\quad$ **if** $(VN \geq k \vee HN \geq k) \wedge ((2 * Area(Node)) \geq l)$ **then**
10 $\quad\quad$ **if** $(VN \geq k \wedge HN \geq k \wedge HN \leq VN) \vee VN < k$ **then**
11 $\quad\quad\quad$ $CR \leftarrow \{HNode, Node\}$;
12 $\quad\quad$ **else**
13 $\quad\quad\quad$ $CR \leftarrow \{VNode, Node\}$;
14 $\quad\quad$ $Risk \leftarrow P(A_t|O_{loc} = CR)$; $Safety(CR) = 1 - Risk$;
15 $\quad\quad$ **if** $Safety(CR) \geq t$ **then**   /* Check safety. */
16 $\quad\quad\quad$ **return** CR;   /* Return one CR */
17 $\quad\quad$ **else**   /* Search its parent recursively. */
18 $\quad\quad\quad$ $CR \leftarrow SCR(k, l, t, Node.ParentNode)$;
19 $\quad$ **else**   /* Search its parent recursively. */
20 $\quad\quad$ $CR \leftarrow SCR(k, l, t, Node.ParentNode)$;

---

maps. Finally, for t-safety about location semantics, the function $Safety(Node)$ refers to $1 - Pr(A_t|O_{loc} = Node)$.

In addition, the computation of $P(A_t|O_{loc})$ can be divided into two phases for reducing its time cost since a region O_{loc} are divided into a set of distinct spatial cells and for $A \in \{A_t, A_f\}$, $P(O_{loc}|A) = \sum_{l \in O_{loc}} P(l|A)$. First, the off-line phase can calculate these prior beliefs $P(l|A)$ for each basic cell $l \in V_l$. Second, Algorithm 1 can obtain $P(A_t|O_{loc})$ with linear complexity $O(m)$ where m is the number of cells in the region O_{loc}, using the prepared prior beliefs from the off-line phase. This way can help to achieve the high processing performance on spatial cloaking and POI forwarding in the real-time LBS environment.

5   Experiments

We implement the proposed solution using JAVA and run it in the experiment platform which is a laptop with a quad-core 2.4Ghz Intel i7 CPU and 16G RAM. The experimental dataset from MNTG[5] holds trajectory data of about 1000 users who move along the real road networks of Beijing on 20 continuous timestamps (from 0 to 19) . All raw locations lie in a rectangle region about $67km^2$ and are indexed by the n-height full Quad-Tree structure [2] where 4^n leafs divide the region into 4^n cells which refer to atomic regions and the default height is 4.

To build the adversary's background knowledge, we extract the real POI dataset from a popular electronic map web site 'map.baidu.com', including about 8700 POI entries in this experimental region. Next, we explore 12 chosen semantics labels which are s_1=hospitals, s_2=nurseries, s_3=restaurant, s_4=hotels, s_5=bank, s_6=malls, s_7=offices, s_8=houses, s_9=school, s_{10}=museums, s_{11}=parks and s_{12}=others. The default privacy profile (k, l, t) is $(10, 2, 0.9)$ and additionally the default value of the total risk belief $Pr(A_t)$ is set to 0.05.

5.1 Evaluating Privacy Risks on Location Semantics

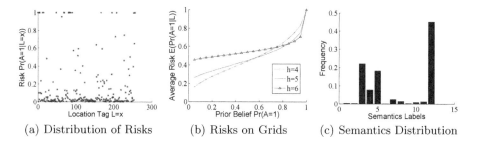

(a) Distribution of Risks   (b) Risks on Grids   (c) Semantics Distribution

Fig. 4. Risk Evaluation for Location Privacy

Experiments in Figure 4 explore the degree of the location privacy risks that mobile users leak their current locations to untrusted LBS servers on partitioned maps annotated by semantic information. Each location point refers to a POI record labeled by a meaningful string according to public real maps and POI databases, and therefore the leakage of location coordinates via a request leads to the leakage of the corresponding meaningful labels.

Figure 4(a) displays a distribution of the belief $Pr(A_t|O_{loc} = x)$ on $4^4 = 256$ cells in the 4-height full Quad-Tree structure. Intuitively, each location-based request involves a piece of risky semantic information. Generally, the majority of these cells have low risks for the perspective of user privacy. e.g. Mobile users visit in locations of public places like offices and malls. And there are some high-sensitive cells which refers to restricted regions such as hospitals and military areas. This distribution relies on two factors: First, the inherent semantic feature of a POI database or a real map expresses as the belief $Pr(O_{loc}|s \in V_s)$; Second, the adversary's intention refers to $Pr(s \in V_s|A_t)$.

Figure 4(b) shows the curves of average risk by adjusting the prior belief $P(A_t)$. Under the distinct values of $P(A_t) \in [0, 1]$, we count up the mathematical expectation (Average Risk) of $Pr(A_t|O_{loc} = x)$ for all cells. Three curves are under different Quad-Tree partitioning [2] configurations whose heights are 4, 5 and 6 respectively. More accurate location information (i.e. more finer granularity and higher Quad-Tree) leads to more privacy leakages and a higher privacy risk level. There is a positive correlation between the risk at a cell $Pr(A_t|O_{loc} = x)$ and the prior belief $P(A_t)$ which refers to the estimated total risk. Specially, when $P(A_t)$ approximates 1, the privacy disclose event on any location is inevitable with the probability that is close to 1.

Figure 4(c) demonstrates the distribution of these 12 semantics labels over the POI dataset. The majority of POI entries have low risks for user privacy and by contrast POI entries with two high risk semantics labels, s_1=hospitals and s_2=kids, take over 0.32% and 0.18% respectively. Clearly, high risky POI entries are sparse in a real-world maps. As a result, (k, l, t)-SCR can be constructed with an accepted success ratio to satisfy its custom privacy conditions.

5.2   Cloaking Published Locations

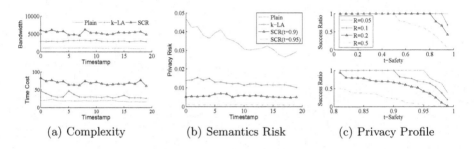

(a) Complexity   (b) Semantics Risk   (c) Privacy Profile

Fig. 5. Performance of Location Anonymization

By comparing existing location cloaking methods, experiments in Figure 5 demonstrates that the proposed location cloaking method is feasible and practical. First, the label 'Plain' means the straightway method that the location cloaking server is only a simple proxy to forward requests from mobile clients to LBS servers by replacing an exact location with a spatial cell. Next, the label 'k-LA' is the popular location k-anonymity method (NewCasper[2]) which generalizes an extended rectangular region under the k-anonymity metric. Finally, the label 'SCR' represents our solution that can guard against the Location Semantics Attack.

Figure 5(a) compares complexity on time and communication. The straightway method 'Plain' has the lowest cost on both execution time and downloaded data amount. And SCR possesses slightly more costs for controlling privacy risks under location semantics than location k-anonymity. Thus additional costs of SCR are still affordable.

As shown in Figure 5(b), the proposed method can control privacy risks on location semantics by checking $Pr(A_t|O_{loc})$ of all cloaking regions. The straightway method labeled by 'Plain' has high risks on location semantics disclosure. Next, location k-anonymity and SCR hold similar performance of privacy preservation but SCR builds safer cloaking regions than other two methods. On two

SCR curves of t=0.9 and t=0.95, for a higher safety threshold t, this method reduces privacy risks by generalizing exact locations into larger regions.

Figure 5(c) shows effects of privacy profiles by comparing success ratios of building SCRs. For specific values of the total risk belief $P(A_t)$ denoted by R=0.05,R=0.1,R=0.2, and R=0.5, the ratios drop significantly after horizontal lines which refer to 100% cloaking success, when the required safety thresholds t increasing gradually. As a result, visiting high-risk LBSs especially, we have to trade off the required safety and the cloaking success ratio.

6 Related Works

When publishing a dataset where each object holds generally one identifier and multiple attributes, the adversary can re-identify objects because of the possible uniqueness of attribute values in spite of removing identifiers. For this, k-anonymity[3][6] ensures that at least k objects are indistinguishable in an anonymity set. l-Diversity[4] requires that the number of different attributes which each object in an anonymity set associates with is more than at l. t-Closeness [7] guarantees that an anonymity set is statistically similar under the probability metric such as Earth-Mover-Distance.

Previous techniques of location privacy protection employed two basic ideas, cryptography and anonymization. Wernke et al.[8] survey research works on attacking and protecting location privacy. Cryptography-based methods[9][10] can give strong privacy assurance but need extremely intensive resources. By comparison, location anonymization (e.g. spatial cloaking) can achieve enough privacy assurance under appropriate resources.

Location k-anonymity[11] from Gruteser et al. generalizes an exact location into a region which holds at least k requests, extended from k-anonymity. Plenty of solutions such as CliqueCloak[12], HilbertCloak[13] and NewCasper[2] have adopted location k-anonymity in the last decade. Following t-closeness[7], Lee et al. introduce a location anonymization method which constructs θ-Secure Cloaking Area[1] after extracting semantics information from staying duration. Shokri et al. introduced a Markov Chain based approach[14] to measure location privacy.

Additionally, location semantics mining is a hot topic in mobile Internet. Parent et al.[15] review various methods which model and mine semantics information on trajectory data.

7 Conclusion

This paper investigated privacy protection against Location Semantics Attacks. To solve this problematic issue, we introduce the Location Semantics Risk Graph model to evaluate privacy risks about the dependence of location coordinates and sensitive semantics information, using Bayesian inference. And next we proposed a spatial cloaking algorithm under this model. Finally, experiments demonstrate that this solution can achieve a better privacy guarantee than existing schemes.

Acknowledgments. This work is partially supported by the HGJ National Significant Science and Technology Projects under Grant No. 2012ZX01039-004-009, Key Lab of Information Network Security, Ministry of Public Security under Grant No.C11606, the National Natural Science Foundation of China under Grant No. 61170263.

References

1. Lee, B., Oh, J., Yu, H., Kim, J.: Protecting location privacy using location semantics. In: Apté, C., Ghosh, J., Smyth, P. (eds.) KDD, pp. 1289–1297. ACM (2011)
2. Mokbel, M.F., Chow, C.Y., Aref, W.G.: The new casper: query processing for location services without compromising privacy. In: Dayal, U., Whang, K.Y., Lomet, D.B., Alonso, G., Lohman, G.M., Kersten, M.L., Cha, S.K., Kim, Y.K. (eds.) VLDB, pp. 763–774. ACM (2006)
3. Sweeney, L.: k-anonymity: A model for protecting privacy. International Journal of Uncertainty, Fuzziness and Knowledge-Based Systems **10**(5), 557–570 (2002)
4. Machanavajjhala, A., Gehrke, J., Kifer, D., Venkitasubramaniam, M.: l-diversity: privacy beyond k-anonymity. In: Liu, L., Reuter, A., Whang, K.Y., Zhang, J. (eds.) ICDE, p. 24. IEEE Computer Society (2006)
5. Mokbel, M.F., Alarabi, L., Bao, J., Eldawy, A., Magdy, A., Sarwat, M., Waytas, E., Yackel, S.: MNTG: an extensible web-based traffic generator. In: Nascimento, M.A., Sellis, T., Cheng, R., Sander, J., Zheng, Y., Kriegel, H.-P., Renz, M., Sengstock, C. (eds.) SSTD 2013. LNCS, vol. 8098, pp. 38–55. Springer, Heidelberg (2013)
6. Sweeney, L.: Achieving k-anonymity privacy protection using generalization and suppression. International Journal of Uncertainty, Fuzziness and Knowledge-Based Systems **10**(5), 571–588 (2002)
7. Li, N., Li, T., Venkatasubramanian, S.: t-closeness: Privacy beyond k-anonymity and l-diversity. In: Chirkova, R., Dogac, A., Özsu, M.T., Sellis, T.K. (eds.) ICDE, pp. 106–115. IEEE (2007)
8. Wernke, M., Skvortsov, P., Dürr, F., Rothermel, K.: A classification of location privacy attacks and approaches. Personal and Ubiquitous Computing **18**(1), 163–175 (2014)
9. Papadopoulos, S., Bakiras, S., Papadias, D.: Nearest neighbor search with strong location privacy. PVLDB **3**(1), 619–629 (2010)
10. Paulet, R., Kaosar, M.G., Yi, X., Bertino, E.: Privacy-preserving and content-protecting location based queries. IEEE Trans. Knowl. Data Eng. **26**(5), 1200–1210 (2014)
11. Gruteser, M., Grunwald, D.: Anonymous usage of location-based services through spatial and temporal cloaking. In: Siewiorek, D.P. (ed.) MobiSys. USENIX (2003)
12. Gedik, B., Liu, L.: Protecting location privacy with personalized k-anonymity: Architecture and algorithms. IEEE Trans. Mob. Comput. **7**(1), 1–18 (2008)
13. Kalnis, P., Ghinita, G., Mouratidis, K., Papadias, D.: Preventing location-based identity inference in anonymous spatial queries. IEEE Trans. Knowl. Data Eng. **19**(12), 1719–1733 (2007)
14. Shokri, R., Theodorakopoulos, G., Boudec, J.Y.L., Hubaux, J.P.: Quantifying location privacy. In: IEEE Symposium on Security and Privacy, pp. 247–262. IEEE Computer Society (2011)
15. Parent, C., Spaccapietra, S., Renso, C., Andrienko, G.L., Andrienko, N.V., Bogorny, V., Damiani, M.L., Gkoulalas-Divanis, A., de Macêdo, J.A.F., Pelekis, N., Theodoridis, Y., Yan, Z.: Semantic trajectories modeling and analysis. ACM Comput. Surv. **45**(4), 42 (2013)

Big Data

SALA: A Skew-Avoiding and Locality-Aware Algorithm for MapReduce-Based Join

Ziyu Lin[1(✉)], Minxing Cai[1], Ziming Huang[1], and Yongxuan Lai[2]

[1] Department of Computer Science, Xiamen University, Xiamen, China
{ziyulin,caiminxing,ziminghuang}@xmu.edu.cn
[2] School of Software, Xiamen University, Xiamen, China
laiyx@xmu.edu.cn

Abstract. MapReduce is a parallel programming model, which is extensively used to process join operations for large-scale dataset. However, traditional MapReduce-based join is not efficient when handling skewed data, because it can lead to partitioning skew, which further results in longer response time of the whole join process. Additionally, some newly proposed methods usually involve large amounts of intermediate results over the network in the shuffle phase of Mapreduce-based join, which may consume a lot of time and cause performance degradation. Here a novel algorithm called SALA is proposed, which employs volume/locality-aware partitioning instead of hash partitioning for data distribution. Compared with other existing join algorithms, SALA has three typical advantages: (1) makes sure that the data is distributed to reducers evenly when the input datasets are skewed, (2) reduces the amount of intermediate results transferred across the network by utilizing data locality, and (3) does not make any modification of the MapReduce framework. The extensive experimental results show that SALA not only achieves better load balance but reduces network overhead, and therefore speeds up the whole join process significantly in the presence of data skew.

1 Introduction

MapReduce is an efficient programming model from Google for large-scale data processing in domains such as search engine, data mining and machine learning. MapReduce is extensively used to process the join operation for large-scale dataset, and various join algorithms have been proposed to implement join operation in MapReduce environment [3].

Traditional MapReduce-based join algorithms, however, are suffering performance degradation when handling skewed data , because they use hash partitioning to distribute data that can lead to partitioning skew. Partitioning skew will bring some problems. On one hand, join algorithms have to take longer time to deal with load imbalance caused by partitioning skew. On the other hand,

---

Supported by the Natural Science Foundation of China under Grant No. 61303004 and 61202012, and the Natural Science Foundation of Fujian Province under Grant No.2013J05099.

J. Li and Y. Sun (Eds.): WAIM 2015, LNCS 9098, pp. 311–323, 2015.
DOI: 10.1007/978-3-319-21042-1_25

large amounts of intermediate results have to be moved from mappers to reducers over the network, thus introducing extra network overhead. As a result, join processing upon skewed data consumes more time.

Some methods such as SAND [2] and LEEN [2], have been proposed to solve the problem of data skew in MapReduce-based join, which adopt partition schemes considering the key's frequency distribution. However, SAND does not take into account the network overhead. LEEN not only solves load imbalance but also reduces network transmission, however, it changes the internal implementation scheme of Hadoop and ignores the advantage of overlapping [1] between the shuffle and map phases.

To overcome the above deficiency, we proposed SALA (Skew-avoiding and Locality-aware) MapReduce-based join algorithm, which uses volume/locality-aware partitioning to distribute data and does not make any modification of the MapReduce framework. Our approach firstly obtains the distribution information of key's frequency and location through data sampling. Based on this distribution information, SALA is able to guarantee the uniform distribution of data even when skewed data exist, so as to effectively avoid partitioning skew. At the same time, SALA reduces the amount of data transferred over network by utilizing the data locality feature of MapReduce, i.e., assigning keys to the nodes on which most of the intermediate results are located. This significantly improves the efficiency of the whole join operation.

In summary, we make the following major contributions:

- A novel algorithm called SALA is proposed to handle skewed data in MapReduce-based join. It not only achieves better load balance but reduces shuffled data over the network, thus resulting in significantly performance improvement.
- Volume/locality-aware partitioning scheme is proposed to distribute data, which is easy to implement without any modification of the MapReduce framework.
- We carry out extensive experiments and the results show the efficiency of SALA in the presence of data skew.

The rest of this paper is organized as follows. Sect.2 briefly introduces MapReduce-based join, and then we investigate the problem of data skew in Sect.3. Sect.4 presents the detail of the SALA. Extensive experimental results are reported in Sect.5. Related work is reviewed in Sect.6. Finally, we conclude the paper and discuss our future work in Sect.7.

2   MapReduce-Based Join

MapReduce-based join algorithms can be classified into two categories: map-side join and reduce-side join. For map-side join, the smaller input dataset is placed on each mapper and join operation only needs to be executed in the map phase to get the final results. Instead, a reduce-side join is carried out on the reduce phase. First, the map function takes the input dataset from DFS(Distributed File

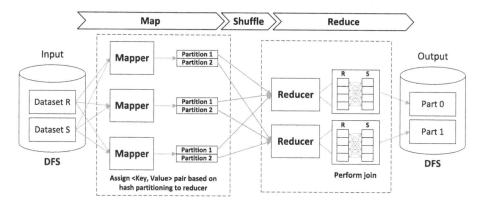

Fig. 1. The process of repartition join with the dataset R and S

System), and generates key-value pairs with the form of *(Key, Value)* as intermediate results, wherein *Key* represents join attribute. These intermediate results are to be assigned to reducers using hash partitioning. Second, in the shuffle phase, the reducers are notified to pull partitions across the network. Finally, the reduce function performs join operation with *(Key, list(Value))* pairs, wherein *list(Value)* is a list of values associated with the given key *Key*, and writes the final results to DFS. Fig.1 shows the process of a join operation between the dataset *R* and *S* with a typical reduce-side join algorithm, which is called repartition join[3].

Map-side join algorithms are more efficient than reduce-side join algorithms, because they produce the final results in map phase without shuffling data across the network. However, they can be used only in particular circumstances, i.e., one of the input datasets must be small enough to be buffered in memory of nodes. Reduce-side join algorithms are commonly used because they have fewer restrictions on input datasets. Therefore we focuse on the problem of data skew in reduce-side join.

3   The Problems in MapReduce-Based Join

MapReduce-based join algorithms sometimes suffer performance degradation from partitioning skew and heavy network overhead.

3.1   Partitioning Skew

Traditional join algorithms use hash partitioning to distribute data. Hash partitioning, the default partitioning function used in MapReduce model, is based on key hashing: *hash(Key) mod R*, wherein *R* is the number of reducers, which can allow data to be distributed uniformly when there are no skewness in the input datasets. In practice, however, partitioning skew tends to occur in processing skewed data and cause load imbalance, which means large amounts of data are distributed on only a few nodes. Because the larger the volume of partition is, the longer time it takes to process data. In addition, the response time of a MapReduce job is dominated by the slowest reduce instance. So partitioning skew results in longer response time of MapReduce-based join on the whole.

According to the process of repartition join shown in Fig.1, to which node the intermediate results will be distributed, is determined by the partitioning function. Therefore, the key factor to achieve load balance in MapReduce-based join operation lies in whether or not the partitioning function can guarantee uniform distribution of data.

3.2 Heavy Network Overhead

Apart from partitioning skew, network overhead is another non-negligible problem. Large amounts of intermediate results are produced and need to be transferred across the nodes through network, which may consume a lot of network resources and result in longer execution time. For Hadoop, it runs mappers on those machines where splits of input datasets are located, so as to avoid network overhead. Most existing MapReduce-based join algorithms, however, does not take full advantage of such data locality feature in the reduce phase, and as a result, lots of intermediate results have to be transferred over network during the shuffle phase. In addition, transmission for skewded partitions may also introduce extra network overhead, because they have more data to transfer than non-skewed partitions. What's more, the reduce phase only can start after the shuffle phase completes, so network overhead is to increase the response time of the whole join operation.

Therefore, with evenly distribution of partitions, data transmission time tends to be equal among various partitions. In addition, applying the data locality feature in the reduce phase, can also reduce the amount of the transferred data and further improve the performance of join operation.

4 SALA Join Algorithm

To solve the above problem, we propose SALA join algorithm to handle skewed data and reduce network overhead. In this section, we first present an overview of SALA, and then present the volume/locality-aware partitioning used in SALA in detail. Also a example is discussed to compare SALA join with repartition join. Finally, we propose a cost model to analyze the performance of our algorithm.

4.1 Overview

We propose SALA join algorithm to handle skewed data. The core idea of SALA join is to distribute intermediate results based on the distribution information of key's frequency and location. With volume/locality-aware partitioning scheme, SALA join is able to not only handle skewed data but also reduce network overhead.

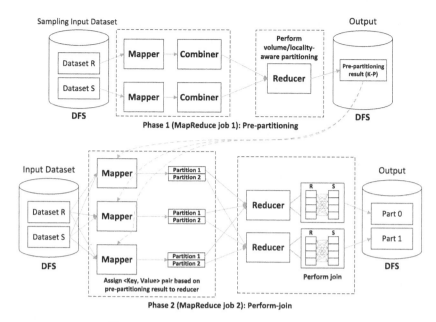

Fig. 2. The process of SALA join algorithm

Fig.1 shows the traditional process of MapReduce-based join, and the join process with SALA is shown in Fig.2. The main difference between the two algorithms is that SALA adds an additional MapReduce job to obtain key's distribution information. The SALA join includes two phases:

1. Phase 1: sample the input dataset and pre-compute the data to get the partitioning results, represented as *K-P*. *K-P* is a mapping array, and each of the array element is a map between a key and the partition that the key is assigned to.
2. Phase 2: perform the actually join operation. The join process is similar with repartition join, except that SALA join directly partitions the intermediate results according to *K-P* instead of using hash partitioning.

Since phase 2 of SALA is similar with repartition join, so we mainly focus on phase 1, i.e., the pre-partitioning process.

4.2   The Pre-partitioning Process

The pre-partitioning process is to pre-compute the sample input dataset to get *K-P*. and includes three phases - map phase, combine phase and reduce phase:

– Map phase: process the sample input dataset and take the join attribute as the *Key*. The output will be *(Key, (node, 1))*, wherein *node* represents the node on which the data is located and the number 1 represents the frequency of this key.

- Combine phase: the combine task will count the frequency of each key on the current node and the output will be *(Key, (node, sum))*, i.e., outputting the total frequency of each *Key* on the node.
- Reduce phase: volume/locality-aware partitioning is employed to get the pre-partitioning result *(Key, Partition)*, i.e., *K-P*.

Volume/locality-aware partitioning plays an important role in SALA join, so we discuss it in more detail below.

4.3   Volume/Locality-Aware Partitioning

Volume/locality-aware partitioning can not only deal with partitioning skew to achieve load balance, but also reduce the data transferred over network. Assuming that the data volume is M (which can be represented by the rows of the input dataset) and the number of nodes is N. In order to achieve load balance, the volume of data distributed to each node should be close to $\frac{M}{N}$. To reduce data transferred over network, volme/locality-aware partitioning makes full use of data locality feature by adopting greedy selection strategy as follows:

1. Each key value is distributed in higher priority to the node on which most intermediate results of this key are located.
2. First process the key value which has larger size of intermediate results.

Volume/locality-aware partitioning involves the following two steps:

1. Preparing step:
   (a) Compute the total rows of intermediate results of each key value in all nodes and write it as T_{key}.
   (b) Extract all *(Key, node, sum)* tuples from $(Key, list(node, sum))$ paris and store them in list L, meanwhile, put all key values into set K. After that, sequence all tuples in L in descending order based on the size of *sum*.
2. Partitioning step:
   (a) Traverse list L and process each tuple *(Key, node, sum)*. We use P_{key} to represent the partitioning result of each key value, which means to which node the key value should be distributed, and use V_{node} to record the volume of data that has been distributed to the node at present. If P_{key} is null, then determine whether or not $V_{node} + T_{key} \leq \frac{M}{N}$. If it is true, let $P_{key} = node$ and $V_{node} = V_{node} + T_{key}$.
   (b) Lastly, there may be some key values in K which have not been partitioned. In this case, find out the smallest V_{node}, to which the minimum volume of data is distributed at present, and then P_{key} will be the *node* that refers to V_{node}.

Algorithm 1 in Fig.3 formally describes volume/locality-aware partitioning. Due to the random sampling method used in the pre-partitioning process, there

Algorithm1: Volume/Locality-aware Partitioning
Input: pairs of $(Key, list(node, sum))$;
             $M \leftarrow$ rows of input dataset; $N \leftarrow$ the number of nodes;
Output: partitioning results K-P

1. $T \leftarrow$ total rows of intermediate results in all nodes for each key value;
2.   traverse the input and put all $(Key, node, sum)$ into L, put all Key into K;
3.   initialize the list P and V;
4.   **for** each $(Key, node, sum) \in L$ **do**
5.   **if** $P[Key]$ is $null$ and $V[node] + T[key] \leq \frac{M}{N}$ **then**
6. $P[Key] = node$;
7. $V[node] = V[node] + T[Key]$;
8.   **endif**
9.   **endfor**
10. **for** each $Key \in K$ **do**
11.   **if** $P[Key]$ is $null$ **then**
12. $node \leftarrow$ the node that refers to minimum $V[node]$ in V;
13. $P[Key] = node$;
14. $V[node] = V[node] + T[Key]$;
15.   **endif**
16. **endfor**
17. return P as K-P;

Fig. 3. The algorithm of volume/locality-aware partitioning

are some key values which may not be counted in. Therefore, when the intermediate results are partitioned in the perform-join process, key values which have been counted in will be partitioned according to the K-P, while key values which have not been counted in will still be partitioned by hash partitioning. Given that key values which have not been counted in only involve a small part of all key values, they will have negligible impact on the data distribution.

4.4   Example

Taking the following join operation for example: $R \overset{R.a=S.a}{\bowtie} S$. Assuming that there are 3 nodes in the cluster and the input data volume of each node is the same, i.e., 70, but with skewed data. Fig.4(a) shows the intermediate results produced in the map phase, in which, each row represents one key group *(Key, volume)*, wherein *volume* is the data volume of this key value on the present node.

The partitioning results of repartitioning join and SALA join are shown in Fig.4(b) and Fig.4(c) respectively. According to Fig.4(b), partitioning skew happens in repartition join. Too much data are distributed to $Node_3$, almost four times of that distributed to $Node_1$. Therefore, load imbalance appears. However, as Fig.4(c) shows, SALA join algorithm has achieved better load balancing, and at the same time, the overall network overhead has reduced by 36% compared with repartition join.

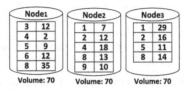

(a) Distribution of intermediate results

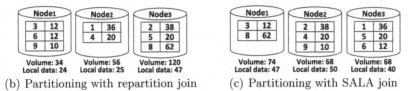

(b) Partitioning with repartition join   (c) Partitioning with SALA join

Fig. 4. Partitioning results of various methods

With SALA join algorithm, the volume of data distributed to each node will tend to be equal and load balance is therefore achieved. Further, because each key value is first distributed to the node on which most of its intermediate results are located, the overall volume of data to be transferred over network is remarkably reduced and the performance of join operation is improved.

4.5 Cost Model

As shown in Fig.1, the whole processing time of the traditional reduce-side join algorithm includes three parts: processing time of map phase, transmission time of shuffle phase and processing time of reduce phase. For convenience, we use the following notations in Table 1:

Table 1. Table of notations

Notation	Meaning
t_m	processing time for a record of input datasets in map phase
t_s	transmission time for a record in shuffle phase
t_r	processing time for a record in reduce phase
M	total rows of input dataset
M_s	total rows of sampling input dataset
N	the number of nodes
B	average available bandwidth of nodes
L	data locality of partitions
s	skewness of input dataset

Because the response time of a MapReduce job is determined by the slowest reduce instance, we can estimate the response time by the reducer which is allocated the most volume of data, represented as R. Therefore, the cost model for a traditional reduce-side join algorithm is as follows:

$$T_{tra} = t_m \cdot \frac{M}{N} + \frac{R \cdot (1 - L)}{B} + t_r \cdot R \tag{1}$$

With uniform distribution of data, R_e tends to be $\frac{M}{N}$. In the case of partitioning skew, however, R_s tends to be:

$$R_s = M \cdot s + \frac{M \cdot (1 - x)}{N} \qquad (2)$$

The key values are $K=\{k_1, k_2, \ldots k_n\}$, and F_k represents the frequency of key value k on nodes. With hash partitioning, the data locality L_{tra} is $\frac{\sum_{k=k_1}^{kn} mean(F_k)}{M}$. Therefore, the cost model for a traditional reduce-side join algorithm in the case of partitioning skew can further be written as:

$$T_{tra} = t_m \cdot \frac{M}{N} + \frac{R_s \cdot (1 - L_{tra})}{B} + t_r \cdot R_s \qquad (3)$$

With SALA join, the data locality L_{sala} tends to be $\frac{\sum_{k=k_1}^{kn} max(F_k)}{M}$. SALA join guarantees the uniform distribution of data, but needs an additional pre-partitioning process, and the required time of pre-partitioning process is represented as T_{pre}. Therefore the cost model for SALA join is:

$$T_{sala} = T_{pre} + t_m \cdot \frac{M}{N} + \frac{R_e \cdot (1 - L_{sala})}{B} + t_r \cdot R_e \qquad (4)$$

Thus, SALA join algorithm is superior to traditional reduce-side join algorithm when satisfying the following condition:

$$T_{sala} - T_{tra} < 0 \Rightarrow T_{pre} < \frac{R_s \cdot (1 - L_{tra}) - R_e \cdot (1 - L_{sala})}{B} + t_r \cdot (R_s - R_e) \quad (5)$$

As can be seen from Eq.(5), SALA join performs better when the decreased of time results in from avoiding solving the partitioning skew is greater than the time used to process pre-partitioning. We can therefore employ Eq.(5) in optimal query plan selection. Here, according to many experiments, T_{pre} tends to be $0.23 \times t_m \times \frac{M}{N}$ and t_r tends to be $0.69 \times t_m$. We take $N=5$ and $s=10\%$, then $L_{sala} = 2.94 \times L_{tra}$ and $R_s = 1.4 \times R_e$, so the Eq.(5) is satisfied, as is shown in Eq.(6). Also with the case of greater data skewness and the lower available bandwidth, SALA join will performs much better.

$$T_{pre} = 0.23 \cdot t_m \cdot \frac{M}{N} < \frac{0.4 + 1.54 \cdot L_{tra}}{B} \cdot \frac{M}{N} + 0.28 \cdot t_m \cdot \frac{M}{N}$$
$$\implies -0.05 \cdot t_m - \frac{0.4 + 1.54 \cdot L_{tra}}{B} < 0 \qquad (6)$$

5   Empirical Study

In this section, we conduct experiments to verify the efficiency of our approach. We mainly use the response time of join operation to demonstrate performance difference in the case of data skew. We compare SALA join with the repartition join algorithm [3] and SAND join algorithm [2], because repartition join is extensively used, and SAND join is a typical join algorithm to deal with skewed data.

5.1   Environmental Setup

Our experiments run on AliCloud (Alibaba Cloud Computing) with a 6-node cluster running native Hadoop 2.4.1, where there are 1 master node scheduling the task and 5 slave nodes taking charge of both storage and computation. Each node has two Xeon 2.3Ghz CPUs, 4GB memory and 60GB disk drive. HDFS block size is set to be 128MB and each node is configured to run one reducer task.

We use TPC-H to generate the input dataset and take the following query in our experiments:

select * from CUSTOMER C join ORDER O on C.CUSKEY = O.CUSKEY

In order to control data skewness, we randomly choose a portion of the input dataset *ORDER* and change its *CUSKEY* to the same value. For example, if the skewness is 10%, it means that we change 10% rows of the input dataset *ORDER* to have the same value in the join attribute *CUSKEY*. Finally, we generate 20 million records for query with various degree of data skewness.

5.2   Partitioning Effectiveness

Firstly, our concern is whether or not SALA join can effectively solve the partitioning skew problem. As analysis in Sect.2 has suggested that the key factor of load balancing is uniform distribution of data, we can therefore evaluate the capability of a join algorithm to handle skewed data by the value of *max-reducer-input*, i.e., the maximum volume of data distributed to any reducer. According to Fig.5(a), as the degree of skewness increases, repartition join concentrates a large amount of data on hot nodes, while both SALA join and SAND join can guarantee the uniform distribution of data.

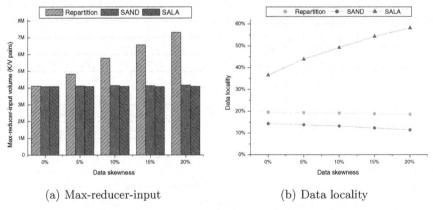

(a) Max-reducer-input   (b) Data locality

Fig. 5. Partitioning with three join algorithms

Meanwhile, we use data locality to represent the volume of intermediate results that do not need to be transferred over network. From Fig.5(b), we can see that the data locality for SALA is much larger than the data locality for both repartition and SAND methods. Because the larger the data locality is, the less the volume of data required to be transferred across the network, thus less data needs to be transferred in SALA join algorithm than in both repartition and SAND methods.

5.3   Response Time

Fig.6(a) shows comparison between response time used to complete the given join operation under different degree of data skewness. The performance of repartition join is the best in the case of no or little data skewness, and the reason is that both SALA join and SAND join require additional MapReduce job to obtain frequency distribution of key values. However, with the increase of data skewness degree, the response time of repartition join increases almost linearly. The reason is that as the degree of data skewness increases, data will concentrates on hot nodes as Fig.5(a) shows, which increases the completion time of the overall join operation. However, both SALA and SAND can guarantee load balance, and therefore the response time remains steady with the increase of skewness degree. Most importantly, SALA join algorithm performs better than others when skewed data exist, because SALA not only achieves load balance but also reduces network overhead, thus speeding up the join operation process with the increase of data skewness degree.

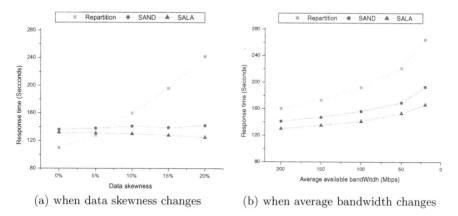

(a) when data skewness changes   (b) when average bandwidth changes

Fig. 6. Response time for three join algorithms

Fig.6(b) shows the variation of response time under different bandwidths when the degree of data skewness is 10%. It can be seen that as the average available bandwidth reduces, the problem of network overhead becomes prominent. It is because that the lower the bandwidth is, the longer time it will cost to complete network transmission. By taking full advantage of data locality feature, the minimum volume of data is transferred with SALA join algorithm, and therefore SALA is preferable in the case of low bandwidth.

6   Related Work

In recent years, various approaches have been proposed to deal with skewed data in MapReduce-based join, such as [3,11,10,2,6,4,9,8]. The research work in [12] has demonstrated that the default hash partitioning function in Hadoop is not efficient for the skewed data and may lead to load imbalance of reducers.

The partitioning skew problem due to data skew can be solved by making a good partition scheme based on the key's frequency distribution, while sampling is a common way to obtain key's frequency [12,2]. The SAND join alogoritm [2] uses simple range partitioning for data distribution to achieve load balancing. Yujie Xu *et al.* proposed two partition schemes, namely cluster combination optimization and cluster partition combination based on random sampling results, to handle slight skew and heavy skew respectively.

Reducing the volume of data transferred across the network is an efficient way to further improve the performance of data-intensive join operation. Based on the priori knowledge of skewed key values, PRPD join geography proposed in [11] keeps all skewed rows locally to reduce the data volume transferred among nodes over network. LEEN scheme presented in [6] partitions the intermediate results based on key's frequency and the fairness score that is calculated after the shuffle phase. However, LEEN scheme changes the internal implementation of Hadoop and overlooks the advantage of overlapping between the shuffle and map.

An alternative strategy to mitigate skew is dividing the workload into fine-grained computation tasks and then scheduling them dynamically at runtime [5,7,9]. SkewTune [7] first identifies the task with the greatest expected remaining processing time when a node in the cluster becomes idle. The unprocessed data of this unfinished task is then repartitioned in a way that fully utilizes the computing power of cluster nodes.

7   Conclusion

In this paper, we propose SALA join algorithm, using volume/locality-aware partitioning to distribute intermediate results. On one hand, SALA guarantees the uniform distribution of data based on key's distribution information and therefore avoids partitioning skew problem. On the other hand, SALA takes full advantage of the data locality feature to reduce the volume of data transferred across the network. Experiments show that SALA is efficient to deal with skewed data. Our future work includes further improving performance of SALA join and extending volume/locality-aware partitioning to other MapReduce applications.

References

1. Ahmad, F., Lee, S., Thottethodi, M., Vijaykumar, T.N.: Mapreduce with communication overlap (marco). J. Parallel Distrib. Comput. **73**(5), 608–620 (2013)
2. Atta, F., Viglas, S.D., Niazi, S.: Sand join - a skew handling join algorithm for google's mapreduce framework. In: 2011 IEEE 14th International Multitopic Conference (INMIC), pp. 170–175, December 2011
3. Blanas, S., Patel, J.M., Ercegovac, V., Rao, J., Shekita, E.J., Tian, Y.: A comparison of join algorithms for log processing in mapreduce. In: Proceedings of the ACM SIGMOD International Conference on Management of Data, SIGMOD 2010, Indianapolis, Indiana, USA, June 6–10, 2010, pp. 975–986 (2010)
4. Bruno, N., Kwon, Y.C., Wu, M.-C.: Advanced join strategies for large-scale distributed computation. PVLDB **7**(13), 1484–1495 (2014)
5. Dhawalia, P., Kailasam, S., Janakiram, D.: Chisel: a resource savvy approach for handling skew in mapreduce applications. In 2013 IEEE Sixth International Conference on Cloud Computing, Santa Clara, CA, USA, June 28 – July 3, 2013, pp. 652–660 (2013)
6. Ibrahim, S., Jin, H., Lu, L., Wu, S., He, B., Qi, L.: LEEN: locality/fairness-aware key partitioning for mapreduce in the cloud. In: Proceedings of the Cloud Computing, Second International Conference, CloudCom 2010, November 30 – December 3, 2010, Indianapolis, Indiana, USA, pp. 17–24 (2010)
7. Kwon, Y.C., Balazinska, M., Howe, B., Rolia, J.A.: Skewtune in action: Mitigating skew in mapreduce applications. PVLDB **5**(12), 1934–1937 (2012)
8. Kwon, Y.C., Ren, K., Balazinska, M., Howe, B.: Managing skew in hadoop. IEEE Data Eng. Bull. **36**(1), 24–33 (2013)
9. Lynden, S.J., Tanimura, Y., Kojima, I., Matono, A.: Dynamic data redistribution for mapreduce joins. In: IEEE 3rd International Conference on Cloud Computing Technology and Science, CloudCom 2011, Athens, Greece, November 29 – December 1, 2011, pp. 717–723 (2011)
10. Yu, X., Kostamaa, P.: Efficient outer join data skew handling in parallel DBMS. PVLDB **2**(2), 1390–1396 (2009)
11. Xu, Y., Kostamaa, P., Zhou, X., Chen, L.: Handling data skew in parallel joins in shared-nothing systems. In: Proceedings of the ACM SIGMOD International Conference on Management of Data, SIGMOD 2008, Vancouver, BC, Canada, June 10–12, 2008, pp. 1043–1052 (2008)
12. Xu, Y., Zou, P., Qu, W., Li, Z., Li, K., Cui, X.: Sampling-based partitioning in mapreduce for skewed data. In: ChinaGrid Annual Conference (ChinaGrid), 2012 Seventh, pp. 1–8, September 2012

Energy-Proportional Query Processing
on Database Clusters

Jiazhuang Xie[1], Peiquan Jin[1,2(✉)], Shouhong Wan[1,2], and Lihua Yue[1,2]

[1] School of Computer Science and Technology,
University of Science and Technology of China, Hefei 230027, China
jpq@ustc.edu.cn
[2] Key Laboratory of Electromagnetic Space Information,
Chinese Academy of Sciences, Hefei 230027, China

Abstract. Energy efficiency has been a critical issue in database clusters. In this paper, we present an energy-proportional database cluster and propose an energy-proportional query processing approach to reduce the energy consumed by database clusters while keeping high time performance. Particularly, we introduce a *query stream buffer* on top of a database cluster and propose an *unbalanced load allocation* algorithm to distribute workloads among the cluster so as to realize better energy proportionality. Further, we present an adaptive algorithm to turn on/off nodes according to workload changes. With this mechanism, we can reduce energy consumption while keeping high time performance for query processing on database clusters. We build a prototype database cluster and use the TPC-H benchmark to compare our proposal with three baseline methods, where different query patterns are used. The results suggest the superiority of our proposal in energy savings and time performance.

Keywords: Energy proportionality · Database cluster · Query processing

1 Introduction

Big data has introduced many challenges to traditional database systems, e.g., reducing energy consumption, improving time performance over big data, etc. In order to cope with the storage and performance needs in big data management, database clusters involving several small servers can offer better time performance for storing and querying big data [1, 2]. Moreover, the size of such a database cluster can be adjusted to workloads so that we can dynamically turn on/off the nodes to obtain better energy proportionality [3].

Energy proportionality means that the energy consumption of a system is proportional to the workloads running on it [3]. This issue was first studied by Google researchers in 2007 [3], where they tested the energy consumption and server utilizations for 5,000 servers running Google services. Consequently, they found that generally all the nodes only used about 20% of their CPU capabilities. In other words, we need not run all the servers to provide services in most cases; thus some servers can be turned off to save energy.

© Springer International Publishing Switzerland 2015
J. Li and Y. Sun (Eds.): WAIM 2015, LNCS 9098, pp. 324–336, 2015.
DOI: 10.1007/978-3-319-21042-1_26

However, it is not a trivial task to realize an energy-proportional database cluster. The most critical challenge is that we have to keep high time performance when making some nodes power-off. Therefore, we have to find a best trade-off between time performance and energy savings. Obviously, the best solution has to adapt to workload changes. For this purpose, effective algorithms regarding load allocation and node activation/deactivation should be proposed.

Aiming for providing an energy-proportional database cluster to meet the urgent needs in big data management, in this paper we first propose a new architecture for energy-proportional database clusters, and then present new algorithms involving load allocation and node activation/deactivation to deal with the key issues in such a cluster. In summary, we make the following contributions in this paper:

(1) We introduce a new architecture for energy-proportional database clusters. It is a hybrid architecture combining the share-nothing and share-disk architectures. Furthermore, we introduce a query stream buffer on top of the hybrid architecture to cache new incoming queries, so that we are able to keep high time performance even when the cluster is over-loaded. (**Section 3**)

(2) We propose an unbalanced load allocation algorithm to distribute workloads among the nodes in the cluster so as to realize better energy proportionality than traditional load balance algorithms for clusters. (**Section 4**)

(3) We present new algorithms for node activation/deactivation that can adapt to work-load changes. (**Section 5**)

(4) We build a prototype database cluster consisting of five nodes and conduct comparative experiments using the TPC-H benchmark on PostgreSQL, with various metrics, varying numbers of query streams, and different query patterns. (**Section 6**)

2   Related Work

The researches on energy-aware data management are mainly based on two viewpoints, namely single-server-based and cluster-based. So far, most previous work focuses on the single-server-based energy-efficient approaches [4-6], which aims to save the energy of a single server by optimizing the traditional database algorithms such as buffer management, query processing, and indexing. For example, in the literature [4], the authors propose the QED method to reduce the energy costs in query processing. They delay the queries and place them into a queue according to their arrival time. When the queue reaches a certain threshold, all the queued queries are clustered into some small groups, such that the queries in each group can be evaluated together. They demonstrate that this approach is more energy-efficient than traditional query processing methods.

Although single-server-based energy-efficient approaches are helpful to reducing the total energy costs of a database system, many studies have shown that even a stand-by server can consume a lot of energy, which implies that the single-server-based energy optimization can only bring limited benefits. Thus, recently many researchers consider the cluster-based situation and propose to construct energy-proportional clusters. With the rapid development of big data researches, the cluster-based research has received more and more attention from both academia and industries [7, 8].

In [1], the authors propose WattDB, a prototype cluster supporting energy proportionality on commodity hardware. WattDB uses the share-storage architecture and all the data are placed in a single node, which may lead to single-point failure. In [8], the authors conduct an experimental study to reveal the inefficiency of current parallel data processing technologies in scalability and energy efficiency. Based on the experimental results, they present some guiding principles for building energy-efficient clusters. The most critical issue in the cluster-based energy proportionality is how to control the power states of nodes [9-12]. In the literature [11, 12], the researchers propose a predictive approach to predict the workload trend and then to turn on or off some nodes if necessary. They use various predictive methods including Last Arrival, MWA, Exponentially Weighted Average, and LR, to predict the future request rate for a website cluster, and then accordingly add or remove servers from a heterogeneous pool.

Our node activation/deactivation depends on more than the prediction result. Specifically, in addition to the prediction result, we also consider the change patterns of workloads so as to determine the right time to turn on /off nodes.

3   Architecture of the Database Cluster

3.1   General Idea

We propose a new architecture for energy-proportional database cluster, as shown in Fig. 1. It consists of a control node and several backend nodes, and the backend nodes are divided into share-disk nodes and share-nothing nodes.

The motivation of our design is two-folds. First, the architecture should support energy proportionality, which implies that it must allow the scale-in property, i.e., removing nodes from the cluster. This is intrinsic for energy proportionality, because the key idea of energy proportionality is to let some nodes power-off in case of low workloads. As the share-disk architecture is appropriate for scale-in design. Therefore, we first use the share-disk architecture in Fig. 1. However, the share-disk architecture is not suitable for storage-volume extension. Hence, we further introduce the share-nothing architecture for organizing the storage nodes, as shown in the bottom part of Fig. 1. Thus, the share-nothing nodes can be regarded as storage nodes, forming a distributed storage center, where each node can have their own storage media such as HDD and SSD (Solid State Drives). On the other hand, the share-disk nodes are used to process queries. In general, the share-disk nodes can be turned on/off while the share-nothing nodes should be kept active.

The control node has two main modules: *Load Allocation* and *Node Activation/Deactivation*. The Load Allocation module is used to distribute the user queries to share-disk nodes. The *Node Activation/Deactivation* is designed to turn on and turn off nodes when necessary. Both the two modules run on top of a *Query Stream Buffer (QSB)*. The *QSB* buffer is introduced to cache the information about new incoming queries, which offers necessary information for load allocation and node activation/deactivation. The details of the *QSB* buffer are presented in Section 3.2. We will discuss the algorithms of *Load Allocation* and *Node Activation/ Deactivation* in Section 4 and 5, respectively.

For the purpose of energy proportionality, the control node will change the power states of share-disk nodes according to workload changes. Basically, a share-disk node has the states shown in Fig. 2. A share-disk node is initially at the *standby* state. This state only requires very little energy for supplying the power to keep network adapter alive so that the control node can turn on the node when needed by using the Wake-on-LAN (WOL) technology [12]. When the node is turned on, it goes into the *power up* state. At this state, the node should complete its boot sequence and then start up the database system. Then the node changes into the *idle* state to be ready to accept query requests. A node with workload running is at the *load* state. At this state, the energy consumption is approximately linear proportional to the usage of CPU. Only the *idle* nodes and *load* nodes can accept query requests, so we also call the nodes at these two state active nodes. When a share-disk node receives shutdown request from control node, it moves to the *power off* state. A node at this state firstly shut down the database system and then executes the shutdown procedure. The node changes into *standby* state again when it finishes the procedures.

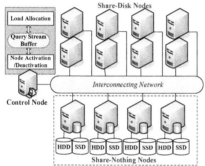
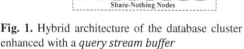
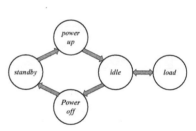

Fig. 1. Hybrid architecture of the database cluster enhanced with a *query stream buffer*

Fig. 2. State transition of a backend node

3.2 Query Stream Buffer

The *Query Stream Buffer (QSB)* is an enhanced module in the control node to cache the new incoming query streams when the cluster is overloaded. It is motivated by the following observations:

(1) We found that the execution time of a query stream is linearly growing with the increase of concurrent number of queries when the nodes are overloaded (details can be found in Section 6). Without *QSB*, an overloaded node will continue to receive new queries, which will in turn worsen the response time for the queries running in the node.

(2) We found that the powerful nodes with high processing capability usually finish the tasks quickly and are very likely to turn to the *idle* state. On the other hand,

the weak nodes with low computing resources tend to be overloaded. Without *QSB*, the queries will be issued to the nodes running fewer queries when the cluster is overloaded. Consequently, each node will have the similar number of queries. However, since the processing capability varies from node to node, the powerful nodes can quickly finish the queries while the weak nodes may be overloaded. To this extent, it is appropriate to use a buffer to cache the query streams, so that when a powerful node finishes its jobs, it can get a new query from the buffer rather than turns to the *idle* state. Hence, we can better balance the workloads among the share-disk nodes.

(3) When the current workload is becoming heavy, we need to wake up some nodes in the *standby* state. However, a *standby* node needs some time to become ready for receiving queries. Thus, it is better to cache the queries (using *QSB*) during the wake-up time period of the *standby* nodes, and send the queries to those nodes when they are ready. Otherwise, the query streams during the wake-up time period will be sent to current active nodes, which will lead to more overloaded nodes and worsen the time performance of the cluster.

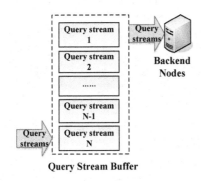

Fig. 3. The schematic diagram of *Query Stream Buffer*

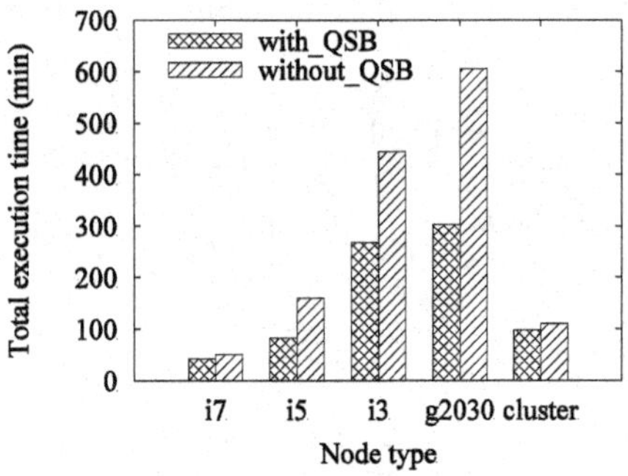

Fig. 4. The time performance of *QSB* for a single node and a cluster

Figure 3 shows the basic structure of *QSB*. Note that *QSB* only works when the workload exceeds the current capacity of the active nodes. We use the *FIFO* mechanism to maintain the buffered query streams. In particular, new query streams will be placed in the queue tail of *QSB*. When a node is ready to receive new queries, the control node will check whether the buffer contains queries to be processed and move the query in the head of the queue to the available node.

The benefits of the *Query Stream Buffer* are three-folds. First, the control node can cache the extra queries in *QSB* and the share-disk nodes can be kept in an efficient mode and will not be overloaded. Second, as *QSB* caches the extra query steams temporarily, the powerful nodes will not be *idle* when the buffer is not empty. Thus, we can let the powerful nodes in the *load* state and improve the overall time performance of the cluster. Third, by caching the new queries during the wake-up time of *standby* nodes, we can avoid the overloading of current nodes, because the new queries can be distributed to the newly wake-up nodes.

We conduct an experiment to verify the efficiency of *QSB*. In this experiment, two new query streams are sent to the control node every 20 seconds. The control node allocates all query streams to available share-disk nodes and we compare two different methods when allocating the workloads: *with QSB* and *without QSB*. For the method *with QSB*, the control node caches the extra streams when the cluster is overloaded. For the method *without QSB*, the control node directly allocates the query streams to the share-disk nodes when it receives new query streams. Figure 4 shows the total execution time of forty query streams for a single node and the cluster. For the method *with QSB*, the execution time of a query stream includes the time that *QSB* consumes and the execution time. We can notice that *QSB* can effectively reduce the execution time for different kinds of nodes. As a result, *QSB* improves the time performance up to 49.7% compared with the method without *QSB*.

4   Load Allocation

In this section, we propose our load allocation algorithm that utilizes the *QSB* buffer. Differing from traditional load balance design for clusters, we adopt the load unbalancing idea in our load allocation algorithm. The load unbalancing design will make full use of the current active nodes and therefore let more nodes to be *idle*. These idle nodes can be set to the *standby* state with some appropriate node activation/deactivation algorithm, and thereby we can reduce the energy consumption of the cluster.

Algorithm 1. *Load_Allocation*

Input: q: a new query stream to be allocated
Output: none

1.   Sort the share-disk nodes w.r.t. capability;
2.   **for** i =1 to |nodes| **do**
3.   **if** $node_i$ .state = *idle* or *load* **then**
4.   **if** $node_i$ can accommodate q **then**
5.   selected_node:= i;
6.   break;
7.   **if** i > |nodes| **then**   selected_node:= -1
8.   **if**   selected_node <> -1 **then**   Allocate q to $node_i$
9.   **else**
10.   Put q into the tail of *QSB*;
11.   **if** *QSB* is full **then**   *judge_open*();

End *Load_Allocation*

Algorithm 1 shows the load allocation procedure. When the control node receives a new query stream, it first checks all active share-disk nodes to see if any of them can accommodate this new query. If one of these nodes has less query streams running than its capacity, the control node will send the new query stream to it. If there are no active nodes that can take over this query stream, either because there are a few standby nodes or all the share-disk nodes are in use, the control node puts the new query stream into *QSB*. Next, we invoke the function *judge_open* to see whether the cached query streams exceed the predefined size of *QSB*.

The proposed load allocation algorithm differs from the previous work. First, it makes good use of the *QSB* buffer that we prove to be effective for time performance. Second, we associate the load allocation with the node activation so that every new query stream has the chance to invoke the node activation. On the contrary, most of the previous load allocation algorithms do not consider the node activation. Finally, according to our algorithm, a new query stream will be cached instead of being allocated to a share-disk node when the load is over the current capacity. The query streams can be issued to any of the share-disk nodes later when they have finished at least one query streams. In this way, the workload can be balanced among all the nodes according to their execution speeds. In other words, our load allocation algorithm combines the advantages of load balancing and unbalancing policies.

5   Node Activation and Deactivation

5.1   Node Activation

The node activation algorithm needs to make two decisions: (a) when to activate nodes and, (b) how to predict the future workload with the help of *QSB* and determine the number of nodes to be turned on.

Regarding the first issue, we first classify workloads into two different types according to their change patterns. The first type of workload is rapidly changing, while the second type is gradually changing. In turn, we propose two methods to decide when to turn on nodes w.r.t. the different changing patterns of the workload:

(1) For the rapidly changing workloads, we propose the *judge_open* algorithm (it has been used in the load allocation algorithm shown in Fig. 5). The *judge_open* algorithm detects whether it is necessary to turn on some standby nodes and activate nodes after that. This algorithm runs on top of *QSB*. Once the number of the cached query streams in *QSB* reaches the size of *QSB*, which means that the workload exceeds the current capacity, the *judge_open* algorithm first predicts the workload in the future of the time when a new active node can get ready. Based on the predicted future workload, the control node determines whether to turn on additional nodes and how many nodes should be turned on. If the prediction result still outstrips the size of *QSB*, control node determines to turn on a few nodes, the number of which is determined by the prediction result and the capacities of the nodes. If the prediction result is less than the size of *QSB*, we do nothing because the result means that the cluster can handle the load in the future, it is better not to turn on nodes for saving more energy. To ensure high time performance, the powerful nodes in the cluster are first selected as new active nodes.

(2) For gradually changing workloads, we have to consider a new method rather than the *judge_open* algorithm. Since *judge_open* is triggered only when the number of the cached query streams exceeds the size of *QSB*, it cannot effectively handle the situation that the workload is gradually changing. In this situation, when the workload finally exceeds the current capacity of the cluster, the number of the cached query streams in *QSB* will be still less than the size of *QSB*. Therefore, we propose a *time-out*

scheme to deal with the gradually changing workloads. We first set a timer that starts whenever the first query stream is put into *QSB*. When the last query stream is moved out from *QSB*, the timer stops and resets to zero. If the aggregated time duration of the timer exceeds a threshold called *Timer_powerup*, the control node activates a standby node to enhance the capability of the cluster.

Regarding the second issue, namely future workload prediction, we adopt a simple way to predict the future workload to reduce the time overhead of the prediction procedure. Suppose that at time *T* we start to predict the future workload and it costs time *Tpowerup* to get a *standby* node be ready to accept queries. Since the boot-up stage of a node is very quick, we can assume that the workload feature during the boot-up period is similar to that in the last period of *Tpowerup*. Thus, we predict the future workload by the following Formula (1).

$$Load_{T+T_{powerup}} = N_{cached\_streams} + N_{newqueries} - N_{completedqueries} \quad (1)$$

Here, *Load* represents the size of *QSB* at the time *T+ Tpowerup*. $N_{cached\_streams}$ represents the current number of the query streams in *QSB*. $N_{newqueries}$ represents the number of the new query streams and $N_{completedqueries}$ represents the number of the completed query streams during *Tpowerup*. We also use a sample thread to record the useful information about the cluster during every *Tpowerup*.

5.2   Node Deactivation

Similarly, we also adopt the *time-out* policy for node deactivation. Due to the load allocation algorithm, the workload is more likely to concentrate on a few active nodes. Consequently, the remaining active nodes will stay in the *idle* state and can be turned off to save energy consumption. Unlike most previous algorithms, we allow each active node in the cluster to decide independently whether to turn itself off or not by setting a timer for each active node. Particularly, we set timers for active nodes. An active node can be shut down as soon as its timer keeps going and exceeds a threshold *Timer_shutdown*. Similarly, the timer starts when the node's state changes into *idle* and resets to zero when the node receives a new query stream.

6   Performance Evaluation

6.1   Experiment Setup

Our prototype database cluster consists of a control node and four backend nodes, running PostgreSQL 9.2.0. All components are interconnected using the TP-LINK Gigabit switcher. The control node can turn on the *standby* nodes by using the *Wake-on-Lan* technology [12]. Each backend node is with an Intel CPU (i3, i5, i7, and G2030), a hard disk of Seagate ST1000DM003 1TB 7200RPM, and a memory of Kingston 8GB DDR3. The energy related information of the backend nodes are shown in Table 1, where the nodes are represented by their CPU type.

Table 1. Energy information of the four backend nodes

Node Power	i7 (Node 1)	i5 (Node 2)	i3 (Node 3)	G2030 (Node 4)
Standby power (Watt)	1.5	1.3	1.5	1.3
Idle power (Watt)	35.3	35.5	36	33
Peak power(Watt)	107	78	58	51
Boot time (S)	30	30	30	30

We take queries from the well-known TPC-H benchmark as the workload for our cluster. There are 22 queries and 2 updates in TPC-H. In order to keep the data in different backend nodes consistent, we eliminate the update statements and use the 22 query statements to generate query streams. A query stream consists of 22 query statements and each query statement is randomly generated by official tools provided by TPC-H. Besides, the sequence of 22 queries in a query stream is randomly arranged.

To observe the effect of energy proportionality, we need to run the workload for a period of time and dynamically turn on/off nodes. To the best of our knowledge, there is no such long-term workload benchmark for database clusters. Thus, we manually create three types of workload patterns in our experiment. The three types of workload patterns are *Fixed Type*, *Poisson Flow*, and *Step Type*. For the *Fixed Type*, the number of queries changes regularly. For the *Poisson Flow*, the number of query streams generated each time satisfies the Poisson distribution. Both the *Fixed Type* and *Poisson Flow* are changing dramatically. In addition, we also create the *Step Type*, in which the number of queries is gradually changing. The total number of query streams of the above three patterns is 268, 302, and 680 respectively.

In addition, we compare our proposal with three baseline methods: (1) *AlwaysOn* [2], which always leaves the backend nodes active regardless of the workload change. (2) *Reactive* [2, 15], which reacts to the current number of query streams, attempting to keep the right number of active nodes at time t. (3) *LR-Predictive* [9, 11], which attempts to predict the future workload by using the linear regression method. It adjusts the number of active nodes according to the prediction result and the current capacity. For this policy, we consider a time window of 100 seconds (ten sample times).

6.2   Proportionality in Node Activation

In this section, we show the advantages of our proposal in node activation. For each workload running, we sample every 10 seconds to record the number of active backend nodes and concurrent query streams.

We show the performance of our proposal in Fig. 5. In our algorithm, there are two tunable parameters named *Timer_powerup* and *Timer_shutdown*. We set the *Timer_powerup* to 10s and *Timer_shutdown* to 40s. We will discuss the effect of these two parameters later.

Our proposal adopts the load unbalancing policy so that it can shut down unnecessary nodes quickly when needed. For instance, at the beginning of the *Step Type* workload, our algorithm finishes shutting down the unneeded nodes at the time 40s (e.g.,

Timer_shutdown), while *LR-Predictive* needs 200s to complete this procedure. Further, our algorithm can effectively reduce the misjudgment phenomenon, which is more noticeable in the *Step Type* workload. We can find from Fig. 5 that our design introduces rare misjudgments and the number of active nodes is pretty close to the current workload. The major reason is that we adopt the time-out principle to conduct the decision of node activation. Moreover, our proposal can avoid the dramatic degradation of time performance with the help of *QSB*.

To quantitatively analyze the effect of our algorithm in achieving the proportionality, we compute the ideal number of active nodes corresponding to the workload at each sample time. Then, we compare the actual conditions with the ideal ones. Figure 6 shows the average distance between the actual number and the ideal number of the active nodes, where our algorithm is denoted as *TAS*. The Y-axis in Fig. 6 represents the average difference between the actual number and the ideal number of the active nodes when finishing the workload. We can see that our proposal obtains a value more close to the idle situation, compared with the other three algorithms. There is an exception that the *Reactive* policy has the least difference for the *Fixed Type* workload pattern. That is mainly because that it turns on/off the nodes immediately when the workload decreases while our algorithm has to wait for a time period. However, this will also lead to some drawbacks, as we have discussed before.

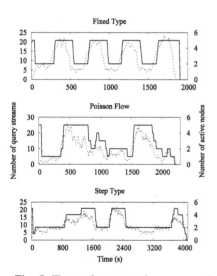

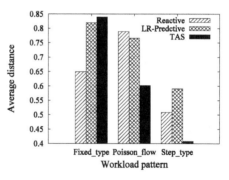

Fig. 5. The performance of our proposal under three workload patterns

Fig. 6. The average distance between the actual and ideal number of the active nodes

6.3   Time Performance

In this section, we compare the total execution time of query streams. The results are shown in Fig. 7, where our algorithm is denoted as *TAS*. As Fig. 7 shows, our algorithm (*TAS*) is superior to *Reactive* and *LR-Predictive* regarding time performance. Figure 8 shows the execution time of our proposal on the *Step Type* workload pattern.

Remember that there are totally 680 query streams in the *Step Type* workload. The dotted line represents the maximal execution time of a query stream that users can tolerate. We can find that the execution time of a large portion (83.23%) of query streams is less than the maximal tolerance time. We also notice that the query streams sent to the newly active nodes cost more time. For a newly active node, none of the pages in the node are cached in the buffer. Thus, we have to read pages from disks when starting to process queries in those newly active nodes.

The performance degradation percentage compared to *AlwaysOn* is shown in Table 2. Comparing to *Alwayson*, the other three algorithms are slower in processing the query streams to some extent. Considering that we have determined the capacity of a node by *P_degradation*, i.e., 30%, only our proposal meets the performance requirement.

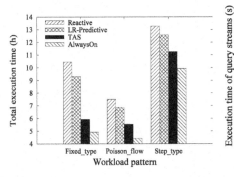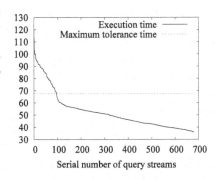

Fig. 7. Total execution time of the query streams

Fig. 8. The execution time of our proposal on the *Step Type* workload

Table 2. Performance degradation percentage

	Reactive	*LR-Predictive*	*TAS*
Fixed Type	113.03%	89.95%	20.63%
Poisson Flow	69.62%	55.24%	25.40%
Step Type	33.64%	26.66%	11.82%

Table 3. Energy saving percentage

	Reactive	*LR-Predictive*	*TAS*
Fixed Type	9.77%	7.52%	12.03%
Poisson Flow	20.69%	16.55%	21.38%
Step Type	16.90%	15.86%	18.62%

6.4   Energy Savings

Table 3 shows the energy saving percentage compared to *Alwayson*, and Fig.9 shows the total energy consumption. We collect the energy consumption via power meters attached to each backend node. When finishing the execution of the workload, we immediately read the values of energy consumption from the screens of the power meters.

As Fig. 9 shows, our method (*TAS*) consumes the least energy for all kinds of workload patterns. Regarding the energy savings, our method saves up to 21.38% energy compared to *Alwayson* under the *Poisson Flow* workload pattern. The ability to save energy not only depends on the energy proportional algorithms, but also the low-utilization period in workload and energy parameters of each node. So far, we can find

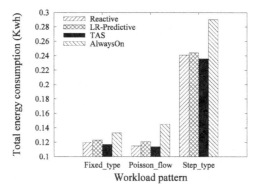

Fig. 9. Total energy consumption of different methods

that our design basically reaches the goal of energy proportionality and saves energy efficiently under the premise of ensuring the query execution performance.

7   Conclusion

In this paper, we construct an energy-proportional database cluster enhanced with a *Query Stream Buffer* on top of a hybrid architecture involving the share-disk and share-nothing architecture. We also present new algorithms involving load allocation and node activation/deactivation to deal with the key issues in such a cluster. We finally build a prototype database cluster consisting of five nodes and conduct comparative experiments using the TPC-H benchmark on PostgreSQL, with various metrics, varying numbers of query streams, and different query patterns. The results demonstrate that our proposal is superior to its competitors and provides new insights for storing and querying big data.

Acknowledgements. This work is supported by the National Science Foundation of China (61379037, 61472376, & 61272317) and the OATF project funded by University of Science and Technology of China.

References

1. Schall, D., Hudlet, V.: WattDB: an energy - proportional cluster of wimpy nodes. In: Proc. of SIGMOD, pp. 1229–1232 (2011)
2. Jin, Y., Xing, B., Jin, P.: Towards a benchmark platform for measuring the energy consumption of database systems. In: Proc. of DTA, pp. 385–389 (2013)
3. Barroso, L., Hölzle, U.: The case for energy-proportional computing. IEEE Computer **40**(12), 33–37 (2007)
4. Lang, W., Patel, J.M.: Towards eco-friendly database management systems. In: CIDR (2009)
5. Graefe, G.: Database servers tailored to improve energy efficiency. In: Proc. of EDBT Workshop SETDM, pp 24–28 (2008)

6. Yang, P., Jin, P., Yue, L.: Exploiting the performance-energy tradeoffs for mobile database applications. Journal of Universal Computer Science **20**(10), 1488–1498 (2014)
7. Wang, X., Liu, X., Fan, L., Huang, J.: Energy-aware resource management and green energy use for large-scale datacenters: a survey. In: Patnaik, S., Li, X. (eds.) Proceedings of International Conference on Computer Science and Information Technology. Advances in Intelligent Systems and Computing, vol. 255, pp. 555–563. Springer, Heidelberg (2014)
8. Lang, W., Harizopoulos, S., Patel, J., et al.: Towards energy-efficient database cluster design. Proceedings of the VLDB Endowment **5**(11), 1684–1695 (2012)
9. Nathuji, R., Kansal, A., Ghaffarkhah, A.: Q-clouds: managing performance interference effects for QoS-aware clouds. In: Proc. of EuroSys, pp. 237–250 (2010)
10. Leite, J., Kusic, D., Mossé, D., et al.: Stochastic approximation control of power and tardiness in a three-tier web-hosting cluster. In: Proc. of ICAC, pp 41–50 (2010)
11. Krioukov, A., Mohan, P., Alspaugh, S., et al.: Napsac: design and implementation of a power-proportional web cluster. Computer Communication Review **41**(1), 102–108 (2011)
12. Horvath, T., Skadron, K.: Multi-mode energy management for multi-tier server clusters. In: Proc. of PACT, pp. 270–279 (2008)

SparkRDF: In-Memory Distributed RDF Management Framework for Large-Scale Social Data

Zhichao Xu[1,2], Wei Chen[1,2(✉)], Lei Gai[1,2], and Tengjiao Wang[1,2]

[1] Key Laboratory of High Confidence Software Technologies (Peking University),
Ministry of Education, Beijing, China
[2] School of Electronics Engineering and Computer Science,
Peking University, Beijing 100871, China
{xuzhichao,pekingchenwei,lei.gai,tjwang}@pku.edu.cn

Abstract. Considering the scalability and semantic requirements, Resource Description Framework (RDF) and the de-facto query language SPARQL are well suited for managing and querying online social network (OSN) data. Despite some existing works have introduced distributed framework for querying large-scale data, how to improve online query performance is still a challenging task. To address this problem, this paper proposes a scalable RDF data framework, which uses key-value store for offline RDF storage and pipelined in-memory based query strategy. The proposed framework efficiently supports SPARQL Basic Graph Pattern (BGP) queries on large-scale datasets. Experiments on the benchmark dataset demonstrate that the online SPARQL query performance of our framework outperforms existing distributed RDF solutions.

Keywords: RDF · SPARQL · Social networks · Query processing

1   Introduction

With the rapid development of web social network applications such as Facebook, Twitter and Microblog, a large number of users linked data are generated. The characteristics of such data are large volume and complicated structure. So how to effectively manage OSN data is a hot topic in academic and industrial research. The scalability and flexibility of RDF, which is designed for Semantic Web [1], are ideal for modeling this kind of data. The projects of Friend of a Friend (FOAF) [2] and Semantically-Interlinked Online Communities (SIOC) [3] have illustrated this tendency. The Simple Protocol and RDF Query Language (SPARQL) [4] can express BGP queries for RDF, which can be directly applied to the OSN subgraph query. In general, the nature of RDF model makes it suitable for large-scale complex OSN management.

Figure 1 illustrates an example for a fraction of OSN graph representing relations between users and User Generated Contents. Query which finds pairs of users in a path of friend relationship which user1 likes the blog1 that created by user2 is expressed in SPARQL as:

© Springer International Publishing Switzerland 2015
J. Li and Y. Sun (Eds.): WAIM 2015, LNCS 9098, pp. 337–349, 2015.
DOI: 10.1007/978-3-319-21042-1_27

Select ?user1, ?company, ?user2 Where {
          ?user1   follows   ?user2 .   // triple pattern #1
          ?user1   worksFor ?company .   // 2
          ?user2   creatorOf blog1 .   // 3
          ?user1   likes blog1   // 4
}

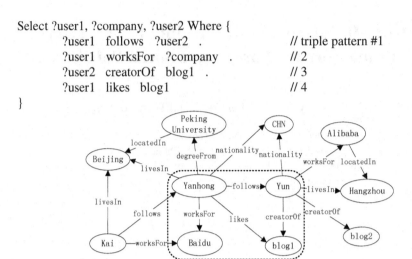

Fig. 1. An example for a fraction of social network

This query returns the results collection of <user1, company, user2> triples which match all the triple patterns. This is also a typically SPARQL BGP query. In Figure 1 we show the target subgraph by dotted line. We observe that BGPs can easily express the query for OSN chains of relationships.

A number of RDF data management systems have been developed in the past few years. The standalone systems like [5-7] store and retrieve all the RDF data in a single node. However, they are limited in storage and query processing capacity. Some works have introduced distributed framework e.g. Hadoop [13] for querying large-scale data. In [8], RDF datasets are stored in Hadoop file system (HDFS), but it has to traverse a mass of triples to select the right triple which matches the pattern because of its lack of index. H2RDF [9] utilizes the MapReduce [10] framework to process the iterative joins, but the intermediate results must be written back to the disk, which leading to low efficiency.

According to the existing works, we need to solve the two problems simultaneously. First, how to create effective index on the triples to select the query-related RDF data in a distributed store. Second, how to keep the intermediate results in memory to improve the query efficiency.

In this paper, we put forward a hybrid two-layer RDF management framework for answering SPARQL BGP queries on large-scale datasets. The contributions of our work are summarized as follows:

1. We present a three-table indexing schema for storing RDF data implemented in HBase [11], which allows bulk loading of MapReduce jobs to import and index large-scale RDF datasets.
2. We propose a pipelined in-memory approach to process SPARQL BGP iterative joins based on scalable and distributed Spark [12] Resilient Distributed Dataset (RDD) [14]. We adopt a heuristics to determine the join order.

3. We conduct comprehensive experiments on benchmark datasets to demonstrate the performance of our index schema and the efficiency of pipelined in-memory joins.

The rest of the paper is organized as follows. Section 2 reviews the related work. In Section 3, we present our framework architecture which is made up of storage schema and query strategy. Section 4 reports the experimental results. Finally we conclude the paper and give directions for the future studies in Section 5.

2   Related Works

Based on the implementation infrastructure and query mechanism, existing RDF data management systems can be divided into two categories, one is centralized systems based on the standalone mode, and the other is distributed systems.

Centralized Systems. Jena [15] is a semantic web framework that first developed by IBM. It utilizes a relational model to manage data, in other words, all data are stored in a single three-column table. Because of its limited triple store schema, Jena has very poor query performance for large-scale datasets. But it provides users abundant APIs to parse the SPARQL query and inference.

Hexastore [6] materializes six indices on triples to answer the triple patterns, one for each possible permutation of subject-predicate-object values. These permutations are spo, sop, pso, pos, osp and ops. But in our work, we show that only three indices are needed to answer all the triple patterns.

RDF-3X [5] is considered the state-of-the-art system in centralized RDF data stores. It employs indices similar to Hexastore as well as histograms, summary statistics and query optimizations to enable high query performance. However, the performance of RDF-3X is limited by the standalone main memory required to perform joins, presenting problems with large input.

Other representative centralized systems include BitMat [16], Sesame [7] and Virtuoso [17]. However, all the aforementioned systems run on a single machine, which limiting their storage and processing capacity.

Distributed Systems. Jaeseok et al. [9] introduces us an iterative MapReduce method to process SPARQL BGP query. The RDF triples are stored in HDFS in an N-Triples format file. MR selection phase obtains RDF triples which satisfy at least one triple pattern and MR join phase merges matched triples into a matched graph. The main problem of this system is that MR selection must examine all of the RDF triples in HDFS whenever a SPARQL query is issued. For this reason, the construction of an index can be considered.

H2RDF [18] presents an architecture that RDF triples are stored in HBase and queried by MapReduce framework. In storage layer, RDF data are indexed into three tables. The SPARQL queries are executed in centralized mode or in distributed mode decided by the estimated join cost based on the index statistics. H2RDF adopts the advantages in centralized systems described above, but the fact that in distributed mode the intermediate results of iterative joins must be written back to the disk may lead to low query efficiency.

Trinity [19] is a distributed, in-memory system. It uses graph exploration instead of join operations and greatly boosts SPARQL query performance. The main drawback of Trinity is that its performance is limited by the main memory of the cluster, since all the triples need to be loaded in main memory. It is not trivial to manage all data in memory, as distributed shared memory increases the complexity for maintenance.

For OSN queries, they focus on efficient graph traversal, which need special efforts on querying processing. All existing works lack the guarantee for the performance of such queries.

3   A Two-Layer RDF Management Framework

A straightforward method to improve query performance is to load all query-related data into main memory. As the basic primitives of a query are selection and join, considering large-scale OSN data, such implementation brings two challenges. First, how to manage all RDF data which are query-related to be efficiently selected in a distributed massive data store. Second, how to implement distributed in-memory join.

To solve these challenges, we propose a two-layer RDF management architecture. Figure 2 presents its overview. We utilize HBase as the distributed indexing and storage substrate (disk layer) for large-scale RDF datasets. RDF data is imported into HBase through a bulk loading process. When users submit a SPARQL query to the system, it is first parsed by the Jena parser [15] that checks the query syntax and creates the query algebra. Then our join planner generates the join order based on the algebra and triple patterns. We pipeline the iterative joins in Spark RDD (main memory layer) and output the final result.

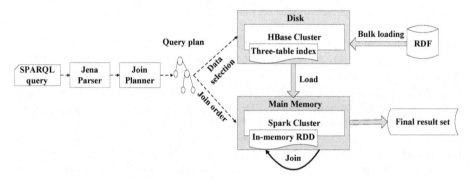

Fig. 2. Two-layer RDF management architecture of our approach

As discussed above, the challenges for efficient query implementation in distributed RDF management is focus on the strategies used in selection and join primitives. In this section, we describe our method for the index schema and join execution.

3.1   Index Schema

An RDF triple consists of a subject, predicate and object. We materialize three permutations of s-p-o values of RDF triples, these permutations are SPO, POS and OSP, which are named based on the components order of the triples stored. These three of all six permutations are necessary and sufficient to answer each possible triple pattern efficiently by only a table range scan. Table 1 shows how all eight triple patterns can map to a table scan of the three tables. Triple Pattern No.1 and No.5 are very unusual in common queries, when this happens we can pick any one of the three tables. The most frequently used triple patterns are No.2, No.4 and No.8, which are bounded with predicate, leaving the subject or the object unbounded.

Table 1. Triple patterns mapped to table scans

No.	Triple Pattern	Table to Scan
1	subject, predicate, object	ANY
2	?subject, predicate, object	POS
3	?subject, ?predicate, object	OSP
4	?subject, predicate, ?object	POS
5	?subject, ?predicate, ?object	ANY
6	subject, ?predicate, object	OSP
7	subject, ?predicate, ?object	SPO
8	subject, predicate, ?object	SPO

To be able to provide efficient indices access in a distributed environment we store the three tables using HBase and thus achieve the desired index scan and search capabilities. HBase is an open-source implementation of Google Bigtable [20]. A data row in HBase consists of a sortable row key and an arbitrary number of columns, which are further grouped into column families. A B+ tree-like index on row key is provided by HBase by default. All the above features can be taken into account for designing RDF storage schema.

We propose a new approach of storing RDF triples in HBase. In these three tables, data are stored in row keys and column names. Each table has only one column family "CF" for now and all columns belong to this column family. Take table SPO for example, the subject and predicate are stored in the row key with the delimiter "|" denoting the boundary between terms, leaving the object stored in the column name which called column qualifier in HBase. Table POS and OSP are in the similar form.

We show through an example how RDF triples are stored in our approach. Table 2 shows several example triples taken from Figure 1 in Introduction section.

Table 2. Sample RDF triples

Subject	Predicate	Object
Yanhong	likes	blog1
Yanhong	worksFor	Baidu
Yun	worksFor	Alibaba
Yun	creatorOf	blog1
Yun	creatorOf	blog2

So the logic storage structure of the triples described above in HBase SPO table is shown in Figure 3, with the timestamps omitted. Table POS and OSP are in the similar form.

The advantages of our indexing schema can be outlined as follows:

1. The multi-valued properties are well handled. There are a number of multi-valued properties in RDF datasets like Yun described above who creates more than one blog. Our approach enables users to manage multi-valued properties just like single-valued ones.
2. As many as possible fields are put into row key which can be indexed. We put two thirds of the triple into row key and leave the last part in column qualifier. We have taken a compromise in maintaining the index size. The index content can be fully utilized in search of one or two variables with our approach.

Row Key	Column Family (CF)	
<Yanhong>l<likes>	CF:<blog1>	
<Yanhong>l<worksFor>	CF:<Baidu>	
<Yun>l<creatorOf>	CF:<blog1>	CF:<blog2>
<Yun>l<worksFor>	CF:<Alibaba>	

Fig. 3. Sample RDF triples in HBase table

In order to handle large-scale RDF datasets we use a MapReduce bulk loading process to minimize the I/O and network operations. We launch MapReduce jobs to create HFiles (the HBase file format) directly and then loaded them into HBase tables instead of calling HBase APIs for each triple insertion. The import procedure consists of two MapReduce jobs for each of the three tables. The first job parses the original triple and generates a key-value pair in HFile. The second job loads the HFiles from HDFS to HBase table through LoadIncrementalHFiles [11].

3.2   Join Execution

SPARQL is the standard query language for RDF, which is based on graph pattern matching [4]. In this part, we focus on parallel processing BGP, which is a set of triple patterns.

In the example query showed in Introduction section, the SPARQL BGP contains four triple patterns which have two shared variables user1 and user2. User1 and user2 are the join keys, and company is a non-join key. The query is first parsed by Jena ARQ [15], including checking the query syntax, replacing the prefix in the triples and then creating the query algebra. As RDF has a fixed simple data model, it is not unusual that a BGP has two or more shared variables. We apply a heuristics to select join key variables which decides the join order. We select a join key according to the number of related triple patterns greedily until every related triple pattern is participated in a join iteration. This strategy is good for selectivity and pipelining the subsequent join iterations.

Algorithm 1. Matching a triple pattern over a table

Input: a triple pattern
Output: matched table scan
1:    // check for SPO table
2:    **if** tp.sp is bounded ∧ tp.op is a variable **then**
3:        tableName = SPO
4:        **if** tp.pp is bounded **then**
5:            rowKey = tp.sp + tp.pp
6:            type = 1
7:        **else** rowKey = tp.sp
8:        **end if**
9:    **end if**
10: // check for POS and OSP table (we omit the detail)
11: // set the range of scan
12: scan.setStartRow(rowKey)
13: **if** type == 1 **then**
14:     scan.setStopRow(rowKey)
15: **else** scan.setStopRow(rowKey + ~)
16: **end if**
17: **return** scan

When it comes to join execution, we must select the join data first. According to the index schema, we can get all the scans matched the related triples. The matching process is outlined in Algorithm 1. It receives one of the SPARQL BGP query pattern, outputs the matched scan for HBase table. In lines 1-9 we check if the triple pattern is No.7 or No.8 in Table 1 for SPO table. Due to the space limit, we omit the detail in line 10 as it does the similar check for POS and OSP table. In lines 11-16 we set the range of the scan.

According to the join order created by heuristics, we pipeline all the iterative joins in distributed main memory, without writing the intermediate results to the disk. We choose to implement the iterative join process on top of Spark [12]. Spark is a large-scale distributed processing framework specifically targeted at iterative workloads. It utilizes a functional programming paradigm, and applies it on large clusters by providing a fault-tolerant implementation of distributed datasets called RDD. Especially for iterative processing, the opportunity to store the data in main memory can significantly speed up processing. We pick Spark as the underlying framework, because of its nature of in-memory computing.

Algorithm 2 is the pseudo-code of the pipelined in-memory iterative join process. The input is the list of execution sequence of join keys, and the output is the final query result. In line 1 we first initialize the midRDD as the first scan to join, and we keep the intermediate results in RDD in main memory, in order to avoid disk I/O in line 8.

For line 6 in Algorithm 2, Spark provides API to load the whole HBase table into RDD, but in this way the index cannot be used. We imitate the method in which Hadoop reads HBase query results as input. Scans are first encoded to bytes, and then transformed into RDD as TableInputFormat. In this way data are transformed locally from HBase to RDD, reducing unnecessary network I/O.

Algorithm 2. Iterative joins in RDD

Input: list of execution sequence of join keys L
Output: final query result
1:   initialize midRDD as the first scan
2:   **for** joinKey in L **do**
3:       get all triple pattern numbers related to joinKey in nums
4:       **for** num in nums **do**
5:           get the scan according to triple[num]   //Algorithm 1
6:           transform scan to newRDD
7:           resRDD = newRDD.join(midRDD)
8:           midRDD = resRDD.cache
9:       **end for**
10: **end for**
11: resRDD = midRDD
12: **return** resRDD

In the whole iterative join process, the data structure in RDD keeps in binary tuple format, which contains a key part and a value part. With the SPARQL query example given in the Introduction section, let us consider example data of triple pattern #1 and #2 as follows:

Original data format for triple pattern #1:

(<Yanhong>, <follows>, <Yun>)
(<Kai>, <follows>, <Yanhong>)

Original data format for triple pattern #2:

(<Yanhong>, <worksFor>, <Baidu>)
(<Yun>, <worksFor>, <Alibaba>)
(<Kai>, <worksFor>, <Baidu>)

When they are got from HBase tables, they are transformed to key-value format according to which variable is the join key. The join key variable with its value is as the key part, separated by delimiter "|". And the value part consists of all the variables with their values like the key part, separated by delimiter ",". Here variable user1 is the join key, and user2 and company are non-join key variables in this iteration, so the key-value formats as the input of joins are as follows:

Input format for triple pattern #1:

(user1|<Yanhong> - user1|<Yanhong>, user2|<Yun>)
(user1|<Kai> - user1|<Kai>, user2|<Yanhong>)

Input format for triple pattern #2:

(user1|<Yanhong> - user1|<Yanhong>, company|<Baidu>)
(user1|<Yun> - user1|<Yun>, company|<Alibaba>)
(user1|<Kai> - user1|<Kai>, company|<Baidu>)

For convenience we separate the key part and value part with delimiter "-", it actually does not exist. In this format we can easily hash join the tuples by RDD join transformation and output in the same format. After join triple pattern #1 and #2 with input above, the intermediate result are as follows:

Output format:

(user1|<Yanhong> - user1|<Yanhong>, user2|<Yun>, company|<Baidu>)
(user1|<Kai> - user1|<Kai>, user2|<Yanhong>, company|<Baidu>)

So the intermediate result can turn into the input of next iteration of joins directly. If a different join key occurs in the next iteration, we just need to replace the key part with the new join key variable with its value. After the last iteration, the final query result will be found in the value part of the RDD.

If the query does not have any shared variables, means that there is no need to join, there will not be any Spark jobs. The result of the triple patterns get from the HBase tables will be directly outputted as the final result.

4   Evaluation

In this section, we present a performance evaluation of our framework. We first present our experimental environment, next the alternative systems we evaluated for comparison, then the benchmark dataset with which we experimented. Finally we present our evaluation results.

4.1   Experiments Set-Up

Cluster Configuration: Our experimental environment consists of a variable number of worker nodes and a single node as the Hadoop, HBase and Spark master. The master node and the worker nodes have the same configuration, which contains two AMD Opteron™ 4180 6-core CPUs at 2.6GHz, 48GB of RAM and 9TB disk. We utilize Hadoop v1.0.4, HBase v0.94.20 and Spark v1.0.0 respectively with default settings.

Baseline Frameworks: We compare the performance of our framework against Iterative Mapreduce (IterMR for short) [9] and H2RDF [18]. We reimplement the method in IterMR and utilize the source code open-sourced by the authors of H2RDF.

Dataset Used: In our experiments with SPARQL BGP query processing, we use Lehigh University Benchmark (LUBM) [21] synthetic dataset. The LUBM dataset generator simulates the network in academic domain, enabling variable numbers of triples by varying the number of university entities. We create 5 datasets with 1k, 5k, 10k, 20k, 30k universities respectively, with random seed 0. The statistics of the datasets are in Table 3. In this evaluation, LUBM (n) indicates that the dataset contains n universities.

Table 3. Dataset statistics

Dataset	# of triples	Raw size
LUBM (1k)	110 million	26 GB
LUBM (5k)	560 million	133 GB
LUBM (10k)	1.2 billion	271 GB
LUBM (20k)	2.5 billion	540 GB
LUBM (30k)	3.7 billion	809 GB

The raw data format from LUBM generator is OWL/XML, we first use Jena rdfcat toolkit to transform it to N-Triple format. LUBM also includes 14 standard queries, some of which require inference to answer. We remove the hierarchy of the predicate by querying for explicit predicate with no subclasses or subproperties.

4.2   Experiments Results

We first recorded the data importing time with the increase of the size of dataset. The results are shown in Figure 4. Note that the IterMR method achieved the minimum time in data importing phase, mainly because it does not create any index on the dataset. For our framework we not only import the datasets to HDFS, but also bulk load them into three HBase tables to create index on row keys. H2RDF additionally brings in data compression and statistics computing in this phase, which we will consider in our future work to refine the join order.

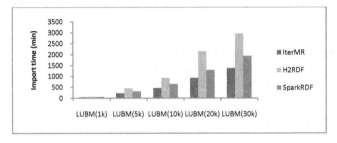

Fig. 4. Data importing time

We divide the 14 LUBM standard queries into two categories: high-selective queries and low-selective queries. High-selective query means the query has small input for the iterative joins, while the input of low-selective query is relatively large. Due to lack of space, we choose to present results of Queries 1,2,3,6 and 9 in this section. Query 1 and 3 represent the high-selective queries while queries 2, 6 and 9 stand for low-selective queries. These queries provide good mixture of high-selective and low-selective structures and multiple types of joins, which cover most variations of all the LUBM test queries. Also these queries can highlight the different characteristics of our framework and compared systems. We test the five queries with LUBM (30k) dataset and the queries execution time are showed in Figure 5. Each query experiment we have run 5 times and take the average execution time.

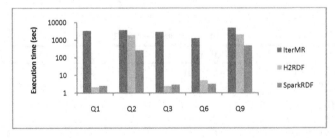

Fig. 5. LUBM execution time

We see that IterMR performed all five queries over 1000 seconds mainly due to its lack of index. In each query it must traverse all the dataset in HDFS to select the triples matched at least one triple pattern in BGP. H2RDF performs quite well in high-selective queries (Q1, Q3) but relatively poorly in low-selective and complex queries (Q2, Q9). The main reason is that H2RDF executes high-selective queries in centralized mode while low-selective queries in distributed mode based on the cost model. SparkRDF performs nearly one order of magnitude better than H2RDF in Q2 and Q9, which demonstrates the effectiveness of our in-memory pipelined join method. Q6 contains no join process, and SparkRDF performs a little better than H2RDF because we do not need to decode literals. Due to the time to setup a spark job, SparkRDF performs a little slower but comparably than H2RDF in high-selective queries. From the results, we observe that SparkRDF works relatively well through the comparative experiment, especially in low-selective and complex queries. We will introduce the method of cost model in our future work.

We evaluate the scalability property of our framework using different number of worker nodes and scale of the dataset. We choose Q2 as the representative for low-selective queries because it is one of the most complicated queries tested, requiring a triangular join, and Q1 as the representative for high-selective queries due to its bounded object. The scalability evaluation results are presented in Figure 6.

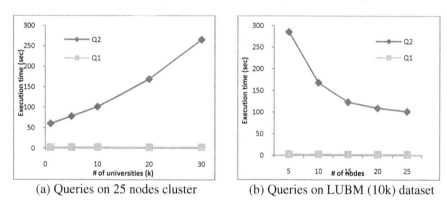

(a) Queries on 25 nodes cluster   (b) Queries on LUBM (10k) dataset

Fig. 6. Query scalability for different number of universities and nodes

Figure 6(a) shows the query execution time as the dataset increases. We clearly notice that the Q2 execution time is almost linear to the size of input data while the Q1 execution time grows slightly. In Figure 6(b), we measure the scalability of the system as the number of worker nodes increases. The tests are executed using the LUBM (10k) dataset. For Q2, the join execution is highly scalable with great speed up by adding more nodes in the front, and then the effect gradually abates due to the fact that the query cannot fully utilize the cluster resources. In the meanwhile the Q1 execution time grows still slightly. The main reason for the different results between Q1 and Q2 is that the input and result size of low-selective queries (Q2) depends on the dataset size, which directly affecting the query execution time, while the input and result size of high-selective queries (Q1) is almost bounded. From the above, we can see that the query processing of our framework is highly scalable, especially for complicated low-selective queries.

5   Conclusions and Future Works

In this paper, we present a two-layer RDF management framework capable of storing and querying enormous amount of RDF data. We have proposed a three-table index schema to store RDF dataset based on HBase which can answer arbitrary pattern of SPARQL BGP queries. A heuristic approach is applied to decide the join order in join planer. We have also proposed the pipelined in-memory distributed join processing method implementing on Spark. In this way we directly pipeline the intermediate join result into next join iteration in main memory, avoiding unnecessary disk I/O. SparkRDF is preferable and more efficient if the query includes low-selective joins.

In the future, we would like to extend the framework in several directions. First, we will store the statistics about the dataset in data loading phase. Dynamic programming can be applied to generate better join order based on the statistics. Second, we will introduce cost estimation model to decide whether the query executed in distributed mode or standalone mode, since the high-selective queries works better in standalone mode. Finally we will handle OWL reasoning and more complex SPARQL patterns.

Acknowledgements. This research is supported by Natural Science Foundation of China (Grant No.61300003), Specialized Research Fund for the Doctoral Program of Higher Education (Grant No.20130001120001) and Ministry of Education & China Mobile Joint Research Fund Program (MCM20130361) .

References

1. Semantic Web. http://www.w3.org/standards/semanticweb/
2. FOAF-project. http://www.foaf-project.org/
3. SIOC project. http://rdfs.org/sioc/spec/
4. SPARQL Query Language for RDF. http://www.w3.org/TR/rdf-sparql-query/
5. Neumann, T., Weikum, G.: RDF-3X: A RISC-Style Engine for RDF. Proceedings of the VLDB Endowment 1(1), 647–659 (2008)
6. Weiss, C., Karras, P., Bernstein, A.: Hexastore: sextuple indexing for semantic web data management. In: PVLDB, pp. 1008–1019 (2008)
7. Sesame. http://www.openrdf.org
8. Husain, M., McGlothlin, J., Masud, M., Khan, L., Thuraisingham, B.: Heuristics-Based Querying Processing for Large RDF Graphs Using Cloud Computing. IEEE Transactions on Knowledge and Data Engineering 23, 1312–1327 (2011)
9. Myung, J., Yeon, J., Lee, S.: SPARQL basic graph pattern processing with iterative MapReduce. In: Proceedings of MDAC, pp. 6:1–6:6 (2010)
10. Dean, J., Ghemawat, S.: MapReduce: simplified data processing on large clusters. In: Proceedings of the 6th Conference on Symposium on OSDI, vol. 6, p. 10 (2004)
11. Kellerman, J.: HBase: Structured storage of sparse data for hadoop (2009). http://hbase.apache.org/
12. Zaharia, M., Chowdhury, M., Franklin, M., Shenker, S., Stoica, I.: Spark: cluster computing with working sets. In: Proceedings of the 2nd USENIX Conference on Hot Topics in Cloud Computing (2010)
13. Hadoop. http://hadoop.apache.org/

14. Zaharia, M., Chowdhury, M., Das, T., Dave, A., Ma, J., McCauley, M., Franklin, M., Shenker, S., Stoica, I.: Resilient distributed datasets: a fault-tolerant abstraction for in-memory cluster computing. In: Proceedings of the 9th USENIX Conference on NSDI (2012)
15. Jena. https://jena.apache.org/
16. Atre, M., Srinivasan, J., Hendler, J.: BitMat: a main-memory bit matrix of RDF triples for conjunctive triple pattern queries. In: ISWC (2008)
17. Erling, O., Mikhailov, I.: Virtuoso: RDF support in a native RDBMS. In: Semantic Web Information Management, pp. 501–519 (2009)
18. Papailiou, N., Konstantinou, I., Tsoumakos, D., Koziris, N.: H2RDF: adaptive query processing on RDF data in the cloud. In: Proc. of WWW, pp. 397–400 (2012)
19. Zeng, K., Yang, J., Wang, H., Shao, B., Wang, Z.: A distributed graph engine for web scale RDF data. In: PVLDB, pp. 265–276. VLDB Endowment (2013)
20. Chang, F., Dean, J., Ghemawat, S., Hsieh, W., Wallach, D., Burrows, M., Chandra, T., Fikes, A., Gruber, R.: Bigtable: a distributed storage system for structured data. In: Proceedings of the 7th USENIX Symposium on OSDI, pp. 305–314 (2006)
21. Guo, Y., Pan, Z., Heflin, J.: LUBM: A benchmark for OWL knowledge base systems. J. Web Semantics 3, 158–182 (2005)

Distributed Grid-Based K Nearest Neighbour Query Processing Over Moving Objects

Min Yang, Yang Liu$^{(\boxtimes)}$, and Ziqiang Yu

School of Computer Science and Technology, Shandong University, Jinan, China
{yangm1022,zqy800}@gmail.com, yliu@sdu.edu.cn

Abstract. K-nearest neighbour (k-NN) queries over moving objects is a classic problem with applications to a wide spectrum of location-based services. Abundant algorithms exist for solving this problem in a centralized setting using a single server, but many of them become inapplicable when distributed processing is called for tackling the increasingly large scale of data. To address this challenge, we propose a distributed grid-based solution to k-NN query processing over moving objects. First, we design a new grid-based index called Block Grid Index (BGI), which indexes moving objects using a two-layer structure and can be easily constructed and maintained in a distributed setting. We then propose a distributed k-NN algorithm based on BGI, called DBGKNN. We implement BGI and DBGKNN in the commonly used master-worker mode, and the efficiency of our solution is verified by extensive experiments with millions of nodes.

1 Introduction

Given a set of N_p moving objects in a two dimensional region of interest, at time t, let $O(t) = \{o_1, o_2, ..., o_{N_p}\}$, each object $o(t)$ can be be represented by a triple $\{o_{id}, (o_x, o_y), (o'_x, o'_y)\}$, where o_{id} is the identifier of the object, (o_x, o_y) is the position at time t and (o'_x, o'_y) represents the previous position of o. Each query in the query set $Q(t) = \{q_1, q_2, ..., q_{N_q}\}$ is a query point in the same region, which can be represented by (q_x, q_y), and N_q is the number of queries in this set. The problem we study in this work is to get the k nearest neighbours (k-NNs) of each query in real-time. We adopt the snapshot semantics, i.e., the answer of $q(t)$ is only valid for the positions of the objects at time $t - \Delta t$, where Δt is the latency due to query processing. Apparently, minimizing this latency Δt is critical in our problem and is the main objective of this work. To make our approach more general, we do not make any assumptions on the movement patterns of the objects, i.e., the objects can move without any predefined pattern.

Such k-NN queries based applications in a wide array of location-based services (e.g., location-based advertising). There have already been many algorithms to process k-NN queries over moving objects. However, most of them are based on a centralized setting with a single server and is thus usually only suitable for applications with a limited data size. On the other hand, we are experiencing a rapid growth in the scale of spatio-temporal data due to the

© Springer International Publishing Switzerland 2015
J. Li and Y. Sun (Eds.): WAIM 2015, LNCS 9098, pp. 350–361, 2015.
DOI: 10.1007/978-3-319-21042-1_28

increasing prevalence of positioning devices, such as GPS trackers and smart phones. The abundance of such data and the increase in the workload in the location-based applications has rendered the centralized solutions inapplicable in many cases, and more scalable solutions are in order.

To address this challenge, we propose a distributed solution to process the k-NN queries. We first design a grid-based index called Block Grid Index (BGI), which can be constructed and maintained efficiently in a distributed setting. BGI is designed as a two-layer structure. The top layer is a grid structure, i.e. it partitions the region of interest into a grid of equal-sized cells without overlap. Each cell in the grid is in charge of indexing the moving objects within itself. The bottom layer is established based on the first layer and consists of a set of blocks, which also constitute a non-overlapping partitioning of the interest region. Each block of the bottom layer corresponds to one or more cells from the top layer and the number of objects within the block cannot be greater/less than the user-specified maximum/minimum thresholds. As objects move, the blocks can be split/merged when the number of objects in the blocks goes out the range we set. The proposed structure is particularly suitable for the distributed setting as it can achieve fast maintenance due to its simple structure.

We propose an algorithm DBGKNN for distributed k-NN query processing based on BGI, which guarantees returning the query results with only two iterations. Given a query q, DBGKNN can directly locate the blocks that are guaranteed to contain at least k neighbours of q based on BGI. Then the algorithm chooses a set of objects that are closest to q from candidate blocks and identifies the k-th nearest neighbour in these selected objects. Using this neighbour as a reference point, it can determine a search region and compute the final k-NNs by calculating the distances between q and the objects in this region. We implement BGI and DBGKNN in a master-worker mode, which can be easily deployed to distributed setting, such as Storm and S4. We conducted extensive experiments to evaluate the performance of our solutions.

Our main contributions can be summarized as follows.

· We propose BGI, a new grid-based index, for supporting k-NN search over moving objects in a distributed setting.

· We develop DBGKNN, a distributed k-NN search algorithm based on BGI. Guaranteeing to return the results with only two iterations, DBGKNN has a superior and more predictable performance than other grid-based approaches.

The rest of the paper is organized as follows. Section 2 provides an overview of related work. Section 3 introduces the BGI index structure. Section 4 presents the DBGKNN algorithm. Experimental results are presented in Section 5. Section 6 concludes this paper.

2   Related Work

As a fundamental operation, k-NN query processing has been intensively studied in recent years. Early k-NN search algorithms are for the case where both the query point and the data points are static. [4] solves this problem using the

R-tree associated depth-first traversal and branch-and-bound techniques. An incremental algorithm using traversed R-tree is developed in [2]. k-NN queries over moving objects have also been considered. The first algorithm for continuous nearest neighbour queries is proposed in [7]. It handles the case that only the query object is moving, while the data objects remain static. An improved algorithm was proposed by [8] which searches the R-tree only once to find the k nearest neighbours for all positions along a line segment.

Existing k-NN search methods can also be classified based on the structure of the index used. Tree-based approaches and grid-based approaches are both widely used. Tree-based approaches mostly are variants of the R-tree. The first algorithm of k-NN was based on R-tree as aforementioned. TPR-tree is used to index moving objects and filter-and-refine algorithms are proposed to find the k-NNs [1,3,5]. The B+-tree structure is employed by [9] to partition the spatial data and define a reference point in each partition, then index the distance of each object to the reference point to support k-NN queries.

The grid index partitions the region of interest into equal sized cells, and indexes objects and/or queries (in the case of continuous query answering) in each cell respectively [6,10,11]. Most of these approaches are designed for the centralized setting, and cannot be directly deployed on a distributed cluster.

3   Block Grid Index

Most existing grid-based algorithms of k-NN search follow the similar thought: (1)location the query object. (2)enlarge the search region iteratively to get the k objects near query object. (3)find the farthest object to query object in(2). (4)taking the distance between farthest object in (2) and the query object as the radius, the query object as the circle center, draw a circle. (5)get the k-NNs from the objects fall in the circle. In step(2), the number of iterations needed to get k objects is unknown. If we implement this kind of algorithm on a master-worker mode, let master node maintain the grid index, worker nodes store the data belongs to each grid cell. Then we will face the uncertain times communication between master and workers, which will lead to low performance. Aim to implement k-NN algorithm on distributed system, we design the BGI, a main-memory index structure to meet these requirements.

3.1   Structure of BGI

Without loss of generality, we assume that all objects exist in the $[0,1)^2$ unit square, through some mapping of the interest region. BGI is designed into a two-layers structure(See Fig.1). The top layer uses a grid structure, which partitions the unit square into a regular grid of cells of equal size δ. Each cell is denote by (i, j), corresponding its row and column indices. Given a query $q(t)$, we can directly know that it falls into the cell (i, j), if $i \leq q(t)_x \leq (i+1)$ and $j \leq q(t)_y \leq (j + 1)$. In the top layer, each cell only contains the id of block which it belongs to. The bottom layer is a set of blocks, which consisting of data located in the corresponding cells. Objects stored in block are organized by cells's boundary. Briefly, we take cell as the smallest unit to partition (without overlap) the region of interest, and objects are partitioned by cell size.

B is the set of blocks, which can be denoted as $b_i (1 \leq i \leq N_b)$ (where N_b is the number of blocks) and also can be represented by $\{b_{id}, CL(b_{id})\}$, where b_{id} is the unique identifier of b_i, and $CL(b_{id})$ is a list of cells that b_i contains. In the bottom layer, each cell is represented by $\{(i, j), OL(i, j)\}$, which has a new element $OL(i, j)$ to store the objects which fall into the cell. The blocks are non-overlapping and every object must fall into one cell of one block.

We require every block to contain at least ξ and at most θ objects, i.e., $\xi \leq N_b \leq \theta$ for all block b_i. N_b is the number of objects in block. When objects are updated, blocks split or merged as needed to meet this condition. We call ξ and θ the minimum and maximum threshold of a block respectively. Typically $\xi \ll \theta$. In the rare case where the total number of objects N_p is less than ξ, the minimum threshold requirement cannot be satisfied. This is handled as a special case in query processing. To simplify our discussion, we assume without loss of generality that at any time the total number of objects $N_p \geq \xi$.

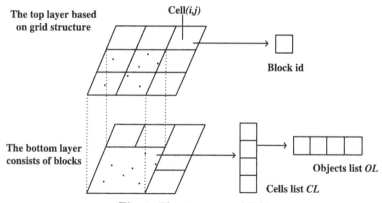

Fig. 1. The structure of BGI

3.2 Insertion

When an object comes, BGI gets the block id where the object located according to its coordinates. i.e., object o located in cell(i, j), o is inserted into block b_i if cell (i, j) has b_i's id. The insertion is done by appending o_{id} into the object list $OL(i, j)$. Initially, there is only one block covering the whole region of interest. When an object o is inserted into a block b_i, b_i will be split if the number of objects in it exceeds the maximum threshold. A split method split b_i and generate two new ones that hold approximately the same number of objects adapt to the data distribution. In this method, we first find the left bottom cell which has objects in block b_i. Then we get the nearest grids to the left bottom cell. Finally, the number of cells we select is more than $N_b/2$. These cells are moved to a new block. Once b_i is split, some cells move to a new block, the information in grid index need to be update. As shown in Fig.2.

3.3   Deletion

When an object disappears or moves out of a block, it has to be deleted from the block that currently holds it. To delete an object o, we need to determine which block currently holds it, which can be done directly using BGI. After deleting an object, if the block b_i has less than ξ objects, it will be merged with an adjacent block. The block b_i needs to send message to the top layer. Then top layer can choose which block that has the most common edges with b_i, denoted by b_j. Next, two blocks merge together and cells which fall into b_i with updating the corresponding information. In case, the number of objects in the resulting block exceeds the threshold θ, triggering another split. However, since in general $\xi \ll \theta$, such situations rarely happen and their impact on the overall performance is minimal.

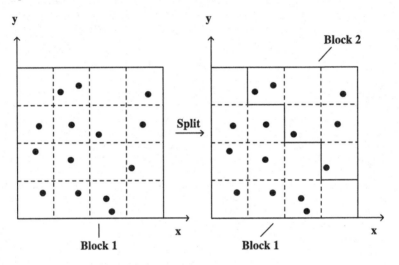

Fig. 2. The split of b_i

3.4   Analysis of the BGI Structure

Time Cost of Maintaining BGI.

Theorem 1. *Let N_p, N_b and N_c be the number of objects, blocks and cells respectively, and assume that the objects are uniformly distributed. T_{insert}, T_{delete}, T_{split}, and T_{merge} are the time costs of the insert, delete, split and merge operations respectively, and $a_i(i = 1, \cdots, 6)$ are constants.*

$$T_{insert} \approx a_0 \frac{N_c}{N_b}, T_{delete} \approx a_1 \log \frac{N_c}{N_b} + a_2 \frac{N_p}{N_c}, T_{split} \approx a_3\theta, T_{merge} \approx a_4\xi \quad (1)$$

Proof.   For an insert operation, we need to locate which cell o falls in and appending it to the end of the cell's object list. We have to spent some time on finding the right cell in a block to complete this operation. Therefore, $T_{insert} \approx a_0 log \frac{N_c}{N_b}$. To delete an object from a cell, we need to locate the cell and then remove the object from its object list. The costs of these two operations are $T_delete \approx a_1 log\frac{N_c}{N_b} + a_2\frac{N_p}{N_c}$. For a split operation, we need to first identify the cells have half of objects, which takes linear time with respect to the number of objects: θ. For a merge operation, we need to verify the block to merger, which requires to traverse the old block, i.e, $a_4\xi$.   □

Advantages of BGI. BGI has the following advantages.

· Parallelizable: BGI's partitioning strategy makes it easy to be deployed in a distributed system. The blocks do not overlap, making it possible to perform query processing in parallel.

· Scalable and Light-weight: Since the top layer, which is a grid structure that only needs to store the cells' boundary and the block id the cell belongs to, the capacity of BGI is directly proportional to the number of servers, lending it well to large-scale data processing.

· Efficient: Having a minimum threshold for each block makes it possible to directly determine the blocks that contain at least k neighbours of a given query, without invoking excessive iterations.

4   A Distributed Block Grid k-NN Search Algorithm

4.1   The DBGKNN Algorithm

The distributed k-NN search algorithm (DBGKNN) we proposed based on BGI follows a filter-and-refine paradigm. Given a query q, (1)the algorithm identify the blocks which are guaranteed to contain at least k neighbours of q through the first layer grid index. (2)corresponding blocks return at least k objects near q. (3)compute the k-th nearest neighbour of query in return objects set. (4)taking the distance between this neighbour and query as the radius, q as center, draw a circle. (5)among objects fall in the circle, compute the k-NN of query object. The algorithm is presented in Algorithm 2. Now we present the details of the algorithm. Without loss of generality, we assume that $N_p \geq k$ where N_p is the number of objects.

Calculating Candidate Blocks. For a given query q, DBGKNN can directly identify the set of blocks that are guaranteed to contain k neighbours of q, called the candidate blocks. First, get which cell the query q falls into. As we partition the region of interest using grid, it is easy to locate which cell contains q according to $q's$ coordinates. We denote the cell as c_q. Second, identify the candidate blocks. We locate a rectangle R_0 centred at the cell c_q, with some size such that R_0 encloses cells falling into at least $\Gamma \geq k/\xi$ candidate blocks. Γ denote the number of candidate blocks and satisfy $\Gamma \cdot \xi \geq k$. This way, we can guarantee that there are at least k neighbours in the candidate blocks.

Fig.3 gives an example, blocks are represented by solid lines. Query q is located in $cell(2,3)$, and R_0 with size 1 encloses cells belongs to there different

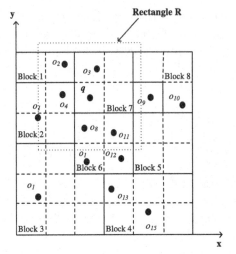

Fig. 3. Determining the set of candidate blocks

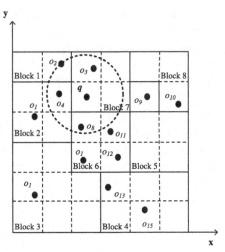

Fig. 4. An example of finding 3-NN using DBGKNN

blocks, b_4, b_6, b_7. Assume $\xi = 3, k = 6$, now $\Gamma = 3, \Gamma \cdot \xi \geq k$, the candidate blocks $\{b_4, b_6, b_7\}$ are identified.

The algorithm of determining the candidate blocks DCS is shown in Algorithm 1. This procedure describes the details of DCS and can be implemented on the master-workers setting as shown in fig.5. BGI is maintained in a distributed fashion by multiple workers, where each worker is responsible for a set of blocks. The master is the entry point for the queries. It maintains the top layer, which only record the block id of each cell. When the master receives a query q, it can immediately determine the candidate blocks by running DCS, and then send q to the workers that hold the candidate blocks.

Determining the Final Search Region. After the candidate blocks are determined, we send query q to the candidate blocks. Then every block return ξ objects which are closest to q. Then we can identify a supporting object o, which is the k-th closest to q in the return of candidate blocks. Let the distance between o and q be r_q. The circle takes (q_x, q_y) as the center and r_q as the radius is thus guaranteed to cover the k-NNs of q. Next, we identify the set of cells that intersect with this circle, and search k-NNs of q in these cells. Fig.4 shows an example, where the query q is a 3-NN query and let $\xi = 1$. We find the supporting object o_5 in its candidate blocks $\{b_1, b_2, b_6, b_7\}$ and set the radius r_q which equals the distance between q and o_5. The circle C_q is guaranteed to contain the 3-NNs of q. After scanning all objects that are located within C_q, we find that the 3-NNs are o_3, o_4 and o_8.

Fig. 5 shows this step in the master-workers setting. Master sends q to workers who holds candidate blocks. Then these workers send objects near q to calculation worker. Next, calculation worker send the circle C_q to master and identify the final set of cells C which intersected with C_q. Then master sends C to the blocks holding cells in C. Finally, k nearest neighbours are chosen from each cell in C (or all the objects in the cell if it contains less than k objects) and sent to calculation worker, where the final k-NN of q are computed.

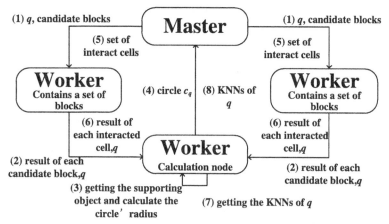

Fig. 5. Processing queries on the master-workers model

4.2 Analysis of the DBGKNN Algorithm

Time Cost of the DBGKNN Algorithm.

Theorem 2. *Let N_p, N_b and N_c be the number of objects, blocks and grid cells respectively, and assume that the objects are uniformly distributed. For a given k-NN query q, the query processing time (without considering the communication cost) by DBGKNN is $T_{query} = T_d + T_c + T_l$, where $T_d \approx a_1$, $T_c \approx a_2 \cdot \xi \frac{N_p}{N_b} + a_3 \cdot k \cdot \log k$, $T_l \approx a_4 \cdot N_c \cdot \frac{k}{N_p} \cdot \log k$, and $a_i(i = 1, \cdots, 4)$ are constants.*

Proof. T_d is the time of determining the candidate blocks, T_c is the time of obtaining the circle C_q, and T_l be the time of searching k-NNs from the set of cells covered by C_q. The time of finding the candidate blocks is a constant, for we can get the set of blocks directly through the grid index. Therefore, $T_d \approx a_1$. To compute the circle C_q, we need time $a_2 \cdot \xi \frac{N_p}{N_b}$ to find the ξ closest objects (to q) from each candidate blocks. Obtaining the radius of the circle C_q then takes time $a_3 \cdot k \cdot \log k$. Therefore, $T_c \approx a_2 \cdot \xi \sqrt{N_p N_b} + a_3 \cdot k \cdot \log k$. Finally, as we assume a uniform distribution of the data, the expected area of C_q is k/N_p. Thus, the time of obtaining the k-NNs is $T_l \approx a_4 \cdot N_c \cdot \frac{k}{N_p} \cdot \log k$. □

Effects of ξ and θ. The minimum threshold ξ influences the frequency of the merge operation. We assume that the N_p objects are uniformly distributed in a unit square for simplicity. When the number of objects in block is lower than ξ, the merge operation is running. Thus, when ξ increases, the probability of merge operations comes higher. However, ξ can not be too small. In the candidate blocks

---

Algorithm 1. DCS Algorithm

Input:

The query $q(q_x, q_y), BGI, \delta, \xi, k$;

Output:

The set of candidate blocks, C_b;

1: get the cell c which q located in through BGI. Put the block id of cell into C_b.
2: **while** $(|C_b| \cdot \xi < k)$ **do**
3:     make a rectangle R_0 centreed at cell c with size l.
4:     **if** the block of cells that fall into R_0 is not in C_b **then**
5:         add the block id into C_b.
6:     **else**
7:         l=l+1;
8:     **end if**
9: **end while**
10: **return** C_b

Algorithm 2. DBGKNN Algorithm

Input:

The query $q(q_x, q_y)$; BGI; The cell size, δ; The set of candidate blocks, C_b; The minimum threshold of block, ξ; The maximum threshold of block, θ;

Output:

k nearest neighbours of q;

1: $C_b = DCS(q_x, q_y, BGI, \delta, \xi)$;
2: Find nearest ξ objects in every candidate block, and put them into O_c.
3: Compute the supporting object o to q which is the k-th nearest object as in O_c.
4: Taking q as centre and the distance between o and q as the radius, draw a circle c_q.
5: Let Υ be the set of cells which interacts with circle c_q.
6: Find k-NNs from the objects covered by cells in Υ;
7: Return k-NNs;

---

notify stage, we need to meet the condition of $\xi \cdot \Gamma \geq k$, if ξ is too small, then we need to enlarge the rectangle R_0 to get more candidate blocks. The maximum threshold θ affects the splitting of blocks. when θ decreases, more blocks need to split. Meanwhile, high θ means every block has a high number of objects, which may influences the performance of circle computation.

Advantages of DBGKNN. The most notable advantage of DBGKNN is that it minimize the probability that the master becomes a bottleneck, for the master only maintenance a grid structure to index objects and store the block id that the object belongs to. Given a query q, DBGKNN gets the k-NNs of q in two steps, first directly determining the candidate blocks using BGI, and then identifying the final set of cells to search by computing the circle C_q. This is highly beneficial when the algorithm is running in a distributed system.

Scalability of DBGKNN. DBGKNN is easily parallelizable and scales well with respect to the number of servers to handle increases in data volumes. These blocks in BGI in general reside on different servers, and the process of searching them for the k-NNs can take place simultaneously on individual servers. More processing power can be obtained by simply adding more servers to the cluster.

5 Experiments

We implement experiments to evaluate BGI and DBGKNN. We mainly test the performance of BGI by changing the parameters. For DBGKNN, we implement a distributed grid-based search algorithm according to [3]. We take this grid-based search algorithm as the baseline, and compare a lot between them. Every experiment is repeated ten times, and the average values are recorded as the final results.

5.1 Experimental Setup

The experiments are conducted on a Dell R210 server, which has a 2.4GHz Intel processor and 8GB of RAM. We use the local mode to simmulate the master-worker mode. We simulate three different datasets for our experiments. The first dataset (UD) is consisting of the objects that follow a uniform distribution. In the second dataset (GD), 70% of the objects follow the Gaussian distribution, and the rest objects are uniformly distributed. In the third database, objects follow the Zipf distribution. All the objects are normalized to a unit square.

5.2 Experiment Performance

We first test the time of building BGI with changing θ, and the number of objects using different database. Fig. 6 shows the time of building BGI as we vary θ. Fig.8 is the time of index construction with data in UD as we change the number of objects. In our study, the number of θ is approximately reversely proportional to the account of spilt operation. The lower *theta* is, the higher split operations are, which means a longer build time. θ cannot be overly large. In extreme cases, when θ is too large, the total number of blocks may be 1, which means we will run the index like single server. A high θ will also increase the time of query processing. The index build time increases almost linearly with the increasing number of objects. As objects changing, the more split and merge operations happen, leading the maintenance time increase. Data in Zipf and GD need more split and merge operations, so they take a higher cost of building index. Fig.7 shows the the query time of different database changing θ. Data follow different distributions perform similarly and the query time of them all increase along with θ getting higher. A higher θ means that the average number of objects in one block is higher, which brings more time for calculation in one block.

Fig.8 compares our algorithm with the baseline method varying the number of objects. Baseline method build index very fast and the time it costs varies

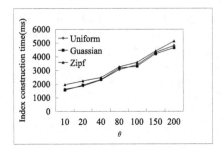

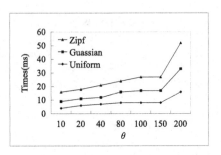

Fig. 6. Index construction time with different θ

Fig. 7. Query time with different θ

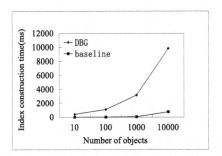

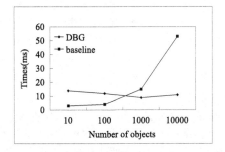

Fig. 8. Index construction time comparison with different number of objects

Fig. 9. time comparison with different number of objects

little when the number of objects changes. In fig.9, we show the query time between two algorithms. The query time of baseline method increases rapidly along with increasing objects number, comparing that our algorithm performs stable. When the number of objects becoming higher, the baseline method suffer from the communication cost of iterations. DBGKNN only need two iterations to get the result so that it is little unaffected by objects number.

In the above experiments, we find that although DBGKNN takes time to build index, it performs better in query processing than the baseline method. The parameter θ in DBGKNN matters the index build time and query processing time. A high θ means more split operations, which leads to fast query processing time and high cost on index maintains. Therefore, we need to choose the optimal value of θ according to the actual conditions. In summary, DBGKNN is more suitable for large volumes of objects in distributed system.

6   Conclusions

The problem of processing k-NN queries over moving objects is fundamental in many applications. The large volume of data and heavy query workloads call for new scalable solutions. We propose a distributed grid index BGI and a distributed k-NN search algorithm DBGKNN to address this challenge. Based on

BGI, we present DBGKNN that can directly determine a region that is guaranteed to contain the k-NNs for a given query with only two iterations. This has a clear cost benefit when compared with existing approaches, such as grid-based methods, which require an uncertain number of iterations. Extensive experiments confirm the superiority of the proposed method. For future work, we would like to explore how to evaluate continuous k-NN queries over moving objects. For a given k-NN query q, it is very possible that its result (a list of objects) remains relatively stable when objects move with reasonable velocities. Therefore, it is promising to investigate how the k-NN results can be incrementally updated as objects move.

Acknowledgments. This work was supported in part by the 973 Program (2015CB352500), the National Natural Science Foundation of China Grant (61272092), the Shandong Provincial Natural Science Foundation Grant (ZR2012FZ004), the Science and Technology Development Program of Shandong Province (2014GGE27178), the Taishan Scholars Program and NSERC Discovery Grants.

References

1. Chaudhuri, S., Gravano, L.: Evaluating top-k selection queries. In: VLDB, vol. 99, pp. 397–410 (1999)
2. Hjaltason, G.R., Samet, H.: Distance browsing in spatial databases. ACM Transactions on Database Systems (TODS) **24**(2), 265–318 (1999)
3. Raptopoulou, K., Papadopoulos, A., Manolopoulos, Y.: Fast nearest-neighbor query processing in moving-object databases. GeoInformatica **7**(2), 113–137 (2003)
4. Roussopoulos, N., Kelley, S., Vincent, F.: Nearest neighbor queries. In: ACM Sigmod Record, vol. 24, pp. 71–79. ACM (1995)
5. Seidl, T., Kriegel, H.-P.: Optimal multi-step k-nearest neighbor search. In: ACM SIGMOD Record, vol. 27, pp. 154–165. ACM (1998)
6. Šidlauskas, D., Šaltenis, S., Jensen, C.S.: Parallel main-memory indexing for moving-object query and update workloads. In: Proceedings of the 2012 ACM SIGMOD International Conference on Management of Data, pp. 37–48. ACM (2012)
7. Song, Z., Roussopoulos, N.: K-Nearest neighbor search for moving query point. In: Jensen, C.S., Schneider, M., Seeger, B., Tsotras, V.J. (eds.) SSTD 2001. LNCS, vol. 2121, pp. 79–96. Springer, Heidelberg (2001)
8. Tao, Y., Papadias, D., Shen, Q.: Continuous nearest neighbor search. In: Proceedings of the 28th International Conference on Very Large Data Bases, pp. 287–298. VLDB Endowment (2002)
9. Yu, C., Ooi, B.C., Tan, K.-L., Jagadish, H.: Indexing the distance: an efficient method to knn processing. In: VLDB, vol. 1, pp. 421–430 (2001)
10. Yu, X., Pu, K.Q., Koudas, N.: Monitoring k-nearest neighbor queries over moving objects. In: Proceedings of 21st International Conference on Data Engineering, ICDE 2005, pp. 631–642. IEEE (2005)
11. Zheng, B., Xu, J., Lee, W.-C., Lee, L.: Grid-partition index: a hybrid method for nearest-neighbor queries in wireless location-based services. The VLDB JournalThe International Journal on Very Large Data Bases **15**(1), 21–39 (2006)

An Efficient Block Sampling Strategy for Online Aggregation in the Cloud

Xiang Ci and Xiaofeng Meng[✉]

School of Information, Renmin University of China, Beijing, China
{cixiang,xfmeng}@ruc.edu.cn

Abstract. As the development of social network, mobile Internet, etc., an increasing amount of data are being generated, which beyond the processing ability of traditional data management tools. In many real-life applications, users can accept approximate answers accompanied by accuracy guarantees. One of the most commonly used approaches is online aggregation. Online aggregation responds aggregation queries against the random samples and refines the result as more samples are received. In the era of big data, more and more data analysis applications are migrated to the cloud, so online aggregation in the cloud has also attracted more attention. There can be a huge difference between the number of tuples in each group when dealing with group-by queries. As a result, answers of online aggregation based on uniform random sampling can result in poor accuracy for groups with very few tuples. Data in the cloud are usually organized into blocks and this data organization makes sampling more complex. In this paper, we propose an efficient block sampling which can exactly reflect the importance of different blocks for answering group-by queries. We implement our methods in a cloud online aggregation system called COLA and the experimental results demonstrate our method can get results with higher accuracy.

Keywords: Online aggregation · Block sampling · Cloud computing

1 Introduction

Data-driven activities are rapidly growing in various applications, including Web access logs, sensor data, scientific data, etc. Distilling the meaning from these data has never been in such urgent demand. All these big data have brought in great challenges to traditional data management in terms of both data size and significance. As the growth of sheer volume of data, performing analysis and delivering exact result on big data can be extremely expensive, sometimes even impossible. For example, suppose one petabyte data are stored in the database and we want to find a particular record from the database. If indices have been built on the attribute we want to query, the answer will be returned soon. But if we do not build indices, how long will it take? The only way we can do without indices is full scan. Now for the fastest Solid State Drives (SSD) with disk scanning speed of about 6GB/s, a linear scan of SSD with one petabyte data will take

© Springer International Publishing Switzerland 2015
J. Li and Y. Sun (Eds.): WAIM 2015, LNCS 9098, pp. 362–373, 2015.
DOI: 10.1007/978-3-319-21042-1_29

about 166666 seconds, i.e. 2777 minutes, 46 hours, or 1.9 days. It is definitely unacceptable. Fortunately, in many real-life applications, people just want to obtain a bird's eye view of the whole dataset. This situation has brought more attention to the already-active area of Approximate Query Processing (AQP).

OLA is one of the most widely used AQP techniques and our work also focuses on OLA. OLA is first proposed in the area of relational database management system (RDBMS). The basic idea behind OLA is to estimate the result by sampling data and the approximate answer should be given some accuracy guarantees. Usually, we use confidence to measure the accuracy. In many scenarios, group-by queries play an important role: data are divided into groups and aggregated within these groups. Uniform random sampling is appropriate only when the utility of the data to the users mirrors the data distribution and less effective for group-by queries because of the problem of "small group". For example, consider a column R with 1000 tuples of which 99% have value 1, while the remaining 1% of the tuples have value 100. If we want to estimate the Sum of the two groups individually by 100 samples, accurate results can be obtained only when there are 99 samples with the value of 1 and 1 sample with the values of 100. If two or more value of 100 are sampled, the estimation will be great error. Here, the group with value of 100 can be considered as "small group". The problem of "small group" can significantly influence the accuracy of the results.

In recent years, with the development of cloud computing, OLA in the cloud has drawn more attention. Especially it can reduce the economic cost of users on the typically pay-as-you-go cloud systems and increase the overall throughput of the cloud system. Although OLA techniques have been extensively studied for RDBMS and OLA is very suitable for cloud environments, there are still challenges to adapt them to the cloud when group-by queries are invloved. First, data in the cloud are usually organized and processed in blocks which may contain thousands of tuples. The accuracy of OLA highly depends on the sampling. When data are organized in blocks, the minimum sampling unit is block. So the layout of the data in block, i.e., the way by which tuples are grouped into blocks, will greatly influence the accuracy. Second, we need to change the sampling method to make the samples reflect the distribution of the original data. Third, many cloud systems, such as Hadoop and Hyracks, are MapReduce-oriented. The naive MapReduce framework is batch processing, which is opposite to OLA. The Reduce phase of MapReduce can not start until the completion of the Map phase while OLA needs pipelining between different phases.

Motivated by above requirements and challenges, we try to solve the problem of "small group" for online aggregation in the Cloud. The specific contributions we make in this paper are:

- We introduce a stage of preprocessing which can randomize the data blocks efficiently and obtain relative data information.
- We propose an adaptive block sampling method which can solve the problem of "small group".

– We implement our method in COLA, a system for Cloud Online Aggregation. And experiments of the proposed technique are reported in terms of performance and accuracy.

The remainder of the paper is organized as follows. In Section 2, we summarize the related work. We provide an overview of our problem in Section 3. In Section 4, we describe the stage of preprocessing. Our adaptive block sampling is presented in Section 5. In Section 6, we use COLA to implement our approach. Experimental results are presented in Section 7, followed by conclusions.

2   Related Work

In general, our work in this paper is related to two fields: online aggregation and data sampling. OLA was first introduced in RDBMS [1], which focuses on single-table queries involving "group by" aggregations. The work in [2] improves the approach in [1] by providing the large-sample and deterministic confidence interval computing methods in the case of single-table and multi-table queries. The query processing and estimate algorithms for OLA were studied in the context of joins over multi-tables [3]. A family of join algorithms called ripple joins were presented in [3]. Wu et al. [6] proposed a new OLA system called COSMOS to process multiple aggregation queries efficiently. All the work above is in the context of RDBMS. In fact, these centralized OLA methods or systems can not be extended to MapReduce-based cloud systems straightforward. Hadoop Online Prototye (HOP) [8] is a modified version of the original MapReduce framework, which is proposed to construct a pipeline between Map and Reduce. COLA [11,28] realizes the estimation of confidence interval based on HOP. A Bayesian framework based approach is used to implement OLA over MapReduce [13] based on the open source project Hyracks [9]. The approach's estimation method is complex, which is hard to be implemented in the MapReduce framework. In the work of [15], the authors focus on the optimization for running OLA over MapReduce-based cloud system.

Sampling has a long history in database. There are two levels of sampling unit in the existing sampling techniques: row-level sampling [19,21] and block-level sampling [20,22]. Row-level sampling provides true uniform-randomness, which is the basis of many approximate algorithms. The work in [20] proposes statistical estimators with block-level samplings. Several special sampling approaches for group-by queries are also proposed. A method called Outlier Indexing [23] is proposed to improve sampling-based approximations for aggregation queries. Congressional sampling [24] stratifies the database by considering the set of queries involving all possible combinations of grouping columns and provides general-purpose synopses. Another approach to solve the small group problem has been developed by Babcock et.al. [25] and is called small group sampling. It generates multiple sample tables and selects an adequate subset of them at query evaluation time. Philipp et.al. [26] proposed a novel sampling scheme for constructing memory-bounded group-aware sample synopses. All above sampling approaches for group-by queries are row-level sampling. In the context of online aggregation in the cloud, existing work of OLA over MapReduce adopts random sampling [8,13], and no special sampling techniques have been proposed.

3 Overview

In this section, we firstly formalize our problem that we study in the paper. This paper mainly focuses on online aggregation for single table, so we consider a relation R and queries like

SELECT $op(exp(t_{ij}))$, col FROM R WHERE $predicate$ GROUP BY col

where op is the operation of Sum, other operator (Count, Avg) can be dealt similarly. exp is an arithmetic expression of the attributes in R, $predicate$ is an arbitrary predicate involving the attributes, and col is one or more columns in R. Because all the data are stored in HDFS, the data unit is block while its counterpart in RDBMS is tuple. t_{ij} represents the j-th tuple in block i.

Figure 1 shows the basic architecture of our system. Online aggregation in the cloud is that when users request an aggregation query, system returns an approximate result within the prescribed level of accuracy against random samples. The result is refined as more samples are received. In this way, users can terminate the running queries prematurely if an acceptable estimate arrives quickly. Usually, confidence interval and confidence level are adopted to measure the accuracy of current result. If users do not terminate the query actively, all the data will be scanned, and the processing is just the same as common aggregation query. The whole data processing is implemented in the cloud.

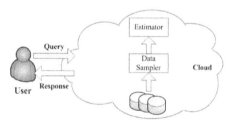

Fig. 1. Basic Architecture

As shown in Figure 1, there are two major issues we should consider when implementing online aggregation with group-by clause in the cloud: (1) How to sample data efficiently and, (2) How to implement the method in the cloud.

We will in turn detail our solutions for the two issues in subsequent sections.

4 Preprocessing

It is extremely difficult to execute the queries online without any assumption or preprocessing. We do not add constraints to the original data, so we need preprocessing to the data. During the stage of preprocessing, two important things should be completed: randomization of data blocks and acquisition of necessary data information. The performance of online aggregation highly depends on the sampling, and the accuracy of sampling depends on the data distribution. If the data are fully random, the result will be good. Unfortunately, this is not always true for real world data. For tuple-level sampling, there are several ways to achieve random sampling from disk. If the data are distributed randomly, we just need sequential access. Otherwise, we must do random disk access to obtain random sampling. For block-level sampling, the case can be worse. We still can not get random sampling by random disk access because the data layout inside the block may be not random. Besides, completely random disk access can be five orders of magnitude slower than sequential access [27]. If we want to ensure samples from different groups can match the data distribution, some additional data information must be obtained. In this section, we introduce one approach to randomize the blocks and obtain the data information simultaneously.

Usually, online aggregation assume the dataset is static, we also reserve the assumption in this paper. Here, static dataset does not mean we never update the dataset. In real-life applications, if the dataset is relatively static or update rarely, we can view it as a static dataset. The cost of preprocessing can be amortized by subsequent massive queries. This stage can be done by one MapReduce job and the basic procedure is as follows:

1) Get current number of different blocks on HDFS from namenode, denoted by N.

2) Initiate the MapReduce job and then Map tasks read data from HDFS.

3) Read value of the key-value pair and record the frequency of different values within each column. The concept of column here is similar to the column of RDBMS. In fact, there is no column in the data block, but different values in one line which are seperated by some predefined characters can be viewed as different columns. Then Map tasks generate a random number between 1 and N, and assign the random number to the header of the key of current key-value pair.

4) Reduce tasks receive the key-value pairs and write them to different blocks according to the header of the key. At the same time, reduce tasks merge the frequency of different values from all the map tasks. These frequencies and corresponding location (i.e. the data block they belong to) will be stored on the master as data information.

This method of data redistribution is easy to be realized and can be used in any MapReduce-based cloud system without any modification. Its performance will be demonstrated in the subsequent experiments.

5   Query Processing

Unlike query processing in RDBMS, OLA must return the estimated result continuously with accuracy guarantees. In this section, we will elaborate the query processing of OLA, which includes a new adaptive block sampling and the estimation of results.

5.1 Adaptive Block Sampling

Uniform random sampling is the most used sampling method of online aggregation. If the data is distributed evenly, random sampling is effective. Unfortunately, in the real world, many datasets are skew and the problem of "small group" is not uncommon. Block sampling is more complex than random sampling because the data within a block may be correlative. The basic processing unit of block sampling is block. If the block is inappropriate to a certain "small group", we still need to scan all the data in that block, this is really wasteful. In this section, we propose a new adaptive block sampling which can avoid above drawbacks.

For a relation R with two columns A, B, let B be the column used for grouping and A be the aggregation attribute. Suppose B can be divided into n groups: $\{b_1, b_2,..., b_n\}$. Since we have known the frequencies of values in different groups during the stage of preprocessing, it is easy to compute the ratio between these n groups, which can be described as follow:

$$b_1 : b_2 : ... : b_n = r_1 : r_2 : ... : r_n \tag{1}$$

The basic steps of our adaptive sampling is described as follows:

- **step1**: Map tasks read the data blocks sequentially. Data blocks have been randomized during the stage of preprocessing, so read blocks sequentially can get random blocks instead of randomly read;
- **step2**: Read key-value pair within the block one by one, and decide to accept or reject current key-value pair by using the method that is similar to acceptance/rejection sampling. The probability of acceptance is decided by column B. If current key-value pair belongs to group b_i, its probability of acceptance p_i is as follows:

$$p_i = \frac{r_i}{\sum_{i=1}^{n} r_i} \tag{2}$$

- **step3**: Compute the current ratio between different groups in the sample. If a certain group is unseen or its ratio is highly under the normal level, we set the probability of acceptance is 1. This means the group is accepted directly as soon as it appears next time.
- **step4**: Compute the terminal condition. We find it is not necessary to process the entire block to get the approximate answers. We define three parameters α, β and γ. Here α means the ratio of data we have processed within current block, β means the ratio of groups we have not seen yet, and γ means the upper bound of the amount of data which will be processed within a block. If α percent of the data have been processed and still β percent of the groups have not seen, this data block will be discarded; If the data block is not discarded, the processing of current data block will stop in two situations: 1) If all the groups have been seen and their ratios approximately match the true ratios, stop processing current block; 2) If γ percent of the data have been processed, we also stop. α, β and γ can be obtained by multiple tests and will be given in the subsequent experiments.

- **step5**: Repeat step 1 to 4 until users terminate the process actively. If users do not terminate, all the data block will be processed.

5.2   Estimation and Confidence Interval

After the sampling procedure, we can get a sampling S_k with size n_k for a particular group k, and set $exp_k = \sum_{i=1}^{n_k} exp_k(t_{ij})$. The estimation of aggregation result is given by

$$\mu = \frac{N}{n_k} * exp_k \tag{3}$$

According to the previous equation, the final aggregation result can also be considered as the average of Y, where $Y = N * exp_k$. The sampled data are retrieved in random order, so Y are identically distributed and independent to each other. According to the central limited theorem, the average of Y in samples obeys the normal distribution, so we can obtain the half-width interval with specified confidence level $100p\%$: $\varepsilon_n = z_p \sigma_n / \sqrt{n}$, where z_p is the p-quantile in the standard normal distribution, and σ_n is the standard deviation of n_k varibles in the sample. The final $100p\%$ confidence interval of the aggregation result is $[\mu - \varepsilon_n, \mu + \varepsilon_n]$.

6   Online Aggregation in the Cloud

Cloud is different from RDBMS, and the major problem of online aggregation in the cloud is that naive MapReduce does not support pipeline operations. Several online aggregation systems have tackled the problem of pipeline, e.g., HOP, COLA. We finally choose COLA, which is described in the paper [11], rather than HOP, to implement our method. Our method is easier to be implemented on COLA because it supports more operations for online aggregation. The basic steps of online aggregation with our method is described as follows:

- **step1**: The user sends a query to the master and the master determines the location (i.e. which slaves have relative data) which will involve the query;
- **step2**: Read data from relative slaves and get the initial samples using our adaptive block sampling;
- **step3**: Compute the estimation and confidence interval;
- **step4**: Output the result and continue sampling;
- **step5**: Repeat step 1 to 4 until users terminate the process actively. If users do not terminate, all the data block will be processed.

Our method primarily focuses on the query on a single table, so the implementation only involves one MapReduce job. In the Map function shown in Algorithm 1, tuples of every block are filtered out according to the predicate and transformed into key-value pairs.

Algorithm 1. Map function

Input:Object t;
Output:Text key, Text value;
1:**if** t satisfies the predicate **then**
2:calculate the probability of acceptance;
3:**if** accept **then**
2:   key.set(t.tuple.lang);
3:   value.set(t.tuple.size);
4:**end if**
5:output.collect(key,value);

The Reduce function is executed each time the estimate is invoked, so it is important to make the computing process incremental and make use of the results of the previous Reduce function. The Reduce function is described in Algorithm 2.

Algorithm 2. Reduce function

Input:Text key,Iterator(Text)values;
Output:aggregation estimate μ, confidence interval ε;
1://n_k: number of tuples processed by the reducer;
2://sum_i: sum of the variables in the last iteration;
3://$quadratisum_i$: quadratic sum of the variables in the last iteration
4:**while** values.hasNext() **do**
5:   Text it=values.getNext();
6:   sum+=it.get_first();
7:   $quadraticsum$+=it.get_second();
8:**end**
9:sum+=sum_i;
10:quadraticsum+=$quadratisum_i$
11:variance=quadraticsum/n-sum*sum/n*n;
12:μ=sum/n;
13:$\varepsilon = z_p sqrt\,(variance)\,/sqrt\,(n)$

---

7 Performance Evaluation

7.1 Experiment Overview

The testbed is established on a cluster of 11 nodes connected by a 1Gbit Ethernet switch. One node serves as master, and the remaining 10 nodes act as slave nodes. Each node has a 2.33GHz quad-core CPU and 7GB of RAM, and the disk size of each node is 1.8TB. We set the block size of HDFS to 64MB.

Naive MapReduce dose not support pipeline operations, so we use COLA, an online aggregation system which is developed based on HOP. We implement our methods described above for COLA, and all the following experiments are conducted on the improved COLA.

In the experiment, we evaluate our approach on a synthetic dataset. We constrcut a relation R which contains three attributes A, B and C. C is used for grouping. We generate about 20GB data and C contains 10 groups. Three of the ten groups are relatively small, and they are defined as "smalll group" in this paper. Table 1 summarizes settings used in the experiments:

Table 1. Settings used in the experiments

Parameter	Values
Number of Computer Node	11
Data size	20G
Platform	COLA
Map task num. per Computer Node	4
Reduce task num. per Computer Node	2

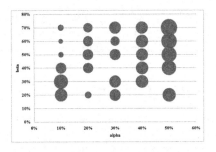

Fig. 2. Parameters: α and β

All the data are stored in HDFS, and we test online aggregation queries of Sum with following example queries Q.

Q= SELECT Sum(A), C FROM R GROUP BY C

In subsquent experiments, we set the confidence level to 95%. The accuracy of estimated aggregation result is measured by *relative_error*, which is computed by following equation:

$$relative\_error = \frac{|estimateValue - actualValue|}{actualValue} \qquad (4)$$

7.2 α, β and γ

These three parameters are very important for our sampling. We want to make our samlping faster, so α can not be too large. At the same time, we need to ensure the accuary of the result, so β can not be too small. We set α to 10%, 20%, 30%, 40% and 50% respectively. For each α, we set β to 20%, 30%, 40%, 50%, 60% and 70%, so we have thirty kinds of combination. We record the time when the relative_error is no more than 5%. In order to choose the parameters accurately, one small group in C is used to show the result. Figure 2 illustrates the result. X-axis means α, Y-axis means β. The area of the bubbles in the figure represents the time. The smaller the bubble is, the shorter the time it takes. We

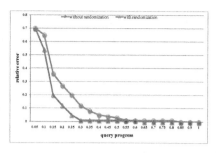

Fig. 3. Effect of Data Randomization

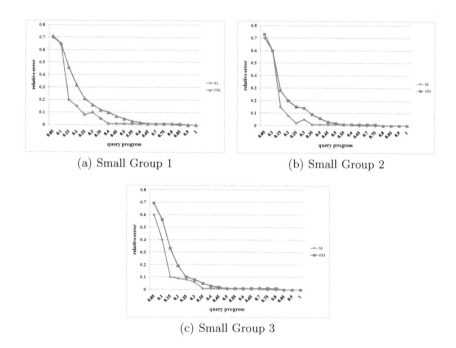

(a) Small Group 1   (b) Small Group 2

(c) Small Group 3

Fig. 4. Query Error of Small Group

can know that when the α is 10% and β is 60%, we can get the shortest running time when the accuracy of relative_error is also satisfied.

γ can not be too small or too large. In our multiple experiments, we found that 55% is a good choice. If γ is smaller than 55%, it is hard to get accurate results. If this value is larger than 55%, longer time will be taken and it is wasteful.

Subsquent experiments all adopt above α, β and γ.

7.3   Effect of Data Randomization

We compare the trends of relative_error between original data and data after randomization. We only choose one large group from the ten groups in C to show the result due to limited space. Figure 3 shows that our method of data randomization really takes effect. We can get the accurate result earlier than the dataset without data randomization.

7.4   Query Error of Small Group

Our method mainly focuses on the problem of small group, so we compare the trends of query error of the three small groups in our experiment. Because there is no work of block sampling for the problem of small group, we just compare our method with simple random sampling. S1, S2 and S3 are the results of small group 1, 2 and 3 by using our adaptive block samping. OS1, OS2 and OS3 are the results of small group 1, 2 and 3 by using simple random sampling. Group 1 is the samllest one of the three groups while group 3 is the largest.

Figure 4 displays the query results of group 1, group 2 and group 3 respectively. From all these figures, our method can get high accuary with less time. We can also find that the samller the group is, the better our method performs.

8   Conclusion

"small group" is a very important problem for online aggregation. Uniform random sampling is not suitable for this problem. We propose a new adaptive block sampling to solve the problem of "small group". This method can adjust the probability of acceptance dynamically and the experiments have shown its effieciency.

Acknowledgments. This research was partially supported by the grants from the Natural Science Foundation of China (No. 61379050,91224008); the National 863 High-tech Program (No. 2013AA013204); Specialized Research Fund for the Doctoral Program of Higher Education(No. 20130004130001), and the Fundamental Research Funds for the Central Universities, and the Research Funds of Renmin University(No. 11XNL010).

References

1. Hellerstein, J.M., Haas, P.J., Wang, H.J.: Online aggregation. In: SIGMOD Conference, pp. 171–182 (1997)
2. Haas, P.J.: Large-sample and deterministic confidence intervals for online aggregation. In: 9th IEEE International Conference on Scientific and Statistical Database Management, pp. 51–62. IEEE Press, New York (1997)
3. Haas, P.J., Hellerstein, J.M.: Ripple Joins for online aggregation. In: SIGMOD Conference, pp. 287–298 (1999)

4. Luo, G., Ellmann, C.J., Haas, P.J., Naughton, J.F.: A scalable hash ripple join algorithm. In: SIGMOD Conference, pp. 252–262 (2002)
5. Jermaine, C., Dobra, A., Arumugam, S., Joshi, S., Pol, A.: A disk-based join with probabilistic guarantees. In: SIGMOD Conference, pp. 563–574 (2005)
6. Wu, S., Ooi, B.C., Tan, K.: Continuous sampling for online aggregation over multiple queries. In: SIGMOD Conference, pp. 651–662 (2010)
7. Wu, S., Jiang, S., Ooi, B.C., Tan, K.: Distributed online aggregation. presented at PVLDB, pp. 443–454 (2009)
8. Condie, T., Conway, N., Alvaro, P., Hellerstein, J.M., Gerth, J., Talbot, J., Elmeleegy, K., Sears, R.: Online aggregation and continuous query support in MapReduce. In: SIGMOD Conference, pp. 1115–1118 (2010)
9. Borkar, V.R., Carey, M.J., Grover, R., Onose, N., Vernica, R.: Hyracks: A flexible and extensible foundation for data-intensive computing. In: ICDE, pp. 1151–1162 (2011)
10. Böse, J.-H., Andrzejak, A., Högqvist, M.: Beyond online aggregation: parallel and incremental data mining with online Map-Reduce. In: 2010 Workshop on Massive Data Analytics on the Cloud, pp. 1–6 (2010)
11. Shi, Y., Meng, X., Wang, F., Gan, Y.: You can stop early with COLA: online processing of aggregate queries in the cloud. In: CIKM, pp. 1223–1232 (2012)
12. Gan, Y., Meng, X., Shi, Y.: COLA: A cloud-based system for online aggregation. In: ICDE, pp. 1368–1371 (2013)
13. Pansare, N., Borkar, V.R., Jermaine, C., Condie, T.: Online aggregation for large mapreduce jobs. In: PVLDB, pp. 1135–1145 (2011)
14. HKalavri, V., Brundza, V., Vlassov, V.: Block sampling: efficient accurate online aggregation in mapreduce. In: CloudCom, vol. (1), pp. 250–257 (2013)
15. Wang, Y., Luo, J., Song, A., Dong, F.: Partition-Based Online Aggregation with Shared Sampling in the Cloud. J. Comput. Sci. Technol., 989–1011 (2013)
16. Qin, C., Rusu, F.: Parallel online aggregation in action. In: SSDBM, p. 46 (2013)
17. Wang, Y., Luo, J., Song, A., Dong, F.: OATS: online aggregation with two-level sharing strategy in cloud. In: Distributed and Parallel Databases, pp. 1–39 (2014)
18. Wu, M., Jermaine, C.: Guessing the extreme values in a data set: a bayesian method and its applications. VLDB J., 571–597 (2009)
19. Antoshenkov, G.: Random sampling from pseudo-ranked B+ trees. In: VLDB, pp. 375–382 (1992)
20. Chaudhuri, S., Das, G., Srivastava, U.: Effective use of block-level sampling in statistics estimation. In: SIGMOD Conference, pp. 287–298 (2004)
21. Olken, F., Rotem, D.: Random sampling from database files: a survey. In: SSDBM, pp. 92–111 (1990)
22. Haas, P.J., Koenig, C.: A Bi-level bernoulli scheme for database sampling. In: SIGMOD Conference, pp. 275–286 (2004)
23. Chaudhuri, S., Das, G., Datar, M., Motwani, R., Narasayya, V.R.: Overcoming limitations of sampling for aggregation queries. In: ICDE, pp. 534–542 (2001)
24. Acharya, S., Gibbons, P.B., Poosala, V.: Congressional samples for approximate answering of group-by queries. In: SIGMOD Conference, pp. 487–498 (2000)
25. Babcock, B., Chaudhuri, S., Das, G.: Dynamic sample selection for approximate query processing. In: SIGMOD Conference, pp. 539–550 (2003)
26. Rsch, P., Lehner, W.: Sample synopses for approximate answering of group-by queries. In: EDBT, pp. 403–414 (2009)
27. Jacobs, A.: The pathologies of big data. Commun. ACM, 36–44 (2009)
28. COLA, http://idke.ruc.edu.cn/COLA/

Computing Probability Threshold Set Similarity on Probabilistic Sets

Lei Wang, Ming Gao, Rong Zhang$^{(\boxtimes)}$, Cheqing Jin, and Aoying Zhou

Institute for Data Science and Engineering and Shanghai Key Lab for Trustworthy
Computing, East China Normal University, No. 3663, Zhongshan Rd. (N),
Shanghai 200062, China
SeiLWang@163.com, {mgao,rzhang,cqjin,ayzhou}@sei.ecnu.edu.cn

Abstract. Currently, the computation of set similarity has become an increasingly important tool in many real-world applications, such as near-duplicate detection, data cleaning and record linkage, etc., in which sets often are uncertain due to date missing, imprecise and noise, etc. The challenge of evaluating similarity between probabilistic sets mainly stems from the exponential blowup in the number of possible worlds induced by uncertainty. In this paper, we define the probability threshold set similarity (**PTSS**) between two probabilistic sets based on the possible world semantics and propose an exact solution to compute **PTSS** via the dynamic programming. To speed up the computation of the probability threshold set query (**PTSQ**), we derive an efficient and effective pruning rule for **PTSQ**. Finally, we conduct extensive experiments to verify the effectiveness and efficiency of our algorithms using both real and synthetic datasets.

1 Introduction

It is critical to identify the similarity of sets in numerous applications, such as near-duplicate detection [21][7], data cleaning [5], and record linkage [10], etc. Jaccard Coefficient (Jac.) is a common metric to evaluate similarity between two sets. Let A and B be two sets, Jaccard Coefficient between sets A and B, $\mathrm{Jac}(A, B)$, is defined as shown in Equation 1.

$$\mathrm{Jac}(A, B) = \frac{|A \cap B|}{|A \cup B|} \tag{1}$$

In these applications, sets often are uncertain due to many factors, including data missing, imprecise and noise, etc. However, it is different and difficult to evaluate the similarity between two probabilistic sets. Lian and Chen define the similarity between two probabilistic sets associated with both the set-level and element-level models [17], where the main limitations include the scalability, computational complexity and the efficiency of their proposed pruning rules [11]. Different from the existing work [11], we define the probability threshold set similarity (**PTSS**) between two probabilistic sets based on the possible world

© Springer International Publishing Switzerland 2015
J. Li and Y. Sun (Eds.): WAIM 2015, LNCS 9098, pp. 374–386, 2015.
DOI: 10.1007/978-3-319-21042-1_30

semantics. Compared with Lian and Chen's work, we exploit an exact solution to compute **PTSS** via the dynamic programming. In addition, people may be interested in whether the value of **PTSS** is larger than a pre-defined threshold. We therefore define a probability threshold set query (**PTSQ**) and drive a more efficient and effective pruning rule to speed up the processing of probability threshold set query (**PTSQ**). In summary, our model achieves a good balance between the model complexity and efficiency of similarity retrieval algorithms. The contributions of this paper are summarized as follows.

- We propose an exact solution to compute **PTSS**. The similarity value can be calculated in both polynomial time and space. Our proposed solution is a dynamic programming-based algorithm and avoids to expand all possible worlds of two probabilistic sets.
- We derive an efficient and effective pruning rule to speed up **PTSQ**. The pruning rule is not only applied in linear time-cost and constant space-cost, but also filters out a lot of probabilistic sets by analyzing the upper bound of **PTSS**.
- We conduct extensive experiments on both synthetic and real probabilistic datasets, and illustrate the efficiency and effectiveness of our proposed exact solution for **PTSS** and the pruning rule for **PTSQ**.

The remainder of paper is organized as follows. Section 2 introduces some preliminary knowledge such as data models, definitions of **PTSS** and **PTSQ**. The exact solution for **PTSS** is introduced in Section 3. In Section 4, we devise a quick-test approach to process **PTSQ**. Furthermore, we report the experimental results upon synthetic and real datasets in Section 5. In the last two sections, we review some related work and conclude the paper.

2 Preliminaries

We model a probabilistic set $\mathcal{A}$ in a discrete domain $\mathcal{D}$ as

$$\mathcal{A} = \left\{ a_i : p_{a_i} | a_i \in \mathcal{D}, \forall i \in [1, n] \right\} \tag{2}$$

where $a_i \in D$ and $p_{a_i} \in (0, 1]$ for $1 \leq i \leq n$, each element a_i associates with a probability p_{a_i}, and any two elements in the set are independent. Note that $\sum_{w \in \mathcal{W}} \Pr[w] = 1$. In fact, many applications indicate that the model is useful even it is simple [11][8]. [11] performs the set similarity query upon the same probabilistic set model. In relational probabilistic database, a relation can be modeled as a probabilistic set if each tuple of the relation is treated as an element of the probabilistic set [8].

Possible world space $\mathcal{W}$ contains a large number of possible worlds [9], where each possible world w ($w \in \mathcal{W}$) only contains a subset of elements from $\mathcal{A}$. The probability of a possible world is computed as $Pr[w] = \Pi_{t \in w} p_t \Pi_{t \notin w} (1 - p_t)$, where the probability, $\Pr[w]$, is calculated as the product of the existing probabilities of all of elements in w and the non-existing probabilities of elements out of w.

2.1 Formulation

Consider two independent probabilistic sets $\mathcal{A}$ and $\mathcal{B}$ as follows.

$$\mathcal{A} = \{a_1 : p_{a_1}, a_2 : p_{a_2}, \cdots, a_n : p_{a_n}\},$$
$$\mathcal{B} = \{b_1 : p_{b_1}, b_2 : p_{b_2}, \cdots, b_m : p_{b_m}\}.$$

The joint possible worlds of two probability sets, $\mathcal{W}(\mathcal{A}, \mathcal{B})$, combines all of possible worlds of both $\mathcal{A}$ and $\mathcal{B}$. The associated probability of a joint possible world $(w_a, w_b) \in \mathcal{W}(\mathcal{A}, \mathcal{B})$ $(w_a \in \mathcal{W}(\mathcal{A})$ and $w_b \in \mathcal{W}(\mathcal{B}))$ is $\mathsf{Pr}[w_a] \cdot \mathsf{Pr}[w_b]$.

Example 1. Given two probabilistic sets, $\mathcal{A} = \{1 : 1, 2 : 0.7\}$ and $\mathcal{B} = \{1 : 0.9, 3 : 0.5\}$, Table 1 enumerates all of joint possible worlds over $\mathcal{W}(\mathcal{A}, \mathcal{B})$. For example, the 3rd possible world consists of $\{1^{(\mathcal{A})}, 2^{(\mathcal{A})}\}$ and $\{3^{(\mathcal{B})}\}$. The probabilities of former and latter are $0.7(= 1 \times 0.7)$ and $0.05(= (1 - 0.9) \times 0.5)$, respectively. Thus, the existential probability of the joint possible world is $0.035(= 0.7 \times 0.05)$.

For each joint possible world (w_a, w_b), Jaccard Coefficient between w_a and w_b can be computed as shown in Equation 1. All of the similarities over $\mathcal{W}(\mathcal{A}, \mathcal{B})$ follows a discrete distribution. In [11], they define both **ES** and **CS** corresponding to the mean and *minconf*−quantile of the distribution, respectively. Both **ES** and **CS**, being statistics over the distribution of similarity values, may not capture the complete characteristics of the distribution. In this paper, we define the **Probability Threshold Set Similarity (PTSS)** as follows.

Definition 1 (PTSS). *Let ϕ be a similarity threshold. The **Probability Threshold Set Similarity (PTSS)** of two probabilistic sets, $\mathcal{A}$ and $\mathcal{B}$, is computed as follows*

$$PTSS(\mathcal{A}, \mathcal{B}, \phi) = Pr[w \in \mathcal{W} \mid Jac(w(\mathcal{A}), w(\mathcal{B})) \geq \phi]. \tag{3}$$

PTSS is the probability of Jaccard Coefficient of all of joint possible worlds $(w_a, w_b) \in \mathcal{W}(\mathcal{A}, \mathcal{B})$ being greater than a pre-defined threshold ϕ. The larger value of **PTSS** indicates that it is more confidential to believe that Jaccard Coefficient of a joint possible world of $\mathcal{W}(\mathcal{A}, \mathcal{B})$ is larger than ϕ.

Example 2. In Table 2, we also show the different similarity values of $\mathcal{A}$ and $\mathcal{B}$ given in Example 1. The **ES** value is the average of Jac. in Table 1 weighted by their respective probabilities. The **CS** value is the 50%−quantile of the similarity distribution over $\mathcal{W}(\mathcal{A}, \mathcal{B})$. In the 3rd row of Table 2, **PTSS** is 0.585 (= 0.315 + 0.135 + 0.135) because Jac. of the 2nd, 5th and 6th joint possible worlds in Table 1 are not less than 0.5.

In Table 2, we can also observe that **PTSS** can give us a more detailed picture of the distribution between two probabilistic sets by varying the values of ϕ.

In some applications, we may also be interested in whether the value of **PTSS** is greater than a pre-defined τ or not. Formally, we define a **Probability Threshold Set Query (PTSQ)** in terms of **PTSS** as follows.

Definition 2 (PTSQ). *Let ϕ and τ be two pre-defined parameters. The* **Probability Threshold Set Query (PTSQ)** *of two probabilistic sets, $\mathcal{A}$ and $\mathcal{B}$, is defined as follows*

$$PTSQ(\mathcal{A}, \mathcal{B}, \phi, \tau) = \mathbb{I}_{PTSS(\mathcal{A},\mathcal{B},\phi) \geq \tau}, \tag{4}$$

where $\mathbb{I}_{PTSS(\mathcal{A},\mathcal{B},\phi) \geq \tau}$ is an indicator function and defined by

$$\mathbb{I}_{PTSS(\mathcal{A},\mathcal{B},\phi)} = \begin{cases} 1, & PTSS(\mathcal{A}, \mathcal{B}, \phi) \geq \tau \\ 0, & PTSS(\mathcal{A}, \mathcal{B}, \phi) < \tau \end{cases} \tag{5}$$

That is, **PTSQ** is a boolean query which return 1 (**true**) when **PTSS** is greater than τ, otherwise 0 (**false**). For example, when $\phi = 0.3, \tau = 0.8$, $\mathcal{A}$ and $\mathcal{B}$ given in Example 1, **PTSQ**$(\mathcal{A}, \mathcal{B}, \phi, \tau)$ returns 1 because **PTSS**$(\mathcal{A}, \mathcal{B}, \phi)$ is larger than 0.8.

Table 1. Jac. of possible worlds in $\mathcal{W}(\mathcal{A}, \mathcal{B})$ **Table 2.** Different similarities of $\mathcal{A}$ and $\mathcal{B}$

w_a	w_b	$Pr[(w_a, w_b)]$	Jac.
$\{1^{(A)}, 2^{(A)}\}$	$\{1^{(B)}, 3^{(B)}\}$	0.315	0.33
$\{1^{(A)}, 2^{(A)}\}$	$\{1^{(B)}\}$	0.315	0.5
$\{1^{(A)}, 2^{(A)}\}$	$\{3^{(B)}\}$	0.035	0
$\{1^{(A)}, 2^{(A)}\}$	$\emptyset$	0.035	0
$\{1^{(A)}\}$	$\{1^{(B)}, 3^{(B)}\}$	0.135	0.5
$\{1^{(A)}\}$	$\{1^{(B)}\}$	0.135	1.0
$\{1^{(A)}\}$	$\{3^{(B)}\}$	0.015	0
$\{1^{(A)}\}$	$\emptyset$	0.015	0

metric	similarity
$ES(\mathcal{A}, \mathcal{B})$ [11]	0.46
$CS(\mathcal{A}, \mathcal{B}, minconf = 0.5)$ [11]	0.5
$PTSS(\mathcal{A}, \mathcal{B}, \phi = 0.5)$	0.585
$PTSS(\mathcal{A}, \mathcal{B}, \phi = 0.3)$	0.9

2.2 Important Symbols

Let k denote the number of identical elements between $\mathcal{A}$ and $\mathcal{B}$. Without loss of the generality, we assume the first k elements in $\mathcal{A}$ and $\mathcal{B}$ are identical, i.e., $a_i = b_i \equiv c_i$ for $1 \leq i \leq k$. Consequently, probabilistic sets $\mathcal{A}$ and $\mathcal{B}$ can be redescribed below.

$$\mathcal{A} = \{c_1 : p_{a_1}, \cdots, c_k : p_{a_k}, a_{k+1} : p_{a_{k+1}}, \cdots, a_n : p_{a_n}\}$$
$$\mathcal{B} = \{c_1 : p_{b_1}, \cdots, c_k : p_{b_k}, b_{k+1} : p_{b_{k+1}}, \cdots, b_m : p_{b_m}\}$$

Let $\mathcal{A}[i, j]$ (or $\mathcal{B}[i, j]$) be the i-th to the j-th elements of $\mathcal{A}$ (or $\mathcal{B}$), $\mathcal{A}_f^l$ (or $\mathcal{B}_f^l$, $1 \leq l \leq k$) be the first l elements of $\mathcal{A}[1, k]$ (or $\mathcal{B}[1, k]$), i.e., the first l common elements, and $\mathcal{A}_t^l$ (or $\mathcal{B}_t^l$) be the first l elements of $\mathcal{A}[k + 1, n]$ (or $\mathcal{B}[k + 1, m]$), i.e., the first l distinct elements. For convenience, $\mathcal{A}_t^{n-k}$ and $\mathcal{B}_t^{m-k}$ are shorted in $\mathcal{A}_t$ and $\mathcal{B}_t$, respectively. Similarly, $\mathcal{A}_f^k$ and $\mathcal{B}_f^k$ are shorted in $\mathcal{A}_f$ and $\mathcal{B}_f$, respectively. Some important symbols are listed in Table 3.

Table 3. Important symbols

Symbol	Description
$\mathcal{A}, \mathcal{B}$	Probabilistic sets
n, m	The number of elements in $\mathcal{A}$ or $\mathcal{B}$
w	A joint possible world instance
$w(\mathcal{A})$ $(w(\mathcal{B}))$	A set of elements in w coming from $\mathcal{A}$ ($\mathcal{B}$)
$\mathcal{A}[i,j], \mathcal{B}[i,j]$	The i-th to the j-th elements from $\mathcal{A}$ or $\mathcal{B}$
$\mathcal{A}_f^l, \mathcal{A}_t^l$	First l elements in $\mathcal{A}[1,k]$ or $\mathcal{A}[k+1,n]$
$\mathcal{A}_f, \mathcal{A}_t, \mathcal{B}_f, \mathcal{B}_t$	Abbr. of $\mathcal{A}_f^k, \mathcal{A}_t^{n-k}, \mathcal{B}_f^k$ and $\mathcal{B}_t^{m-k}$
ϕ, τ	Parameters used in **PTSS** or **PTSQ**

Table 4. $H[i,j]$

	$j=0$	$j=1$	$j=2$	$j=3$
$i=0$	0	0.015	0.05	0.035
$i=1$	0	0.135	0.45	0.315

3 The Computation of PTSS

The naive approach for exactly computing **PTSS** is to explore all of joint possible worlds based on its definition. However, the number of joint possible worlds is exponential to the sizes of two probabilistic sets. The key observation for an efficient solution is that **PTSS** only involves a few key statistics which can be computed by the dynamic programming.

3.1 Overview

As shown in Equation 6, we observe that only the joint possible worlds, whose Jaccard Coefficients are larger than given threshold ϕ, can contribute to the value of **PTSS**.

$$\text{PTSS}(\mathcal{A}, \mathcal{B}, \phi) = \sum_{(w_a, w_b) \in \mathcal{W}(\mathcal{A}, \mathcal{B})} \text{Pr}[w] \cdot \mathbb{I}_{Jac(w_a, w_b) \geq \phi}, \tag{6}$$

where $\mathbb{I}_{Jac(w_a, w_b) \geq \phi}$ is an indicator function that returns 1 if $Jac(w_a, w_b) = \frac{|w_a \cap w_b|}{|w_a \cup w_b|} \geq \phi$, otherwise 0.

In terms of Equation 6, the union size of a joint possible world ($\leq n + m - k$) cannot be larger than $\frac{i}{\phi}$ if we fix the intersection size to i. After we partition the joint possible worlds of $\mathcal{A}$ and $\mathcal{B}$ into the equivalent classes based on their intersection and union sizes, the value of **PTSS** can be computed as

$$\text{PTSS}(\mathcal{A}, \mathcal{B}, \phi) = \sum_{i=1}^{k} \sum_{j=i}^{\min(\lfloor \frac{i}{\phi} \rfloor, n+m-k)} H[i,j], \tag{7}$$

where $H[i,j] = \sum_{|w_a \cap w_b| = i \wedge |w_a \cup w_b| = j} \text{Pr}[w_a]\text{Pr}[w_b]$.

Example 3. Consider again the two probabilistic sets in Example 1, $H[i,j]$ can be easily computed and shown in Table 4. For computing $PTSS(\mathcal{A}, \mathcal{B}, 0.5)$ ($\phi = 0.5$), we only need to consider the second row of Table 4. When we fix $i = 1$ (i.e., intersection size), j (i.e., union size) cannot be larger than 2 ($\leq \frac{i}{\phi}$). Thus, $PTSS(\mathcal{A}, \mathcal{B}, 0.5) = 0.45 + 0.135 = 0.585$.

3.2   The Computation of $H[i, j]$

Now, we illustrate how to compute $H[i, j]$ in polynomial time via the dynamic programming. Since the intersection size only relies on the common elements of $\mathcal{A}$ and $\mathcal{B}$, we consider the common elements and the distinct elements separately.

For the common elements, we construct a two-dimensional table $F^l[i, j]$, $0 \leq i \leq j \leq l \leq k$, which corresponds to probability $\Pr[|w(\mathcal{A}^l_f) \cap w(\mathcal{B}^l_f)| = i \wedge |w(\mathcal{A}^l_f) \cup w(\mathcal{B}^l_f)| = j]$. Finally, $F^k[i, j]$ is the F we need for the common elements. For $\forall l \in [1, k]$, we have

$$
\begin{aligned}
F^l[i, j] = {} & F^{l-1}[i-1, j-1] p_{a_l} p_{b_l} \\
& + F^{l-1}[i, j-1]((1 - p_{a_l}) p_{b_l} + p_{a_l}(1 - p_{b_l})) \\
& + F^{l-1}[i, j](1 - p_{a_l})(1 - p_{b_l})
\end{aligned}
\tag{8}
$$

In Equation 8, four possible cases, which regard the occurrence of the common elements in $\mathcal{A}$ and $\mathcal{B}$, are considered. For example, the final part is for the case that the l-th common element does not appear in either $\mathcal{A}$ or $\mathcal{B}$. Initially, we set $F^0[0, 0] = 1$, otherwise $F^0[i, j] = 0$. We also return $F^{l-1}[i, j] = 0$ if condition $0 \leq i \leq j \leq l \leq k$ is violated.

If the current element is a distinct element, we similarly construct an one-dimensional table $G^l_{\mathcal{X}}[i]$ ($\mathcal{X}$ can be $\mathcal{A}$ or $\mathcal{B}$), which is probability $\Pr[|w(\mathcal{X}^l_t)| = i]$.

$$
G^l_{\mathcal{X}}[i] =
\begin{cases}
G^{l-1}_{\mathcal{X}}[i-1] p_l + G^{l-1}_{\mathcal{X}}[i](1 - p_l), & 0 < i \leq l \leq |\mathcal{X}_t|; \\
G^{l-1}_{\mathcal{X}}[i](1 - p_l), & i = 0; \\
0 & otherwise.
\end{cases}
$$

where $p_l = p_{a_l}$ if $\mathcal{X}$ is $\mathcal{A}$, otherwise $p_l = p_{b_l}$.

Finally, $H[i, j]$ can be computed as in Equation 9.

$$
H[i, j] = \sum_{j_c=i}^{min\{j,k\}} F^k[i, j_c] \cdot \sum_{j_a=0}^{min\{j-j_c, n-k\}} G^{n-k}_{\mathcal{A}}[j_a] \cdot G^{m-k}_{\mathcal{B}}[j - j_c - j_a],
\tag{9}
$$

where j_c is the union size of $\mathcal{A}_f$ and $\mathcal{B}_f$, j_a is the size of $\mathcal{A}_t$. Assume $n \geq m \geq k$, the time complexity of computing $H[i, j]$ is $O(k^3 + n^2)$, and the space complexity is $O(kn)$.

4   Answering a PTSQ

Given parameters ϕ and τ, we can address a **PTSQ** by directly computing **PTSS**$(\mathcal{A}, \mathcal{B}, \phi)$, and comparing the value with τ. However, the drawback is to compute the similarity value between every probabilistic set in a database and the query set. In this section, we derive an efficient and effective pruning rule by comparing the upper bound of **PTSS** with τ.

4.1   Analysis

The upper bound of **PTSS** is given in Theorem 1.

Theorem 1. *Let w denote a randomly selected possible world upon two probabilistic sets $\mathcal{A}$ and $\mathcal{B}$. Let M_1 and M_2 be two random variables, such that $M_1 = |w(\mathcal{A}) \cap w(\mathcal{B})|$ and $M_2 = |w(\mathcal{A}) \cup w(\mathcal{B})|$. For $\forall \phi \in (0,1)$, we have:*

$$Pr[M_1 \geq \phi M_2] \leq \min_{0 \leq c \leq 1} \left(\frac{E(M_1)}{\phi(1-c)E(M_2)} + \exp \frac{-c^2 E(M_2)}{2} \right). \qquad (10)$$

Proof. Let events H_1 and H_2 be

$$H_1 : M_1 \geq \phi M_2,$$
$$H_2 : M_2 \geq (1-c)E(M_2).$$

We decompose the event H_1 into two disjoint events $H_1 \cap H_2$ and $H_1 \cap \overline{H_2}$. For $\forall c \in [0,1]$, we bound their probabilities as follows.

$$Pr[H_1 \cap H_2] \leq Pr[M_1 \geq \phi(1-c)E(M_2)] \leq \frac{E(M_1)}{\phi(1-c)E(M_2)} \quad \text{(Markov inequality)}$$

$$Pr[H_1 \cap \overline{H_2}] \leq Pr[M_2 < (1-c)E(M_2)] \leq \exp \frac{-c^2 E(M_2)}{2} \quad \text{(Chernoff inequality)}$$

Thus, we can obtain Inequation 10.

4.2   Pruning Rule

Theorem 1 indicates an upper bound of **PTSS** due to the following equation.

$$Pr[M_1 \geq \phi M_2] = Pr[\frac{M_1}{M_2} \geq \phi] = Pr[Jac(\mathcal{A}, \mathcal{B}) \geq \phi].$$

Given a query probabilistic set q, we can filter out a probabilistic set $\mathcal{A}$ and do not need to compute the value of **PTSS** if the upper bound of $Pr[|w(q) \cap w(\mathcal{A})| \geq \phi|w(q) \cup w(\mathcal{A})|]$ is smaller than τ.

In Inequality 10, the values of $E(M_1)$ and $E(M_2)$ can be computed as follows.

$$E(M_1) = \sum_{i=1}^{k} p_{a_i} p_{b_i};$$

$$E(M_2) = E\big(|w(\mathcal{A})| + |w(\mathcal{B})| - |w(\mathcal{A}) \cap w(\mathcal{B})|\big)$$

$$= \sum_{i=1}^{n} p_{a_i} + \sum_{i=1}^{m} p_{b_i} - \sum_{i=1}^{k} p_{a_i} p_{b_i}.$$

It is cheap to compute the upper bound since the complexities of time and space for computing $E(M_1)$ and $E(M_2)$ are $O(m+n)$ and $O(1)$, respectively.

5   Performance

In this section, we begin to evaluate the effectiveness, efficiency and scalability of the proposed algorithm upon both synthetic and real datasets. All programs were implemented in Java, and were conducted on a dual core 64-Bit processor with 3.06 and 3.06 GHz CPUs respectively, and 128GB of RAM.

- *SYN-N*: this is a synthetic probabilistic dataset, which is a collection of probabilistic sets of N elements. The collection is induced by an original probabilistic set, whose elements randomly sample from a large universal of elements, each element associates a probability generated from a uniform distribution within the range of $[v, 0.9]$ with default value 0.2 for v. We generate a controlled probabilistic set collection by adding "noise" to the original set. We randomly remove γ percent of elements from the set, then add the same amount of random elements from the universe and assign probabilities to the added elements. Varying γ from 1% to 100%, we obtain a collection containing 100 probabilistic sets. For **PTSQ**, we consider the original probabilistic set as a query set, and the collection as a database.
- *pDBLP-N*: it is a real probabilistic dataset, which is a collection of proba-bilistic sets of N elements. The collection is also induced by an original proba-bilistic set which involves an author from DBLP. Its elements randomly select N 3−grams of the publication titles of the author, a 3−gram e associates a probability which is derived from Sigmoid function $p(e) = \frac{2}{1+exp(-c(e))} - 1$, where $c(e)$ is the frequency of e in all of publications of the author. Using the manner for constructing *SYN-N*, we construct the real probabilistic dataset.

We adopt four performance measures: (i) the elapsed time; (ii) the pruning time; (iii) the memory usage and; (iv) the pruning ratio. We measure the elapsed time and the pruning time in milliseconds, the memory usage presenting the total amount of memory used by an algorithm, and the pruning ratio is defined as $\frac{N-C}{T}$, where N, C and T are the sizes of database, candidate probabilistic sets and probabilistic sets violated the query condition.

In [17], they propose a pruning rule combined two techniques to speed up **PTSQ**, denoted as **PUBP**. We treat **PUBP** as a baseline, and compare it with our proposed pruning rule, denoted as **PTSQ**. Note that **PUBP**-r (**PUBP**-s) presents the experimental results of **PUBP** on real dataset *pDBLP-N* (synthetic dataset *SYN-N*). For **PTSS** and **PTSQ**, the notations are similar. Unless spec-ified, otherwise the default parameter settings are: $\phi = 0.5, \tau = 0.8$ and $v = 0.2$.

5.1   Efficiency of the Exact Solution

In this subsection, we begin to evaluate the efficiency of the exact solution pro-posed in Section 3.

Figure 1 illustrates the memory usage of the exact solution for computing **PTSS** on both *SYN-N* and *pDBLP-N*, where N is from 100 to 5,000. The memory usage goes up as the size of a probabilistic set increases. Note that

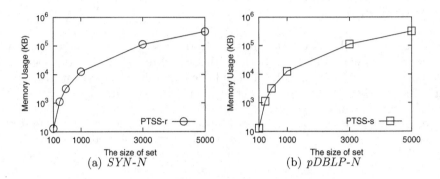

Fig. 1. The memory usage of the exact solution

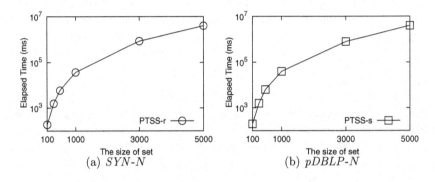

Fig. 2. The elapsed time of exact solution

the memory usage exceeds 1G bytes when there are merely five thousands of elements in two probabilistic sets.

Figure 2 illustrates the elapsed time of the exact solution for computing **PTSS** on both *SYN-N* and *pDBLP-N*. The elapsed time increases as the size of a probabilistic set increases.

5.2   Comparison with PUBP [17]

Here, we mainly compare **PTSQ** with **PUBP** as shown in Figures 3 and 4.

Figure 3 shows the scalability of **PUBP** and our proposed pruning rule for **PTSQ** on both *SYN-N* and *pDBLP-N*. We observe that our proposed pruning rule outperforms **PUBP** significantly. That is due to the factor that **PUBP** requires to construct a 2D table in memory, so that it has complexity $O(n^2 + m^2)$ for both time and space. Contrarily, our method only needs one scan of the probabilistic sets. Thus, the time and space complexities are $O(m + n)$ and $O(1)$, respectively. It means that our proposed pruning rule can support the larger probabilistic sets.

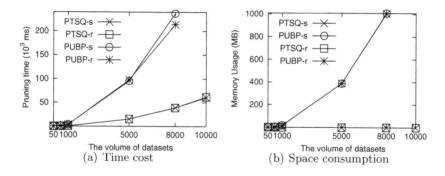

(a) Time cost   (b) Space consumption

Fig. 3. The scalability of our proposed pruning rule VS. **PUBP**

Figure 4 illustrates that the pruning ratios of **PUBP** and our proposed pruning rule on both *SYN-N* and *pDBLP-N*. In Figures 4(a) and (c), the pruning ratios go up as the value of ϕ increases. This is due to the fact that the larger value of ϕ results in a smaller value of **PTSS**. Comparing to τ, a smaller **PTSS** further leads to a higher pruning ratio. In Figures 4(b) and (d), the pruning ratios go down for **PUBP** when the set sizes increase. This is due to the factor that two probabilistic sets with larger sizes have larger upper bound given by **PUBP**. On the contrary, the upper bound given by **PTSQ** goes down as the sizes of two probabilistic sets increase. From Figure 4, we can conclude that our proposed approach outperforms **PUBP** significantly.

6   Related Work

Similarity Search returns objects in a set of data collection which are similar or close to the query object, such as top-k, K-NN, skyline, and range query, etc. In the determinate scenario, there are too many works focused on this topic [2][13][3], etc. This topic is hot in the uncertain data[12][19] [24][6][15]. The key point is how to evaluate an object similar or close to the other ones in different semantics. In another words, it is significant to evaluate the similarity or distance between two uncertain objects. For two images or PDFs presented by histogram, we can measure their distance by Earth Mover's Distance (EMD)[20]. For the string set data, the paper [14] proposed expected edited distance to measure the distance between two probabilistic strings. And the paper [17] proposed probability threshold set similarity for Jaccard index.

Similarity Join is based on the similarity metric between two objects stored in the relations. As a fundamental query, similarity join has been extensively studied in many applications with different similarity metrics, such as the Euclidean distance between two spatial objects [4], the Edit distance for two strings [22], the Earth Mover's Distance between two PDFs[23], the Jaccard similarity and Hamming distance for two sets [1], etc. The most common technique is the index-based join solution which constructs index structure built on both

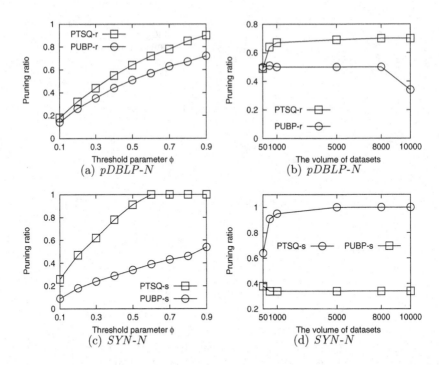

Fig. 4. The pruning ratios of PTSQ and PUBP

relations. Actually, this technique employs the framework of pruning and verification, in the stage of pruning, some indexes, inverted indexes and approximate computing the distance between two objects. However, the problem becomes more complex in the uncertain data. There are some works, which studied the similarity join over probabilistic data, such as Euclidean distance between spatial objects [18], probabilistic distance functions for uncertain data[16], expected edit distance for probabilistic strings [14], Jaccard distance for probabilistic sets[17]. In the work of answering string similarity join over string-level and character-level uncertain model. And the work of [17] responded the set similarity join over set-level and element-level uncertain model. However, for uncertain set, the paper[17] was the only work related to it, which studied the PUBP over probabilistic sets. Furthermore, only the second pruning rule proposed in that paper could work for PUBP between two probabilistic sets. What's more important, that paper did not exactly compute the similarity between two probabilistic sets.

7   Conclusion

The importance of probabilistic similarity queries has been recognized in many applications, such as near-duplicate detection, data cleaning and record linkage, etc. In this paper, we present an exact solution for computing the probabilistic threshold set similarity and propose an efficient and effective pruning rule for speeding up the probabilistic threshold set query. The experimental results on both synthetic and real probabilistic datasets show the effectiveness and efficiency of our proposed solutions. Future work includes handling streaming data, supporting other uncertain data models.

Acknowledgments. This work is supported by the National Basic Research Program (973) of China (No. 2012CB316203) and NSFC under Grant No. 61232002 and 61402177.

References

1. Arasu, A., Ganti, V., Kaushik, R.: Efficient exact set-similarity joins. In: VLDB, pp. 918–929,(2006)
2. Bharambe, A.R., Agrawal, M., Seshan, S.: Mercury: supporting scalable multi-attribute range queries. In: SIGCOMM, pp. 353–366 (2004)
3. Börzsönyi, S., Kossmann, D., Stocker, K.: The skyline operator. In: ICDE, pp. 421–430 (2001)
4. Brinkhoff, T., Kriegel, H.-P., Seeger, B.: Efficient processing of spatial joins using r-trees. In: SIGMOD Conference, pp. 237–246 (1993)
5. Chaudhuri, S., Ganti, V., Kaushik, R.: A primitive operator for similarity joins in data cleaning. In: ICDE, p. 5 (2006)
6. Cheema, M.A., Lin, X., Wang, W., Zhang, W., Pei, J.: Probabilistic reverse nearest neighbor queries on uncertain data. IEEE Trans. Knowl. Data Eng. **22**(4), 550–564 (2010)
7. Chum, O., Philbin, J., Isard, M., Zisserman, A.: Scalable near identical image and shot detection. In: Proc. of CIVR (2007)
8. Dalvi, N.N., Suciu, D.: Efficient query evaluation on probabilistic databases. The VLDB Journal **16**(4), 523–544 (2007)
9. Dalvi, N.N., Suciu, D.: Management of probabilistic data: foundations and challenges. In: Proc. of ACM PODS, pp. 1–12 (2007)
10. Dong, X.L., Halevy, A.Y., Yu, C.: Data integration with uncertainty. In: VLDB, pp. 687–698 (2007)
11. Gao, M., Jin, C., Wang, W., Lin, X., Zhou, A.: Similarity query processing for probabilistic sets. In: 29th IEEE International Conference on Data Engineering, ICDE 2013, Brisbane, Australia, April 8–12, pp. 913–924 (2013)
12. Hua, M., Pei, J., Zhang, W., Lin, X.: Ranking queries on uncertain data: a probabilistic threshold approach. In: SIGMOD Conference, pp. 673–686 (2008)
13. Ilyas, I.F., Aref, W.G., Elmagarmid, A.K.: Supporting top-k join queries in relational databases. VLDB J. **13**(3), 207–221 (2004)
14. Jestes, J., Li, F., Yan, Z., Yi, K.: Probabilistic string similarity joins. In: SIGMOD Conference, pp. 327–338 (2010)
15. Jin, C., Yi, K., Chen, L., Yu, J.X., Lin, X.: Sliding-window top-k queries on uncertain streams. PVLDB **1**(1), 301–312 (2008)
16. Kriegel, H.-P., Kunath, P., Pfeifle, M., Renz, M.: Probabilistic similarity join on uncertain data. In: Li Lee, M., Tan, K.-L., Wuwongse, V. (eds.) DASFAA 2006. LNCS, vol. 3882, pp. 295–309. Springer, Heidelberg (2006)

17. Lian, X., Chen, L.: Set similarity join on probabilistic data. In: Proc. of VLDB (2010)
18. Ljosa, V., Singh, A.K.: Top-k spatial joins of probabilistic objects. In: ICDE, pp. 566–575 (2008)
19. Pei, J., Jiang, B., Lin, X., Yuan, Y.: Probabilistic skylines on uncertain data. In: VLDB, pp. 15–26 (2007)
20. Rubner, Y., Tomasi, C., Guibas, L.J.: The earth mover's distance as a metric for image retrieval. International Journal of Computer Vision 40(2), 99–121 (2000)
21. Theobald, M., Siddharth, J., Paepcke, A.: Spotsigs: robust and efficient e detection in large web collections. In: Proc. of ACM SIGIR, pp. 563–570 (2008)
22. Xiao, C., Wang, W., Lin, X.: Ed-join: an efficient algorithm for similarity joins with edit distance constraints. PVLDB 1(1), 933–944 (2008)
23. Xu, J., Zhang, Z., Tung, A.K.H., Yu, G.: Efficient and effective similarity search over probabilistic data based on earth mover's distance. PVLDB 3(1), 758–769 (2010)
24. Yi, K., Lian, X., Li, F., Chen, L.: A concise representation of range queries. In: ICDE, pp. 1179–1182 (2009)

Region-aware Top-k Similarity Search

Sitong Liu$^{(\boxtimes)}$, Jianhua Feng, and Yongwei Wu

Department of Computer Science and Technology, Tsinghua University,
Beijing, China
liu-st10@mails.tsinghua.edu.cn, {fengjh,wuyw}@tsinghua.edu.cn

Abstract. Location-based services have attracted significant attention for the ubiquitous smartphones equipped with GPS systems. These services (e.g., Google map, Twitter) generate large amounts of spatio-textual data which contain both geographical location and textual description. Existing location-based services (LBS) assume that the attractiveness of a Point-of-Interest (POI) depends on its spatial proximity from people. However, in most cases, POIs within a certain distance are all acceptable to users and people may concern more about other aspects. In this paper, we study a region-aware top-k similarity search problem: given a set of spatio-textual objects, a spatial region and several input tokens, finds k most textual-relevant objects falling in this region. We summarize our main contributions as follows: (1) We propose a hybrid-landmark index which integrates the spatial and textual pruning seamlessly. (2) We explore a priority-based algorithm and extend it to support fuzzy-token distance. (3) We devise a cost model to evaluate the landmark quality and propose a deletion-based method to generate high quality landmarks (4) Extensive experiments show that our method outperforms state-of-the-art algorithms and achieves high performance.

1 Introduction

With the popularity of global position systems (GPS) in smartphones, location-based services (LBS) have recently attracted significant attention from both academic and industrial communities. These services generate large amounts of spatio-textual data which contain both geographical location and textual description. Traditional LBS services assume that people are more interested in the POIs around them. That is, the attractiveness of a POI depends on its spatial proximity from people. However, in most cases, POIs within a certain distance are all acceptable to users and they may concern more about other aspects. For example, a hospital around 2km may be more attractive than closer ones if it is more accordance with users' requirements.

In this paper, we study a region-aware top-k similarity search problem, which, given a set of spatio-textual objects, a spatial region and several tokens, finds k most textual-relevant objects falling in this region. This problem can be applied to many existing LBS systems. For example, in Google map, users are allowed to adjust their interesting regions through "zoom-in", "zoom-out" and "move" operation. No matter which position and size the selected region is, we always

© Springer International Publishing Switzerland 2015
J. Li and Y. Sun (Eds.): WAIM 2015, LNCS 9098, pp. 387–399, 2015.
DOI: 10.1007/978-3-319-21042-1_31

display k objects which locate in this region and are most relevant to the query tokens. Different from traditional spatial-keyword search problem, consider the possibility that users' input may contain some typo-errors or misspellings, we use a fuzzy-token metric [10] to measure the textual relevancy. To the best of our knowledge, none of current solutions can efficiently support both top-k requirement and fuzzy-token distance. [1,14] focus on range-based spatial search and [5,11,13] focus on top-k textual similarity search. Although we can combine these two kinds of methods to support our problem (Section 2.3), they are quite inefficient since they use the spatial and textual pruning separately. LBAK-tree [1] and IRtree [4] focus on spatial keyword search. They embed textual index into each node of R-tree. However, LBAK-tree cannot address "top-k" and IRtree cannot support "fuzzy" (e.g., edit distance). To address the limitation, we propose a hybrid-landmark index (HLtree) which generates high quality landmarks to dynamically divide objects into hierarchical clusters. We further devise a priority-based method to efficiently support fuzzy-token distance.

We summarize our main contributions as follows: (1) We propose a hybrid-landmark index which integrates the spatial and textual pruning seamlessly. (2) We explore a priority-based algorithm and extend it to support fuzzy-token distance. (3) We devise a cost model to evaluate the landmark quality and propose a deletion-based method to generate high quality landmarks. (4) Extensive experiments show that our method outperforms state-of-the-art algorithms and achieves high performance.

The rest of this paper is organized as follows. We formulate our problem in Section 2 and introduce HLtree in Section 3. Section 4 discusses the landmark selection. We show experiments in Section 5 and make a conclusion in Acknowledgement.

2 Preliminaries

We first formulate the problem of region-aware top-k similarity search in Section 2.1, and then show some related work in Section 2.2. We extend state-of-the-art algorithms to support our problem in Section 2.3.

2.1 Problem Statements

Consider a collection of objects $\mathcal{R} = \{r_1, r_2, \ldots, r_{|\mathcal{R}|}\}$. Each object $r \in \mathcal{R}$ includes a spatial location $\mathcal{L}_r$ and textual description $\mathcal{T}_r$, denoted by $r = \{\mathcal{L}_r, \mathcal{T}_r\}$. We use the coordinate of an object to describe its spatial location, denoted by $\mathcal{L}_r = [\mathcal{L}_r.x, \mathcal{L}_r.y]$. And we use a set of tokens to capture the textual description, denoted by $\mathcal{T}_r = \{t_1, t_2, \ldots, t_{|\mathcal{T}_r|}\}$, which describes an object (e.g., Gym, Hotel, Restaurant) or users' interests (e.g., jogging, yoga, pilates). A region-aware top-k similarity query $q = \{\mathcal{M}_q, \mathcal{T}_q, k\}$ includes a spatial region $\mathcal{M}_q$, textual description $\mathcal{T}_q$ and a top-k parameter k. The spatial region can be retrieved through the screen of a computer or a phone. Since users can adjust the position and size of the region using "zoom-in", "zoom-out" and "move" operation, we only consider the textual similarity of objects falling in this region.

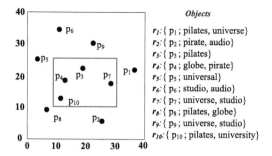

Fig. 1. An example of spatio-textual objects

Inspired by Wang et al [10], we combine the character-based distance and token-based similarity to quantify the textual distance. There are two advantages: (1) Character-based distance (e.g., edit distance) can satisfy error-tolerant requirement. Sometimes query may contain typos and misspellings. (2) Token-based similarity (e.g., overlap) helps us find the potential relationship of tokens regardless of the order. Each token can be considered as a separate condition (e.g, Jogging, Yoga). By mapping tokens between the query and data, we are able to measure how well the user's requirements are satisfied by a record. Thus, in this paper, we combine the edit distance and the overlap metric as a fuzzy-token distance (Definition 1). Based on this definition, we define the region-aware top-k similarity search problem (Definition 2).

Definition 1 (Fuzzy-token distance). *Given two sets of tokens $\mathcal{T}_r = \{t_1, t_2, \ldots, t_m\}$ and $\mathcal{T}_s = \{s_1, s_2, \ldots, s_n\}$. For each token $t_i \in \mathcal{T}_r$, we find its most similar token s_j in $\mathcal{T}_s$ which has the minimum edit distance, i.e., $\forall s \in \mathcal{T}_s, Ed(t_i, s_j) \leq Ed(t_i, s)$. For all those token pairs (t_i, s_j), we take the sum of these edit distance as the fuzzy-token distance between $\mathcal{T}_r$ and $\mathcal{T}_s$, denoted by $Dis_{ft}(\mathcal{T}_r, \mathcal{T}_s)$.*

Definition 2 (Region-aware top-k similarity search). *Given a collection of objects $\mathcal{R}$ and a query $q = \{\mathcal{M}_q, \mathcal{T}_q, k\}$, a region-aware top-k similarity search finds a subset of objects $\mathcal{S} = \{r_1, r_2, \ldots, r_k\}$ which satisfy: (1) $\mathcal{S} \subseteq \mathcal{R}$ and $|\mathcal{S}| = k$, (2) $\forall r \in \mathcal{S}$, we have $\mathcal{L}_r \in \mathcal{M}_q$ (3) $\forall r_i \in \mathcal{S}, \forall r_j \in (\mathcal{R} - \mathcal{S}) \cap \mathcal{M}_q$, we have $Dis_{ft}(\mathcal{T}_{r_i}, \mathcal{T}_q) \leq Dis_{ft}(\mathcal{T}_{r_j}, \mathcal{T}_q)$.*

Example 1. Consider the ten spatio-textual objects in Figure 1. Suppose query q is $\{[(10, 8), (30, 24)], (\text{"piate"}, \text{"universt"}), 1\}$. Then objects $\{r_3, r_4, r_7, r_{10}\}$ fall in the query region. We compare their fuzzy-token distance. Take r_{10} as an example. Its token set $\mathcal{T}_{r_{10}}$ is $\{\text{"pilates"}, \text{"university"}\}$. For the first query token "piate", we calculate its edit distance to each token in $\mathcal{T}_{r_{10}}$. Then we have $Ed(\text{"piate"}, \text{"pilates"}) = 2$ and $Ed(\text{"piate"}, \text{"university"}) = 9$. Thus, its best match token is "pilates" since they have the minimum edit distance 2. For the second query token "universt", its best match token in $\mathcal{T}_{r_{10}}$ is "university" where $Ed(\text{"universt"}, \text{"university"}) = 3$. Thus, we add them to get its fuzzy-token distance $Dis_{ft}(\mathcal{T}_q, \mathcal{T}_{r_{10}}) = 2 + 3 = 5$. Similarly, we have $Dis_{ft}(\mathcal{T}_q, \mathcal{T}_{r_3}) = 10$, $Dis_{ft}(\mathcal{T}_q, \mathcal{T}_{r_4}) = 11$ and $Dis_{ft}(\mathcal{T}_q, \mathcal{T}_{r_7}) = 10$. Thus, r_{10} is the top-1 result of q.

2.2   Related Work

Range-Based Spatial Search. Many well-known data structures such as R-tree [8] can address this problem. It organizes objects in a hierarchical way. When a query q comes, it iteratively visits nodes from the root using a top-down pattern. At each node, only those child nodes whose MBRs have intersection with $\mathcal{M}_q$ can be added to the visiting list for further extension.

Spatial Keyword Search. There are many studies on spatial keyword search [1,4,7,14,16]. [3] gives an overall experimental analysis. [4,7] focus on the top-k topic. They embed inverted index or signature files to every node of R-tree. [1,14,16] focus on the range topic. [16] combines the inverted lists and Rtree in different orders. [1,14] support fuzzy search. These methods cannot address our problem directly. For example, the language model in IRtree based on the exact match between tokens. It cannot support fuzzy search at the character level (e.g., edit distance).

Top-k Textual Similarity Search. Though many works have studied threshold-based string similarity search [2,9,15], few focus on the "top-k" topic [5,11,13]. Two kinds of methods are explored recently: (1) Trie-based method [5] utilizes a trie index to avoid calculating edit distance repeatedly for tokens sharing the same prefix. It first builds a trie for all the tokens and then incrementally search the trie nodes. At the beginning, we find all the trie nodes n which can completely match any query prefix p, i.e., $Ed(n,p) = 0$ and $p \in \mathcal{T}_q$. At each step, we iteratively extend pairs (n,p) to get the pairs (n',p') satisfying $Ed(n',p') = Ed(n,p) + 1$. (2) [11,13] propose a q-gram based method. We define "q-gram" in Definition 3. For example, the 3-gram set of "Jogging" is {"Jog","ogg","ggi","gin","ing"}, denoted by $Q_{\text{"Jogging"}}$. According to Lemma 1, edit distance evaluation can be easily transformed to counting the overlap of q-grams [13]. Thus, they organize q-grams using an inverted index and only objects shareing at least $|\mathcal{T}_q| - q + 1 - q \times \tau$ q-grams with the query can become candidates.

Definition 3 (q-gram). *Given a token t and a positive integer q, a q-gram of t is a pair (i,s), where s is a q-length substring which starts from position i in t, i.e., $s = t[i, i+q-1]$. By sliding a window of length q over the characters of t, we get a q-gram set for token t, denoted by Q_t.*

Lemma 1. *Consider two token s and t, if $Ed(s,t) < \tau$, we have $|Q_s \cap Q_t| \geq \max(|s|, |t|) - q + 1 - q \times \tau$.*

2.3   Straightforward Solutions

Based on the two prevalent techniques in Section 2.2, we embed the trie tree and q-gram based inverted index into Rtree to address our problem: (1) Rtree+trie: We first build a R-tree for the spatial components of all the objects. Then in each leaf node, we build a trie tree for all the objects falling in this region. Given a query q, we first traverse Rtree to locate those leaf nodes which have overlap with $\mathcal{M}_q$ and then search each embedded trie to find top-k similar objects. Finally we combine these candidates to get final results. (2) Rtree+q-gram: Similar to the

---

Algorithm 1. HLtree construction $(\mathcal{R}, M)$

Input: $\mathcal{R}$: An object set

$\qquad M$: The capacity limitation of each node

Output: *root*: A constructed HLtree

```
1  begin
2      root ← a new HLtree node; Q ← an empty queue;
3      Q.push(⟨root, R⟩);
4      while Q is not empty do
5          ⟨n, R_n⟩ ← Q.top(); Q.pop();
6          if |R_n| < M then
7              BuildLeafAndQgram(R_n);

8          else
9              n.L ← GenerateLandmark(R_n);
10             P ← |n.L| empty object lists;
11             for each r ∈ R_n do
12                 l_r ← SelectLandmark(n.L);
13                 add r to P_{l_r};
14                 update l_r;

15             for each l ∈ n.L do
16                 n_l ← a new tree node;
17                 n.child.add(n_l)
18                 Q.add(⟨n_l, P_l⟩);

19     return root;
20 end
```

Fig. 2. Construction of a hybrid-landmark tree

former method, we build a q-gram based inverted index instead of trie. Given a query q, we search R-tree using $\mathcal{M}_q$ to locate the leaf nodes. We then divide $\mathcal{T}_q$ into q-gram sets and probe the corresponding inverted lists. By utilizing the heap structure, we can quickly calculate q-gram overlaps and select the objects appearing in more than $|\mathcal{T}_q| - q + 1 - q \times \tau$ inverted lists as candidates.

3   Hybrid-Landmark Tree

In this section, we introduce a hybrid-landmark tree called HLtree. We first introduce the basic idea in Section 3.1 and then show the construction of HLtree in Section 3.2. The query processing methods will be presented at Section 3.3.

3.1   Basic Idea

The two methods in Section 2.3 have several drawbacks: (1) Trie-based method incrementally extends the pair $Ed(n, p) = x$ to get the pairs $Ed(n', p') = x + 1$. Suppose the query is "restroom" and current edit distance threshold is 3. Even if $Ed(\text{"restaurant"}, \text{"restroom"}) = 5 > 3$, the algorithm still need to visit its prefix node "restaur" since $Ed(\text{"restaur"}, \text{"restr"}) = 2 < 3$. Thus, all the prefixes whose edit distance is less than the k_{th} result will be visited. (2) Q-gram based method uses $Q_{\mathcal{T}_q}$ to probe the inverted index. Though some q-gram sets have few intersection with $Q_{\mathcal{T}_q}$, they may still be visited to exactly count the overlap. (3) For both methods, the spatial and textual components are organized separately which may involve large numbers of false-positives. When locating leaf nodes with the query region, if the intersection part is too small compared to the leaf

MBR, the algorithms will waste much time in calculating the textual similarity for non-intersecting candidates. Besides, if tokens distribute uniformly, then each leaf node contains similar textual index. When the query region is large, it will repeatedly calculate textual similarity for every intersecting leaf node. However, these calculations can be reduced to only once if we utilize the textual index at a higher level. To overcome these drawbacks, we propose a hybrid-landmark based index called HLtree. The basic idea is to dynamically select a set of spatial-textual landmarks which can best represent the distribution of objects. At each level, we divide objects into partitions according to the hybrid proximity to the landmarks. In this way, similar objects can be mapped into the same group while dissimilar ones are put separately. Given a query q, we only need to compare q with these selected landmarks to avoid visiting extremely dissimilar groups. We first introduce the concept of the hybrid landmark.

Definition 4 (Hybrid Landmark). *Given a partition p with a set of objects $\mathcal{R}_p$, its corresponding hybrid landmark L_p is a triple $\langle t, mbr, rads \rangle$ where t is a selected token, mbr is the MBR of $\mathcal{R}_p$ and $rads$ is the minimum and maximum edit distance of all the objects in $\mathcal{R}_p$ compared with t, i.e., $L_p.rads = [\min_{r \in \mathcal{R}_p} (L_p.t, \mathcal{T}_r), \max_{r \in \mathcal{R}_p} (L_p.t, \mathcal{T}_r)]$, denoted by $[L_p.rads.l, L_p.rads.r]$.*

3.2 Construction of HLtree

HLtree Construction. We build HLtree in a top-down pattern. As Algorithm 1 shows, we use a queue Q to maintain current generated partitions (nodes) and then iteratively divide them into smaller ones. At the beginning, all the objects $\mathcal{R}$ are taken as a whole partition (the root node). We then add the root node and its corresponding object list to Q for further partition (Line 3). At each step, we retrieve the top element $\langle n, \mathcal{R}_n \rangle$ from Q where $\mathcal{R}_n$ is the object list corresponding to node n. We use M to limit the capacity of each node. If $\mathcal{R}_n$ contains less than M objects, we mark n as a leaf node and build a q-gram based inverted index (Line 7). Otherwise, we generate a set of high quality landmarks $n.\mathcal{L}$ which can best represent the distribution of $\mathcal{R}_n$ (The details of generation method will be presented in Section 4). We initiate $|n.\mathcal{L}|$ empty object lists to keep newly-generated partitions. For each object r in $\mathcal{R}_n$, we select the nearest landmark l_r and add r to its corresponding object list $\mathcal{P}_{l_r}$ (Line 11-14). Finally, for each landmark l in $n.\mathcal{L}$, we create a new node and take it as a child of n. We also add pair $\langle n_l, \mathcal{P}_{l_r} \rangle$ to queue Q for further partition (Line 15-18).

Q-gram Prefix Based Index Construction. A main challenge is to avoid completely counting the overlaps when searching the q-gram based inverted lists. Inspired by [12], we explore a q-gram prefix based method. Consider two tokens $\mathcal{T}_r$ and $\mathcal{T}_s$ with q-gram sets $Q_{\mathcal{T}_r} = \{g_1, g_2, \ldots, g_{|\mathcal{T}_r|-q+1}\}$ and $Q_{\mathcal{T}_s} = \{g'_1, g'_2, \ldots, g'_{|\mathcal{T}_s|-q+1}\}$. Suppose q-grams have beed sorted according to a global ordering in advance (e.g., tf-idf). As Lemma 1 shows, if $Ed(\mathcal{T}_r, \mathcal{T}_s) \leq \tau$, we have $|Q_{\mathcal{T}_r} \cap Q_{\mathcal{T}_s}| \geq \max(|\mathcal{T}_r|, |\mathcal{T}_s|) - q + 1 - q \times \tau$. Even if we cut off the last $|\mathcal{T}_r| - q - q \times \tau$ q-grams from $Q_{\mathcal{T}_r}$ and the last $|\mathcal{T}_s| - q - q \times \tau$ q-grams from $Q_{\mathcal{T}_s}$, the remaining part of $Q_{\mathcal{T}_r}$ and $Q_{\mathcal{T}_s}$ should still have intersection to

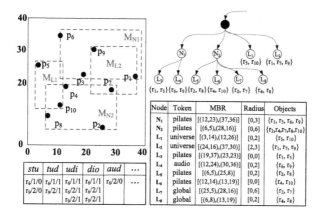

Fig. 3. Construction of a HLtree

preserve the overlap constraint, i.e., $Q_{\mathcal{T}_r}[1..q\tau+1] \cap Q_{\mathcal{T}_s}[1..q\tau+1] \neq \phi$. Thus, if $Q_{\mathcal{T}_r}[1..q\tau+1] \cap Q_{\mathcal{T}_s}[1..q\tau+1] = \phi$, $Ed(\mathcal{T}_r, \mathcal{T}_s)$ must be larger than τ. Then for any q-gram prefixes $Q_{\mathcal{T}_r}[1..i]$ and $Q_{\mathcal{T}_s}[1..j]$, if $Q_{\mathcal{T}_r}[1..i] \cap Q_{\mathcal{T}_s}[1..j] = \phi$, we can deduce a lower bound of τ as Lemma 2. Based on this analysis, we develop our q-gram based inverted index. Given an object r, for each token t in $\mathcal{T}_r$, we generate its q-gram set $\mathcal{Q}_t = \{g_1, g_2, \ldots, g_{|t|-q+1}\}$. For each q-gram $g_i \in \mathcal{Q}_t$, we add triple $(rid, tid, \lceil \frac{i-1}{q} \rceil)$ to the inverted lists corresponding to g_i. rid is the id of record and tid is the token position in $\mathcal{T}_r$. $\lceil \frac{i-1}{q} \rceil$ indicates the minimum bound of edit distance if no candidates have intersection with $\mathcal{Q}_t[1..i-1]$ (Lemma 2).

Definition 5 (Q-gram prefix). *Given a q-gram sets $Q_{\mathcal{T}_r} = \{g_1, g_2, \ldots, g_{|\mathcal{T}_r|-q+1}\}$. Any subset $\{g_1, g_2, \ldots, g_i\}$ is called a q-gram prefix of $Q_{\mathcal{T}_r}$, denoted by $Q_{\mathcal{T}_r}[1..i]$.*

Lemma 2. *Consider two q-gram prefixes $Q_{\mathcal{T}_r}[1..i] = \{g_1, g_2, \ldots, g_i\}$ and $Q_{\mathcal{T}_s}[1..j] = \{g'_1, g'_2, \ldots, g'_j\}$. If $Q_{\mathcal{T}_r}[1..i-1] \cap Q_{\mathcal{T}_s}[1..j-1] = \phi$, we have $\tau \geq \lceil \frac{\max(i-1, j-1)}{q} \rceil$.*

Example 2. Consider objects in Figure 1. Suppose the node capacity is 3 ($M = 3$). At first, all the objects are taken as a whole partition. We generate four landmarks $\{\langle$ "pilates", $M_{N_1}, [0,3]\rangle, \langle$ "pilates", $M_{N_2}, [0,6]\rangle, \langle$ "universe", $M_{L_1}, [0,2]\rangle, \langle$ "universe", $M_{L_2}, [2,3]\rangle\}$. For each object, we find proper landmarks for all its tokens. Take r_7 as an example, $\mathcal{T}_{r7} = \{$ "universe", "studio"$\}$ and p_{r7} locates in MBR M_{N_2} and M_{L_2}. Since $Ed($ "universe", $N_2.t) = 7$ and $Ed($ "universe", $L_2.t) = 0$, we map $(r_7,$ "universe"$)$ to node L_2. Similarly, we map $(r_7,$ "studio"$)$ to N_2. After allocation, L_1 and L_2 contain no more than 3 objects and are taken as leaf nodes. For each token in leaf node, we generate its q-gram set and update the inverted lists. For example, $Q_{r6}[$ "studio"$]$ contains the q-gram "stu", then we add $\langle r_6/1/0 \rangle$ to the inverted list of "stu" where 1 is the token position in r_6 and 0 is the estimated edit distance in Lemma 2. For nodes contain more than 3 objects, we continue this procedure.

3.3   Query Processing of Hybrid-Landmark Tree

We first discuss the algorithm for the query containing only one token and then extend it to support the query with multiple tokens.

Algorithm for Single-token Query. Similar to all the hierarchical structures, HLtree explores a priority-first traverse strategy. A priority queue Q is used to keep the priority of each node. At first, Q is empty and we add $\langle root, 0 \rangle$. We then iteratively retrieve the top element n from Q. If n is a leaf node, we search its q-gram based inverted index. Otherwise, for each child node, we check its MBR intersection and estimate its visiting priority. Child nodes which pass the filtering rules are added to Q for further extension. Since spatial constraint can be checked in $O(1)$ time, we use the textual distance as the visiting priority. Because T_q only contains one token, then the textual metric is simplified to edit distance which follows the triangle inequality: given three tokens T_a, T_b, T_c, we have $|Ed(T_a, T_b) - Ed(T_b, T_c)| \leq Ed(T_a, T_c) \leq Ed(T_a, T_b) + Ed(T_b, T_c)$. Thus, in partition l_i, only the token t satisfying $|Ed(l_j.t, T_q) - Ed(T_q, t)| \leq Ed(t, l_j.t) \leq Ed(l_j.t, T_q) + Ed(T_q, t)$ can be the answer. We use the current k_{th} best token to estimate $Ed(T_q, t)$. That is, only the tokens between $\max\{0, Ed(l_j.t, T_q) - k_{th}\}$ and $\min\{l_j.rads.r, Ed(l_j.t, T_q) + k_{th}\}$ need to be visited. We can deduce a tighter pruning. Consider two landmarks l_i and l_j. If T_q is quite closer to $l_i.t$ than to $l_j.t$, i.e., $Ed(T_q.t, l_i.t) \ll Ed(T_q.t, l_j.t)$, then the token near T_q cannot appear in partition l_j since it is only mapped to the nearest landmark. Thus, we can deduce a safety edit distance bound for each partition l_j as Lemma 3. Once the current k_{th} result is less than $\max_i\{\frac{Ed(T_q, l_j.t) - Ed(T_q, l_i.t)}{2}\}$, we can prune the whole partition l_j.

Another problem is how to efficiently find results in leaf n. As Lemma 2 shows, each q-gram prefix corresponds to an edit distance lower bound. We utilize this property to avoid visiting tokens with low values. Suppose the query q-gram set Q_{T_q} is $\{g_1, g_2, \ldots, g_{|T_q|-q+1}\}$. Consider a q-gram starting from position i, if its previous q-grams $\{g_1, g_2, \ldots, g_{i-1}\}$ have intersection with any top-k result, we can find it when visiting the inverted lists of $g_1, g_2, \ldots, g_{i-1}$. Otherwise, the edit distance of unvisited objects can not be less than $\lceil \frac{i-1}{q} \rceil$. Thus, we use a priority queue Q_g to keep candidate q-grams and their estimated values. At first, for each q-gram in Q_{T_q}, we add $\langle g_i, \lceil \frac{i-1}{q} \rceil \rangle$ to Q_g as a non-visited q-gram. At each time, we retrieve the q-gram g_t with minimum estimated value from Q_g. If g_t is from the query, then the next inverted list has not been explored yet. We find the inverted list corresponding to g_t and add the first element to Q_g. Otherwise, we calculate its real edit distance and update current results. Then we update Q by adding the next element in the same inverted list.

Lemma 3. *Consider any two landmarks l_i, l_j and a query q. Token t which satisfies $Ed(T_q, t) < \frac{Ed(T_q, l_j.t) - Ed(T_q, l_i.t)}{2}$ can not appear in landmark l_j. That is, we can take $\max_i\{\frac{Ed(T_q, l_j.t) - Ed(T_q, l_i.t)}{2}\}$ as a lower bound of partition l_j.*

Algorithm for Multiple-token Query. Inspired by the TA algorithm [6], we extend our framework to support query with multiple tokens. Suppose the query is $T_q = \{t_1, t_2, \ldots, t_m\}$. Based on the method above, for each token t_i, we incrementally find the next object r which is closest to t_i under the edit distance metric. That is, we can dynamically defer an implicit visiting list for each token t_i according to the edit distance proximity, denoted by $\mathcal{T}_q = \{\mathcal{O}_1, \mathcal{O}_2, \ldots, \mathcal{O}_m\}$. Each object r in $\mathcal{O}_i$ corresponds a score $Ed(\mathcal{T}_r, t_i)$. We then alternatively scan these object lists and combine the candidates. The method is similar to the previous one (search in q-gram based inverted lists). We use a priority queue Q_{ta} to keep the processing objects in each list and their corresponding scores. At the beginning, we add the first object in each list $\mathcal{O}_i$ and its similarity to Q_{ta}. We then iteratively retrieve the top element from Q_{ta} and update the results by calculating its fuzzy-token distance. For each popped token, we incrementally find its next closest object and add it to Q_{ta}. Notice that we do not need to calculate the entire ordering of each token. Similar to TA, once the sum of visiting frontiers (i.e, elements in Q_{ta}) is larger than the current k_{th} result, we can stop the algorithm. For constraint of space, we omit the details.

4   High Quality Hybrid-Landmark Selection

We first analyze the effectiveness of a landmark selection. At each node n, we either take n as a leaf node or divide it into smaller parts. The benefit of dividing n depends on the ratio of the query time reduction and the additional space consumption (Equation 1). If n is a leaf node, the query time at n is to search its q-gram inverted index, i.e., $c_R * |\mathcal{R}(n)|$ where c_R is the cost to refine an object and $|\mathcal{R}(\cdot)|$ is the quantity of visiting objects. Otherwise, suppose n is divided into m parts, denoted by $\{n_1, n_2, \ldots, n_m\}$. Then the query time contains two parts: (1) prune child nodes using landmarks (i.e., $c_F * m$); (2) search the inverted index of non-filtered child nodes $(sum_i(p(n_i) * c_R * |\mathcal{R}(n_i)|))$ where $p(n_i)$ is the probability to visit n_i. The analysis of the additional space consumption is similar. For constraint of space, we omit this explanation ($\mathcal{I}(\cdot)$ is the number of objects in the inverted index. c_L is the space cost of a landmark and c_I is the space cost of an object in the inverted index). We use the tightness degree of spatial and textual components to measure the probability a child node will be visited. The intuition is that, the closer a newly-generated partition is, the more precise it can be used in the filter stage. Thus, we formulate $p(n_i)$ as $\frac{|l_{n_i}.mbr|}{|l_n.mbr|} * \frac{|l_{n_i}.rads|}{|l_n.rads|}$.

$$\mathcal{B}(n) = \frac{\Delta time}{\Delta space} = \frac{c_R * |\mathcal{R}(n)| - (c_F * m + \sum_i (p(n_i) * c_R * |\mathcal{R}(n_i)|))}{c_I * |\mathcal{I}(n)| - (c_L * m + \sum_i (c_I * |\mathcal{I}(n_i)|))} \qquad (1)$$

To simplify the analysis, we make two reasonable deductions: (1) we map each token only to one child, then the extra space cost turns to $c_L * m$ since $|\mathcal{I}(n)| = \sum_i |\mathcal{I}(n_i)|$; (2) suppose $|\mathcal{R}(\cdot)|$ is proportional to the size of the inverted index, then we have $|\mathcal{R}(n)| = \sum_i |\mathcal{R}(n_i)|$. Thus, the time cost $c_R * |\mathcal{R}(n)| - \sum_i (p(n_i) * c_R * |\mathcal{R}(n_i)|)$ equals to $c_R * \sum_i ((1 - p(n_i)) * |\mathcal{R}(n_i)|)$. Therefore, the algorithm performance is only affected by three factors: $p(n_i)$, m and $|\mathcal{R}(n_i)|$. For simplicity, we fixed m and $|\mathcal{R}(n_i)|$ can be estimated as $\frac{|\mathcal{I}(n)|}{m}$. Then we only need to improve $p(n_i)$, i.e., to make objects within each child node close in both spatial location and textual distance. Traditional methods (e.g., k-means, hierarchical

Table 1. Dataset Statistics

	# of objects	# of tokens	# of average tokens	Data size	Index size
USA	1 Million	383481	16.94	140.8M	40.28M
Twitter	1 Million	313439	7.72	37.57M	75.5M

clustering) are infeasible in our problem since: (1) "k-means" cannot update the cluster center under the edit distance metric; (2) "hierarchical clustering" needs to calculate the edit distance for every object pair.

A natural idea is to alternatively divide the spatial and textual components to generate landmarks with proper granularity. We can random select tokens and divide their tokenMBRs into partitions. However, this method is quite inefficient since: (1) the selected token may not best fit the distribution. (2) mapping each token to the closest partition needs to calculate its distance from all the landmarks which leads to heavily overloads. Thus, we propose a deletion-based method. The idea is twofold: (1) we can use the inverted index to avoid calculating textual distance for dissimilar pairs; (2) instead of using the entire token, we select a substring to represent the textual center. Given a token t, we delete any i characters to generate its i-deletion set, denoted by $d_i(t)$. For example, $d_0($"pizza"$) = \{$"pizza"$\}$ and $d_1($"pizza"$) = \{$"izza","pzza","piza","pizz"$\}$. At each node n, we first build an inverted index using the 0-deletion of all its tokens. For each inverted list l_t corresponding to the token t, if l_t contains more than $\frac{|\mathcal{I}(n)|}{m}$ objects, we calculate its MBR $\mathcal{M}_t$ and divide it into four regions (like quad-tree). Then we have four spatial regions, each centered with token t and radius $[0,0]$. If objects in any new generated region $\mathcal{M}'$ exceed $\frac{|\mathcal{I}(n)|}{m}$, we take $\langle \mathcal{M}', t, [0,0] \rangle$ as a selected landmark. For the remaining tokens, we repeat this step by adding deletion sets $d_1(n), d_2(n)$ and so on. After processing 3 deletions, the textual parts are quite dissimilar, we then partition the remaining objects only according to the spatial region.

5   Experimental Study

5.1   Experimental Settings

Datasets. We conducted extensive experiments on two datasets: Twitter and USA. Table 1 summarizes these two datasets. The Twitter dataset is a real dataset. We crawled 1 million tweets with location and textual information from Twitter[1]. The USA dataset is a synthetic dataset which randomly combines the Points of Interests (POIs) in US and the publications in DBLP.

Experimental Environment. All the algorithms were implemented in C++ and run on a Linux machine with an Intel(R) Xeon(R) CPU E5-2650 @ 2.00GHz and 48GB memory. The algorithms were complied using GCC 4.8.2.

Parameter Setting. Unless stated explicitly, parameters were set as follows by default: $k = 20$ and the average number of query tokens is 4.

---

[1] http://twitter.com

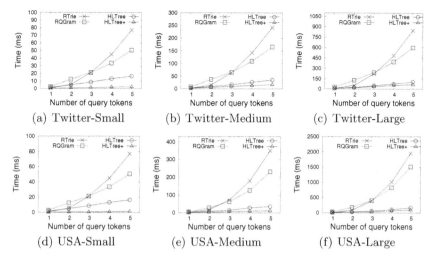

Fig. 4. Evaluation on number of query tokens

5.2 Evaluating Queries with Single and Multiple Tokens

In this section, we evaluated our methods HLtree and HLtree+ by comparing them with two existing solutions Rtree+Trie algorithm (RTrie) and Rtree+Qgram algorithm (RQGram). HLtree randomly selected tokens to generate hybrid landmarks while HLtree+ explored a deletion-based method as Section 4. We suppose that there are three kinds of query requirements corresponding to the different size of the query regions: (1)Small Size: region like a university; (2)Medium Size: region like a district; (3)Large Size: region like a city. Thus, we randomly selected 1000 objects from USA and Twitter as the center points and then extended them by 1496×1496, 20000×20000, 125040×125040 square meters respectively. In each evaluation, we executed 1000 queries and compared the average processing time. Experimental results show that our methods outperform the state-of-art algorithms and achieve high performance on all the evaluations.

Evaluation on the Number of Query Tokens. To evaluate the performance under different number of query tokens, we fixed k to 20 and varied query tokens from 1-5. The results are shown in Figure 4. Take Twitter as an example, HLtree+ achieved the best performance. It was 4-10 times faster than HLtree and 2-30 times faster than RTrie and RQGram. When we increased the query tokens, RTrie and RQGram increased sharply while HLtree+ changed very slowly. The reason is that the TA Algorithm in HLtree+ helps quickly locate similar candidates when dealing with queries with multiple tokens. Figure 4(b) and Figure 4(c) had similar performance.

Evaluation on k. To evaluate k, we varied k from 1 to 500. The result is shown in Figure 5. For constraint of space, we only show the result of USA. HLtree+ was 8-15 times faster than HLtree, was 40-75 times faster than RTrie and was 10-19 times faster than RQGram. Notice that when k increased, HLtree+, HLtree

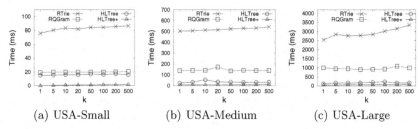

Fig. 5. Evaluation on parameter k

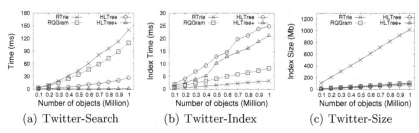

Fig. 6. Evaluation on time and index scalability

and `RQGram` almost kept a straight line while `RTrie` increased obviously. The reason is that for `RTrie`, all the prefixes whose edit distance is less than the current k_{th} candidate should be visited.

Evaluation on Scalability. We evaluated the time and index scalability. We varied object size from 0.1-1 million. As Figure 6(a) shows, `RTrie` and `RQGram` increased drastically while `HLtree` and `HLtree+` almost achieved a linear scalability. Figure 6(b) shows the index time. The memory of `HLtree+` is larger than `HLtree` because of the intermediate deletion-based inverted index (Figure 6(c)).

Acknowledgement. This work is supported by National High-Tech $R\&D$ (863) Program of China (2012AA012600), the NSFC project (61272090, 61472198), and the Chinese Special Project of Science and Technology (2013zx01039-002-002).

References

1. Alsubaiee, S., Behm, A., Li, C.: Supporting location-based approximate-keyword queries. In: GIS, pp. 61–70 (2010)
2. Bayardo, R.J., Ma, Y., Srikant, R.: Scaling up all pairs similarity search. In: WWW, pp. 131–140 (2007)
3. Chen, L., Cong, G., Jensen, C.S., Wu, D.: Spatial keyword query processing: An experimental evaluation. PVLDB **6**(3), 217–228 (2013)
4. Cong, G., Jensen, C.S., Wu, D.: Efficient retrieval of the top-k most relevant spatial web objects. PVLDB **2**(1), 337–348 (2009)
5. Deng, D., Li, G., Feng, J., Li, W.-S.: Top-k string similarity search with edit-distance constraints. In: ICDE, pp. 925–936 (2013)
6. Fagin, R., Lotem, A., Naor, M.: Optimal aggregation algorithms for middleware. In: Proceedings of the Twentieth ACM (2001)

7. Felipe, I.D., Hristidis, V., Rishe, N.: Keyword search on spatial databases. In: ICDE, pp. 656–665 (2008)
8. Guttman, A.: R-trees: A dynamic index structure for spatial searching. In: SIGMOD Conference, pp. 47–57 (1984)
9. Li, C., Lu, J., Lu, Y.: Efficient merging and filtering algorithms for approximate string searches. In: ICDE, pp. 257–266 (2008)
10. Wang, J., Li, G., Feng, J.: Fast-join: An efficient method for fuzzy token matching based string similarity join. In: ICDE, pp. 458–469 (2011)
11. Wang, X., Ding, X., Tung, A.K.H., Zhang, Z.: Efficient and effective knn sequence search with approximate n-grams. PVLDB **7**(1), 1–12 (2013)
12. Xiao, C., Wang, W., Lin, X., Shang, H.: Top-k set similarity joins. In: ICDE, pp. 916–927 (2009)
13. Yang, Z., Yu, J., Kitsuregawa, M.: Fast algorithms for top-k approximate string matching. In: AAAI (2010)
14. Yao, B., Li, F., Hadjieleftheriou, M., Hou, K.: Approximate string search in spatial databases. In: ICDE, pp. 545–556 (2010)
15. Zhang, Z., Hadjieleftheriou, M., Ooi, B.C., Srivastava, D.: Bed-tree: an all-purpose index structure for string similarity search based on edit distance. In: SIGMOD Conference, pp. 915–926 (2010)
16. Zhou, Y., Xie, X., Wang, C., Gong, Y., Ma, W.: Hybrid index structures for location-based web search. In: CIKM, pp. 155–162 (2005)

A Partition-Based Bi-directional Filtering Method for String Similarity JOINs

Ying Huang, Baoning Niu[(✉)], and Chunhua Song

School of Computer Science and Technology, Taiyuan University of Technology,
79 West Yingze Street, Taiyuan 030024, China
huangying0242@link.tyut.edu.cn,
{niubaoning,songchunhua}@tyut.edu.cn

Abstract. A string similarity join finds similar string pairs from two sets of strings, which is frequently found in many applications, such as duplicate detection, data integration and cleaning. Various algorithms have been proposed to address its efficiency issues. Partition-based filtering methods, such as Pass-JOIN, are promising, which quickly screens out possible similar string pairs by searching partitioned parts of a string in another string, in order of increasing length, and then performs similarity verification base on edit-distance. We notice that, filtering with different direction produces different candidate sets, which motivate us using a bi-directional filtering mechanism. This paper proposes a novel bi-directional filtering mechanism to enhance the filtering capability, which pipelines filtered results in forward direction to the process of backward filtering. The substring selection method of Pass-JOIN is adapted for the backward filtering. Experimental results show that the proposed bi-directional filtering algorithm outperforms the origin algorithm on real-world datasets.

Keywords: String similarity joins · Bi-directional filtering method · Partition-based scheme

1 Introduction

A string similarity join finds similar string pairs from two sets of strings, with the similarity value of the pairs above a given threshold, which can be quantified by a specified similarity metric. String similarity joins are expensive and yet frequently found in many real-world applications, such as duplicate detection, entity resolution, pattern recognition, and data integration and cleaning. Hence, the challenge is how to perform the similarity join in an efficient and scalable way.

Intuitively, a string similarity join needs to enumerate every string pairs and calculate their similarity values, which bears a prohibitively $O(n^2b^2)$ time complexity for string sets with n strings and string length less than b. Since the cost of calculating the similarity value is unavoidable, the primary approach is to avoid enumerating all the string pairs. Existing methods can be broadly classified into two categories [1]. The first category adopts a trie-based framework [2-4], which represents a string as a

© Springer International Publishing Switzerland 2015
J. Li and Y. Sun (Eds.): WAIM 2015, LNCS 9098, pp. 400–412, 2015.
DOI: 10.1007/978-3-319-21042-1_32

tree path from the root node to the leaf node. If the prefixes of a string pair violate the similarity value constraint, the string pair must be dissimilar and pruned to gain performance improvement. The second category employs a filter-verification framework [5-8]. At the filter step, an effective filtering algorithm prunes large numbers of dissimilar pairs and generating a set of candidate pairs. Most of existing filtering algorithms employ a signature-based technique, which generates signatures for each string such that if two strings are similar, their signatures overlap. At the verification step, similarity values of candidate pairs are computed to get the final similar pairs. Since it is expensive to compute the similarity value, filter-verification framework tries to minimize the size of the candidate set. However the filtering algorithm itself has overhead, there exists a tradeoff between the filtering power and the filtering cost.

The efficiency of the two category methods has to do with the distribution of string length. The trie-based framework is inefficient for long strings for two reasons. First, it is expensive to traverse a deep trie. Second, the number of strings with a common prefix is inversely proportional to the average length of strings, which weakens the pruning power. The filter-verification framework, on the other hand, is inefficient for the datasets with short strings. The reason is that it is difficult to extract high-quality signatures for short strings, which results in a large number of candidates to be verified.

Pass-Join [5] is one of the filter-verification methods, which is efficient for both short and long strings by taking a partition-based approach for filtering. In the filter step, Pass-Join maintains an inverted index of string segments. It sorts strings in the ascending order of length, and for each string in the order checks if its substrings are similar to the indexed segments. If it is similar, the pairs consisting of the current string and the indexed strings are marked as possible candidates, otherwise, pruned. And then, the current string is segmented and indexed.

We notice that, filtering in the descending order of length produces a different candidate set, which motivate us to infer that whether the filtering power can be further strengthened if a bi-directional filtering mechanism is used. As the results of experimentally proving the hypothesis, this paper claims the following contributions.

1) A novel bi-directional filtering mechanism is proposed to improve partition-based filter-verification framework by enhancing its filtering capability.
2) The multi-match-aware substring selection method [5] is redesigned and applied to the backward filtering.
3) A set of experiments on real-world datasets are conducted to show the proposed bi-directional filtering algorithm outperform the origin algorithm.

The rest of this paper is organized as follows. Section 2 presents preliminaries and backgrounds about the partition-based filter-verification framework. Section 3 introduces the proposed partition-based bi-directional filtering method. We propose to effectively pipeline way and modify the multi-match-aware substring selection method for the backward filtering in Section 3.3 and Section 3.4 respectively. Experimental results are provided in Section 4. Related work is covered in Section 5 and we make a conclusion in Section 6.

2   Preliminaries and Backgrounds

In this section, we formally define the string similarity join problem with edit-distance constraints as Definition 1, and introduce the background knowledge about the partition-based filter-verification framework.

Definition 1 (String Similarity Joins). Given two sets of strings R and S, and an edit-distance threshold τ, a string similarity join finds all similar string pairs $<r, s> \in R \times S$ satisfied $ED(r, s) \le \tau$. $ED(r, s)$ is the value of real edit-distance between string r and s.

In this paper, we only focus on the circumstance of $R = S$, and $R \ne S$ is similar to it. For example, consider the strings in Table 1. If the given edit-distance threshold is $\tau = 3$, then $<s_1, s_2>$ and $<s_4, s_5>$ will be considered as the similar pairs since their edit distance is not larger than τ.

Table 1. A collection of strings

Id	Strings	Length
1	Similarity	10
2	Similar	7
3	Diminishing	11
4	Conference	10
5	Confidence	10

Partition-base filtering is based on Lemma 1 [5].

Lemma 1. Given a string r with $\tau+1$ segments and a string s, if s is similar to r within an edit-distance threshold τ, s must contain a substring which matches a segment of r based on the pigeonhole principle.

Given two strings r and s, and an edit-distance threshold τ, we partition r into $\tau+1$ disjoint segments. If s has a substring matching a segment of r, we need verity the pair $<r, s>$ with edit-distance constraint. Otherwise s must be dissimilar to r, thus we can prune the pair $<r, s>$.

2.1   Partition Scheme

Given a string, many plans could partition the string into $\tau+1$ segments. Intuitively, the shorter a segment of r is, the higher the probability that a substring matches the segment and the more strings will be taken as r's candidates, thus the pruning power is weaker. The even-partition scheme can make each segment have nearly the same length. Consider a string r with length $|r|$. In even partition scheme, each segment has a length (l_i) of $\lfloor |r|/(\tau+1) \rfloor$ or $\lfloor |r|/(\tau+1) \rfloor + 1$. Let $k = |r| - \lfloor |r|/(\tau+1) \rfloor \times (\tau+1)$, the last k segments have length $\lfloor |r|/(\tau+1) \rfloor + 1$, and the first $\tau+1-k$ ones have length $\lfloor |r|/(\tau+1) \rfloor$ [5]. For example, for s_1 in Table 1 and $\tau = 3$, the length of s_1 ($|s_1|$) is 10, k equals 2, and s_1 has four segments {"si", "mi", "lar", "ity"}.

2.2   The Partition-Based Filtering Mechanism

For ease of presentation, we first introduce the notations used in the paper.

L_l^i: denote the inverted index for the i-th segment of strings with length of l.

$L_l^i(w)$: denote the set of all strings with length of l and whose the i-th segment is w.

$W(s,l,i)$: denote all substrings of s, whose substrings can match the i-th segment of strings with length of l.

$W(S,l,i)$: denote all substrings of all strings whose substrings can match the i-th segment of strings with length of l.

We first sort strings by their length in ascending order. Then we visit strings according to the sorted order. The visited string with length of l is partitioned to $\tau+1$ segments and the segments are inserted into the inverted index L_l^i ($1 \leq i \leq \tau+1$). Table 2 lists the inverted index of all strings in Table 1 for $\tau = 3$. Consider the current string s with length $|s|$. Based on length filtering, we check whether the strings in L_l^i ($|s|-\tau \leq l \leq |s|$, $1 \leq i \leq \tau+1$) are similar to s. If it is, s should contain a substring matching a segment in L_l^i.

Substring Selection. A straightforward method might be enumerating all substrings of s, and checks each of them whether it appears in L_l^i. In fact, it is not necessary to check all of them. Instead this framework only selects some substrings by using the multi-match-aware substring selection method [5] described in detail in Section 3.4. For each selected substring $w \in W(s,l,i)$, we check whether it appears in L_l^i. If so, for each $r \in L_l^i(w)$, $<r, s>$ are added to the candidate set $C(s)$.

Verification. To verify whether string r is similar to s, a straightforward method computes their real edit distance by using the dynamic programming algorithm. However this method is rather expensive. Instead, this framework utilizes the extension-based verification method which partitions r and s into three parts, the matching parts ($r_m = s_m$), the left parts r_l, s_l (on the left side of the matching part), and the right parts r_r, s_r. Formally, if $ED(r_l, s_l) \leq i-1 \cap ED(r_r, s_r) \leq \tau+1-i$, then the pair passes the verification, we add the pair into the result set [5].

Table 2. The inverted index of all strings in Table 1

L_l^i	L_7^1	L_7^2	L_7^3	L_7^4	L_{10}^1	L_{10}^2			L_{10}^3			L_{10}^4			L_{11}^1	L_{11}^2	L_{11}^3	L_{11}^4
segment	s	im	il	ar	si	co	mi	nf	lar	ere	ide	ity	nce		di	min	ish	ing
	↓	↓	↓	↓	↓	↓	↓	↓	↓	↓	↓	↓	↓		↓	↓	↓	↓
strings	2	2	2	2	1	4	1	4	1	4	5	1	4		3	3	3	3
						5		5					5					

3   Partition-Based Bi-directional Filtering Method

In this section, we describe a novel partition-based bi-directional filtering method to improve partition-based filter-verification framework by enhancing the filtering capability.

3.1 Motivations

We propose the partition-based bi-directional filtering method based on two considerations.

First, the effectiveness of partition-based filtering in descending order of string length (backward filtering) is not worse than that in ascending order of string length (forward filtering). As introduced in Section 2, the forward filtering uses the segments of short strings to select the substrings of current string, and, on the contrary, the backward filtering uses the segments of long strings to select the substrings of current string. When we partition a string into $\tau+1$ segments, the length of segments of short strings is not large than that of long strings. Intuitively, the longer a segment is, the lower the probability that a substring matches the segment and the more dissimilar string pairs can be eliminated. That is to say backward filtering should outperform forward filtering.

Second, the candidate set of forward filtering is different from that of backward filtering. A segment of string r matches a substring of string s, which cannot guarantee the opposite is founded, because the substring of s that matches a segment of r may be partitioned into two segment of s. For example, consider the strings r = "similar" and s = "dimension". Suppose $\tau = 3$. We partition r into four segments ("s", "im", "il", "ar"). s has a substring that matches the second segment ("im") of r. When we partition s, we get four segments ("di", "me", "ns", "ion").There is no any substring of r matching segments of s, because the substring ("im") of s is now partitioned into two segments ("di", "me"). Only if one string has a substring that matches a segment of another string, they can be added to the candidate set. This proves filtering with different direction can produce different candidate sets.

According to the property of partitioning, no matter for forward filtering or backward filtering, the similar string pairs will always appear in the candidate set. Although the string pair $<r, s>$ is in the candidate set of forward filtering, it does not appear in the candidate set of backward filtering. We can conclude that r and s are not similar.

This comes to the partition-based bi-directional filtering, which we name it as Bi-Filtering JOIN. We first perform forward filtering to generate candidate set C_1,we then perform backward filtering to generate candidate set C_2, and finally we compute the intersection of C_1 and C_2 to get the final candidate set C.

3.2 The Framework of Bi-Filtering JOIN

In this section, we introduce the framework of Bi-Filtering JOIN, which is divided into six steps. We give the pseudo-code of our algorithm as Algorithm 1.

Step 1: Sorting. Strings are sorted in ascending order by string length and the strings with same length are sorted in alphabetical order (line 2).

Step 2: Substring selection. Consider the current string s with length $|s|$. Based on length filtering, we utilize the segment information of the visited strings satisfying $|s|-\tau \leq l \leq |s|+\tau$ to select substrings of the current string, add them into $W(s,l,i)$ (line 3-line 6), and save the selected substrings' information of s ($W(s,l,i)$) into substrings' information of all strings ($W(S,l,i)$) (line 7). If $|s|-\tau \leq l \leq |s|$, the selected substrings of

s are prepared for matching a segment in inverted index L_l^i ($|s|-\tau \leq l \leq |s|$, $1 \leq i \leq \tau+1$) in forward filtering. If $|s| \leq l \leq |s|+\tau$, the selected substrings of s are prepared for matching the segment of visiting string in the future in backward filtering.

Step 3: Forward filtering. For each selected substring $w \in W(s,l,i)$, we check whether it appears in L_l^i. If it does, for each $r \in L_l^i(w)$, $<r, s>$ joins the initial candidate set $C_1(s)$ (line 8-line 11).

Step 4: Building inverted index. The segments of string s need be used for backward filter, so we partition s into $\tau+1$ segments and insert the segments into inverted index $L_{|s|}^i$ ($1 \leq i \leq \tau+1$) (line 12).

Step 5: Backward filtering. We pipeline the initial candidate set $C_1(s)$ from forward filtering to perform the backward filtering, and generate final candidate set $C_2(s)$ (line 13).

Step 6: Verification. We verify whether a candidate pair $<r, s>$ in the final candidate set $C_2(s)$ is an answer by computing their edit distance. If the pair passes the verification, it is added to the result set (line 14).

The function SubstringSelection selects all substrings for the current string by using the multi-match-aware substring selection method. We redesigned it for backward filtering and explained in detail in Section 3.4, the function ReverseFilter performs pipelined backward filtering and described in Section 3.3, The function Verification [5] verifies whether a candidate pair is satisfy the similarity constraint by computing the edit distance of two strings.

```
Algorithm 1: Bi-Filtering( S, τ )
Input: S: A collection of strings
       τ: A given edit-distance threshold
Output: A = { (s∈S, r∈R) | ED(r,s)≤τ}
1   Begin
2     Sort S first by string length and second in
        alphabetical order
3     for s ∈ S
4       for 1 ≤ i ≤ τ+1
5         for |s|-τ ≤ l ≤ |s|+τ
6           W(s,l,i) = SubstringSelection( s,l,i )
7           Add W(s,l,i) into W(S,l,i)
8           if |s|-τ ≤ l ≤ |s| then
9             for w ∈ W(s,l,i)
10              if w is in L_l^i then
11                Add L_l^i(w) into C_1(s)
12        Partition s into τ+1 segments and add s_i into L_{|s|}^i
13        C_2(s) = ReverseFilter( s,C_1,W )
14        Verification( s,C_2,τ )
15  End
```

3.3   Pipelined Filtering

A full backward filtering is not necessary, because those pairs eliminated by forward filtering should not be processed again. This reminds us of pipelining the candidate set from forward filtering to backward filtering as shown in Fig. 1. It consists of two steps and the pseudo-code is shown as Algorithm 2.

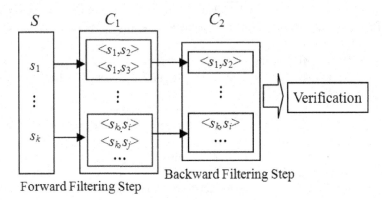

Fig. 1. The process of pipelined filtering

Step 1: Perform forward filtering to generate the initial candidate set $C_1(s)$ of the current string s.

Step 2: Perform the backward filtering step on the initial candidate set $C_1(s)$. For each initial matching string $c \in C_1(s)$ of the current string s, we check whether it has substring to matches a segment of s. If the segment s_i of s appears in $W(c,|s|,i)$, $<c, s>$ will be put in s' final candidate set $C_2(s)$, otherwise be pruned (line 2-line 5).

```
Algorithm 2: ReverseFilter( s,C₁,W )
Input: s: the current string
       C₁: the initial candidate set of s
           after forward filtering
       W: substring information
Output: C2: the final candidate set of s
            after backward filtering
1  Begin
2    for c ∈ C₁(s)
3      for 1 ≤ i ≤ τ+1
4        if sᵢ is in W(c,|s|,i) then
5          Add c into C₂(s)
6  End
```

3.4   Redesigned Substring Selection Function

The muti-match-aware substring selection method used in Pass-JOIN determines the start position of a substring supposes that the strings processed in ascending order of string length, which is designed for forward filtering. For backward filtering the strings are processed in descending order of string length. The muti-match-aware substring selection method should be redesigned.

Multi-Match-Aware Substring Selection Method for Forward Filtering. Given two strings r and s, $|r| = l$ and $|s|-\tau \leq l \leq |s|$, we partition r into $\tau+1$ segments. For each segment of r, we select some substrings of s and check whether they match the i-the segment of r. The length of selected substring is equal to the length of the i-th segment of r is l_i. The start position p of the substring can be computed using the start position p_i of the i-th segment of r, which is between $pf^i_{min} = \max(1, p_i-(i-1), p_i+\Delta-(\tau+1-i))$ and $pf^i_{max} = \min(|s|-l_i+1, p_i+(i-1)), p_i+\Delta+(\tau+1-i))$ where Δ is the length difference of the two strings [5]. For example, suppose $\tau = 3$, consider string r = "similarity" with four segments {"si", "mi", "lar", "ity"} and string s ="diminishing". The selected substrings of s are shown in Table 3.

Table 3. The selected substrings of s

i	segement (r)	$l_i(r,s)$	$p_i(r)$	$pf^i_{min}(s)$	$pf^i_{max}(s)$	substrings (s)
1	si	2	1	1	1	di
2	mi	2	3	2	4	im, mi, in
3	lar	3	5	5	7	nis, ish, shi
4	ity	3	8	9	9	ing

Multi-Match-Aware Substring Selection Method for Backward Filtering. Given two strings r and s, $|r| = l$ and $|s|-\tau \leq l \leq |s|+\tau$. we partition r into $\tau+1$ segments. If $|s|-\tau \leq l \leq |s|$, the computation of the start position of s' substring matching the i-th segment of r is the same of the method for forward filtering. If $|s| \leq l \leq |s|+\tau$, the computation is as shown in Fig. 2. We partition r and s into three parts, the matching parts $(r_m = s_m)$, the left parts r_l and s_l, and the right parts r_r and s_r. If $\Delta_l \geq i$, then $d_l = ED(r_l, s_l) \geq \Delta_l \geq i$, $d_r = ED(r_r, s_r) \leq \tau-d_l \leq \tau-i$ when s is similar to r. As r_r contains $\tau+1-i$ segments, s_r must contain a substring matching a segment in r_r based on the pigeon-hole principle, which can be proved similar to Lemma 1. In this way, we can discard s_m and use the next matching segment. Thus let $\Delta_l=|p-p_i| \leq i-1$, the minimal start position is $pb^l_{min_i} = \max(1, p_i-(i-1))$ and the maximal start position is $pb^l_{max_i} = \min(|s|-l_i+1, p_i+(i-1))$. Similarly, let $\Delta_r = |p-(p_i-\Delta)| \leq \tau+1-i$, the minimal start position is $pb^r_{min_i} = \max(1, p_i-\Delta-(\tau+1-i))$ and the maximal start position is $pb^r_{max_i} = \min(|s|-l_i+1, p_i-\Delta+(\tau+1-i))$. In summary, the start position p of substring is between $pb^i_{min} = \max(pb^l_{min_i}, pb^r_{min_i})$ and $pb^i_{max} = \min(pb^l_{max_i}, pb^r_{max_i})$ where Δ is the length difference of the two strings.

Based on above analysis, we give the pseudo-code of the redesigned substring selection algorithm as algorithm 3. It utilize the segment information of the strings with length in $|s|-\tau \leq l \leq |s|+\tau$ to determine the start position and length of the substrings of current string, and adds the substrings into the s $(W(s,l,i))$(line 2-line 7).

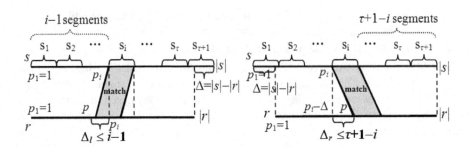

Fig. 2. Redesigned substring selection method

```
Algorithm 3: SubstringSelection( s,l,i )
Input:  s: The current visit string
        l: The length of string
        i: The id of segment
Output: W(s,l,i): The selected substrings'
                  information of s
1  Begin
2    if |s|-τ ≤ l ≤ |s| then
3      for p ∈ (pf^i_min, pf^i_max)
4        Add the substring of s with start position p
           and length l_i (s[p,l_i]) into W(s,l,i)
5    if |s| ≤ l ≤ |s|+τ then
6      for p ∈ (pb^i_min, pb^i_max)
7        Add the substring of s with start position p
           and length l_i (s[p,l_i]) into W(s,l,i)
8  End
```

4   Experiments

To evaluate Bi-Filtering JOIN, we designed a set of experiments to compare the performance of Bi-Filtering JOIN with the performance of Pass-JOIN with different data sets and varying edit-distance thresholds.

Experiment Setup: All the algorithms are written in C++, and complied by GCC 4.9.1. We run our programs on an Ubuntu machine with Intel Celeron E3500 2.7GHz processors and 4GB memory.

Dataset: We use the two real datasets Word and DBLP. Word is a dataset with short strings and consisted of 146,033 different English words, whose average length, maximal length and minimal length are 8, 30, and 1 respectively. DBLP is a dataset with long strings and included publication titles and author names. It contains 1,385,925 records and its average length, maximal length and minimal length are 105, 1626, 1 respectively.

4.1   The Size of Candidate Sets

The sizes of the candidate sets generated by Bi-Filtering JOIN and Pass-JOIN with different data sets and varying edit-distance threshold are shown in Fig. 3(a) and 3(b) respectively. Note that the vertical-axis is in logarithm scale.

Several observations can be made:

- The size of the real result grows modestly when the edit-distance threshold increases.
- Two algorithms generate more candidate pairs with the increase of the edit-distance threshold. Bi-Filtering JOIN has a decent reduction on the candidate size of Pass-JOIN, as the backward filtering prunes many candidates. The process of forward filtering in Bi-Filtering JOIN is the same as the process of filtering in Pass-JOIN, so the difference between candidate sets is the size of dissimilar string pairs pruned by backward filtering of Bi-Filtering JOIN. When $\tau = 3$, for the Word dataset, the size of candidate set generated by Bi-Filtering JOIN is 629 million pairs less than that by Pass-JOIN, which is the number of dissimilar string pairs pruned by backward filtering.
- For different data sets, filtering capability of the backward filtering varies, but the backward filtering always can further reduce the size of candidate set, which generated by Bi-Filtering JOIN is less than a half of that by Pass-JOIN for Word dataset, and is at least 10% smaller than that of Pass-JOIN for DBLP dataset.

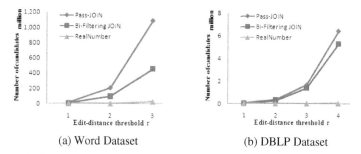

(a) Word Dataset   (b) DBLP Dataset

Fig. 3. Number of candidates with varying edit-distance threshold

4.2   Response Time

The Response times of Bi-Filtering JOIN and Pass-JOIN with different data sets and varying edit-distance threshold are shown in Fig.4(a) and4(b) respectively. The observations follow:

- The general trend is that the response time increases with the increase of the edit-distance threshold. One of the reasons is that inverted index lists and substring lists are getting longer with the increase of edit-distance threshold, thus increasing the matching time.

- The larger size of candidate set needs the more processing time. Because the candidate sizes on Word dataset are much larger than it on DBLP dataset as shown in Fig. 3, the runtime is also larger.
- In all the settings, Bi-Filtering JOIN is the more effective than Pass-JOIN. Although the bi-directional filtering has more overhead than forward only filtering, the backward filtering can reduces the size of candidate set, and effectively cut down the time spent on verification step, which is much greater than the overhead time. So the total response time of Bi-Filtering JOIN is much less than that of Pass-JOIN.
- The more string pairs pruned on the backward filtering by BiFilter-join, the more the response time reduced. When $\tau = 3$, as shown in Fig. 3, the backward filtering pruned about 600 million and 1 million string pairs in Word and DBLP dataset respectively, which result in a reduction of 200 seconds for the response time on Word dataset, and 10 seconds on the DBLP dataset.

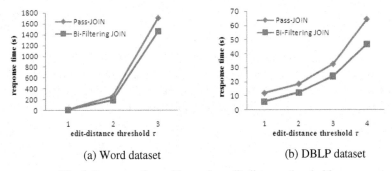

(a) Word dataset   (b) DBLP dataset

Fig. 4. Response time with varying edit-distance threshold

5   Related Work

The filter-verification framework focuses on two topics, filtering techniques and similarity metrics.

Filtering techniques include signature-based approaches and partition-based approaches. Gram-based approaches [6-9] are the common signature-based approaches. It transforms strings into grams and use common grams to filter similar pairs. Filtering techniques include count filtering [9-11] which mandates that two similar string s and r must share at least $(\max(|s|, |r|) - q + 1) - q\tau$ common q-grams, position filtering [6, 7] which mandates that s and r must share at least $(\max(|s|, |r|) - q + 1) - q\tau$ matching positional q-grams, length filtering [5, 10] which mandates that $\|s| - |r\| \le \tau$, and content filter [6] which select a probing window and look into the contents of both strings within the probing window to lower bound of the edit distance. Partition-based approaches [5, 12, 13] partitioned strings into segments and proved that if two strings are similar then they must share some common segments, such as Pass-JOIN [5], which quickly screens out possible similar string pairs by searching partitioned parts of a string in another string and then performs similarity verification.

Similarity metrics can be divided into two categories: The first category is set-based similarity metrics [7-9, 14], which rely on the overlap of two string sets and string set sizes to compute the similarity. The more overlap the more similar. The well-known set-based metrics include Overlap, Jaccard, Cosine, and Dice. Another is character-based similarity metrics, which treat the string as a character sequence. The representative character-based metric is edit distance [5, 6, 15, 16].

6   Conclusions

This paper proposes a partition-based bi-directional filtering method for string similarity joins by exploiting a novel bi-directional filtering mechanism to improve partition-filter-based framework and enhance the filtering capability, thus reduce the time spent in verification step. We develop a pipelined filtering mechanism, which pipeline the results of forward filtering to the process of backward filtering, and redesigned the multi-match-aware substring selection method for backward filtering. Experimental results show that our bi-directional filtering algorithm outperform the origin algorithm on real-world datasets.

Our future work includes how to further reduce the filtering cost while guaranteeing the filtering power, and how to make use of the matching segments to speed up verification.

Acknowledgments. This work is supported by The National Key Technology Support Program of China (grant No. 2012BAH04F02).

References

1. Jiang, Y., Li, G., Feng, J.: String similarity joins: an experimental evaluation. In: Proceedings of the 40th International Conference on VLDB (2014)
2. Wang, J., Feng, J., Li, G.: Trie-Join: efficient trie-based string similarity joins with edit-distance constraints. In: Proceedings of the 36th International Conference on VLDB (2010)
3. Ji, S., Li, G., Li, C., Feng, J.: Efficient interactive fuzzy keyword search. In: Proceedings of the 18th International Conference on WWW, pp. 433–439 (2009)
4. Li, G., Ji, S., Li, C., Feng, J.: Efficient fuzzy full-text type-ahead search. In: Proceedings of the 37th International Conference on VLDB Journal, pp. 617–640 (2011)
5. Li, G., Deng, D., Wang, J., Feng, J.: Pass-Join: a partition-based method for similarity joins. In: Proceedings of the 38th International Conference on VLDB (2012)
6. Xiao, C., Wang, W., Lin., X.: Ed-Join: an efficient algorithm for similarity joins with edit distance constraints. In: Proceedings of the 34th International Conference on VLDB Endowment (2008)
7. Xiao, C., Wang, W., Lin., X.: Efficient similarity joins for near duplicate detection. In: Proceedings of the 17th International Conference on WWW (2008)
8. Wang, J., Li, G., Feng, J.: Can we beat the prefix filtering? An adaptive framework for similarity join and search. In: SIGMOD Conference, pp. 85–96 (2012)
9. Li, C., Lu, J., Lu, Y.: Efficient merging and filtering algorithms for approximate string searches. In: ICDE Conference, pp. 257–266 (2008)

10. Gravano, L., Ipeirotis, P.G., Jagadish, H.V., Koudas, N., Muthukrishnan, S., Srivastava, D.: Approximate string joins in a database (Almost) for free. In: Proceedings of the 28th International Conference on VLDB, pp. 491–500 (2001)
11. Sarawagi, S., Kirpal, A.: Efficient set joins on similarity predicates. In: SIGMOD Conference, pp. 743–754 (2004)
12. Wang, W., Xiao, C., Lin, X., Zhang, C.: Efficient approximate entity extraction with edit distance constraints. In: SIGMOD Conference, pp. 759–770 (2009)
13. Wang, J., Li, G., Feng, J.: Fast-Join: an efficient method for fuzzy token matching based string similarity join. In: ICDE Conference, pp. 458–469 (2011)
14. Bayardo, R.J., Ma, Y., Srikant, R.: Scaling up all pairs similarity search. In: Proceedings of the 16th International Conference on WWW (2007)
15. Deng, D., Li, G., Hao, S., Wang, J., Feng, J.: Massjoin: a mapreduce-based method for scalable string similarity joins. In: IEEE 30th International Conference, pp. 340–351 (2014)
16. Jiang, Y., Deng, D., Wang, J., Li, G., Feng, J.: Efficient parallel partition-based algorithms for similarity search and join with edit distance constraints. In: EDBT Workshop (2013)

Fast Multiway Maximum Margin Clustering Based on Genetic Algorithm via the NystrÖm Method

Ying Kang[1,2], Dong Zhang[3,4], Bo Yu[1(✉)], Xiaoyan Gu[1],
Weiping Wang[1], and Dan Meng[1]

[1] Institute of Information Engineering, Chinese Academy of Sciences, Beijing, China
{kangying,yubo}@iie.ac.cn
[2] University of Chinese Academy of Sciences, Beijing, China
kangying@iie.ac.cn
[3] State Key Laboratory of High-end Server and Storage Technology, Jinan, China
zhangdong@inspur.com
[4] Inspur Group Corporation Ltd., Jinan, China

Abstract. Motivated by theories of support vector machine, the concept of maximum margin has been extended to the applications in the unsupervised scenario, developing a novel clustering method—maximum margin clustering (MMC). MMC shows an outstanding performance in computational accuracy, which is superior to other traditional clustering methods. But the integer programming of labels of data instances induces MMC to be a hard non-convex optimization problem to settle. Currently, many techniques like semi-definite programming, cutting plane etc. are embedded in MMC to tackle this problem. However, the increasing time complexity and premature convergence of these methods limit the analytic capability of MMC for large datasets. This paper proposes a fast multiway maximum margin clustering method based on genetic algorithm (GAM3C). GAM3C initially adopts the NystrÖm method to generate a low-rank approximate kernel matrix in the dual form of MMC, reducing the scale of original problem and speeding up the subsequent analyzing process; and then makes use of the solution-space alternation of genetic algorithm to compute the non-convex optimization of MMC explicitly, obtaining the multiway clustering results simultaneously. Experimental results on real world datasets reflect that GAM3C outperforms the state-of-the-art maximum margin clustering algorithms in terms of computational accuracy and running time.

Keywords: Maximum margin clustering · NystrÖm method · Genetic algorithm

1 Introduction

Clustering is a fundamental research for deep learning on large datasets, which are derived from many domains like information retrieval, bioinformatics and social media analysis etc. Given a dataset, the aim of clustering is to group data points with similar properties into one same cluster [1]. Over the past decades, except for the

© Springer International Publishing Switzerland 2015
J. Li and Y. Sun (Eds.): WAIM 2015, LNCS 9098, pp. 413–425, 2015.
DOI: 10.1007/978-3-319-21042-1_33

familiar methods like K-Means [2], spectral clustering [3] etc., plenty of novel clustering approaches have been proposed. Inspired by the preponderance of support vector machine (SVM) which is successfully applied to many research areas, the concept of maximum margin has been extended to the unsupervised learning scenario, forming a new clustering method—maximum margin clustering (MMC) [4]. Lots of applications show that MMC often outperforms other traditional clustering algorithms.

Unfortunately different from SVM, MMC needs to identify an optimal labeling of data instances which results in the maximum margin cutting hyperplanes between different clusters. The additional integer programming of data-instance labels induces the original optimization problem to be non-convex [4]. To settle the optimization problem on non-convex condition, many schemes have been employed in MMC successively, such as semi-definite programming (SDP) [5], alternating optimizing [6], cutting plane [7] and so on. However, the computational complexity of each of these methods is greater than $\Theta(n^2)$ at least, which limits the scalability of MMC for analyzing large datasets. Moreover, there's no further work to prove how and how fast these methods converge to the optimal solutions [8].

In light of the above presented deficiencies, this paper proposes a fast multiway maximum margin clustering method based on genetic algorithm (GAM3C) for large datasets. GAM3C firstly employs the Nyström method to generate a low-rank approximate kernel matrix in the dual form of MMC, to reduce the complexity of original problem and accelerate the subsequent clustering process, promoting the scalability of MMC for large datasets. Secondly, by means of embedding genetic algorithm (GA) into the optimization procedure of MMC and alternating the solution space by its crossover and mutation operators, GAM3C avoids the premature convergence which is caused by local minima and outputs the multiway clustering results simultaneously, improving the computational accuracy of MMC. Experiments on real datasets show that the performances of GAM3C are superior to the state-of-the-art maximum margin based clustering algorithms, especially in analyzing large datasets.

2   Related Work

MMC was firstly proposed by Xu et al. [4] to tackle the binary clustering problems, and then developed to be a multiway clustering technique [9]. Due to involving the identification of cluster labels, the computation of MMC becomes a non-convex optimization problem. Therefore, MMC was transformed into a semi-definite programming problem. Although the following generalized maximum margin clustering (GMMC) was proposed by Valizadegan et al. [10] to reduce the number of variables of MMC, the expensive computation makes MMC or GMMC impractical.

To solve the non-convex optimization problem, Zhang et al. [11] adopted an alternating scheme SVR with Laplacian loss to avoid premature convergence. However, this algorithm is time-consuming, and how fast it converges is uncertain. Zhao et al. [12] used the cutting plane to relax constraints of MMC and solved the non-convex optimization problem by means of constrained concave-convex procedure (CCCP). Then Wang et al. [8] proposed an advanced maximum margin clustering algorithm

(CPMMC), which decomposed the original clustering into a series of convex sub-problems and settled them by a cutting plane algorithm. These two approaches provide efficient solutions for multiway MMC clustering, but the optimal results from a sequence of decomposed convex sub-problems are prone to get stuck in local minima.

Recently, Gieseke et al. [13] exhibited a fast evolutionary maximum margin clustering method (containing EMMC and FEMMC), which speeds up the computational procedure by sampling a subset from either the data label set or the Lagrangian multiplier set to approximate. However, this method is just a binary clustering. Moreover, owing to lack of basic cluster information, the aimless approximation degrades the entire accuracy of clustering results.

In the applications of analyzing large dataset, many approaches [14][15] make use of the NystrÖm method to extend their scalability. The NystrÖm method has been proved to be an efficient technique for matrix approximation, which samples a low-rank matrix in terms of some selecting schemes to reduce the complexity of original problem. Motivated by these applications, this paper employs the NystrÖm method to MMC for an approximate computation firstly; then based on the obtained approximate matrix, searches for the optimal solutions from the non-convex problem explicitly based on genetic algorithm. Most important, this is a multiway clustering method, and the local minima can be avoided efficiently by alternation of solution space.

3   Preliminaries

3.1   The NystrÖm Method

The NystrÖm method is originally used to solve the numerical approximation of integral equations in the following form [14]:

$$\int K(x,y)\phi(y)p(y)dy = \lambda\phi(x) \tag{1}$$

where p is a probability density function, K is a kernel function, λ and $\phi(x)$ are the eigenvalue and eigenvector of K respectively. To approximate the integral in Eq.(1), sample q interpolation points $\{x_1, x_2, ..., x_q\}$ drawn from p, and the approximate result by the empirical average is:

$$\frac{1}{q}\sum_{j=1}^{q} K(x, x_j)\tilde{\phi}(x_j) \simeq \lambda\,\tilde{\phi}(x) \tag{2}$$

where $\tilde{\phi}(x)$ is an approximate value of $\phi(x)$ in Eq.(1), and choose x in Eq.(2) from $\{x_1, x_2, ..., x_q\}$ as well to generate an eigen-decomposition $\overline{K}\overline{U} = q\overline{\Lambda}\overline{U}$, $\overline{K} = \{K(x_i, x_j)|\, i,j = 1,2, ..., q\}$. So any eigenvector $\tilde{\phi}_i(x_j)$ and eigenvalue $\tilde{\lambda}_i$ in Eq.(1) can be approximated by $\overline{U} = \{\bar{u}_{ij}\}_{i,j=1}^{q}$ and $\overline{\Lambda} = \{\bar{\lambda}_{ii}\}_{i=1}^{q}$:

$$\tilde{\phi}_i(x_j) \simeq \sqrt{q}\bar{u}_{ij}, \ \tilde{\lambda}_i \simeq \bar{\lambda}_{ii}/q \tag{3}$$

The eigenvector of any point x can be also approximated by the eigenvectors of interpolation points in $\{x_1, x_2, ..., x_q\}$, because the kth eigenvector at an unsampled point x can be computed by:

$$\tilde{\phi}_i(x) \simeq \frac{1}{q\lambda_i}\sum_{j=1}^{q} K(x, x_j)\tilde{\phi}_i(x_j) \simeq \frac{\sqrt{q}}{\lambda_{ii}}\sum_{j=1}^{q} \bar{u}_{ij}K(x, x_j) \tag{4}$$

3.2   Multiway Support Vector Machine

Mathematically, given a instance set $X = \{x_1, ..., x_n\}$, $x_i \in R^m$ and the corresponding labels $Y = (y_1, ..., y_n), y_i \in \{1, ..., k\}^n$, SVM defines a matrix $P \in R^{k \times m}$ which consists of k class indicator vectors and the class of instance x can be predicted [16] by:

$$max_r\{P_r * x + 1 - \delta_{y_i,r}\} - P_{y_i} * x \tag{5}$$

where P_r is the rth row of P (P_r^T is the indicator vector), $\delta_{y_i,r}$ is equal 1if $y_i = r$ and 0 otherwise. The inner-product $P_r * x$ is a metric of similarity, and the gained highest similarity score with x from each row of P means the class affiliation of x. Sum up all instances in X, an upper bound on the empirical loss is obtained as follows:

$$\epsilon_T(P) \le \frac{1}{n}\sum_{i=1}^{n}[max_r\{P_r * x_i + 1 - \delta_{y_i,r}\} - P_{y_i} * x_i] \tag{6}$$

Furthermore, if the training set X is linearly separable, there exists a matrix P which satisfies the following constraint:

$$\forall i, r \quad P_r * x_i - \delta_{y_i,r} - P_{y_i} * x_i \le -1 \tag{7}$$

Therefore, the aim of multiway SVM is to seek such matrix P that the distances from support vectors to class indicator hyperplanes $f_r(x) = P_r * \phi(x) + b_r$, $(r = 1, ..., k)$ are maximized, where ϕ is a feature projecting function.

4   Our Approach

4.1   Multiway Maximum Margin Clustering

In essence, MMC is an extension of SVM to the unsupervised learning scenario. Differing from theories of SVM [17], the label vector Y is often not in hand, so the formulized definition of multiway MMC is to compute the following minima:

$$min_{P,\xi,Y} \frac{1}{2}\|P\|_2^2 + C\sum_{i=1}^{n}\xi_i \tag{8}$$

$$subject\ to: \delta_{y_i,r} + \left(P_{y_i} * \phi(x_i) - P_r * \phi(x_i)\right) \ge 1 - \xi_i, \xi_i \ge 0,$$

$$\sum_{i=1}^{n}\delta_{y_i,r}y_i \le s, \quad Y = (y_1, ..., y_n) \in \{1, ..., k\}^n, \quad r = 1, ..., k$$

where $C > 0$ is a regularization constant which trades off the empirical risk and the model complexity, $\|P\|_2^2 = \|(P_1^T, ..., P_k^T)^T\|_2^2 = \sum_{r,j} p_{rj}^2$, $r = 1, ..., k$, $j = 1, ..., m$, $\{\xi_i\}_{i=1}^n \ge 0$ are slack variables. $\sum_{i=1}^{n}\delta_{y_i,r}y_i \le s$ is a cluster balance constraint which prevents all instances from being assigned to one cluster.

The most important thing is that, similar to P and ξ, Y becomes a vector variable which needs to be confirmed by an optimal labeling, resulting in the maximum margin cutting hyperplanes $f_r(x)$ between k clusters. What's more, we can deduce the dual form of optimization in Eq.(8) by a direct calculation. Consider the Lagrangian function of MMC as follows:

$$L(P, \xi, \alpha, Y) = \frac{1}{2}\sum_{r=1}^{k} P_r P_r^T + C\xi^T e$$
$$- \sum_{i=1}^{n}\sum_{r=1}^{k} \alpha_i^r \left(P_{y_i} * \phi(x_i) - P_r * \phi(x_i) + \xi_i + \delta_{y_i,r} - 1\right) \qquad (9)$$

where $\{\alpha_i^r\}_{i,r} \geq 0$ are Lagrangian multipliers. To search for the extreme value of Lagrangian function, we rewrite the following term:

$$\sum_{i=1}^{n}\sum_{r=1}^{k} \alpha_i^r P_{y_i} * \phi(x_i) = \sum_{r=1}^{k}\sum_{i:y_i=r}\sum_{s=1}^{k} \alpha_i^s P_{y_i} * \phi(x_i)$$
$$= \sum_{r=1}^{k} P_r \sum_{i=1}^{n} \delta_{y_i,r} \sum_{s=1}^{k} \alpha_i^s \phi(x_i) \qquad (10)$$

and execute the partial derivative on P_r and ξ_i of Lagrangian function, we can obtain:

$$P_r^* = \sum_{i=1}^{n} \delta_{y_i,r} \sum_{s=1}^{k} \alpha_i^s \phi(x_i) - \sum_{i=1}^{n} \alpha_i^r \phi(x_i),$$
$$subject\ to:\ r = 1, \dots, k,\ \sum_{r=1}^{k} \alpha_i^r = C \qquad (11)$$

thus the following term can be rewritten as:

$$\frac{1}{2}\sum_{r=1}^{k} P_r^* P_r^{*T} = \frac{1}{2}\sum_{r=1}^{k} P_r^* \sum_{i=1}^{n} \delta_{y_i,r} \sum_{s=1}^{k} \alpha_i^s \phi(x_i)$$
$$- \frac{1}{2}\sum_{i=1}^{n}\sum_{r=1}^{k} \alpha_i^r P_r^* * \phi(x_i) \qquad (12)$$

based on Equation (9),(10),(11)and(12), we have:

$$L(P^*, \xi^*, \alpha, Y) = \frac{1}{2}\sum_{r=1}^{k} P_r^* P_r^{*T} + C\xi^T e - \sum_{r=1}^{k} P_r^* \sum_{i=1}^{n} \delta_{y_i,r} \sum_{s=1}^{k} \alpha_i^s \phi(x_i)$$
$$+ \sum_{i=1}^{n}\sum_{r=1}^{k} \alpha_i^r P_r^* * \phi(x_i) - \sum_{i=1}^{n}\sum_{r=1}^{k} \alpha_i^r \left(\xi_i + \delta_{y_i,r} - 1\right)$$
$$= -\frac{1}{2}\sum_{r=1}^{k} P_r^* \sum_{i=1}^{n} \delta_{y_i,r} \sum_{s=1}^{k} \alpha_i^s \phi(x_i) + \frac{1}{2}\sum_{i=1}^{n}\sum_{r=1}^{k} \alpha_i^r P_r^* * \phi(x_i)$$
$$- \sum_{i=1}^{n}\sum_{r=1}^{k} \alpha_i^r \left(\delta_{y_i,r} - 1\right) \qquad (13)$$

finally, the dual of multiway MMC is formulized as:

$$\gamma^* = \min_{\alpha,Y}(-\frac{1}{2}\sum_{r=1}^{k}\sum_{i=1}^{n}\sum_{j=1}^{n} \alpha_i^r \alpha_j^r \left(1 - \delta_{y_i,r}\right)\left(1 - \delta_{y_j,r}\right) K_{ij}$$
$$+ \sum_{i=1}^{n}\sum_{j=1}^{n} \alpha_i^r \alpha_j^r) \qquad (14)$$
$$subject\ to:\ \sum_{r=1}^{k} \alpha_i^r = C, 0 \leq \sum_{r=1}^{k} \alpha_i^{y_i} \leq C,$$
$$Y = (y_1, \dots, y_n) \in \{1, \dots, k\}^n, i = 1, \dots, n$$

where in the dual form of multiway MMC, $K_{ij} = \phi(x_i)^T\phi(x_j)$ is the *(i,j)*th entry of kernel matrix K, which can be explicitly built up by a kernel function. By virtue of the nonlinear function ϕ that maps the instances X into a high (possible infinite) dimensional feature space, the non-separable problem in the original space becomes linear separable in the feature space.

4.2   Approximating Kernel Matrix via the NystrÖm Method

Recall the dual form of multiway MMC in the above subsection. It is obvious that the computation of kernel matrix K is a premise of multiway MMC, but it is often time-consuming and space-occupying in practice. In order to lower the complexity of problem and extend the scalability of multiway MMC model, this paper adopts the NystrÖm method to approximate kernel matrix K, so that the subsequent analyzing procedure of MMC can be simplified. Identical to the interpolation points in the NystrÖm method mentioned in Section 3.2, we generate a low-rank approximate matrix by the intersection of sampled columns and rows.

Definition 1 (*Kernel Matrix Approximation*).

Input: *kernel matrix* $K = \{K_{ij} = \phi(x_i)^T\phi(x_j)\} \in R^{n \times n}, r = rank(K)$.

Output: *low-rank approximate matrix* $\widetilde{K} \in R^{m \times m}$, *a rank parameter* $m \leq r$ & $m \ll n$.

Process: *to approximate* K *efficiently, in terms of the probability distribution* $p_i = K_{ii}^2 / \sum_i K_{ii}^2$, *we sample m rows and m columns from* K *and intersect them to generate* $\widetilde{K}$. *In addition, if* $\varepsilon > 0$ *and* K^* *is the best rank-m approximation of* K, *by choosing* $\Theta(r/\varepsilon^4)$*columns and rows, the expectation of approximate error is:*

$$\left\| K - M\widetilde{K}^+ M^T \right\|_F \leq \left\| K - K^* \right\|_F + \varepsilon \sum_{i=1}^{n} K_{ii}^2 \tag{15}$$

Where $\widetilde{K}^+$ is the best rank-m approximation of $\widetilde{K}$, $M \in R^{n \times m}$ is a matrix formed by the sampled m columns from K. According to the Drinear and Mahoney's theorem [18], this approximation can be completed by $\Theta(n)$ additional occupation of time and space, after passing the data from external storage two times.

4.3   Optimization with Genetic Algorithm

Based on the obtained low-rank approximate matrix $\widetilde{K}$, the computational burden of γ^*can be reduced enormously. But due to the integer programming of labeling Y, the optimization of γ^* becomes a non-convex problem. Therefore, this paper is to embed the genetic algorithm (GA) 19] into multiway MMC to tackle the non-convex problem efficiently.

Definition 2 (*Optimizing* γ^* *via Genetic Algorithm*).

Input: *approximate matrix* $\widetilde{K}$, *initial population* G_0 *composed of s individuals and each of which is a genetic representation of clustering results on* Y, *a threshold T for objective function* γ^*, *iterator i=0 and most iteration times I.*

Output: *cluster indicator matrix* P, *k rows of which indicate k clusters obtained on* $\widetilde{K}$.

Process: *1. generate c novel individuals by crossover and add them to* G_i, $G_i = s + c$;

    2. compute the minima of the convex objective function γ^* *directly based on each definite* Y *in* G_i, *select s individuals with elite value of* γ^* *as parents;*

    3. mutate some individuals in generated parent population in a certain proportion to prevent the optimal value of γ^* *from local minima, and produce the next generation of population* G_{i+1}, $i = i + 1$;

    4. if $\gamma^* \geq T$ *or* $i \geq$ I, *terminate the computing process and output the clustering results with the global minimum value of* γ^*, *else return to step 1.*

Objective Function and Genetic Representation. Objective function is the only guide for optimizing infinite variables which are encoded as individuals of population in GA. Here we select γ^* in Eq.(14) as the objective function of GA, and encode $Y = (y_1, \ldots, y_n) \in \{1, \ldots, k\}^n$ into individuals. Each individual consists of n genes, and if two genes of an individual take a same value, they belong to the same cluster.

Initialization. Initialization is applied to produce s individuals for the initial population, which plays a key role in the method's convergence. This paper adopts a link-edge-weight-based gene representation to initialize Y efficiently.

Crossover. When each individual (or cluster labeling vector Y) of population is determined, the optimization of γ^* becomes a convex problem. We can compute the saddle points of α and the minima of γ^* by partial derivative. Here we adopt uniform crossover and elite selection to create offsprings of next generation of population.

Mutation. Selecting the elite candidate parents for crossover is to guide the objective solution space to a better direction and speed up the velocity of convergence of the algorithm. But sometimes this improvement is prone to incur a premature convergence. Moreover, MMC tends to get stuck in local minima as well. To avoid both above unfortunate situations, mutation operator is added as a key technique to exchange the objective solution space into a global range.

4.4   Extension of Clustering Results to Out-of-Samples

Until now, what has been obtained is the clustering results based on the low-rank approximate matrix $\widetilde{K}$. How to extend it to the unsampled? According to the NystrÖm method in Eq.(9), the cluster affiliation of out-of-samples can be estimated by:

Definition 3   (*Extension to Out-of Samples*).
Input: *cluster indicator matrix P which is computed based on the low-rank approximate $\widetilde{K}$, and each row P_r of P represents a cluster, $r = 1, \ldots, k$.*
Output: *final clustering results on K.*
Process: *if an out-of-sample x belongs to the same cluster with the rth row P_r of P, they are collinear in the $(k-1)$-dimensional subspace spanned by $\{P_r\}_{r=1}^k$, and the affiliated cluster of x is determined by indicator function*
$$\mathcal{H}(x) = \arg\max_r (P_r * \phi(x)) = \arg\max_r \left\{ \sum_i p_{ir} K(\phi(x_i), \phi(x)) + b_r \right\},$$
$i = 1, \ldots, m.$

The collinear property of two co-cluster vectors results from the following proposition: if the term $K(\phi(x_i), \phi(x))$ equals to 1, x are predicted in the same cluster with x_i [20]. The row cluster indicator P_r, which denotes the normal vector of one maximum margin cutting hyperplane, is piecewise constant. In addition, because the cluster indicator matrix P is searched for by optimizing the labeling of Y, P_r has only one direction and no chance to rotate.

4.5   Computational Complexity

Firstly, let's give the implementation of GAM3C in Algorithm 1. By analyzing Algorithm 1 in detail, we deduce Lemma1 as follows:

Algorithm 1. **GAM3C**

Input: instance set $X = \{x_1, \ldots, x_n\} \in R^m$, kernel matrix $K = \{K_{ij} = \phi(x_i)^T \phi(x_j)\} \in R^{n \times n}$, objective function γ^*, initial population G_0, a threshold T, largest number of generations I, mutation rate η, parameters $m, s, c, i, and\ m \ll n, i = 0$.

Output: an optimal labeling of $Y = (y_1, \ldots, y_n) \in \{1, \ldots, k\}^n$

1 Begin
2    $\widetilde{K}$ = intersection of m columns and m rows sampled by probability $p_i = K_{ii}^2 / \sum_i K_{ii}^2$;
3   Initialize $G_0 = \{Y_1, \ldots, Y_s\}$;
4   While ($\gamma^* \leq T$ & $i \leq$ I) do
5   Generate individuals $Y_s, \ldots, Y_{s+c}$ by crossover , $G_i = G_i + \{Y_s, \ldots, Y_{s+c}\}$;
6   Compute γ^* based on each Y in G_i, select s elite individuals as parents;
7   Mutate η individuals of parents and produce the next generation G_i ($i = i + 1$);
8   End while
9   Obtain cluster indicator matrix P based on the optimal value of γ^* of G_i;
                                          // P consists of k normal vectors $\{P_r\}_{r=1}^k$
10   $\mathcal{H}(x) = arg\ max_r(P_r * \phi(x))$;   // x is one out-of-sample instance
11 End

Lemma 1 (Running Time). *The worst case time complexity of GAM3C is $\Theta(n) + \Theta((s + c)mL) + \Theta(sm) + \Theta(km(n - m))$, where s is the basic number of individuals in each generation of population, c is the incremental number of individuals generated by crossover, m is the dimension of matrix $\widetilde{K}$, L stands for the maximum value between the threshold T and the most iteration times I, k is the number of cluster number, and n is the number of instances in the original input dataset X.*

Proof.   The generation of low-rank approximate matrix $\widetilde{K}$ takes time $\Theta(n)$; when turning to the "While" loop, each loop iteration needs time $\Theta(m)$ to compute the objective function γ^* based on each individual in the ith generation, and we select the maxima between T and I to denote the worst case, thus the whole running time of iterative loop is $\Theta((s + c)mL)$; after the loop, we need to spend extra time $\Theta(sm)$ obtaining the cluster indicator matrix P; finally, the consuming time of extension of clustering results based on $\widetilde{K}$ to the out-of-samples is $\Theta(km(n - m))$. Therefore, the total time complexity of GAM3C is $\Theta(n) + \Theta((s + c)mL) + \Theta(sm) + \Theta(km(n - m))$ in the worst case, especially on the condition of $m \ll n$.

5   Experiments

5.1   Datasets and Experiment Setup

In the sequel, we will evaluate the performance of GAM3C in terms of computational accuracy and running time on real world datasets. We choose K-Means (KM) [2] and the state-of-the-art MMC algorithms such as ISVR [11], CPM3C [8], EMMC [13] and FEMMC [13] as the baseline for comparison. Experimental datasets are listed in Table 1 and Table 2 divisively. Datasets in Table1 come from a wide range of repositories[1], while datasets in Table2 are collected from Stanford University's SNAP networks[2]. Obviously, the datasets of Table 2 are all larger than the datasets of Table 1. All of the experiments are performed on a Linux machine with 4Core 2.6GHz CPU and 4G main memory, and the algorithms are implemented in Java.

Table 1. Information of experimental datasets (small-scale)

Data	Size(n)	Feature(F)	Cluster(C)	Sparsity Degree
Letter	1555	16	2	98.9%
Satellite	2236	36	2	100%
Text-1	1981	8014	2	0.70%
Newsgroup	3970	8014	4	0.75%
WK-Texas	814	4029	7	1.97%
RCV1	21251	47152	4	0.16%
MNIST	70000	784	2	19.14%
USUP	3046	256	4	36.24%

Table 2. Information of experimental datasets (large-scale)

Data	Size(n)	Cluster(C)	Average clustering coefficient
Youtube	1,134,890	8,385	0.0808
Amazon	334,863	151,037	0.3967
Orkut	3,072,441	6,288,363	0.1666
DBLP	317,080	13,477	0.6324

Since all the above datasets own their real clusters which can be referred as ground-truth clustering results, we are able to compare the analyzing capability of each algorithm by the following two criteria:

1. Normalized Mutual Information (NMI). NMI [21] is such a metric that it is capable of maintaining the balance between clustering quality and number of clusters.

2. Rand Index (RI). RI [21] is a measure of similarity, the definition of which is: $RI = \frac{TP+TN}{TP+TN+FP+FN}$, where TP, TN, FP, FN respectively denote true positives, true negatives, false positives, false negatives, and their concrete meaning is omitted here.

No matter which criterion is selected, the greater value means the better clustering results.

On account of the premised kernel matrix approximation by the NystrÖm method, the clustering results of GAM3C from different sampling proportion vary. Based on NMI, we firstly measure the variation of computational accuracy of GAM3C on each dataset in Table 2, with an increasing sampling proportion. Moreover, to avoid dropping in local minima, we stress the mutation operator in GAM3C. Thus the parameters of embedded genetic algorithm are set as: crossover rate is 0.9, population size is 500, number of generations is 50, and mutation rate takes value from{0.1, 0.2, 0.3, 0.4, 0.45, 0.5} in turn. The analyzing results are drawn as folding lines in Figure 1.

---

[1] http://mlearn.ics.uci.edu/MLRepository.html
http://people.csail.mit.edu/jrennie/20Newsgroups/
http://www.cs.cmu.edu/~WebKB/
http://www.kernel-machines.org/data.html
http://yann.lecun.com/exdb/mnist/
[2] http://snap.stanford.edu/data/

Secondly, we run all of above algorithms on datasets both in Table 1 and Table 2 to mine their own cluster structures, and the implementation of each algorithm is similar to their derivation. According to the obtained empirical results from the first experiment, we select the better sampling proportion and parameters to build up GAM3C, meanwhile run GAM3C several times for average to enhance its stability. This time we compare the *RI* and running time of GAM3C with all the other algorithms'. The corresponding results are listed respectively in Table 3 and Table 4.

5.2   Experimental Results and Analysis

Figure 1 shows the trend of clustering accuracy *NMI* with an increasing number of interpolation points on four real datasets such as Youtube, Amazon, Orkut and DBLP. Each dataset graph involves six different folding lines, standing for different clustering results in line with six mutation rates. Clearly, no matter which dataset, the initial acute variation (below 1%) of folding lines implies the instability of GAM3C when the sampling proportion is much smaller, but the overall trend of all lines is rising. When the proportion of interpolation points is larger than 1%, all folding lines tend to be flat gradually, indicating that the computational accuracy reaches to its extreme value. On the other hand, from the viewpoint of six different folding lines on each dataset graph, we observe that the clustering results with mutation rate more than 0.4 is superior to the ones with other mutation rate, and the three folding lines with 0.4, 0.45, 0.5 mutation rate are close to one another. So in summary, we can gain a much more ideal computational accuracy by selecting mutation rate 0.4 and proportion of interpolation points 5% or larger.

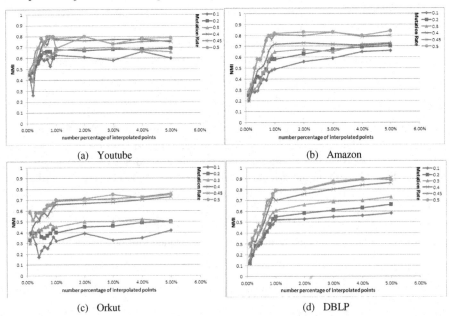

(a) Youtube

(b) Amazon

(c) Orkut

(d) DBLP

Fig. 1. The *NMI* variation trend of GAM3C with an increasing sampling proportion

All measure results of *RI* on all datasets are accumulated in Table 3. Comparing the *RI* between different algorithms, we notice that ISVR, EMMC and FEMMC are only fit for the binary clustering problems, and even in the binary cases, their *RI* are smaller than GAM3C's. When analyzing the multiway clustering problems, the *RI* of CPM3C is much larger than KM; except for an outlier which is caused by the randomness with a probability near to zero, GAM3C performs better than CPM3C. What is stated above demonstrates that the embedded genetic algorithm improves the entire clustering quality of GAM3C. Furthermore, we contrast the running time of each algorithm displayed in Table 4. No matter binary clustering or multiway clustering, the consuming time of GAM3C is the least among all the clustering methods. Although the time complexity of KM and CPM3C is almost linear, GAM3C which is based on the NystrÖm method executes much faster, making it scalable to analyze the cluster structure of larger datasets with more than millions of instances.

Table 3. The *RI* of all algorithms for comparison on various datasets

Data	KM	ISVR	CPM3C	EMMC	FEMMC	GAM3C
Letter	0.709	0.866	0.895	0.890	0.892	**0.901**
Satellite	0.919	0.940	0.968	0.973	0.977	**0.985**
Text-1	0.503	0.939	0.906	0.933	0.927	**0.937**
newsgroup	0.584	-	0.779	-	-	**0.789**
WK-Texas	0.606	-	0.711	-	-	**0.715**
RCV1	0.471	-	0.701	-	-	**0.712**
MNIST	0.809	0.862	0.920	0.934	0.942	**0.945**
USUP	0.932	-	**0.952**	-	-	0.950
Youtube	0.492	-	0.657	-	-	**0.757**
Amazon	0.611	-	0.772	-	-	**0.839**
Orkut	0.419	-	0.603	-	-	**0.716**
DBLP	0.554	-	0.716	-	-	**0.804**

Table 4. Running time of all algorithms on various datasets (in seconds)

Data	KM	ISVR	CPM3C	EMMC	FEMMC	GAM3C
Letter	0.039	1003	0.41	180	29,67	0.016
Satellite	0.084	3190	2.23	398	65.32	0.029
Text-1	30.1	478	9.25	69.7	10.5	1.71
newsgroup	1126	-	108	-	-	11.25
WK-Texas	390	-	5.50	-	-	0.37
RCV1	205332	-	301	-	-	35.42
MNIST	8.35	2572	49.67	301	22.8	0.98
USUP	2.67	-	17.79	-	-	0.096
Youtube	819973	-	1281	-	-	102.3
Amazon	2099	-	345	-	-	87.31
Orkut	1068851	-	1776	-	-	165.2
DBLP	428	-	892	-	-	21.64

6   Conclusion

This paper proposes a fast multiway maximum margin clustering method based on genetic algorithm (GAM3C) for large datasets. GAM3C makes use of the genetic algorithm to solve the non-convex maximum-margin optimization problem explicitly, avoiding the premature convergence efficiently by crossover and mutation operations, and generating the multiway clusters simultaneously. Before analyzing the cluster structure of objective datasets, GAM3C achieves a premised approximation for kernel matrix via the Nyström method, reinforcing its scalability for the further applications in large datasets. Experimental results on real world datasets manifest that, GAM3C outperforms the state-of-the-art MMC algorithms such as ISVR, CPM3C, EMMC, FEMMC and the classical K-Means in terms of computational accuracy and running time, meanwhile exhibiting its scalable analyzing capability.

Acknowledgement. This work is supported by the National Science Foundation of China under grant number 61402473, KeJiZhiCheng Project under grant number 2012BAH46B03, National HeGaoJi Key Project under grant number 2013ZX01039-002-001-001, and "Strategic Priority Research Program" of the Chinese Academy of Sciences under grant number XDA06030200.

References

1. Jain, A., Dubes, R.: Algorithms for clustering data. Englewood Cliffs (1988)
2. Kanungo, T., Mount, D.M., Netanyahu, N.S., et al.: An efficient k-means clustering algorithm. IEEE Trans. on PAMI **24**(7), 881–892 (2002)
3. Ng, A.Y., Jordan, M.I., Weiss, Y.: On spectral clustering: analysis and an algorithm. Advances in NIPS, pp. 849–856 (2001)
4. Xu, L., Neufeld, J., Larson, B., Schuurmans, D.: Maximum margin clustering. In: Advances in NIPS, pp. 1537–1544 (2004)
5. Nesterov, Y., Nimirovskii, A.: Interior-point polynomial algorithms in convex programming. SIAM (1994)
6. Bezdek, J., Hathaway, R.: Convergence of alternating optimization. Journal Neural, Parallel & Scientific Computations **11**(4), 351–368 (2003)
7. Kelley, J.E.: The cutting-plane method for solving convex programs. Journal of SIAM **8**(4), 703–712 (1960)
8. Wang, F., Zhao, B., Zhang, C.S.: Linear time maximum margin clustering. IEEE Trans. on Neural Network **21**(2), 319–332 (2010)
9. Xu, L., Schuurmans, D.: Unsupervised and semi-supervised muliti-class support vector macine. Proc. of NCAI **2**, 904–910 (2005)
10. Valizadegan, H., Jin, R.: Generalized maximum margin clustering and unsupervised learning. In: Advances in NIPS, pp. 1417–1424 (1994)
11. Zhang, K., Tsang, I.W., Kwok, J.T.: Maximum margin clustering made practical. In: Proc. of ICML, pp. 1119–1126 (2007)
12. Zhao, B., Wang, F., Zhang, C.: Efficient multiclass maximum margin clustering. In: Proc. of ICML, pp. 1248–1255 (2008b)
13. Gieseke, F., Pahikkala, T., Kramer, O.: Fast evolutionary maximum margin clustering. In: Proc. of ICML, pp. 361–368 (2009)

14. Xianchao Z., Quanzeng Y.: Clusterability analysis and incremental sampling for nyström extension based spectral clustering. ICDM, pp. 942–951 (2011)
15. Choromanska A., Jebara T., Kim H., et al.: Fast spectral clustering via the nyström method. In: Algorithmic Learning Theory, pp. 367–381 (2013)
16. Crammer, K., Singer, Y.: On the algorithmic implementation of multiclass kerner-based vector machine. Journal of MLR **2**, 265–292 (2001)
17. Lee, C.P., Lin, C.J.: A study on L2-loss (square hinge-loss) mulit-class SVM. Neural Computation **25**(5), 1302–1323 (2013)
18. Drineas, P., Mahoney, M.W.: On the Nyström method for approximating a Gram matrix for improved kernel-based learning. Journal of MLR **6**, 2153–2175 (2005)
19. Pizzuti, C.: GA-Net: A genetic algorithm for community detection in social networks. In: Proc. of ICPPSNX, pp. 1081–1090 (2008)
20. Alzate, C., Suykens, J.A.K.: Multiway spectral clustering with out-of-sample extensions through weighted kernel PCA. IEEE Trans. on PAMI **32**(2), 335–347 (2010)
21. Labatut, V.: Generalized measures for the evaluation of community detection methods. Journal of CoRR (2013)

Short Papers

DTMF: A User Adaptive Model for Hybrid Recommendation

Wenlong Yang, Jun Ma$^{(\boxtimes)}$, and Shanshan Huang

School of Computer Science and Technology, Shandong University,
Jinan 250101, China
{yangwenlonghi,huangshanshansdu}@163.com, majun@sdu.edu.cn

Abstract. Due to the uneven distributions of the ratings and social relationships for each user, two types of above recommendation methods should have varying weights when make recommendations. In this paper, we propose a user adaptive hybrid recommendation model, which dynamically combines a trust-aware based method and low-rank matrix factorization with adaptive tradeoff parameters, named as DTMF. It can utilize the advantages of these two methods and learn combinative parameters automatically. We investigate our model's performance on two social data sets - Epinions and Flixster. Experimental results show that DTMF performs better than other state-of-the-art methods.

Keywords: Recommender systemsa · Trust-aware · Matrix factorization

1 Introduction

In social network sites, as the uneven distributions of social relationships and ratings for each user, we should adopt different recommendation methods. When a user has much more ratings than social relations, we should pay more attention to his/her own interests and weaken the impact of social relations. Nevertheless, when a user has large amount of social relations with others and few ratings, it is more feasible to infer user's interests from that ratings of his social network. With the concerns mentioned above, in this paper we propose a dynamic user adaptive combination model for hybrid recommendation. The experiments conducted on two large data sets show that our method outperforms the state-of-the-art CF methods.

2 Data Description

The Epinions dataset[1] we used was published in the paper [2]. Every user of Epinions maintains a "trust" list which presents a social network of trust relationships between users. Flixster is a social networking site in which users can rate movies and add other users to their "friend" list. For the Flixster dataset[2], it is a recently released social dataset by the authors of [1]. The statistics of the data sets are showed in Table 1.

---

[1] http://www.trustlet.org/wiki/Epinions_datasets
[2] http://www.cs.sfu.ca/~sja25/personal/datasets/

© Springer International Publishing Switzerland 2015
J. Li and Y. Sun (Eds.): WAIM 2015, LNCS 9098, pp. 429–432, 2015.
DOI: 10.1007/978-3-319-21042-1_34

Table 1. Statistics of Epinions and Flixster Data Sets

Dataset	Num. Ratings	Num. Users	Num. Items	Density (%)	Trust Relationships
Epinions	371,909	8,677	31,454	0.14	234,665
Flixster	2M	23,842	33,587	0.25	248,092

3   User Adaptive Hybrid Recommendation Model

Before the introduction of the algorithm, we first give some notations and definition of the task of recommendation. Let $U = \{u_1, u_2, \cdots, u_m\}$ be the user set, $I = \{i_1, i_2, \cdots, i_n\}$ be the item set and $R^{m \times n}$ be the rating matrix where $R_{u,j}$ represents the rating score of user u gives to the item j. In a social trust network $G = (U, E)$ where the vertices are all the users and the edge set E denotes the trust relations between users. Let $T^{m \times m}$ be the trust matrix where $T_{u,v} = 1$ if user u trusts user v, otherwise 0. The task of our model is to predict the missing values in R for the users by employing the trust network and user-item rating matrix.

3.1   Our Dynamic User Adaptive Hybrid Model with Trust Relationships

As far as we know, there is no research has directly integrate the prediction of the trust-aware methods and matrix factorization techniques. Firstly, we integrate trust-aware method and matrix factorization into a simple hybrid model with a constant trade-off parameter α, and its probabilistic function can be defined as:

$$
\begin{aligned}
p(U, V | & R, \sigma_R^2, \sigma_U^2, \sigma_V^2) \\
& \propto \prod_{u=1}^{m} \prod_{i=1}^{n} \left[N(R_{u,i} | \alpha(r_m + U_u^T V_i) + (1 - \alpha)\hat{r}_{u,i}^T, \sigma_R^2) \right]^{I_{u,i}} \\
& \times \prod_{u=1}^{m} N(U_u | 0, \sigma_U^2 I) \times \prod_{i=1}^{n} N(V_i | 0, \sigma_V^2 I)
\end{aligned}
\tag{1}
$$

where $N(x | \mu, \sigma^2)$ is the probability density function of the Gaussian distribution with mean μ and variance σ^2, $\hat{r}_{u,i}^T$ is the predicted rating of user u to item i by a trust-aware method, and $I_{u,i}$ is the indictor function.

However, due the uneven distributions of trust relationships and ratings data for each user, the prediction accuracy should be diverse for each user. Motivated by the aforementioned, we then propose a dynamic user adaptive hybrid model to predict the missing ratings, which combines a trust-aware method and matrix factorization with the user adaptive tradeoff parameters. We call it DTMF, which can dynamically integrate the trust-aware recommendation method and collaborative filtering. The predicted rating for dynamic user adaptive hybrid model can be defined as:

$$\hat{r}_{u,i} = \alpha_u(r_m + U_u^T V_i) + (1 - \alpha_u)\hat{r}_{u,i}^T \tag{2}$$

where α_u is the tradeoff parameter for user u which needs to be learned.

We assume that the prior distribution of α_u is a Gaussian distribution with mean α_0 and variance σ_α^2. And through a Bayesian inference, the posterior probability of the latent factors U and V with above equation can be formulated as:

$$
\begin{aligned}
L &= p(R|U, V, \sigma_R^2, \alpha_u)p(U|\sigma_U^2)p(V|\sigma_V^2)p(\alpha|\sigma_\alpha^2) \\
&= \prod_{u=1}^{m}\prod_{i=1}^{n}\left[N(R_{u,i}|\alpha_u(r_m + U_u^T V_i) + (1 - \alpha_u)\hat{r}_{u,i}^T, \sigma_R^2)\right]^{I_{u,i}} \\
&\quad \times \prod_{u=1}^{m} N(U_u|0, \sigma_U^2 I) \times \prod_{i=1}^{n} N(V_i|0, \sigma_V^2 I) \times \prod_{u=1}^{m} N(\alpha_u|\alpha_0, \sigma_\alpha^2 I)
\end{aligned} \tag{3}
$$

3.2   Parameters Estimation

According to Equation 3, we can maximize the log-posterior of the joint probability with hyper-parameters:

$$
\begin{aligned}
(U, V, \alpha) &= \operatorname{argmin}\left[-\log L\right] \\
&= \arg\min \frac{1}{2}\sum_{u=1}^{m}\sum_{i=1}^{n} I_{u,i}\left\{R_{u,i} - (\alpha_u(r_m + U_u^T V_i) + (1 - \alpha_u)\hat{r}_{u,i}^T, \sigma_R^2))\right\}^2 \\
&\quad + \frac{1}{2}\lambda_U \|U\|_F^2 + \frac{1}{2}\lambda_V \|V\|_F^2 + \frac{1}{2}\lambda_\alpha \|\alpha\|^2
\end{aligned} \tag{4}
$$

where $\lambda_U = \sigma_R^2/\sigma_U^2$, $\lambda_V = \sigma_R^2/\sigma_U^2$, $\lambda_\alpha = \sigma_R^2/\sigma_\alpha^2$, and $\|\bullet\|_F$ denotes the Frobenius norm of a matrix.

4   Experiments and Analysis

4.1   Methodology

To show the performance improvement of our DTMF model, we compare our proposed method with several following methods:

- **Neighbor:** It selects the nearest neighbors to predict the missing values.
- **PMF:** PMF is proposed in [3] by Salakhutdinov and Mnih.
- **Trust:** In the experiments, we set the default trust distance as 4 for two data sets.
- **HybTMF:** This approach formulated as Equation 1 is a simply integrated algorithm.

4.2   Performance Analysis

We report the MAE and RMSE values of all comparison methods on Epinions and Flixster in Table 2. We observe that our method DTMF constantly outperforms HybTMF on both data sets. This demonstrates that DTMF model can learn proper parameters for every user and select the appropriate ways to combine matrix factorization and Trust methods. To show the significance of our performance improvement, we add t-test to test the significance levels of our model compared to other comparison parters and find p-value is less than 0.01 on both data sets.

Table 2. Comparison on MAE and RMSE for DTMF and other recommendation approahces

Data	Train	Metrics	Neighbor	PMF	Trust	HybTMF	DTMF
Epinions	80%	MAE	0.8688	0.8521	0.8223	0.8390	**0.8125**
		RMSE	1.1514	1.0759	1.0754	1.0589	**1.0388**
Flixster	80%	MAE	0.6726	0.6746	0.6752	0.6640	**0.6503**
		RMSE	0.9054	0.9067	0.9014	0.8833	**0.8726**

5   Conclusions

Due to the uneven distributions of ratings and relations, different algorithm may benefit each user to different extents.In this paper, we present a dynamic user adaptive hybrid recommendation model which combines matrix factorization and a trust-aware based method with user adaptive tradeoff parameters. By comparing with static combination parameters, the experimental results verify that user adaptive parameters could improve the prediction accuracy.

Acknowledgments. This work was supported by Natural Science Foundation of China (61272240, 71402083, 6110315).

References

1. Jamali, M., Ester, M.: A matrix factorization technique with trust propagation for recommendation in social networks. In: Proceedings of the Fourth ACM Conference on Recommender Systems, RecSys 2010, pp. 135–142. ACM (2010)
2. Massa, P., Avesani, P.: Trust-aware recommender systems. In: Proceedings of the 2007 ACM Conference on Recommender Systems, RecSys 2007, pp. 17–24. ACM (2007)
3. Mnih, A., Salakhutdinov, R.: Probabilistic matrix factorization. In: Advances in Neural Information Processing Systems, pp. 1257–1264 (2007)

An Adaptive Skew Handling Join Algorithm for Large-scale Data Analysis

Di Wu[1,3], Tengjiao Wang [1,2,3(✉)], Yuxin Chen[2,3], Shun Li[5],
Hongyan Li[2,4], and Kai Lei[1]

[1] School of Electronics and Computer Engineering (ECE),
Peking University, Shenzhen 518055, China
[2] School of Electronics Engineering and Computer Science,
Peking University, Beijing 100871, China
[3] Key Laboratory of High Confidence Software Technologies,
Peking University, Ministry of Education, Beijing 100871, China
[4] Key Laboratory of Machine Perception,
Peking University, Ministry of Education, Beijing 100871, China
[5] University of International Relations, Beijing 100091, China
wudi@sz.pku.edu.cn, {tjwang,chen.yuxin}@pku.edu.cn

Abstract. Join plays an essential role in large-scale data analysis, but the performance is severely degraded by data skew. Existing works can't adaptively handle data skew very well and reduce communication cost simultaneously. To address these problems, we firstly propose a mixed data structure comprising Bloom Filter and Histogram(BFH). Based on BFH, Bloom Filter and Histogram Join(BFHJ) is proposed to handle data skew adaptively. BFHJ can reduce communication cost by filtering unnecessary records. Furthermore, BFHJ adopts a heuristic partitioning strategies to balance workload. Experiments on TPC-H demonstrate that BFHJ outperforms the state-of-the-art methods in terms of communication cost, load balance and query time.

Keywords: Skew handling join · Adaptive · Partitioning strategy

1 Introduction

With the wave of big data coming, join plays an essential role in large-scale data analysis. Unfortunately, traditional RDBMS are no longer practical for such large-scale data. Parallel framework MapReduce[1] has been proven as a powerful approach for large-scale data analysis.

Several well-known join strategies has been implemented on MapReduce. However, data skew happens naturally in many applications. The well-known

This work was supported by Natural Science Foundation of China (Grant No. 61300003), Specialized Research Fund for the Doctoral Program of Higher Education(Grant No. 20130001120001) and Ministry of Education & China Mobile Joint Research Fund Program (MCM20130361).

© Springer International Publishing Switzerland 2015
J. Li and Y. Sun (Eds.): WAIM 2015, LNCS 9098, pp. 433–437, 2015.
DOI: 10.1007/978-3-319-21042-1_35

2/8 law[2] demonstrates this phenomenon. Data skew will cause load imbalance and hot nodes which may overshadow the strengths of parallel infrastructure.

Several improvements have been proposed[3–5] to handle skewness, however, there is still some room for improvement. For example, [3] can't execute the query adaptively. [4,5] fail in balancing the workload very well when data skew is severe. What's more, existing skew join algorithms haven't considered reducing communication cost by filtering records not in the join result.

To overcome above problems, we propose a mixed data structure comprising Bloom Filter and Histogram(BFH). Based on BFH, Bloom Filter and Histogram Join(BFHJ) is proposed to handle data skew adaptively. BFHJ can detect and filter unnecessary records. Furthermore, BFHJ adopts a heuristic partitioning strategies to balance the workload. Experiments on TPC-H demonstrate that BFHJ outperforms the state-of-the-art methods. In the following sections, BFHJ will be presented and evaluated.

2   Bloom Filter and Histogram Join

The intuition of Bloom Filter and Histogram Join(BFHJ) is to handle data skew adaptively and reduce communication cost. This section describes the architecture of this adaptive skew handling join algorithm using MapReduce framework.

2.1   The BFH Data Structure

BFH is a statistical data structure encompassing Bloom filter[6] and equi-width histogram. BFH is built on the join attribute. It combines the space efficiency of Bloom filter and the statistical property of histogram, which will help us estimate the distribution of join attribute before query processing.

There are two key components in BFH: (i) a single-hash-function Bloom filter of size m(bits); (ii) a hash table of counters for each non-zero bit of the Bloom filter, which is considered as an equi-width histogram. Each bin in the histogram is treated as a join part in our algorithm.

2.2   Filter and Partitioning Strategy

Assume that the BFH of relation L and R have been created called bfh_L and bfh_R. A new operation called bitwise multiply is defined for bfh_L and bfh_R to get the distribution of join attribute in the result. In the bitwise multiply operation, the bitmaps are conducted a bitwise-and operation. The corresponding bins between two histograms are conducted a multiply operation.

The new bitmap is considered as a filter. If the value equals 0 in the filter, this indicates the corresponding record is not in the result. Although few records will not be detected, it is a trade-off between space and accuracy.

In the new histogram after the bitwise multiply operation, the value of each bin represents the count of a join part in the join result, which can also be considered as the workload of this join part.

For simplicity, assume that the processing ability of reducers is same. So, the task is to to assign each join part to each reducer, minimize the max workload, make the workload among reducers more balanced. Unfortunately, solving the optimal assignment strategy is NP hard. So, an adaptive heuristics partitioning strategies BFH-S(Sequence) is proposed which is shown in Algorithm 1.

Algorithm 1. BFH-S Partitioning Strategy

1: Input: workload of n join parts $W = \{w_1, w_2, ..., w_n\}$
2: Output: join part sets of r reducers $S = \{S_1, S_2, ..., S_r\}$
3: $L = \{l_1, l_2, ..., l_r\}$
4: $l_k \leftarrow 0, S_k \leftarrow \emptyset (1 \leq k \leq r)$
5: **for** i from 1 to n **do**
6:     $l_j = min\{l_1, l_2, ..., l_r\}$
7:     $S_j = S_j \cup \{i\}$
8:     $l_j = l_j + w_i$
9: **end for**
10: **return** S

In BFH-S, we pick the join part sequentially which is not yet assigned to a reducer, and assign it to the reducer which has the smallest total workload. We repeat these steps until all parts have been assigned.

3   Experiments

All our experiments are run on a 20-nodes cluster of Hadoop, which is an open-source implementation of MapReduce. We compare the performance of Reduce-Side join(RSJ),Pig's skew join(PSJ), Hive's skew join(HSJ) and our BFHJ.

The TPC-H benchmark[7] is used in our experiments. The following query is executed: select * from Customer C, Supplier S where C.Nationkey = S.Nationkey. To control skewness, we randomly choose a portion of data and change Nationkey to some values. To evaluate the BFH filter, 15% records which are not in the join result are randomly selected by changing Nationkey.

3.1   Experimental Results

The BFH size is set to 10000 which is a trade-off between space and capacity. Then BFH uses less than 100kb space. Tab.1 shows the creation cost of BFH on different data scale. From Tab.1, the creation time accounts for less than 5% of the whole query time of BFHJ when skewness is 0.1. It is acceptable because BFH could be reused.

Fig.1 illustrates the influence of join key cardinal and load balance of different algorithms. The data scale is 50G and the skewness is set to 0.1 on both relations.

Fig.1(a) demonstrates the communication cost. The map output records of BFHJ are less than other algorithms since the BFH filter some unnecessary

Table 1. Creation cost of BFH Data Structure

Data scale	20G	50G	100G
Creation time	22.843s	82.172s	201.617s
Query time(skewness = 0.1)	534.969s	1989.631s	4789.676s
Create Time/Query Time	4.27%	4.13%	4.21%

records. If BFH size is fixed, the amount of filtering records decreases with the cardinal increasing. It is because more bits are set to 1 which is a false positive. When the cardinal is 12000, all the bits are set to 1, so the filter ratio is 0.

In Fig.1(b), the query time is decreasing with the cardinal increasing because total join results decrease. Significantly, no matter what the cardinal is, BFHJ is better than other algorithms in terms of query time.

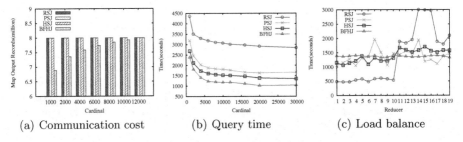

(a) Communication cost   (b) Query time   (c) Load balance

Fig. 1. The influence of join key cardinal and load balance

To evaluate the load balance, we figure out the using time of each reducer when cardinal is 4000. In Fig.1(c), RSJ will cause some hot reducers because of the skew. PSJ's range partition can't handle the skew on both relation. HSJ improves the performance, but the using time are still unbalanced. The results show that BFHJ can achieve the most balanced workload.

4   Conclusion

To handle data skew adaptively, this paper proposes the mixed data structure BFH. Based on BFH, Bloom Filter and Histogram Join(BFHJ) is proposed. BFHJ reduce communication cost by detecting and filtering unnecessary records. BFHJ balance workload by adopting a heuristic partitioning strategies. Experiments demonstrate that BFHJ outperforms the state-of-the-art methods.

References

1. Dean, J., Ghemawat, S.: Mapreduce: simplified data processing on large clusters. Communications of the ACM **51**(1), 107–113 (2008)
2. Walton, C.B., Dale, A.G., Jenevein, R.M.: A taxonomy and performance model of data skew effects in parallel joins. In: VLDB, vol. 91, pp. 537–548 (1991)

3. Thusoo, A., Sarma, J.S., Jain, N., Shao, Z., Chakka, P., Anthony, S., Liu, H., Wyckoff, P., Murthy, R.: Hive: a warehousing solution over a map-reduce framework. Proceedings of the VLDB Endowment **2**(2), 1626–1629 (2009)
4. Atta, F., Viglas, S.D., Niazi, S.: Sand join skew handling join algorithm for google's mapreduce framework. In: 2011 IEEE 14th International on Multitopic Conference (INMIC), pp. 170–175. IEEE (2011)
5. Gates, A.: Programming Pig. O'Reilly (2011)
6. Bloom, B.H.: Space/time trade-offs in hash coding with allowable errors. Communications of the ACM **13**(7), 422–426 (1970)
7. Council, T.P.P.: Tpc-h benchmark specification (2008). Published at http://www.tpc.org/tpch/

NokeaRM: Employing Non-key Attributes in Record Matching

Qiang Yang[1], Zhixu Li[1(✉)], Jun Jiang[1], Pengpeng Zhao[1],
Guanfeng Liu[1], An Liu[1], and Jia Zhu[2]

[1] School of Computer Science and Technology, Soochow University, Suzhou, China
{20144227003,20134527003}@stu.suda.edu.cn,
{zhixuli,ppzhao,gfliu,anliu}@suda.edu.cn
[2] School of Computer Science, South China Normal University, Guangzhou, China
jzhu@m.scnu.edu.cn

Abstract. Record Matching (RM) aims at finding out pairs of instances referring to the same entity between relational tables. Existing RM methods mainly work on key attribute values, but neglect the possible effectiveness of non-key attribute values in RM. As a result, when two instances referring to the same entity do not have similar key attribute values, they are unlikely to be linked as an instance pair. On the other hand, the two instances may share some important non-key attribute values which can also help us identify the relationship between them. With this intuition, we propose to employ non-key attributes in RM. Basically, we propose a rule-based algorithm based on a tree-like structure, which can not only deal with noisy and missing values, but also greatly improve the efficiency of the method by finding out matched instances or filtering unmatched instances as early as possible. The experimental results based on several data sets demonstrate that our method outperforms existing RM methods by reaching a higher precision and recall. Besides, the proposed techniques can greatly improve the efficiency of a baseline.

Keywords: Record matching · Non-key attribute · Algorithm

1 Introduction

RM aims at finding out pairs of instances referring to the same entity across different databases. Most existing RM solutions rely on string similarity metrics to measure the similarity between key attribute values of instances and then make decisions according to a predefined similarity threshold [1]. However, an arbitrary threshold hurts either the matching precision or the recall.

It happens frequently that two instances which do not have similar key attribute values may share some non-key attribute values. Based on the observation, this paper works on employing non-key attributes for RM. We can see that non-key attribute values can help us identify the relationship between two instances from Table 1 and Table 2. Note that our method is orthodox to the existing RM methods based on key attributes. We mainly pay attention on how to use non-key attributes smartly to improve the precision and recall of RM.

© Springer International Publishing Switzerland 2015
J. Li and Y. Sun (Eds.): WAIM 2015, LNCS 9098, pp. 438–442, 2015.
DOI: 10.1007/978-3-319-21042-1_36

Table 1. Example "Cellphones on Sale" Table 1 collected from Tmall

	Product	Manufacturer	Size	RAM	Release	Type	OS	...
t1	w2013	SAMSUNG	3.7 inches	1GB	2013.04	Flip	Android 4	...
t2	8295	Coolpad	4.7 inches	1GB	2013.01	Bar	Android 4.1	...
t3	MXII	MeiZu	4.4 inches	2GB	2012.12	Bar	Flyme 2	...
t4	Ericsson U1	Sony	3.5 inches	512MB	2009.10	Bar	S60V5	...
t5	4S	Apple	3.5 inches	512MB	2011.01	Bar	IOS 5	...
t6	A880	Lenove	6 inches	1GB	2013.04	Bar	Android 4.2	...
t7	Ascend M2	HuaWei	6.1 inches	2GB	2014.03	Bar	Android 4.3	...
t8	S930	Doov	2.8 inches	512MB	2010.09	Slide	-	...
t9	G9098	SAMSUNG	3.67 inches	2GB	2014.09	Bar	Android 4.3	...
t10	8730L	Collpad	5.5 inches	1GB	2012.05	Bar	Android 4.3	...

Table 2. Example "Cellphones on Sale" Table 2 collected from PConline

	Product	Manufacturer	Size	RAM	Release	Type	OS	...
s1	Galaxy w2013	SAMSUNG	3.7 "	1GB	2013-04	Flip	Android 4.0	...
s2	8295	Coolpad	4.7 "	-	2013-01	Bar	Android 4.1	...
s3	Meizu MX2	MeiZu	4.5 "	2.0G	2012-04	-	Flyme 2.0	...
s4	3s	XIAOMI	5.1 "	3G	-	Bar	MIUI V5	...
s5	IPhone 4s	Apple	-	512MB	2011-05	Bar	IOS 5.0	...
s6	A880	Lenove	6.0 "	1G	2013-04	Bar	Android 4.1	...
s7	Mate2	HuaWei	6.1 "	2GB	2014-03	Bar	Android 4.3	...
s8	S930	Doov	2.8 "	512M	2010-09	Slide	-	...
s9	8730L	Coolpad	-	1G	2012-05	Bar	Android 4.3	...

It is non-trivial to employ non-key attributes for RM. Compared to key attribute, non-key attributes can be more noisy and inconsistent. Besides, there are usually a lot more non-key attributes than key attributes, thus RM based on non-key attributes has a significant efficiency problem. We propose to build a specific *Probabilistic Rule-based Decision Tree* with non-key attributes based on their specific abilities: the ability in identifying matched instances referring to the same instances, and the ability in filtering unmatched instances. With the tree, we expect to find out matched instance pairs, as well as filtering unmatched instances as early as possible.

Definition 1. (RM with Non-Key Attributes (NokeaRM)). *Given two tables* $T_1 = \{t_1, t_2, ..., t_n\}$ *and* $T_2 = \{s_1, s_2, ..., s_m\}$ *sharing a set of attributes* $S_{NK} = \{A_1, A_2, ..., A_p\}$, *NokeaRM problem aims at finding a function* $\mathcal{F}(t_i, s_j)$ *based on* S_{NK} *and a threshold* τ, *such that for* $\forall t_i \in T_1$ ($1 \leq i \leq n$), *and* $\forall s_j \in T_2$ ($1 \leq j \leq m$), *they are linked instances referring to the same entity if and only if: (1)* $\mathcal{F}(t_i, s_j) \geq \tau$, *and (2)* $\forall s_k \in T_2$, $\mathcal{F}(t_i, s_j) \geq \mathcal{F}(t_i, s_k)$.

2 A Probabilistic Rule-Based Algorithm

We propose a NokeaRM algorithm based on a probabilistic rule-based decision tree built with non-key attributes. In the following, we first introduce how we build the tree, and then present a tree-based NokeaRM algorithm.

1) Building the Tree: Basically, the Probabilistic Rule-based Decision Tree (or PRTree for short) is built with non-key attributes according to two important properties of each non-key attribute, so-called *Sufficiency* and *Necessity*. One's sufficiency reflects its ability to find out matched instance pairs while one's necessity reflects its ability to reject unmatched instance pairs. Basically, we always prefer to put attribute with the maximum sufficiency or necessity score at the root part, such that it can accept matched pairs or decline unmatched pairs as early as possible. Based on this intuition, we describe how to build the PRTree.

(1) **Root Node:** After estimating the sufficiency and necessity of every non-key attribute, we select the one with the maximum sufficiency or necessity as the root node. The root node has three branches, i.e., *Matched (Y)*, *Unmatched (N)* and *Invalid (Null)*.

(2) **Non-Leaf Node:** To select an attribute for a non-leaf node, we estimate the sufficiency and necessity of all remaining attributes under the condition of all its ancestor nodes, and select the one with the maximum conditional sufficiency or necessity as the node.

(3) **Leaf Node:** Every leaf node outputs "Matched" or "Unmatched", showing that the two instances are matched pair or not.

2) NokeaRM Algorithm based on PRTree: We describe the NokeaRM algorithm based on PRTree. Basically, every instance pair (t, s) visits the PRTree from the root node and stops when they reach a leaf node.Each time the pair (t, s) comes to a node with attribute A_k, we check whether the two instances in the pair share the same attribute value in the node. If yes, we go to the "Matched" child node of the node, otherwise the "Unmatched" child node of the node. But if some value under this attribute in the two instances are missing, we go to the "Invalid" child node.When the pair goes to a leaf node, we will output "Matched" or "Unmatched" according to the property of the leaf node. The confidence of the decision is jointly decided by all the nodes the pair passes by from the root to the current node.

3 Experiments

We have experimented on two real world data sets and one synthetic data set. We compare the precision and recall of our algorithms including **Baseline** and **PRTree (Nokey)**-based methods against a state-of-the-art **Pure Key**-based RM method. Edit distance is the way we use in all algorithms for estimating string similarity. Besides, we also consider a proper way to combine our PRTree-based method with the key-based method and find out that the best way is to also take the key attribute into the tree. So we also use have a **PRTree (Key)**-based method for comparison.

3.1 Comparison with Previous Methods

As shown in Figure 1, we can see that pure key-based method reaches the worst precision and recall than the other methods, while the best performance is reached by Baseline and PRTree method using both key and non-key attributes. The PRTree method without key also has an impressive performance, but without the key attribute we miss some important information for RM. Basically, our method can always improve the precision by nearly 15%, and recall by around 20% on all the three data sets.

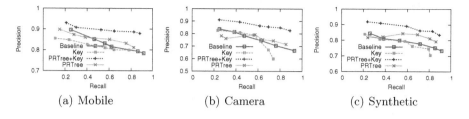

(a) Mobile   (b) Camera   (c) Synthetic

Fig. 1. Comparing the Precision and Recall against Previous Approaches

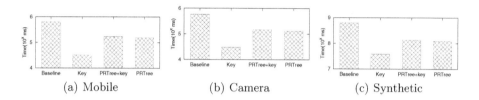

(a) Mobile   (b) Camera   (c) Synthetic

Fig. 2. Comparing the Efficiency of NokeaRM against Previous Approaches

3.2 Scalability

As can be observed in Fig. 2, we compare the cost time for all the four methods on these three datasets. The pure key method uses the least time since it uses the key attribute only, while the PRTree-based method uses 10 times less time cost than the baseline since the PRTree make all decisions as early as possible, which greatly reduces the times of comparing attribute values.

4 Related Work and Conclusions

Record Matching (RM) has been studied extensively [3]. To avoid pairwise comparison, many effective techniques have been proposed. Some work exploits Q-Grams together with inverted indices [2], or prefix-based pruning techniques [4] to filter a large part of unmatched pairs for comparison. We study the problem of RM based on non-key attributes and propose an effective NokeaRM algorithm. Compared to existing methods, we reach a much higher precision and recall on several data sets. As a future work, we consider to involve crowdsourcing in our method for reaching higher precision and recall.

Acknowledgements. This research is partially supported by Natural Science Foundation of China (Grant No. 61472263, 61402313, 61303019).

References

1. Cheatham, M., Hitzler, P.: String similarity metrics for ontology alignment. In: Alani, H., et al. (eds.) ISWC 2013, Part II. LNCS, vol. 8219, pp. 294–309. Springer, Heidelberg (2013)
2. Ukkonen, E.: Approximate string-matching with q-grams and maximal matches. Theoretical Computer Science **92**(1), 191–211 (1992)
3. Verykios, V.S., Elmagarmid, A.K., Houstis, E.N.: Automating the approximate record-matching process. Information Sciences **126**(1), 83–98 (2000)
4. Wang, W., Xiao, C., Lin, X., Zhang, C.: Efficient approximate entity extraction with edit distance constraints. In: Proceedings of the 2009 ACM SIGMOD International Conference on Management of data, pp. 759–770. ACM (2009)

Efficient Foreign Key Discovery Based on Nearest Neighbor Search

Xiaojie Yuan, Xiangrui Cai, Man Yu, Chao Wang,
Ying Zhang$^{(\boxtimes)}$, and Yanlong Wen

College of Computer and Control Engineering, Nankai University,
94 Weijin Road, Tianjin 300071, People's Republic of China
{yuanxj,caixr,yuman,wangc,zhangy,sun}@dbis.nankai.edu.cn

Abstract. With rapid growth of data size and schema complexity, many data sets are structured in tables but without explicit foreign key definitions. Automatically identifying foreign keys among relations will be beneficial to query optimization, schema matching, data integration and database design as well. This paper formulates foreign key discovery as a nearest neighbor search problem and proposes a fast foreign key discovery algorithm. To reduce foreign key candidates, we detect inclusion dependencies first. Then we choose statistical features to represent an attribute and define two attributes's distance. Finally, foreign keys are discovered by finding nearest neighbors of all primary keys. Experiment results on real and synthetic data sets show that our algorithm can discover foreign keys efficiently.

Keywords: Foreign key · Nearest neighbors · Schema

1 Introduction

Schema is the basis to comprehend a database, which is of great significance in data modeling, query optimization, schema matching and indexing, etc. Primary/foreign key relationships play an important role in the schema of relational database. However, many reasons will lead to incompletion of foreign key constraints in a database. For example, the database lacks support for checking foreign key constraints, the designers ignore to check foreign keys on purpose for performance consideration, or some relationships are not known to designers but are inherent in the data. All these reasons are quite frequent met in databases. The absence of foreign keys will lead to poor data quality and thus influence data analysis. When it happens in a database, which contains hundreds of tables, thousands of attributes, countless tuples and lacks of documentation in addition, it is extremely difficult to identify the foreign keys even for an expert.

Surprisingly, little attention was paid to foreign key discovery. Most previous work focuses on detecting INclusion Dependencies(INDs) [1][3]. An IND $A \subseteq B$ demands tuples in attribute A should appear in attribute B. Obviously, it is not sufficient to identify foreign keys. In light of this, some researchers try to

© Springer International Publishing Switzerland 2015
J. Li and Y. Sun (Eds.): WAIM 2015, LNCS 9098, pp. 443–447, 2015.
DOI: 10.1007/978-3-319-21042-1_37

discovery all foreign keys in relations [4][5]. Zhimin C. *et al.* [2] simplify the problem by discovering foreign keys between specified tables. All these work either takes long time or discovers specified foreign keys. This paper focuses on discovering all foreign keys in relations. We formulate foreign key discovery as a nearest neighbor search problem, which reveals the essence of this problem. Considering three similarities between two attributes, we propose a good distance measure and separate foreign keys from other INDs effectively and efficiently.

2   Foreign Key Discovery

We regard foreign key discovery as a nearest neighbor search problem. Given a collection of relations $\mathbf{T}$, a parameter $R > 0$ and a distance metric $dist(*, *)$, the primary key set $\mathbf{P}$ is known, let $\mathbf{C}$ be the set of all attributes in $\mathbf{T}$. The foreign key discovery is constructing a data structure that, for each primary key $P_i \in \mathbf{P}$, reports all $F_{ij} \in \mathbf{C}$ as P_i's foreign key(s) if $dist(P_i, F_{ij}) < R$.

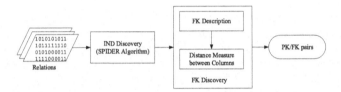

Fig. 1. Overview of foreign key discovery

Figure 1 shows an overview of our algorithm. Similar to previous work, we detect INDs first to reduce candidates. We take SPIDER [1] to detect INDs. To cope with multi-column INDs, we modify the SPIDER algorithm with a level-wise method [3], which is widely used traverse all attribute combinations.

Since the choice of features has significant influence on the achievable performance, we performed an extensive manual study to conclude meaningful features. We sort attributes and represent them by feature vector. We list these features below, each followed by a brief explanation.

- **Distinct Tuples(DT):** The number of distinct tuple in an attribute. Foreign key values often cover a wide range of primary key values.
- **Attribute Name(N):** This feature is used to measure the distance of two attribute names, which reflect the similarity of names.
- **Average(A):** The average of all distinct tuples for a numerical attribute. The averages of a foreign key and its primary key are often very close.
- **Variance(V):** The variance of all distinct tuples for a numerical attribute. The variances are often close between a foreign key and its primary key.
- **Average Length(AL):** The average length of all tuples for a sting attribute. The average lengths should be similar when the values of a foreign key form a non-bias sample of the primary key.

– **Median(M):** The median tuple of all distinct tuples for an attribute. The median of a foreign key should be close to the median of its primary key.

These features mainly cover name similarity and distribution similarity, for the value similarity is considered during IND detection. All these features should be normalized and then we use l_2 to measure two attributes' distance:

$$dist(P, F) = \|(DT_P, N_P, A_P, V_P, AL_P, M_P) - (DT_F, N_F, A_F, V_F, AL_F, M_F)\| \quad (1)$$

where $DT_P, N_P, A_P, V_P, AL_P, M_P; DT_F, N_F, A_F, V_F, AL_F, M_F$ are the features of attribute P and F. We compute all distances of each IND pair. There is a big jump between foreign keys and other INDs for foreign key and primary key have underlying sematic relationship. Our algorithm can detect this big jump point automatically and unitize it as threshold to separate foreign keys from INDs.

3   Experiments

We evaluate our algorithm on two benchmark synthetic databases(TPC-E and TPC-H), a sample database of MySQL(sakila), as well as an information management system database(EMIS). The characteristics of them are shown in table 1, where **T** is the number of non-empty tables, **Avg(C)** and **Max(C)** are the average and maximum number of columns per table, **Avg(R)** and **Max(R)** are the average and maximum number of rows per table, **SC** and **MC** are numbers of single column and multi-column foreign keys.

Table 1. Characteristics of data sets

	T	Avg(C)	Max(C)	Avg(R)	Max(R)	SC	MC
TPC-H	8	8	16	1082504	6000003	9	1
TPC-E	32	6	24	171127	4469625	44	1
sakila	16	6	13	2954	16049	22	0
EMIS	112	9	59	120744	4062145	110	1

Our algorithm is implemented with C#, and performed experiments on an Intel Core i7-2600 3.40GHz with 8GB RAM running SQL Server 2008. We evaluate Precision, Recall and F-measure. The schemas of the databases are the ground truth. Table 2 shows our algorithm's effectiveness for different data sets, where "CP NO." means the big jump point. It is easy to see our algorithm can obtain a good precision and recall.

We compare our algorithm with randomness test [5] in TPC-H and TPC-E, as shown in table 3. The effectiveness of the two method is close. However, we avoid high complexity attribute distance computation.

We test the scalability of our algorithm on four TPC-H instances with size 1MB, 10MB, 100MB, 1GB and 10GB. The running times for each of two phases and the total time are shown in figure 2. Phase 1 stands for IND detection,

Table 2. Effectiveness for each data set

	TPC-H	TPC-E	sakila	EMIS
CP NO.	10	59	22	45
Precison	0.8	0.59	0.77	1
Recall	1	0.78	0.81	0.49
F-measure	0.89	0.67	0.79	0.66

Table 3. Effectiveness comparison with randomness test

		Precison	Recall	F-measure
TPC-H	randomness test	1	1	1
	our method	0.8	1	0.89
TPC-E	randomness test	0.57	0.82	0.67
	our method	0.59	0.78	0.67

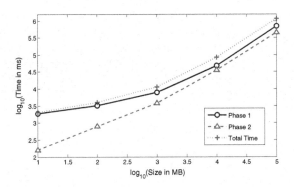

Fig. 2. Scalability Results

and phase 2 stands for foreign key discovery. For the 10GB instance, the total running time is less than 20 minutes, while [5] takes more than 2.5 hours. Our algorithm is more applicable to enterprise-scale data sets.

4 Conclusion

Discovering foreign keys in relations is a very important task with wide applications. But it is very challenging both in effectiveness and efficiency. This paper formulates it as a nearest neighbor search problem and defines a sound distance between attributes. Experiments demonstrate our method obtains high efficiency with little effectiveness loss. One interesting future work is generalizing this work by allowing fuzziness of the column values because data may come from different sources.

Acknowledgement. This work is supported by NSFC under Grant No. 61170184 and 61402243, National 863 Project of China under Grant No. 2015AA011804, National Key Technology R&D Program under Grant No.2013BAH01B05, and Tianjin Municipal Science and Technology Commission under Grant No. 13ZCZDGX02200, 13ZCZDGX01098, 14JCQNJC00200 and 14JCTPJC00543.

References

1. Bauckmann, J., Leser, U., Naumann, F., Tietz, V.: Efficiently detecting inclusion dependencies. In: Proceedings of ICDE, pp. 1448–1450 (2007)
2. Chen, Z., Narasayya, V., Chaudhuri, S.: Fast foreign-key detection in microsoft sql server powerpivot for excel. Proceedings of the VLDB Endowment **7**(13) (2014)
3. De Marchi, F., Lopes, S., Petit, J.M.: Unary & n-ary inclusion dependency discovery in relational databases. Journal of Intelligent Information Systems **32**(1) (2009)
4. Rostin, A., Albrecht, O., Bauckmann, J., Naumann, F., Leser, U.: A machine learning approach to foreign key discovery. In: 12th International Workshop on the Web and Databases, Providence, Rhode Island, USA (2009)
5. Zhang, M., Hadjieleftheriou, M., Ooi, B.C., Procopiuc, C.M., Srivastava, D.: On multi-column foreign key discovery. PVLDB **3**(1), 805–814 (2010)

OntoEvent: An Ontology-Based Event Description Language for Semantic Complex Event Processing

Meng Ma[1(✉)], Ping Wang[2,4], Jun Yang[3], and Chao Li[4]

[1] School of EECS, Peking University, Beijing 100871, China
mameng@pku.edu.cn
[2] National Engineering Research Center for Software Engineering,
Peking University, Beijing 100871, China
pwang@pku.edu.cn
[3] School of Electronic and Computer Engineering, Peking University, Shenzhen 518055, China
yangjvn@pku.edu.cn
[4] School of Software and Microelectronics, Peking University, Beijing 102600, China
li.chao@pku.edu.cn

Abstract. In this paper, we propose an ontology-based event description language, OntoEvent, for semantic event modeling in CEP system. In OntoEvent language, complex events are modeled and described in the form of event ontology. We propose the concept of nature language event constructor and define them by synonym set of WordNet database to describe logical and temporal event relationship in OntoEvent. We demonstrate event ontology examples in application domain of smart home, and our analysis shows that OntoEvent is of rich expressiveness compared to other related languages.

Keywords: Complex event processing · Ontology · Event description language · Semantic · WordNet · Natural language processing

1   Introduction

Complex event processing (CEP) [1] technique is one of the important cornerstones to connect cyber and physical world. Semantic event modeling and processing helps CEP systems to establish a transparent, machine-understandable knowledge discovering process. Some typical semantic complex event processing (SCEP) systems such as SCEPter [2], OECEP [3] and DyKnow [4] are proposed. However, a fundamental problem for SCEP is that no general-purpose semantic model for event construction has been proposed. In this paper, we propose an ontology-based event description language, OntoEvent, for semantic event model in CEP systems. In OntoEvent semantic model, we divide the concepts into two levels: general concepts and domain-specific instances, which promises the OntoEvent can be dynamic extended for different application usage. Complex events are modeled and described in the form of event ontology based on nature language event constructors. Our analysis and preliminary implementation proves that OntoEvent is of rich expressiveness and interoperability between different event description languages. The rest of this paper is organized as follows: Section 2 presents

© Springer International Publishing Switzerland 2015
J. Li and Y. Sun (Eds.): WAIM 2015, LNCS 9098, pp. 448–451, 2015.
DOI: 10.1007/978-3-319-21042-1_38

the semantic models and elaborates the event description language. We demonstrate an event example and discuss the language expressiveness in Section 3. Section 4 concludes the paper and discusses our future works.

2   OntoEvent Language

The semantic model of OntoEvent language defines and describe the concepts and their inter-relationship in complex event processing. It is a two-level framework which includes a set of general upper ontology and interfaces for different application domain ontology extensions. All the concepts in OntoEvent semantic model are grouped into four domains: *Entity*, *Dimension*, *Activity*, and *Service*. The semantic model of OntoEvent is established and implemented in OWL [5] ontology languages. Based on this semantic model, we present a novel ontology-based event description language, OntoEvent. Different from other event description language which we have surveyed in Section 2, OntoEvent language defines complex event as an ontology by its event component and their relationship. The component of a complex event includes primitive event, event constructor, event pattern, attribute, sliding window size and response action. The OntoEvent language is defined as follow:

OntoEvent Language Model. The OntoEvent language model is denoted as $E = (e, c, p, a)$, in which e is the event set, o is the event constructor set, $p = \{hasComponent, hasAttribute, hasSource, hasData, hasWindow, hasAction\}$ is the ontology property set and a stands for the attribute/data set. The pattern, primitive event source, attribute constraints, event window and response action of a certain complex event is defined by the following triples:

Pattern. $\subseteq \{e \cup c\} \times \{hasComponent\} \times \{e \cup c\}$, defines how the event is constructed by primitive events and event constructors.

Source. $\subseteq e \times \{hasSource\} \times a$, defines the primitive event source and indicates their corresponding attributes.

Constraints. $\subseteq \{e \cup c\} \times \{hasData, hasAttribute\} \times a$, defines the attribute constraints of primitive of event constructor.

Window. $\subseteq e \times \{hasWindow\} \times a$, defines the overall processing time limitation on the input event stream.

Action. $\subseteq e \times \{hasAction\} \times a$, defines the response action for the complex event when it is detected.

In OntoEvent language, we introduce the notion of nature language event constructor to express the logic and temporal relationship in event pattern. These constructors are the "keywords" for pattern description in natural English language. In order to support synonym keyword in event description, we introduce WordNet [6], which is a lexical semantic database groups English words into sets of synonyms. With the help of WordNet, we can avoid repetition of similar keyword definitions. Besides, natural or semi-natural description from end-users can be processed by natural language processing (NLP) and mapped into their equivalent constructors. In Table 1, we present and explain some typical representative constructors and their equivalent WordNet synset.

Table 1. Typical Nature Language Event Constructors

Representative Constructor	Equivalent Synset in WordNet		Constructor Semantics
	Type	Synonymy	
and	(adv)	meanwhile, meantime, in the meantime	Conjunction of two event/constructor components.
or	(adv)	either	Disjunction of two event/constructor components.
not	(adv)	no	Negation of event/constructor component.
then	(adv)	subsequently, later, afterwards, after	Sequence of two event/constructor components, representing the two event occurs sequentially.
distance	(n)	distance, space	Describing the temporal distance of two event/constructor components.
within	(adv)	within, in	Describing that event/constructor component occurs within the certain time interval.
keep	(v)	retain, continue, keep, keep on	Describing that one or more event/constructor component occurs and no counter instance occurs within a certain time interval.

3   Application Scenario and Event Example

We illustrate our proposed event description language using event examples in smart home application scenario. Smart home refers to a home environments that are enabled for co-operation of smart objects and systems for ubiquitous interactions [7]. In this environment, we leverage devices such as location sensor, body sensor and other sensing devices to generate primitive event streams.

Natural Language Event Description. This pattern set the status of user as *sleeping* if user enters the bedroom first and **then** heartbeat rates **keep** lower than 70 **within** 30 seconds, while the temporal **distance** between these two events is [600, 1200].

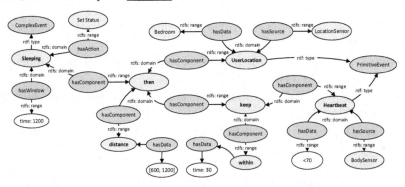

Fig. 1. Event Ontology Example

The event ontology for this event example is depicted in Figure 1. This event describes a temporal-constrained event sequence and iteration consists of two primitive event *UserLocation* and *HeartBeat*. The source of primitive event are declared by attribute *hasSource*. The event constraints are defined by *hasData*. To describe this complex event *Sleeping*, we leverages the constructor *then, keep, within* and *distance* to depict the event logic. Specifically, we use constructor *distance* to describe the temporal distance between two event components of constructor *then*. We use *keep* to describe the event repetition of *HeartBeat* when the reading is lower than 70 and declared the time scope of 30 seconds by *within* constructor.

OntoEvent has rich expressiveness in logical and temporal aspects. Especially, it provides a set of synonym constructor to express the event iteration or status maintenance. The expressiveness of OntoEvent can be extended by defining more event constructors. Because of the essence of ontology, the event ontology can be defined and demonstrated in a user-friendly visualized paradigm. Therefore, as an ontology-based language, it is designed to be a ***middle-language*** to provide interoperability among different event processing systems by language rewriting techniques. Natural language event descriptions can be tranformed into event ontology by NLP. Besides, existing non-semantic event descriptions in existing CEP systems can be transformed into OntoEvent language by event rewriting.

4   Future Works

In our other works, an ontology-based event processing engine have been preliminarily implemented based on OntoEvent language. In this event processing model, we transform event ontology into non-deterministic finite automata-based detection model. A multi-target detection model is established according to the inference relationship between events which is derived from defined event ontologies. In the future work, we plan to establish a natural event query processing approach with the help of NLP, to map natural language query into equivalent event ontology. By this means, ontology-based CEP system can provide a user-friendly interface in complex event querying.

References

1. Cugola, G., Margara, A.: Processing flows of information: From data stream to complex event processing. ACM Computing Surveys (CSUR) **44**, 15 (2012)
2. Zhou, Q., Simmhan, Y., Prasanna, V.: SCEPter: Semantic complex event processing over end-to-end data flows. Technical Report 12-926, Computer Science Department, University of Southern California (2012)
3. Binnewies, S., Stantic, B.: OECEP: enriching complex event processing with domain knowledge from ontologies. In: Proceedings of the Fifth Balkan Conference in Informatics, pp. 20–25. ACM, Novi Sad (2012)
4. De Leng, D., Heintz, F.: Towards on-demand semantic event processing for stream reasoning. In: 17th International Conference on Information Fusion (FUSION), pp. 1–8 (2014)
5. McGuinness, D.L., Van Harmelen, F.: OWL web ontology language overview. W3C recommendation, February 10, 2004
6. Miller, G.A.: WordNet: a lexical database for English. Communications of the ACM **38**, 39–41 (1995)
7. Helal, S., Mann, W., El-Zabadani, H., King, J., Kaddoura, Y., Jansen, E.: The gator tech smart house: A programmable pervasive space. Computer **38**, 50–60 (2005)

Differential Trust Propagation with Community Discovery for Link-Based Web Spam Demotion

Xianchao Zhang, Yafei Feng, Hua Shen, and Wenxin Liang[✉]

School of Software, Dalian University of Technology, Dalian 116620, China
{xczhang,wxliang}@dlut.edu.cn, yafei.feng@gmail.com,
shenhua_as@126.com

Abstract. In this paper, we propose a novel differential trust propagation scheme with community discovery, which can be applied to all kinds of trust propagation algorithms. We first use a random walk-based community discovery algorithm to preselect suspicious communities in which the members are almost spam pages. We then utilize these suspicious communities to limit the across-community-boundary trust propagation. Experimental results on WEBSPAM-UK2007 and ClueWeb09 demonstrate that the proposed penalizing scheme significantly improves the performance of trust propagation algorithms such as TrustRank, LCRank, CPV.

Keywords: Web spam · Community discovery · Differential trust propagation

1 Introduction

Trust propagation algorithms (e.g., TrustRank [1]), which propagate trust of a carefully selected seed set of good pages to the entire Web, have been widely used for spam demotion. However, in most existing algorithms, trust is propagated in non-differential ways. Thus, a differential trust propagation scheme in which a page propagates less trust to a (suspicious) spam neighbor than a normal one is urgently needed [2]. In this paper, we propose a differential trust propagation scheme with community discovery. Firstly, considering known spam pages and reputable pages as seeds, we extract communities using seed-based community identification techniques. We then use the extracted communities to limit across-community-boundary trust propagation. Since most members of the extracted communities are spam pages, the good-to-bad trust propagation can be efficiently limited. Experimental results show that our scheme can significantly improve the performance of trust propagation algorithms such as TrustRank, LCRank, CPV.

W. Liang—This work was supported by NSF of China (No. 61272374,61300190), SRFDP of Higher Education (No.20120041110046) and Key Project of Chinese Ministry of Education (No. 313011).

ⓒ Springer International Publishing Switzerland 2015
J. Li and Y. Sun (Eds.): WAIM 2015, LNCS 9098, pp. 452–456, 2015.
DOI: 10.1007/978-3-319-21042-1_39

2   The Framework

2.1   Extracting Spam Communities

In this section, we propose a community discovery method to aid trust propagation for spam demotion based on a lazy random walk [3]. Given both spam seed set and good seed set, we can use a biased random walk to extract other members within the same community. A formal description of this strategy is presented in Algorithm 1.

Algorithm 1. Spam Community Discovery

Input: Web graph $G = (V, E)$, Adjacency matrix $\mathbf{A}$, Spam seed Set S^-, Reputable seed set S^+
Output: Spam community $\mathcal{C}$
$t = 0$;
while $t < step$ **do**
$\quad | \quad \boldsymbol{p_{t+1}} = \frac{1}{2}(I + AD^{-1})\boldsymbol{p_t}$;
$\quad | \quad \boldsymbol{p'_{t+1}} = \frac{1}{2}(I + AD^{-1})\boldsymbol{p'_t}$;
$\quad | \quad t = t + 1$;

for each page $i \in V$ **do**
$\quad | \quad p_t(i) = p_t(i)/(p_t(i) + p'_t(i))$

normalization $\boldsymbol{p}$;
for each page $i \in V$ **do**
$\quad | \quad$ **if** $p_t(i)$ ranks top k-percentile **then**
$\quad | \quad \quad | \quad$ Add i into $\mathcal{C}$;

return $\mathcal{C}$;

2.2   Trust Propagation with Limitation

Limited TrustRank. In our framework, the trust scores are limited when propagate from non-community members to community members. The formula of Limited TrustRank (L-TrustRank) is:

$$t(p) = \begin{cases} \alpha \cdot \sum_{q:(q,p)\in\mathcal{E}} \frac{\gamma \cdot t(q)}{\omega(q)} + (1-\alpha) \cdot s(p), & \text{if } p \in \mathcal{C} \text{ and } q \notin \mathcal{C}, \\ \alpha \cdot \sum_{q:(q,p)\in\mathcal{E}} \frac{t(q)}{\omega(q)} + (1-\alpha) \cdot s(p), & \text{otherwise.} \end{cases} \quad (1)$$

where α is a decay factor, $\mathcal{C}$ is the spam communities extracted, γ is a penalty factor. $\boldsymbol{s}$ is the normalized trust score vector for the good seed set $\mathcal{S}^+$.

Limited LCRank. Anti-Trust Rank [4], which is broadly based on the same principle as TrustRank, propagates distrust via inverse-links from a seed set of spam pages. The iteration formula of Anti-Trust Rank is:

$$d(p) = \alpha' \sum_{q:(p,q)\in\mathcal{E}} \frac{d(q)}{\iota(q)} + (1-\alpha')s'(p) \quad (2)$$

where s' is the normalized distrust score vector for the bad seed set $\mathcal{S}^-$:

$$s'(p) = \begin{cases} 0, & \text{if } p \notin \mathcal{S}^-, \\ 1/|\mathcal{S}^-|, & \text{otherwise.} \end{cases}$$

We call linear combination of TrustRank and Anti-Trust Rank [5] LCRank for short, which is calculated as $l = \eta \times t - \beta \times d$, where η and β $(0 < \eta < 1, 0 < \beta < 1)$ are two coefficients to give different weights to Limited TrustRank scores t and Anti-Trust Rank scores d in the linear combination. Our Limited LCRank (L-LCRank) algorithm uses L-TrustRank score in the combination.

Limited CPV. CPV [6] assigns two scores to each page called *AVRank (AV for Authority value)* and *HVRank (HV for Hub Value)*. The iteration formulas of Limited CPV (L-CPV) are:

$$\mathbf{av}(p) = \begin{cases} \beta \sum_{q:q \to p} \frac{\gamma \cdot \mathbf{av}(q)}{\omega(q)} + (1 - \beta) \sum_{q:q \to p} \frac{\mathbf{hv}(q)}{\omega(q)}, & \text{if } p \in \mathcal{C} \text{ and } q \notin \mathcal{C}, \\ \beta \sum_{q:q \to p} \frac{\cdot \mathbf{av}(q)}{\omega(q)} + (1 - \beta) \sum_{q:q \to p} \frac{\mathbf{hv}(q)}{\omega(q)}, & \text{otherwise.} \end{cases} \quad (3)$$

$$\mathbf{hv}(p) = \begin{cases} \beta' \sum_{q:p \to q} \frac{\mathbf{hv}(q)}{\iota(q)} + (1 - \beta') \sum_{q:p \to q} \frac{\gamma \cdot \mathbf{av}(q)}{\iota(q)}, & \text{if } p \in \mathcal{C} \text{ and } q \notin \mathcal{C}, \\ \beta' \sum_{q:p \to q} \frac{\mathbf{hv}(q)}{\iota(q)} + (1 - \beta') \sum_{q:p \to q} \frac{\mathbf{av}(q)}{\iota(q)}, & \text{otherwise.} \end{cases} \quad (4)$$

3   Experiments

We conducted experiments on WEBSPAM-UK2007 dataset [7] and TREC Category B of ClueWeb09 dataset [8]. We chose TrustRank [1], LCRank [5] and CPV [6] as the baseline algorithms for comparison with L-TrustRank, L-LCRank and L-CPV. To evaluate the performances of spam demotion algorithms, the set of sites (pages) is split into a number (we use 20 here) of buckets according to PageRank values, then evaluation criteria below are used.

Spam Sites (Pages) in Top-k Buckets: The results on WEBSPAM-UK2007 and ClueWeb09 are shown in Fig. 1 - 3 and Fig. 4 - 6, respectively. It can be obviously seen that our algorithms put fewer spam sites in the top buckets than the baseline algorithms, which is a significant improvement in demoting spam sites.

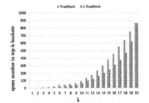 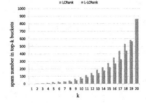 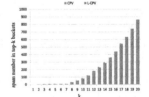

Fig. 1. Spam sites in top-k buckets of TrustRank and L-TrustRank on WEBSPAM-UK2007

Fig. 2. Spam sites in top-k buckets of LCRank and L-LCRank on WEBSPAM-UK2007

Fig. 3. Spam sites in top-k buckets of CPV and L-CPV on WEBSPAM-UK2007

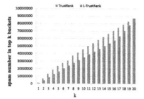

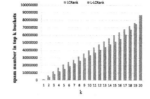

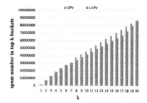

Fig. 4. Spam sites in top-k buckets of TrustRank and L-TrustRank on ClueWeb09

Fig. 5. Spam sites in top-k buckets of LCRank and L-LCRank on ClueWeb09

Fig. 6. Spam sites in top-k buckets of CPV and L-CPV on ClueWeb09

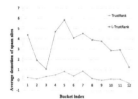

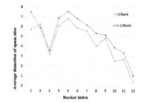

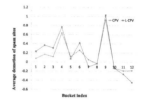

Fig. 7. Average demotion distances of spam sites of TrustRank and L-TrustRank on WEBSPAM-UK2007

Fig. 8. Average demotion distances of spam sites of LCRank and L-LCRank on WEBSPAM-UK2007

Fig. 9. Average demotion distances of spam sites of CPV and L-CPV on WEBSPAM-UK2007

Table 1. SpamFactors on WEBSPAM-UK2007

	TrustRank	LCRank	CPV
Baselines	0.0518	0.0566	0.0573
L-version	**0.0429**	**0.0449**	**0.0557**

Table 2. SpamFactors on ClueWeb09

	TrustRank	LCRank	CPV
Baselines	0.4329	0.4069	0.4015
L-version	**0.1622**	**0.1588**	**0.2331**

Average Spam Sites (Pages) Demotion Distances: This criterion indicates the average demotion distance (how many buckets) of spam sites in the ranking results. Fig. 7, Fig. 8 and Fig. 9 show the average demotion distances of spam site on WEBSPAM-UK2007, which validate the superiority of our algorithms.

SpamFactor: This criterion increases when more spam pages are presented in the top positions of the ranking list. The smaller the SpamFactor is, the more effective the algorithm is. The results shown in Table 1 and Table 2 demonstrate that our algorithms outperform the baseline algorithms.

4 Conclusions

In this paper, we have proposed an integrated framework to combat Link-based Web spam. The framework includes two steps: extracting spam communities and propagating scores with limitation, which is more aligned with the idea of trust propagation, i.e. it only propagates trust to the pages which are really trustworthy. Experimental results indicate that the limited versions of such algorithms as TrustRank, LCRank and CPV can achieve better results for spam demotion.

References

1. Gyöngyi, Z., Garcia-Molina, H., Pedersen, J.: Combating web spam with trustrank. In: VLDB 2004, pp. 576–587 (2004)
2. Zhang, X., Wang, Y., Mou, N., Liang, W.: Propagating both trust and distrust with target differentiation for combating web spam. In: AAAI (2011)
3. Wu, B., Chellapilla, K.: Extracting link spam using biased random walks from spam seed sets. In: Proceedings of the 3rd International Workshop on Adversarial Information Retrieval on the Web, pp. 37–44. ACM (2007)
4. Krishnan, V., Raj, R.: Web spam detection with anti-trust rank. In: AIRWeb 2006, pp. 37–40 (2006)
5. Wu, B., Goel, V., Davison, B.D.: Propagating trust and distrust to demote web spam. In: MTW 2006 (2006)
6. Zhang, Y., Jiang, Q., Zhang, L., Zhu, Y.: Exploiting bidirectional links: making spamming detection easier. In: Proceedings of the 18th ACM Conference on Information and Knowledge Management, pp. 1839–1842. ACM (2009)
7. Yahoo!: Yahoo! research: Web spam collections. http://barcelona.research.yahoo.net/webspam/datasets/ Crawled by the Laboratory of Web Algorithmics. University of Milan (2007). http://law.dsi.unimi.it/
8. Callan, J., Hoy, M., Yoo, C., Zhao, L.: The clueweb09 data set (2009). http://boston.lti.cs.cmu.edu/Data/clueweb09/

Inferring User Preference in Good Abandonment from Eye Movements

Wanxuan Lu and Yunde Jia[✉]

Beijing Laboratory of Intelligent Information Technology, School of Computer Science, Beijing Institute of Technology, Beijing 100081, China
{luwanxuan,jiayunde}@bit.edu.cn

Abstract. Many studies have been done to investigate good abandonment, but only a few have utilized it to improve search engine performance. In this paper, we aim at inferring user preference in good abandonment. Particularly, we use eye movement data to infer which search result has satisfied user's information need in each good abandonment instance. An eye-tracking experiment was conducted to capture user's eye movement data in good abandonment search tasks. These data were transformed into histograms and sequences on which we applied popular machine learning algorithms for the inference. Our results show that the approach can infer user preference with reasonable accuracy.

Keywords: Good abandonment · User preference inference · Eye movement · Search result preference · Eye-tracking

1 Introduction

Search engines have a long history of using user behavior as implicit feedback to improve retrieval performance. Typically, it is effective and efficient to use clickthrough data as implicit feedback, with the assumption that clicked search results are commonly more relevant than non-clicked ones. But, a query may not always be followed by a click. User's inactivity in Web search is called search/query abandonment which has been generally considered as a negative signal. Recent studies (e.g., [1]) report that search abandonment is sometimes a good thing, because user's information need can be satisfied directly by the content of the search engine results page (SERP). Search abandonment in this kind of situation is defined by the notion of good abandonment [2]. Many studies have investigated good abandonment [3][4][5][6][7][8], but only a few have addressed the issue of how good abandonment can be used for search engine improvement [1][5][7][8].

In this paper, we aim at inferring user preference in good abandonment, which can be seen as a next step after successfully identifying good abandonment. To achieve our goal, we propose to use eye-tracking technologies, assuming that user's preference can be revealed by his/her eye movements on the SERP.

© Springer International Publishing Switzerland 2015
J. Li and Y. Sun (Eds.): WAIM 2015, LNCS 9098, pp. 457–460, 2015.
DOI: 10.1007/978-3-319-21042-1_40

2   Method

2.1   Eye-Tracking Experiment

The experiment was conducted to capture user's eye movements in good abandonment search tasks. A task is considered as good abandonment when the information need can be directly satisfied by the content of the SERP. We designed ten tasks within the categories in [2], e.g., the definition of image processing. For each task, we carefully selected the keywords to avoid any Answer (a direct result for topics such as weather) showing up on SERPs, and to make sure the search results satisfying the search goal were not always in top positions. We recruited 30 university students (15 females) who were from a variety of disciplines, with an age range from 19 to 28 ($M = 23.9$, $SD = 2.0$), and were familiar with Web search engines. Eye-tracking was performed using a Tobii T120 eye-tracker. Logging of click and eye movement data were done by the software Tobii Studio. For each participant, the experiment began with a practice. Following the practice there were the 10 search tasks. For each search task, a search engine portal with predefined keywords was shown on the screen. Participants read the keywords and clicked on the search button triggering the SERP to show up. Participants had to find the information within the SERP and click on a "complete" button (a simulation of "closing the browser" [6]). After that, the closed SERP was shown again, and participants need to click on the search result in which he/she found the information to provide a preference feedback.

2.2   Data Transformation and Machine Learning Algorithms

We transformed the raw eye movement data into histograms and sequences. A histogram is a fixed-length feature vector describing the distribution (location wise) of one participant's eye movements in one search task. The histogram has three forms: total fixation count (TFC), total fixation duration (TFD), and TFC+TFD. That is,

$$\mathbf{h}_{TFC} = \{c_1, c_2, \ldots, c_{10}\} \tag{1}$$

$$\mathbf{h}_{TFD} = \{d_1, d_2, \ldots, d_{10}\} \tag{2}$$

$$\mathbf{h}_{TFC+TFD} = \{c_1, c_2, \ldots, c_{10}, d_1, d_2, \ldots, d_{10}\} \tag{3}$$

where c_x and d_x are the fixation count and fixation duration of search result x (10 search results in each SERP), respectively. A sequence is an unfixed-length feature vector presenting one participant's scan path in one search task. The sequence has two forms: fixation location (FL), and fixation location and fixation duration (FL+FD). That is,

$$\mathbf{s}_{FL} = \{l_1, l_2, \ldots, l_n\} \tag{4}$$

$$\mathbf{s}_{FL+FD} = \{(l_1, d'_1)^T, (l_2, d'_2)^T, \ldots, (l_n, d'_n)^T\} \tag{5}$$

where l_x and d'_x are the location and duration of fixation x ($x \in N$), respectively.

After the data transformation, we gave each histogram and sequence a label in light of its corresponding preference feedback. This makes the inference of user preference become a multi-class classification problem. We employed two machine learning algorithms for each data type: Support Vector Machine (SVM) and Random Forests (RF) for histograms; and Hidden Markov Model (HMM) and Hidden-state Conditional Random Fields (HCRF) for sequences.

3 Result

From the eye-tracking experiment, we obtained usable eye movement data for 285 of the 300 tasks. Based on the distribution of participant's preference feedback, a naive majority baseline (MB), assigning all instances to the class of highest frequency, resulted in an accuracy of 30.88%. Besides the MB, we created three criteria to infer user preference: most viewed (MOV), longest viewed (LOV), and last viewed (LAV). The MOV and LOV assume that the preferred search result is the most frequently viewed one, measuring by either fixation count (MOV) or fixation duration (LOV). The LAV assumes that the preferred search result is the last viewed one. We used the metric of accuracy to evaluation the inference of user preference, where $Accuracy = (1 - \frac{misclassified}{all}) \times 100\%$. Our results are summarized in Table 1. All machine learning algorithms and criteria outperform the MB baseline, and machine learning algorithms outperform criteria (except for HCRF which is worse than LAV). Besides, the performances of SVM and RF are better than HMM and HCRF, suggesting that the histogram is more suitable for inferring user preference than the sequence.

Table 1. The summary of inference accuracies

	Method	Inference accuracy	Relative to MB	Relative to LAV
Algorithms	SVM (TFC+TFD)	80.53%	+161%	+13%
	RF (TFC+TFD)	80.26%	+160%	+12%
	HMM (FL)	73.48%	+138%	+3%
	HCRF (FL)	68.95%	+123%	−4%
Criteria	LAV	71.58%	+132%	−
	MOV	54.74%	+77%	−24%
	LOV	54.39%	+76%	−24%
Baseline	MB	30.88%	−	−57%

Based on our approach, it is possible to take advantage of good abandonment instances. For example, clickthrough rate (CTR) is one of the most used features for search engine improvement. Traditionally, CTR is calculated by $CTR = \frac{Clicks}{Impressions}$. Now we know which search result has satisfied user's information need in good abandonment, we can create another CTR or a Good Abandonment CTR (GACTR) by taking inferred user preference into account:

$$CTR' = \frac{Clicks + Preferences}{Impressions} \qquad (6)$$

$$GACTR = \frac{Preferences}{GoodAbandoments} \qquad (7)$$

where *Preferences* is the number of inferred user preferences, and *GoodAbandoments* is the number of good abandonment instances. We believe that the proposed CTR features are more precise than the original one. But further studies are needed to examine the impact of these new features.

4   Conclusion

In this paper, we proposed an approach to infer user preference in good abandonment based on eye movement data. Experimental results demonstrated the effectiveness of our approach. In addition, we discussed how to use the inference result to improve search engine performance by refining the CTR feature.

Acknowledgments. This work was supported in part by the Natural Science Foundation of China under Grant No.61375044, the specialized Research Fund for the Doctoral Program of Chinese Higher Education under Grant No. 20121101110035, and the Specialized Fund for Joint Building Program of Beijing Municipal Education Commission.

References

1. Stamou, S., Efthimiadis, E.N.: Queries without clicks: successful or failed searches?. In: Proceedings of the 2009 SIGIR Workshop on the Future of Information Retrieval Evaluation, pp. 13–14 (2009)
2. Li, J., Huffman, S.B., Tokuda, A.: Good abandonment in mobile and PC internet search. In: Proceedings of the 32nd International ACM SIGIR Conference on Research and Development in Information Retrieval, pp. 43–50 (2009)
3. Chuklin, A., Serdyukov, P.: Good abandonments in factoid queryies. In: Proceedings of the 21st International Conference Companion on World Wide Web, pp. 483–484 (2012)
4. Chuklin, A., Serdyukov, P.: Potential good abandonment prediction. In: Proceedings of the 21st International Conference Companion on World Wide Web, pp. 485–486 (2012)
5. Chuklin, A., Serdyukov, P.: How query extensions reflect search result abandonments. In: Proceedings of the 35th International ACM SIGIR Conference on Research and Development in Information Retrieval, pp. 1087–1088 (2012)
6. Diriye, A., White, R.W., Buscher, G., Dumais, S.T.: Leaving so soon? understanding and predicting web search abandonment rationales. In: Proceedings of the 21st ACM International Conference on Information and Knowledge Management, pp. 1025–1034 (2012)
7. Song, Y., Shi, X., White, R.W., Hassan, A.: Context-aware web search abandonment prediction. In: Proceedings of the 37th International ACM SIGIR Conference on Research and Development in Information Retrieval, pp. 93–102 (2014)
8. Arkhipova, O., Grauer, L.: Evaluating mobile web search performance by taking good abandonment into account. In: Proceedings of the 37th International ACM SIGIR Conference on Research and Development in Information Retrieval, pp. 1043–1046 (2014)

Private Range Queries on Outsourced Databases

Lu Li[1]([✉]), Liusheng Huang[1], An Liu[2], Yao Shen[1], Wei Yang[1],
and Shengnan Shao[3]

[1] School of Computer Science and Technology,
University of Science and Technology of China, Hefei, China
liluzq@mail.ustc.edu.cn
[2] School of Computer Science, Soochow University, Suzhou, China
[3] School of Software Engineering of USTC, Suzhou, China

Abstract. With the advent of cloud computing, data owners could upload their databases to the cloud service provider to relief the burden of data storage and management. To protect sensitive data from the cloud, the data owner could publish an encrypted version of the original data. However, this will make data utilization much harder. In this paper, we consider the problem of private range query. Specifically, the data owner wants to obtain all the data within a query region, while keeping the query private to the service provider. Previous works only provide partial security guarantee, and are inefficient to deal with large scale datasets. To solve this problem, we present a fully secure scheme based on private information retrieval (PIR) and batch codes (BC).

Keywords: Secure range query · Cloud computing · Private information retrieval

1 Introduction

In cloud computing paradigm, data owners could outsource their databases to the service provider, and thus reap huge benefits from releasing the heavy storage and management tasks to the cloud server. However, sensitive data, such as medical or financial records, should be encrypted before being uploaded to the cloud [7]. To preserve the privacy of sensitive data, decryption keys should never be revealed to the cloud server. Unfortunately, this will introduce new challenges to data utilization. Considering the multi-dimensional range queries, which are typical database query operations in real life, lightweight encryption scheme, e.g. block cipher [3], cannot be directly applied for the server to conduct the queries. To enable private queries over encrypted data, Lu [1] proposed **LSE**, a symmetric encryption scheme, which can deal with the private single-dimensional range queries in *logarithmic* time. As a multi-dimensional range query can be decomposed to several single-dimensional queries, **LSE** theoretically can be extended to support private multi-dimensional range queries. However, the extended scheme will cause significant information leakage. Specifically, the cloud server will learn the exact relationship between every single-dimension of the query data and the

© Springer International Publishing Switzerland 2015
J. Li and Y. Sun (Eds.): WAIM 2015, LNCS 9098, pp. 461–464, 2015.
DOI: 10.1007/978-3-319-21042-1_41

outsourced data. Other recent works, e.g. [2], are also suffered from leaking this kind of information to the cloud server.

To prevent the single-dimensional information leakage, Wang *et al.* [4] proposed **Maple** by leveraging Hidden Vector Encryption [5] in a novel way. Although **Maple** can provide stronger privacy guarantee than previous works, it is still not fully secure in the sense of cryptography, e.g., the cloud server can learn the query path as well as which data records are the query results during each query. The cloud server can easily determine the data distribution based on this knowledge through statistical analysis. Besides, this work is inefficient due to the heavy computational operations of cryptography.

To deal with this problem, we present in this paper a fully secure scheme based on private information retrieval (PIR) [8] and batch codes (BC) [6].

2 Approach

To improve computational efficiency and to reduce storage cost, we adopt the block cipher as the underlying encryption scheme. Specifically, we use the AES in counter mode to encrypt the data records before uploading them to the cloud server.

2.1 Basic Scheme

We first give a basic private range query protocol. When the data owner wants to issue a range query, she does not send the server any information about the query region. Instead, she just submits a query requirement. Once receiving a query requirement from the data owner, the cloud server forwards all the encrypted nodes and the topology structure in high levels of the encrypted tree to the data owner. Each node in the bottom level is required to be associated with an identity that contains the indexes of its children nodes in the next level. The data owner first decrypts the root node, and then decides which nodes in the next level should be selected based on the intersection judgement. This process will continue until the data owner obtains the desired indexes of the children nodes of the bottom level. The above process is secure, because the data owner does not send any confidential information to the cloud server. Also, storage and computational overhead can be reduced by restricting the number of levels. Obviously, this process only enables the data owner to perform traversal over limited levels, as transferring the nodes in the next level will incur large communication cost and storage overhead on the data owner. Fortunately, this problem can be resolved by adopting PIR protocol. Recalling that during the above process, the data owner has already known which nodes in the next level should be transferred. In the next step, the data owner and the cloud server could engage in PIR protocol several times to let the data owner extract the information of these nodes, without revealing the required nodes to the cloud server. The data owner then decrypts these nodes and judges which children nodes should be transferred in the next level. This process could continue to the end, and the data owner will obtain all the data records within the query region.

Although the above solution ensures the data owner obtains the correct query result and does not leak much information, it may cause large volume of computation on the server side, because the computational time on the cloud server will increase linearly with respect to the amount of query result. The protocol will suffer from inefficiency in such a case. In the next subsection, we will present an improvement to greatly reduce the cost. Integrating the improvement into the above query protocol, we can easily eliminate the potential security risk by masking the amount of elements issued in each level, and it will not cause significant increase of the overhead.

2.2 Efficient Private Range Query Protocol

As shown in [6], batch code constructions can be used to support multi-round PIR. We now give an example to show how it works. Considering a database with n bits, one can straightforwardly use the above PIR scheme twice to obtain two elements of the database. However, the corresponding computational overhead is $2n$ multiplications at the server. If we partition the database into two parts: L and R containing $n/2$ bits each, and store L on the first bucket, R on the second and $L \oplus R$ on the third. Now one can extract two elements by calling single-bit PIR in each bucket, and the computational overhead at the server is reduced to $3n/2$ multiplication. Also, the database can be partitioned into m parts, i.e. $n = n_1||...||n_m$, where $|n_i| = \lceil \frac{n}{m} \rceil, i = 1, ..., m$. We also append a part $n_{m+1} = n_1 \oplus ... \oplus n_m$. Now it is easy to know that we can retrieve any two bits in n by extracting single bit from each of the $m + 1$ parts, and the computational overhead at the server will be reduced to $(m + 1)n/m$ multiplication.

This process can be recursively applied to support 2^d-bit information retrieval, and the computational overhead at server side is monotonic with respect to m. However, we cannot simply set m the maximum value to reduce the total cost, as this will also increase the corresponding computational overhead at the data owner side and the communication overhead. Fortunately, the optimization coding scheme can be selected easily by considering the overall overhead, and using such a coding scheme will greatly reduce the total cost.

3 Experiments

Our protocols are implemented in C++ using GMP library and tested on two MacBook Pro laptops (2.2GHz CPU and 16GB RAM) connected by a 100 Mbps LAN. The default experimental setting is as follows: the encryption key length is 128, the bit length of N is 1024, the bit length l of each data record is 24, the maximum number of result points k is 64, the number of data records n is 4,000,000. We show in Fig. 1 the computation time and message volume by executing different partition scheme (where m varies from 1 to 13). Clearly, the computation overhead will decrease rapidly, and the transmitted message volume will increase slowly when m varies from 1 to 7. So, we could easily select the best encoding scheme to greatly reduce the total overhead. Based on the optimization partition scheme, even to deal with 10,000,000 data records, our protocol can be finished within 10 minutes, which is much faster than the most secure previous work (Maple requires 928 seconds to deal with 100,000 data records).

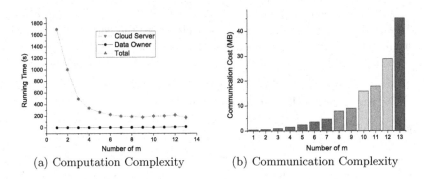

(a) Computation Complexity   (b) Communication Complexity

Fig. 1. Performance evaluation v.s. number of m

4   Conclusions

In this paper, we have studied the problem of private range queries in outsourced database. We have presented a fully secure scheme based on private information retrieval. By adopting batch code, our scheme is much more efficient than state-of-the-art approaches.

References

1. Lu, Y.: Privacy-preserving logarithmic-time search on encrypted data in cloud. In: NDSS (2012)
2. Wang, P., Ravishankar, C.V.: Secure and efficient range queries on outsourced databases using $\hat{R}$ tree. In: ICDE, pp. 314–325 (2013)
3. Katz, J., Lindell, Y.: Introduction to Modern Cryptography (Chapman Hall/Crc Cryptography and Network Security Series). Chapman Hall/CRC (2007)
4. Wang, B., Hou, Y., Li, M., et al.: Maple: scalable multi-dimensional range search over encrypted cloud data with tree-based index. In: ASIACCS, pp. 111–122 (2014)
5. Boneh, D., Waters, B.: Conjunctive, subset, and range queries on encrypted data. In: Vadhan, S.P. (ed.) TCC 2007. LNCS, vol. 4392, pp. 535–554. Springer, Heidelberg (2007)
6. Ishai, Y., Kushilevitz, E., Ostrovsky, R., Sahai, A.: Batch codes and their applications. In: STOC, pp. 262–271 (2004)
7. Armbrust, M., Fox, A., Griffith, R., Joseph, A., et al.: A View of Cloud Computing. Communications of the ACM **53**(4), 50–58 (2010)
8. Kushilevitz, E., Ostrovsky, R.: Replication is not needed: single database, computationally private information retrieval. In: FOCS, pp. 364–373 (1997)

Efficient Influence Maximization Based on Three Degrees of Influence Theory

Yadong Qin, Jun Ma$^{(\boxtimes)}$, and Shuai Gao

School of Computer Science and Technology,
Shandong University, Jinan 250101, China
yadongqinsdu@gmail.com, majun@sdu.edu.cn, gao_shuai@mail.sdu.edu.cn

Abstract. The study on influence modeling is to understand the information diffusion and word-of-mouth marketing. In this paper, based on Three Degrees of Influence theory, we propose a suitable diffusion model named Three Steps Cascade Model (TSCM) to simulate online social network information diffusion process. We focus on the influence maximization problem under TSCM and devise an efficient algorithm to solve this problem. The experiment results on real-networks show the robustness and utility of our approach.

Keywords: Viral marketing · Influence maximization · Three degrees of influence · Social network

1 Introduction

In [1], Nicholas A. Christakis and James H. Fowler proposed Three Degrees of Influence theory (TDI theory for short). Based on their long-term investigation, they found that in social networks, everything we do or say tends to ripple through our network, having an impact on our friends (one degree), our friends' friends (two degrees), and even our friends' friends' friends (three degrees). Our influence gradually dissipates and ceases to have a noticeable effect on people beyond the social frontier that lies at three degrees of separation. Likewise, we are influenced by friends within three degrees but generally not by those beyond.

In this paper, based on TDI theory, we propose a diffusion model named TSCM to simulate information diffusion in real social network. Then we solve influence maximization problem under TSCM efficiently. The rest of this paper is organized as follows. We propose TSCM in section 2. Section 3 gives the formal definition of influence maximization under TSCM. Besides, we devise the TLGreedy to solve this problem efficiently. Section 4 shows our experimental results.

2 Three Steps Cascade Model

Online social network is modeled by a directed graph $G = (V, E)$, where a node $v \in V$ represents the individual of the social network and an edge $(u, v) \in E$

© Springer International Publishing Switzerland 2015
J. Li and Y. Sun (Eds.): WAIM 2015, LNCS 9098, pp. 465–468, 2015.
DOI: 10.1007/978-3-319-21042-1_42

denotes that u can influence v in the network. For every edge $(u, v) \in E$, there is a number $0 \leq p(u, v) \leq 1$ which means the intrinsic strength of the link.

TSCM can be formalized as follows: given a seed set $S \subseteq V$ and three positive real numbers λ_1, λ_2 and λ_3 which represent the cascade decay ratios of step 1, step 2 and step 3 respectively. Let $S_t \subseteq V$ be the set of nodes that are activated at step t with $0 \leq t \leq 3$ and $S_0 = S$. At step $t+1$, every node $u \in S_t$ can activate its out-neighbors $v \in V \backslash \cup_{0 \leq i \leq t} S_i$ with probability $\lambda_{t+1} \cdot p(u, v)$. We assume that $\lambda_1 \geq \lambda_2 \geq \lambda_3$ because of information decaying. The diffusion process ends at the step $S_t = \emptyset$ or $t = 3$. Each node stays as an activated node after it is activated and each activated node only has one chance to activate its out-neighbors at the step right after itself is activated.

The diffusion probability of an edge in TSCM is dependent on the intrinsic strength of the link and the step of propagation. We set $\lambda_4 = 0$ to guarantee the length of influence path within three degrees which is consistent with TDI theory. Given a seed set S, let $\sigma_{TSCM}(S)$ denote the expected number of activated nodes when the diffusion process ends under TSCM.

3   Influence Maximization Under TSCM

Given a social network graph $G = (V, E)$ and a number k, the influence maximization problem under the TSCM model is to find a subset $S^* \subseteq V$ such that $|S^*| = k$ and $\sigma_{TSCM}(S^*) = \max_{S \subseteq V, |S|=k} \{\sigma_{TSCM}(S)\}$. The similar reduction method from set cover problem in [2] is sufficient to show that this optimization problem is NP-hard.

3.1   Our Three Layers Approximation Approach

Given a seed set S and three real numbers λ_1, λ_2 and λ_3 such that $\lambda_1 \geq \lambda_2 \geq \lambda_3$. For arbitrary node $v \in V$, we define $dis(S, v) = \min_{w \in S} dis(w, v)$ where $dis(w, v)$ denotes the shortest graph distance from node w to v. If there is no path from w to v, then we set $dis(w, v) = \infty$. Let $D(S, i) = \{v | dis(S, v) = i, v \in V\}$. If $v \in D(S, i)$, then v belongs to layer i. The basic idea of TLAA is that we only consider the influence from layer i to layer $i + 1$. The initial set $S = D(S, 0)$. Given a seed set S, for each $v \in V$, we define the final activated probability of v as:

$$p(v) = \begin{cases} 1 & v \in D(S,0) \\ 1 - \prod_{u \in D_{i-1}, (u,v) \in E} (1 - p(u) \cdot p(u, v) \cdot \lambda_i) & v \in D(S,i), 0 < i < 4 \\ 0 & v \in D(S,i), i \geq 4. \end{cases} \tag{1}$$

Let $\sigma_{TL}(S)$ denote the final influence spread of S in our TLAA, then we have

$$\sigma_{TL}(S) = \sum_{v \in V} p(v) \tag{2}$$

Breadth-first search (BFS [3]) from S is used to compute $\sigma_{TL}(S)$.

3.2   Efficient Greedy Algorithm on $\sigma_{TL}(\cdot)$

Suppose we have found a set T_m of size m, and we want to find a new node $u \in V \backslash T_m$ to satisfy $u = \arg\max_{v \in V \backslash T_m}(\sigma_{TL}(T_m \cup \{v\}) - \sigma_{TL}(T_m))$. In order to get the influence gain of a node v, the only thing we need to do is to get the sum of incremental value of $p(w)$ with $dis(v, w) < 4$. The activated probability $p(v)$ and $dis(T_m, v)$ of each node v in V are recorded when $\sigma_{TL}(T_m)$ is computed. We can make full use of $p(v)$ and $dis(T_m, v)$ to compute influence gain efficiently. The efficient greedy algorithm for TLAA is shown in Algorithm 1.

Algorithm 1. TLGreedy(G,λ,k)

Input: G : the network; k : size of final seed set ; λ : array of λ_1, λ_2 and λ_3;
Output: seed set S;
1: initial set $S = \emptyset$;
2: **while** $|S| < k$ **do**
3:     select $v = \arg\max_{u \in V}(InfluenceGain(G, S, u, \lambda))$;
4:     $S = S \cup \{v\}$;
5:     update $p(w)$ and record $dis(S \cup \{v\}, w)$ with $dis(v, w) < 4$;
6: **end while**
7: **return** S;

Given a network $G = (V, E)$, let $n = |V|$ and $m = |E|$. Let $G_u = (V_u, E_u)$ denote the subgraph which is induced by $V_u = \{v \mid dis(u, v) < 4\}$, and let $n_{singleMax} = \max_{u \in V}\{|V_u| + |E_u|\}$.

For Algorithm 1, in each iteration of selecting next seed, InfluenceGain (G,S,u,λ) in line 3 spends $O(n_{singleMax})$ to get the influence gain of each node. As a result, when $n_{singleMax} = n+m$, TLGreedy(G, λ, k) spends $O(k \cdot n \cdot (n+m))$.

However, $n_{singleMax} = n + m$ is almost impossible. That is, a node in the social network can not reach all the nodes within three degrees. In this case, TLGreedy(G, P, k) takes $O(n_{singleMax}*n + n_{singleMax}*n + \ldots + n_{singleMax}*n) = O(k \cdot n_{singleMax} \cdot n)$ of time.

4   Experiments

We downloaded HEP-PH and EU-email from site [4] to illustrate that our proposed method is robust for Influence Maximization problem under TSCM.

We use five algorithms of **CELFGreedy** [5], **TLLFGreedy**[1], **TSLF-Greedy**[2], **SingleDiscount** [6] and **Random**[3] to prove that our proposed method can guarantee both accuracy and scalability. For each seed set, we run Monte-Carlo simulation 20000 times to obtain the expected influence spread of

---

[1] Lazy-forward method on TLGreedy.
[2] Let $\sigma_{TS}(S)$ denote the number of nodes that S can reach within three degrees in the graph. TSLFGreedy is the greedy algorithm on $\sigma_{TS}(\cdot)$ with the CELF optimization.
[3] As a baseline comparison, simply select k random nodes.

the heuristic algorithms. Figure 1 and Figure 2 show the experimental results on influence spread using algorithms above. The seed set size k ranges from 1 to 50. We set $\lambda_1 = 1$, $\lambda_2 = 0.5$, $\lambda_3 = 0.1$ and $p(e) = 0.1$ for all $e \in E$.

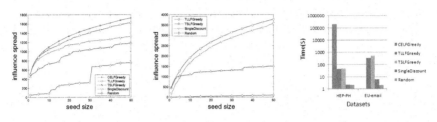

Fig. 1. HEP-PH   **Fig. 2.** EU-email   **Fig. 3.** Running times

As show in Figure 1, the CELFGreedy algorithm provides the best influence spread where the CELFGreedy algorithm is feasible to run. TLLFGreedy provides the largest influence spread except CELFGreedy. In Figure 2, the graph is too large to run CELFGreedy, so it is out of the picture. TLLFGreedy outperforms all other algorithms. Figure 3 shows the running time of different algorithms to get 50 seeds in Figure 1 and Figure 2. The TLLFGreedy is more than three orders magnitude faster than CELFGreedy over HEP-PH. When graph becomes larger, TLLFGreedy is faster than TSLFGreedy.

In conclusion, our proposed method is scalable for large social network while it can guarantee accuracy.

Acknowledgments. This work was supported by Natural Science Foundation of China (61272240, 71402083, 6110315).

References

1. Christakis, N.A., Fowler, J.H.: Connected: The surprising power of our social networks and how they shape our lives. Hachette Digital, Inc. (2009)
2. Kempe, D., Kleinberg, J., Tardos, É.: Maximizing the spread of influence through a social network. In: Proceedings of the Ninth ACM SIGKDD International Conference on Knowledge Discovery and Data Mining, pp. 137–146. ACM (2003)
3. Breadth-first search. http://en.wikipedia.org/wiki/Breadth-first_search
4. Stanford Large Network Dataset Collection. http://snap.stanford.edu/data/
5. Leskovec, J., Krause, A., Guestrin, C., Faloutsos, C., VanBriesen, J., Glance, N.: Cost-effective outbreak detection in networks. In: Proceedings of the 13th ACM SIGKDD International Conference on Knowledge Discovery and Data Mining, pp. 420–429. ACM (2007)
6. Chen, W., Wang, Y., Yang, S.: Efficient influence maximization in social networks. In: Proceedings of the 15th ACM SIGKDD International Conference on Knowledge Discovery and Data Mining, pp. 199–208. ACM (2009)

Bichromatic Reverse kNN Query Algorithm on Road Network Distance

Tin Nilar Win, Htoo Htoo, and Yutaka Ohsawa[✉]

Graduate School of Science and Engineering,
Saitama University, Saitama, Japan
`ohsawa@mail.saitama-u.ac.jp`

1 Introduction

A bichromatic reverse nearest neighbor (BRNN) query has been studied in a road network distance recently, however, existing works for BRNN query have shortcomings in the long processing time. In this paper, we propose a BRNN query algorithm in a road network distance by using the simple materialized path view (SMPV) data structure [1].

When a set of rival objects S and a set of interest objects P are given, and a random query point $q \in S$ is set, BRNN query retrieves reverse nearest neighbors of q, which are interest objects in the set P that are nearest to q among the other rival objects in S. In BRNN query, two sets, P and S, are different data objects. For instance, let P be a set of hospitals, and S be a set of pharmacy shops, and to open a new pharmacy shop q ($\in S$), BRNN query finds all hospitals which are nearest to the new pharmacy shop among other pharmacy shops. In this case, depending on the result of BRNN, the best location to open a pharmacy shop can be decided from several candidate locations.

Fig.1 illustrates an example of BRNN query in Euclidean distance. In this figure, white circles are rival objects ($\in S$), black rectangles are interest objects ($\in P$). Here, p_1 is BRNN of s_1, p_4 is BRNN of s_2 and p_2, p_3 are BRNN of s_3.

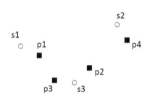

Fig. 1. Example of BRNN query

When k is optionally set to retrieve more than one nearest interest objects in P, kNN ($k > 1$), this type of query is called BRkNN query, and can be defined as the following.

$$\mathrm{BR}k\mathrm{NN}(q, P) = \{p | p \in P, q \in k\mathrm{NN}(p, S)\}$$

where kNN(p, S) is a set of kNN objects in S for p.

© Springer International Publishing Switzerland 2015
J. Li and Y. Sun (Eds.): WAIM 2015, LNCS 9098, pp. 469–472, 2015.
DOI: 10.1007/978-3-319-21042-1_43

In traditional algorithms for BRkNN in road network distance, when k number increases or when rival objects on road network are sparsely located, it takes overlong processing time. In this paper, a fast BRkNN query algorithm is introduced to overcome the deficiency in traditional methods.

2 Proposed Method

2.1 Basic Concepts of BRNN Query

Yiu et al. [2] proposed the Eager algorithm which is a sequential search for monochromatic reverse nearest neighbor (MRkNN). This algorithm is also applicable to BRkNN query. The Eager algorithm is an efficient algorithm to find BRkNN on a large network like a road network, however, when the rival objects are distributed sparsely on the road network or the value of k is large, it takes long processing time. Therefore, this paper proposes an efficient query algorithm that works fast in the difficult conditions mentioned above.

The proposed algorithm overcomes the difficulty by the following two methods; (1) to apply incremental Euclidean restriction (IER) [3] strategy when to decide to expand the region, (2) to suppress the times of checking a road network node whether it is included in the BRkNN region or not by applying the simple materialized path view (SMPV) [1] data.

2.2 BRNN Query on a Road Network

When a set of rival objects S and a set of interest objects P are given and a query object $q(\in S)$ is specified, to find BRkNN in general, the range A in which q is included as kNN is first generated. And then interest objects in P, which lie in the range A, are retrieved. Accordingly, when BRkNN query is in the road network distance, the query retrieves interest objects in P which lie in the range A in road network distances.

Yiu et al.[2] presented the following lemma for an MRkNN query on a road network. This lemma also stands for BRkNN query in the road network distance to prune nodes that are not included in BRkNN of the query object q.

Lemma 1. *Let q be a query point, n be a road network node, and p be a data point that satisfies $d_N(q, n) > d_N(p, n)$. For any data point $p'(\neq p)$ whose shortest path to q passes through n, $d_N(q, p') > d_N(p, p')$. This means that p' is not an RNN of q.*

Fig.2 explains the Lemma 1 with a concrete example using a border node of SMPV structure. In this figure, black circles are border nodes. If the condition $d_N(s_1, b_i) > d_N(b_i, s_2)$ is true for the border node b_i, there is no NN of s_1 on the path passes through b_i. Because s_2 is closer to b_i in this case, s_2 is an NN on all paths which pass through b_i. Therefore, all paths which pass through the border node b_i can be pruned safely if $d_N(s_1, b_i) > d_N(b_i, s_2)$ stands.

This constraint can also be applied to BRkNN query. While retrieving kNNs $(\in S)$ of a border node (b_i), if q is not included in the kNN set of b_i, and then the search for all paths which pass through b_i can be pruned.

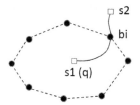

Fig. 2. The example of the Lemma 1

As a similar concept in Yiu et al. [2], kNN search from a border node b_i is called range-NN search. In this search, the range is set to the road network distance from q to b_i, $d_N(q, b_i)$. Then, kNN of b_i in set S within the specified range is searched. If q is not included in kNN of b_i, the search beyond b_i can be discarded and the search is terminated.

If q is included in the NN of b_i, interest objects $p_i (\in P)$ in all adjacent cells to b_i are checked whether these objects are BRkNN of q or not. If the result is BRkNN of q, $p(\in P)$ is added to the result set. This verification takes long processing time when the number of interest objects ($\in P$) for a subgraph is large. Thus, duplicated verification for the same objects within the same subgraphs is avoided.

3   Experimental Evaluation

To evaluate our proposed method comparing with the existing Eager algorithm, we conducted extensive experiments by using the real road map data of Saitama city, Japan with 16,284 road network nodes and 24,914 links. We generated variety of rival(S) and interest(P) data point sets on the road network links by pseudorandom sequences. Both algorithms were implemented in Java and evaluated on a PC with Intel Corei7-4770 CPU(3.4 GHz) and 32GB memory.

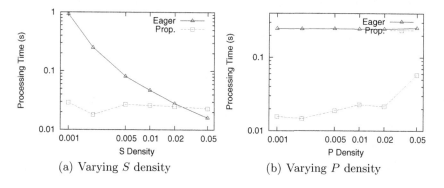

(a) Varying S density   (b) Varying P density

Fig. 3. Comparison for Processing Time

Fig.3 measures the processing time for BR1NN. Fig.3(a) is the result by varying data points density for S when the density of P is set to 0.002, and (b) shows the processing time varying the density of P when the density of S is set to 0.002. In the experiment, the probability density of S or P means the probable existence of objects in S or P at a road network link. For instance, the density 0.01 indicates that a data point exists once every 100 links.

As shown in Fig.3 (a), when the density of S is low, it requires to search in large range, and in such case, Eager algorithm takes very long processing time. Conversely, if the density is dense, the searching area becomes narrow and processing time is faster as a result. Our proposed method showed the stable result that is independent of the density. Consequently, Fig.3 (b) shows that the proposed method outperformed the Eager algorithm in which the processing time is significantly degraded. However, if the density increases, the processing time is also slightly increased in the proposed method. Because when candidates of BRkNN are found from P, it is required to verify whether each candidate in P is truly BRkNN of q or not. If the density of P is dense, the existence of objects in P within candidate subgraphs might be increased and several verification steps leads the increase of processing time.

4   Conclusion

In this paper, we proposed a fast BRkNN query algorithm in the road network distance by using the simple materialized path view (SMPV). With extensive experiments, we showed that the performance of the proposed method comparing to the existing method, Eager algorithm. Especially, the proposed method is 10 to 100 times faster in processing time significantly when the density distribution of S is sparse on a road network. On the other hand, the higher the density distribution of S is, the lesser the searching range is necessary in the real road network. Therefore, Eager algorithm is also efficient for such case. To advance an efficient and adaptive query which is not depending on the density distribution of S is for our future work.

Acknowledgments. The present study was partially supported by the Japanese Ministry of Education, Science, Sports and Culture (Grant-in-Aid Scientific Research (C) 24500107).

References

1. Hlaing, A.T., Htoo, H., Ohsawa, Y., Sonehara, N., Sakauchi, M.: Shortest path finder with light materialized path view for location based services. In: Wang, J., Xiong, H., Ishikawa, Y., Xu, J., Zhou, J. (eds.) WAIM 2013. LNCS, vol. 7923, pp. 229–234. Springer, Heidelberg (2013)
2. Yiu, M.L., Papadias, D., Mamoulis, N., Tao, Y.: Reverse nearest neighbor in large graphs. IEEE Transaction on Knowledge and Data Engineering 18(4), 1–14 (2006)
3. Papadias, D., Zhang, J., Mamoulis, N., Tao, Y.: Query processing in spatial network databases. In: Proc. 29th VLDB, pp. 790–801 (2003)

IIRS: A Novel Framework of Identifying Commodity Entities on E-commerce Big Data

Qiqing Fang[1], Yamin Hu[1], Shujun Lv[1], Lejiang Guo[1], Lei Xiao[1], and Yahui Hu[1,2(✉)]

[1] Air Force Early Warning Academy, Wuhan 430000, China
[2] School of Computer, Wuhan University, Wuhan 430072, China
hyh5800@163.com

Abstract. Identification of the same commodity entities is a major challenge in the heterogeneous multi-source e-commerce of big data. This paper introduces a framework based on Map-Reduce, called IIRS, which is made up of data index, data integration, entity recognition and data sorting. IIRS aims to form the unified model and high efficient commodity information with building an index model based on commodity's attribute/value and constructing a global model map to record commodity's attribute and value, identify the commodity entities in different e-commerce with measuring the similarity of the commodity's identity, and then output the same identity commodity sets and their associated properties organized in the inverted index list. Through an extensive experimental study on real e-commerce dataset on Hadoop, IIRS significantly demonstrates its feasibility, accuracy, and high efficiency.

Keywords: Big data · E-commerce · Entity identification · Map reduce · Normalization identification

1 Introduction

The recent blossom of big data has revolutionized our life by providing everyone with the ease and fun never before. Meanwhile, how to leverage these multi-source heterogeneous, fragmented, various, inconsistent and disruptive e-commerce data for better business intelligence raises a very valuable and challenging topic. In this paper, we aim to harness the power of big data to propose a novel framework based on Map-Reduce, called IIRS, which is made up of data index, data integration, entity recognition and data sorting. The main idea of this framework is to build an index model based on commodity's attribute/value, construct a global model map to record commodity's attribute and value, form the unified model and high efficient commodity information data, measure the similarity of the commodity's identity by the multilayer hierarchical probabilistic model, identify the commodity entities in different e-commerce, and then output the same identity commodity sets and their associated properties organized in the inverted index list.

© Springer International Publishing Switzerland 2015
J. Li and Y. Sun (Eds.): WAIM 2015, LNCS 9098, pp. 473–480, 2015.
DOI: 10.1007/978-3-319-21042-1_44

2   Related Work

The research of identifying the same commodity entity in e-commerce is a typical branch of big data entity recognition. The current methods are mainly based on the similarity function or rules to identify the entity. Herndndez et al.[1], Arasu et al.[2] leveraged the empirical knowledge to define the principles and then solve the problem of entity identity. Fan et at.[3] proposed the entity identity description rules for the first time. Chaudhuri et al.[4], Chen et al.[5] judged the same commodity from the many attributes with clustering and machine learning methods. Singla et al. [6] mentioned a similarity measure with Markov chain. Augsten et al.[7] identified the entity of XML with sliding window. Wang et al.[8] measured the entity with the similar matrix. In this article, we move towards this direction, and study the complex heterogeneous mass Web big data to find the same commodity entity, which is different from the traditional entity identification methods.

3   IIRS Framework

3.1   Index of Commodity Entity of Big Data

For data source S of e-commerce contain a large number of commodity web pages with different classifications, different descriptions of the same commodity entity. For the commodity web detailed pages W have the structure of the web page information (such as columns, page layout) and the detailed content information. Through the analysis of the information extraction and semantic mining, the web object data model can be defined as $W=\{C,O,B\}$.

C denotes the column and structural information of the commodity, which is defined as a no-empty five layer tree while the website data source is a root node, columns and its subtopics are the intermediate nodes, and the web pages are the leaf nodes. O denotes the object of web page. B denotes all of the data item sets in the web pages.

To avoid the NP hard problem in many data sources, we build an inverted index list $SCi=\{A,V,SW\}$ for every attribute value to effectively find the same commodity entity. Here, A is the attribute name of this record, V is its value and SW is the set of commodity web pages including the item$<A,V>$ in the datasets.

3.2   Integration of the Normalized Attribute and Value

Let CA be a set of all attributes, CV be a set of all values CV, H is the weighted edge set between CA and CV with $\omega<A,V>$, while the global model is defined as $G=<CA,CV,H>$.

Definition1. Equivalent value set.

$\bar{V} = \{V_1, V_2, ...\}$ represents the equivalent value set. All elements in the equivalent value set are not only equivalent each other but also the same or similar meaning. As the complexity of the e-commerce platform, for $\forall V_i \in \bar{V}, V_j \in \bar{V}$,

$$\text{satisfy } Sim_{value}(V_i, V_j) \geq \mu_1, \tag{1}$$

where μ_1 is the value threshold.

Definition2. Equivalent attribute set.

$\bar{A} = \{A_1, A_2, ...\}$ represents the equivalent attribute set. All elements in the equivalent attribute set are equivalent each other to describe one of the features on the same commodity. As the diversity and heterogeneity of the e-commerce platform, for

$$\forall A_i \in \bar{A}, A_j \in \bar{A},$$

$$\text{satisfy } Sim_{attr}(A_i, A_j) \geq \mu_2, \tag{2}$$

where μ_2 is the attribute threshold.

The identity of semantics of identity between the attributes is defined as

$$Sim_{attr}(A_i, A_j) = \frac{1}{2} \times Sim_{str}(A_i, A_j) + \frac{1}{2} \times Sim_{range}(A_i, A_j) \tag{3}$$

where Sim_{str} is the similarity of text, $Sim_{range}(A_i, A_j)$ is the value range fit.

3.3   Recognition of Commodity Entity

The information of commodity is composed of many attribute/value items, comparing the similarity and identity of two commodities' pages. And thus the set of data W.B can be rewritten as:

$$W.B = \{D_1 < T, E, V, P, w >, D_2 < T, E, V, P, w >, ..., D_k < T, E, V, P, w >\}, \tag{4}$$

where D_i is the ith item of W_j.B, D_i.T is the type of D_i, D_i.E is the name of ith item of W_j.B, D_i.V is the value of ith item of W_j.B, D_i.P is the credibility of ith item of W_j.B, D_i.w is the weight of ith item of W_j.B while D_i.w $\in (0,1)$.

For a given commodity Wa, identify the possible candidate commodity set.

$$Wa = \{W_1, W_2, ..., W_k\}, \tag{5}$$

where candidate commodity Wi has φk same items with commodity Wa while $\varphi \in (0,1)$.

We use the entity segmentation tools extracting the key terms $K = \{k_1, k_2, ..., k_n\}$, measure the similarity of k_i and items of the set, then get the weighted vector $\omega_K = \{\omega_{k_1}, \omega_{k_2}, ..., \omega_{k_n}\}$ where $\sum_{i=1}^{n} \omega_{k_i} = 1$, and finally calculate the similarity of two commodities' titles.

For $K_a = \{k_{a_1}, k_{a_2}, ..., k_{a_n}\}$ and $K_a = \{k_{b_1}, k_{b_2}, ..., k_{b_n}\}$, their key items vector are respectively $\omega_{K_a} = \{\omega_{a_1}, \omega_{a_2}, ..., \omega_{a_n}\}$ and $\omega_{K_b} = \{\omega_{b_1}, \omega_{b_2}, ..., \omega_{b_n}\}$. They are equal to

$$K_{ab} = K_a \cup K_b = \{k_{a_1}, k_{a_2}, ..., k_{a_n}, k_{ab_1}, k_{ab_2}, ..., k_{ab_m}\}. \tag{6}$$

where K_b-K_a is the key items vector in W_b but not in Wa while the order ab_m keep the same order Kb.

The Tanimoto title similarity of the Wa and Wb is:

$$\text{Sim}_{\text{title}}(\text{Wa}, \text{Wb}) = \frac{\omega'_{Ka} \cdot \omega'_{Kb}}{\left\|\omega'_{Ka}\right\|^2 + \left\|\omega'_{Kb}\right\|^2 - \omega'_{Ka} \cdot \omega'_{Kb}} \tag{7}$$

where ω'_{Ka} and ω'_{Kb} is the related weight vector respectively of Wa and Wb.

If $\text{Sim}_{\text{title}}(\text{Wa}, \text{Wb}) \geq \xi$, then Wa and Wb is the same title. ξ is the same title threshold of commodities.

In the same way, the price distance of two commodities is:

$$\text{Sim}_{\text{price}}(\text{Wa}, \text{Wb}) = \sqrt{1 - \frac{|\text{Wa.price} - \text{Wb.price}|}{Max(\text{Wa.price}, \text{Wb.price})}} \tag{8}$$

If $\text{Sim}_{\text{price}}(\text{Wa}, \text{Wb}) \geq \rho$, then Wa and Wb is the same price. ρ is the same price threshold of commodities.

Finally, the credibility of two commodity Wa and Wb is :

$$\text{Sim}(\text{W.B}, \text{Wi.B}) = \frac{(\text{SIM}_{\text{item}}(\text{Wa.B}, \text{Wb.B}) + \text{SIM}_{\text{title}}(\text{Wa}, \text{Wb}) + \text{SIM}_{\text{price}}(\text{Wa}, \text{Wb}))}{3} \tag{9}$$

3.4   Sorting of the Commodity Entity

```
Algorithm. Entity identification of commodities
input• commodity Wa
output•the same entity commodity set of Wa
Map:
input•key=Wa•value=Wa.B
process•
1. map(key,value)
2.    for each iteminvalue
3.      <A_a1,V_a1>,W_a•Extraction (item)
4.   collect(<A_a1,V_a1>,W_a)
Reduce:
input•key=<W_a>•value=<A_a1,V_a1>
process•
5.reduce(key,value)
6.   SW'•∅
7.   for each avinavList(W_a)
```

```
8.      SW•Search(av)
9.    SW'•Filter(φk,SW1,SW2,...)
10.     for each CommodityinSW'
11.       ifnot SIM(Wa.B,Wi.B) ≥ τ
12.         SW'•Remove(SW', Commodity)
13.     for each CommodityinSW'
14.       ifnot Sim_title(W a, W b) ≥ ξ
15.         SW'•Remove(SW', Commodity)
16.     for each CommodityinSW'
17.       ifnot Sim_price(W a, W b) ≥ ρ
18.         SW'•Remove(SW', Commodity)
19.     for each CommodityinSW'
20.     Stafcom•Commodity
21.     Similarity• Sim(W.B,Wi.b)
22. collect(Stafcom,Similarity)
```

From Algorithm, the department of Map is to build the unified description of commodity information with index and normalize commodities' attributes and values. For the Reduce, we search the same items of commodity set of SW in the inverted index list with Wa, find k items of SW1,SW2,...,SWk from Wa, filter them and get candidate commodity set SW', remove the items mismatched the similarity measure, and finally get the credibility of two commodity Wb and Wi.b and then output them in the inverted index list with the value of Sim(W.B,Wi.b) .

4   Experiment

In order to evaluate our algorithm, we crawled the real commodity data sets on Chinese three mainstream B2C e-commerce platforms and manually annotated part of the sampling data. The details of our data sets are summarized in Table 1.

Table 1. The distribution of data set classification

Type	Jingdong	Yixun	Yihaodian
Mobile phone	12298	11692	10509
Digital products	43910	66521	45393
Computer	13014	18741	9473
Office supply	96518	98379	64852
Household articles	20435	33452	38280
Sports & Outdoor	5260	5214	2448
living goods	16843	24649	11041

4.1   Performance Metrics

To evaluate our results, we show experimental results to evaluate the accuracy and effectiveness of system. Our experiments focused on five important aspects: precision, recall, F1- measure, RT and IS.

(1) Accuracy

The accuracy contains precision, recall and F1-measure. Set the candidate commodity sets SW = {W_1, W_2,...W_n}, RW_k = {W_{k1} and W_{k2},..., W_{km}}is identified the same commodity entities of the kth commodity. According to the k_{m1} error commodities identified and the k_{m2} true commodities digital actually, the average accuracy of the algorithm is:

$$P = \frac{\sum_{k=1}^{n} \frac{k_m - k_{m1}}{k_m}}{n} \qquad (10)$$

The average recall of the algorithm is:

$$R = \frac{\sum_{k=1}^{n} \frac{k_m - k_{m1}}{k_{m2}}}{n} \qquad (11)$$

The average of F1-measure of the algorithm is:

$$F1 = \frac{2 \times P \times R}{P + R} \qquad (12)$$

(2) Efficiency

The efficiency of the algorithm contains RT (Running Time Per 100 Thousand Data) and IS(Increment Speed).

$$RT = 100000 \times \frac{T}{Datasize} \qquad (13)$$

where T means the time of algorithm. Datasize is the size of data set.IS is the slope of the linear approximation of RT.

4.2   Analysis of Experimental Results

Figure 1 shows computing time of each point in the data set under the different methods. Results show that the method based on traditional multi-threading in relational database Oracle is superior to our method under the less amount of data .But as the growth of the amount of data, our method based on the Map-Reduce reflects its stronger advantage than the traditional way.

Table 2 shows that based on the Map-Reduce algorithm in all aspects of running efficiency is superior to traditional methods.

Table 3 shows that our method of precision, recall and F1-measure are better than 360 shopping search while the recall is slightly lower than Huihui shopping assistant. But the accuracy and F1-measure are far higher than Huihui shopping assistant. Because there are many similar but not same commodities of Huihui shopping assistant thus to improve the recall rate, but the sharp decline in accuracy.

From several groups of experiments can be concluded that, our method based on Map-Reduce has high accuracy and efficiency. And even in the larger data set and the more complex data environment, our algorithm's efficiency has good applicability.

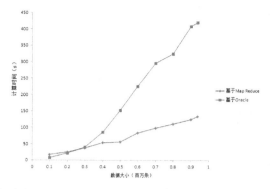

Fig. 1. Time consuming based on Map-Reduce and Oracle

Table 2. The performance comparison based on Map-Reduce and Oracle

	Map-Reduce	Oracle
RT	1420	4465
IS	142.6	534.7

Table 3. the performance comparison of three methods

	Our method	360 shopping search	Huihui shopping assistant
P	0.85	0.82	0.52
R	0.89	0.71	0.91
F1	0.81	0.76	0.66

5   Conclusion and Future Work

Nowadays identification of the same commodity entities is a major challenge in the heterogeneous multi-source e-commerce of big data. This paper introduces a framework based on Map-Reduce, called IIRS, which is made up of data index, data integration, entity recognition and data sorting. Through an extensive experimental study on real e-commerce dataset on Hadoop, our method demonstrates its feasibility, accuracy, and high efficiency. However the more complexity of large e-commerce data environment and the more varieties of different data types, data errors and so on will be the next focus in our future work.

References

1. Herndndez, M.A., Stolfo, S.J.: The merge/purge problem for large databases. SIGMOD **24**(2), 127–138 (1995)
2. Arasu, A., Kaushik, R.: A grammar-based entity representation framework for data cleaning. In: SIGMOD, pp. 233–244 (2009)

3. Wenfei, F., Xibei, J., Jianzhong, L., et al.: Reasoning about record matching rules. VLDB **2**(1), 407–418 (2009)
4. Chaudhuri, S., Ganti, V., Motwani, R.: Robust identification of fuzzy duplicates. In: ICDE 2005, pp. 865–876 (2005)
5. Chen, Z., Kalashnikov, D.V., Mehrotra, S.: Adaptive graphical approach to entity resolution. In: Proc. of the 7th ACM IEEE-CS Joint Conf. on Digital Libraries, New York, pp. 204–213 (2007)
6. Singla, P., Domingos, P.: Entity resolution with markovlogic. In: ICDM, pp. 572–582 (2006)
7. Augsten, N., Bohlen, M., Dyreson, C., et al.: Approximate joins for data—centric XML. In: ICDE 2008, pp. 814–823 (2008)
8. Li, W., Rong, Z., Chaofeng, S., et al.: A Product Normalization Method for E-Commerce. Chinese Journal of Compters **37**(2), 312–325 (2014)

Extracting Appraisal Expressions from Short Texts

Peiquan Jin[1,2(✉)], Yongbo Yu[1], Jie Zhao[3], and Lihua Yue[1,2]

[1] School of Computer Science and Technology,
University of Science and Technology of China, Hefei 230027, China
jpq@ustc.edu.cn
[2] Key Laboratory of Electromagnetic Space Information,
Chinese Academy of Sciences, Hefei 230027, China
[3] School of Business, Anhui University, Hefei 230601, China

Abstract. Short texts such as tweets and E-commerce reviews can reflect people's opinions on interested events or products, which are much beneficial to many applications. However, one opinion word may have different sentiment polarities when modifying different targets. Therefore, in this paper we propose to extract *"appraisal expressions"* that are represented by tuples of (*opinion word*, *target*), indicating an opinion word and the target modified by the word. By extracting appraisal expressions, we can further construct target-sensitive sentiment dictionaries and improve the effectiveness of sentiment analysis on short texts. Consequently, we propose a *filtering-refinement* framework to extract appraisal expressions from short texts. In the *filtering* step, we extract appraisal-expression candidates, and in the *refinement* step, we use SVM to extract appraisal expressions and present a *dependency-grammar-based* approach to automatically label training data. Comparative experiments between our proposal and three baseline methods suggest the superiority and effectiveness of our proposal.

Keywords: Appraisal expression · Sentiment analysis · Short text

1   Introduction

Short texts are very popular in social network and online reviewing platforms. It is beneficial to conduct sentiment analysis on the massive short texts on the Web so that people can know the public opinions on interested events or products.

So far, most sentiment analysis works on short texts are based on sentiment dictionaries maintaining sentiment polarities (e.g., *positive*, and *negative*) of selected opinion words [1-4]. According to these approaches, we can simply detect opinion words and their sentiment polarities by looking up the dictionary. Further, we can determine the final sentiment polarity of a sentence by some scoring methods, which can be classified into two types. One is to count the opinion words of each polarity, respectively, and further determine the sentimental polarity of the sentence [1, 2]. The other type is based on supervised classification [3, 4], which regards the polarity detection of a sentence as a classification task and considers opinion words as features. However, there are two problems existing in previous studies:

© Springer International Publishing Switzerland 2015
J. Li and Y. Sun (Eds.): WAIM 2015, LNCS 9098, pp. 481–485, 2015.
DOI: 10.1007/978-3-319-21042-1_45

Problem 1: It is not appropriate to simply count the number of opinion words to determine the polarity of a sentence, because an opinion word may not contain any kind of sentiment. For example, in the sentence *"well, this engine did a bad work."*, the words *"well"* and *"bad"* will be recognized as a positive opinion word and a negative word because they appear in the sentiment dictionary. However, in this example, *"well"* is rather an auxiliary word and does not bear any sentiment because it does not modify any targets in the sentence. Thus, only opinion words modifying some targets in a sentence can make sense in sentiment analysis.

Problem 2: The sentiment polarity of a sentence is not only determined by the opinion words in the sentence. We have to consider the target modified by an opinion word, because many opinion words have different polarities when modifying different targets. For example, the opinion word *"long"* is positive in the sentence *"the battery life is very long."*, where *"long"* modifies the target *"life"*. However, it is negative in the sentence *"this engine has a long startup time."*, where it modifies the target *"startup time"*. This indicates that it is more meaningful to consider the modification relationship between opinion words and targets when detecting the sentiment polarity of a sentence.

Therefore, in this paper we focus on *"appraisal expressions"* [5, 6], which are formed as pairs of (*opinion word, target*), indicating the modification relationship between an opinion word and its related target. By extracting appraisal expressions, we can further construct target-sensitive sentiment dictionaries to solve the aforementioned problems, and improve the effectiveness of sentiment analysis on short texts. Specially, we propose a *filtering-refinement* framework to extract appraisal expressions from short texts.

2   The Proposed Approach

The framework of our proposed approach to extracting appraisal expressions is shown in Fig. 1.

Our proposed framework contains a *filtering* stage and a *refinement* stage. During the filtering procedure, we use named entity recognition tools to extract pairs of noun phrases and adjectives as appraisal expression candidates. This is to get all the possible appraisal expressions. Compared with existing rule-based approaches that will exclude many correct appraisal expressions, this filtering procedure is beneficial to keep a high recall for appraisal expression extraction. In order to improve precision, we design a refinement procedure that works on the appraisal expression candidates generated by the filtering step. This

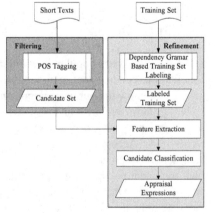

Fig. 1. Framework of the proposed approach

procedure is to remove wrong instances from the appraisal expression candidates. Particularly, we model this problem as a binary classification problem, i.e., classifying appraisal expression candidates into two sets, namely a set of correct expressions and another set of wrong expressions. Unlike traditional classification methods that need manual labeling on training sets, we introduce *dependency grammar* into the labeling on training sets and devise an automatic method to label training sets. Thus, combined with existing classification models (we use SVM in this work), we can provide an adaptive solution for appraisal expression extractions. This is simple because our approach does not depend manual labeling on training sets and can suit for different styles of short texts.

Dependency grammar is used to describe the dependency relationship between words in a sentence [8]. For example, in the sentence "*this car has a fantastic shape*", the word "*fantastic*" has a dependency relationship with "*shape*". There have been many tools that can be used to extract dependency relations of a sentence, such as Stanford Parser. Thus, in our study, we simply use the Stanford Parser for dependency grammar analysis.

The basic idea of dependency-grammar-based training set construction is to utilize the dependency relations in a sentence to automatically identify appraisal expressions. According to the properties of short texts like online reviews, we consider two types of relations to label appraisal expressions, which are "*amod*" and "*nsubj*".

Definition 1 (*amod*). Let w_a be an adjective and w_n be a noun. An *amod* relation between w_a and w_n is valid if there is a modification dependency-relationship between w_a and w_n, indicating that w_a modifies w_n in the sentence.

Definition 2 (*nsubj*). Let w_n be a noun and w_a be an adjective. An *nsubj* relation between w_n and w_a is valid if there is a subject dependency-relationship between w_n and a linking verb v as well as a predicate dependency-relationship between v and w_a.

For example, in the sentence "*this car has a fantastic shape*", there is an *amod* relation between the word "*fantastic*" and "*shape*", because the adjective "*fantastic*" modifies the noun "*shape*", which is represented as a "*mod*" dependency-relationship in dependency grammar. In the sentence "*this car engine is powerful*", there is an *nsubj* relation between the word "*engine*" and "*powerful*", because "*engine*" depends on "*is*" with a dependency relation "*s*" indicating "*engine*" is the subject of "*is*".

3   Performance Evaluation

The data sets are from Task 2 of COAE 2013, including two data sets with one about car domain and the other about camera domain. There are totally 9815 sentences in each data set. We use the Stanford POS Tagger to perform the POS tagging on short texts. In addition, we use the Stanford Parser to get the dependency-grammar relations for each sentence in the data sets. Regarding the classification model, we use SVM to perform the binary classification on appraisal expression candidates. Then, we conduct comparative experiments to compare our proposal with three existing methods, which are *Nearest* [1], *RB* (Rule-Based) [5], and *SPB* (Syntactic-Path-Based) [7]. In the experiments, we use four categories of features for extracting appraisal expressions, namely word features, POS features, syntactic-path features [7], and other features (including count of adjectives and count of nouns and noun-phrases).

We compare our method with the three competitor methods in terms of precision, recall, and F-measure. The results are shown in Table 1 and 2. As shown in Table 1 and 2, our method outperforms all the baseline methods w.r.t. recall and F-measure. In addition, it has a comparable precision with SPB. The nearest method performs poorly, because it lacks deep exploration on the lexical, syntactic, and semantic relationship between words. This indicates that distance-only methods are not appropriate for appraisal expression extraction. Both RB and SPB have a higher precision than the nearest method, since they both employ the exact-matching idea in appraisal expression extraction. However, they both have a low recall because many correct appraisal expressions will be excluded in the result list.

Table 1. Results on the car-domain data set

	Precision	Recall	F-measure
Nearest	0. 6780	0.6519	0.6647
RB	0.7788	0.6942	0.7341
SPB	0.7936	0.7553	0.7740
Our Method	0.7911	0.7904	0.7908

Table 2. Results on the camera-domain data set

	Precision	Recall	F-measure
Nearest	0.6237	0.6783	0.6499
RB	0.7619	0.6640	0.7096
SPB	0.7910	0.7377	0.7634
Our Method	0.7837	0.7825	0.7831

4   Conclusion

Extracting appraisal expressions from short texts is an important issue for sentiment analysis. In this paper, we propose a new approach for appraisal expression extraction, which is based on a *filtering-refinement* framework. Similar with existing methods, we also use a classification model to extract appraisal expressions but we introduce a *dependency-grammar-based* method to automatically label training sets, and thus make the classification more adaptable and scalable to different kinds and sizes of data. We evaluate the performance of our proposal on two real data sets, and the experimental results show that our proposal outperforms its competitors.

Acknowledgement. This paper is supported by the National Science Foundation of China (No. 61379037 and No. 71273010).

References

1. Hu, M., Liu, B.: Mining and summarizing customer reviews. In: Proc. of SIGKDD, pp. 168–177 (2004)
2. Zheng, L., Jin, P., Zhao, J., Yue, L.: Multi-dimensional sentiment analysis for large-scale E-commerce reviews. In: Decker, H., Lhotská, L., Link, S., Spies, M., Wagner, R.R. (eds.) DEXA 2014, Part II. LNCS, vol. 8645, pp. 449–463. Springer, Heidelberg (2014)
3. Zhao, J., Liu, K., Wang, G.: Adding redundant features for CRFs-based sentence sentiment classification. In: Proc. of EMNLP, pp. 117–126 (2008)

4. Jin, W., Ho, H.H.: A novel lexicalized HMM-based learning framework for web opinion mining. In: Proc. of ACL, pp. 465–472 (2009)
5. Bloom, K., Garg, N., Argamon, S.: Extracting appraisal expressions. In: Proc. of HLT-NAACL, vol. 2007, pp. 308–315 (2007)
6. Wilson, T., Wiebe, J.: Annotating attributions and private states. In: Proceedings of the Workshop on Frontiers in Corpus Annotations II: Pie in the Sky, pp. 53–60 (2005)
7. Zhao, Y., Qin, B., Che, W., Liu, T.: Appraisal expression recognition with syntactic path for sentence sentiment classification. International Journal of Computer Processing of Languages **23**(1), 21–37 (2011)
8. Kübler, S., McDonald, R., Nivre, J.: Dependency parsing. Morgan and Claypool Publishers (2009)

Friendship Link Recommendation Based on Content Structure Information

Xiaoming Zhang[✉], Qiao Deng, and Zhoujun Li

State Key Laboratory of Software Development Environment,
Beihang University, Beijing, China
{yolixs,lizj}@buaa.edu.cn

Abstract. Intuitively, a friendship link between two users can be recommended based on the similarity of their generated text content or structure information. Although this problem has been extensively studied, the challenge of how to effectively incorporate the information from the social interaction and user generated content remains largely open. We propose a model (LRCS) to recommend user's potential friends by incorporating user's generated content and structure features. First, network users are clustered based on the similarity of user's interest and structural features. Users in the same cluster with the query user are considered as the candidate friends. Then, a weighted SimRank algorithm is proposed to recommend the most similar users as the friends. Experiments on two real-life datasets show the superiority of our approach.

1   Introduction

Link prediction is the problem of predicting the existence of a link between two entities in an entity relationship graph, where prediction is based on the attributes of the observed links and other information related to the entities [1]. The link may exist between two entities that are either familiar or strange with each other. The problem can be also viewed as a link recommendation problem, where we aim to suggest to the query user a list of people that the user is likely to create new connections to. Link recommendation in social network is closely related to link prediction [4][5], but has its own specific properties. In addition to the structural features, users in social network are associated with other information such as common interest and interaction. Thus, how to incorporate different types of feature to recommend potential links in social network is still a challenge. Another challenge is the sparsity of real social network, which means that the existing links between nodes are only a very small fraction of all potential edges in the network graph. It is not feasible to test every user whether the query user will create connection to. In this paper, we propose to incorporates both of user's text content and structural features to recommend friendship linkages in social networks. First, we cluster all the network users based on the similarity of user's interest and structural features. Then, a weighted graph is constructed for each cluster, and the weighted SimRank algorithm is proposed to estimate the similarity between the query user and each candidate user. Finally, candidates with the greatest similarity values are recommended as the friends of the query user.

© Springer International Publishing Switzerland 2015
J. Li and Y. Sun (Eds.): WAIM 2015, LNCS 9098, pp. 486–489, 2015.
DOI: 10.1007/978-3-319-21042-1_46

2   Friendship Link Recommendation

Our recommendation approach contains two steps. To deal with the sparsity problem, we first filter some of the negative instances by clustering users based on the similarity of structural features and user's interest. Usually, users who have similar interest and structural features are more likely to make friends with each other. Thus, users in the some cluster with the query user compose the candidate list. User's interest is mined from the uploaded micro-blog documents using the topic model. The structural features refer to following relationship, such as follower and followee relationship in microblogging site, and social interaction, such as retweeting, mention, and comment between two users in microblogging site. Then, potential friends are recommended from the candidate list. To accomplish this, a weighted graph is constructed for each cluster, and a weighted SimRank algorithm is proposed to estimate the similarity between users in the same cluster. The users that have the greatest values of similarity with the query user are recommended to create friendship link with the query user.

Usually, user's interest is reflected by the uploaded micro-blogs. We aggregate all the micro-blogs generated by the same individual user into a profile that corresponds to a document in the LDA model [2]. Then, user's interest is modeled by the profile's probability distribution over topics. To measure the similarity of two users' interest, we use JSD (Jensen-Shannon-Divergence) to measure the similarity between their two probability distributions over topic. The similarity of structural features is composed of two components. The first component is the similarity of social link relationships, i.e., followee and follower, and it is estimated as following:

$$Sim_{rel}(u_i, u_j) = (\frac{\left|fee_i \cap fee_j\right|}{\left|fee_i\right| + \left|fee_j\right| - \left|fee_i \cap fee_j\right|} + \frac{\left|fer_i \cap fer_j\right|}{\left|fer_i\right| + \left|fer_j\right| - \left|fer_i \cap fer_j\right|}) / 2 \qquad (1)$$

where fee_i is the set of followees of u_i, and fer_i is the set of followers of u_i.

The second component is a measure of interactions:

$$Sim_{inter}(u_i, u_j) = \frac{r_{ij} + c_{ij} + a_{ij} + r_{ji} + c_{ji} + a_{ji}}{6 \cdot Max} \qquad (2)$$

where r_{ij} is the times that u_i has retweeted u_j's micro-blogs, c_{ij} is the times that u_i has commented on u_j's micro-blogs, a_{ij} is the times that u_i has mentioned u_j, and Max is the largest frequency of these interaction in a given dataset. Then, the similarity between two users is a linear combination of interest similarity and structural similarity.

Then, we introduce a weighted SimRank algorithm [6] to take the strength of relationship into the estimation of similarity. We construct a weighted graph for each cluster, in which each node denotes a user and the weight of an edge denotes the strength of the relationship between this pair of users. The similarity value from a node u_i is not equally propagated to the out-neighbor nodes. Instead, it is propagated based on the normalized probability estimated as following:

$$p(u_i, u_j) = \frac{Sim(u_i, u_j)}{\sum_{u_k \in O(u_i)} Sim(u_i, u_k)} \tag{3}$$

where $O(u_i)$ is the out-neighbor nodes of u_i, and $Sim(u_i, u_j)$ denotes the weight of edge $<u_i, u_j>$. Then the recursive form of the similarity between node u_i and u_j at the $(k+1)^{th}$ iteration is represented as following:

$$s_{k+1}(u_i, u_j) = C \sum_{u_{in} \in I(u_i)} \sum_{u_{jn} \in I(u_j)} p(u_{in}, u_i) p(u_{jn}, u_j) s_k(u_{in}, u_{jn}) \tag{4}$$

Meanwhile, the similarity in the initial state is revised as following:

$$s_0(u_i, u_j) = \begin{cases} 1 & u_i = u_j \\ p(u_i, u_j) & u_j \in O(u_i) \\ 0 & otherwise \end{cases} \tag{5}$$

3   Experiments

To evaluate our approach, two real-life datasets are used in the experiments. The first dataset is collected from Sina Weibo. We sample 10 user ids uniformly at random from the space of Sina Weibo user id numbers, and then we use the width-first search method to visit other users and download the profiles and links of these visited users. The second dataset is published in the SiGKDD Cup Track1 2012, and it is collected from Tecent Weibo. We randomly selected a subset of 51938 users and their links information to build the second dataset. We divided the datasets into 80% for training and 20% for testing. Three methods are used as the baselines. (1) **Content-based** [3]: it applies the topic modeling techniques, specifically LDA, on user's profile data to recommend link based on interest similarity and existing friendships. (2) **Structure-based** [1]: it uses random walk algorithm on an augmented social graph with both attribute and link information to recommend links. (3) **Naïve SimRank** [6]: it uses the Naïve SimRank algorithm on the unweighted graph of user relation to iteratively compute the similarity between nodes and then return the top-k ranked nodes.

Table 1. Precision and Recall of different approaches

dataset	method	P(%)	R(%)
Sina Weibo	Content-based	49.2	40
	Structure-based	71.2	45.8
	Naïve SimRank	70.3	51.9
	LRCS	**75.3**	**56.7**
Tecent Weibo	Content-based	47.1	38.3
	Structure-based	73.4	47.1
	Naïve SimRank	71.8	54.2
	LRCS	**74.9**	**63.4**

We compare the precision and recall of link recommendation of different methods on two datasets, and the results are shown in table 1. The results show that our approach outperforms other approaches obviously. Our approach combines both the interest and structure information to retrieve the potential friends, which is more effective to find the friends that may be neglected by the structure-based or content-based method. Second, we revise the Naïve SimRank algorithm by assigning each edge with a transforming probability and the initial similarity of two neighbouring nodes with a nonzero value. This revision is more coincident with reality, since the effects of different users on a given user are unequal. Thus, the weight SimRank is also more effective than the Naïve SimRank.

4   Conclusion

In this paper, we proposed to recommend friendship link based on content and structural features in a social network. First, we cluster users based on the similarity of structure features and interests mined from the generated text content, which filters the unrelated users and thus reduces the search space. Then, each user cluster is represented by a weighted graph, and the weighted SimRank is proposed to estimate similarity between two users, and the top-ranked users are recommended to be linked by the query user. Experimental results on two real-life datasets demonstrate the superiority of our approach. Our approach can be further improved in several aspects. For example, we can extend our approach to recommend user with other type of objects, such as movie, products, and so on, by including user's relationship with the objects in the approach.

Acknowledgements. This work was supported by the National Natural Science Foundation of China (No. 61202239, No. 61170189, and No. 61370126), the Fundamental Research Funds for the Central Universities (No. YWF-14-JSJXY-16), and the Fund of the State Key Laboratory of Software Development Environment (No. SKLSDE-2015ZX-11).

References

1. Yin, Z., Gupta, M., Weninger, T., Han, J.: A unified framework for link recommendation using random walks. In: International Conference on Advances in Social Networks Analysis and Mining (2010)
2. Blei, D.M., Ng, A.Y., Jordan, M.I.: Latent Dirichlet Allocation. The Journal of Machine Learning Research **3**, 993–1022 (2003)
3. Parimi, R., Caragea, D.: Predicting friendship links in social networks using a topic modeling approach. In: Huang, J.Z., Cao, L., Srivastava, J. (eds.) PAKDD 2011, Part II. LNCS, vol. 6635, pp. 75–86. Springer, Heidelberg (2011)
4. Tang, W., Zhuang, H., Tang, J.: Learning to infer social ties in large networks. In: Gunopulos, D., Hofmann, T., Malerba, D., Vazirgiannis, M. (eds.) ECML PKDD 2011, Part III. LNCS, vol. 6913, pp. 381–397. Springer, Heidelberg (2011)
5. Tang, J., Lou, T., Kleinberg, J.: Inferring social ties across heterogenous networks. In: 5th International ACM Conference on Web Search and Data Mining (2012)
6. Jeh, G., Widom, J.: SimRank: a measure of structural-context similarity. In: 18th ACM SIGKDD International Conference on Knowledge Discovery and Data Mining (2002)

Overlapping Community Detection in Directed Heterogeneous Social Network

Changhe Qiu[1,2,3], Wei Chen[1,2(✉)], Tengjiao Wang[1,2], and Kai Lei[3]

[1] Key Laboratory of High Confidence Software Technologies, Peking University, Ministry of Education, Beijing, China
pekingchenwei@pku.edu.cn
[2] School of Electronics Engineering and Computer Science, Peking University, Beijing 100871, China
[3] Shenzhen Key Lab for Cloud Computing Technology and Applications (SPCCTA), School of Electronics and Computer Engineering, Peking University, Beijing, China

Abstract. In social networks, users and artifacts (documents, discussions or videos) can be modelled as directed bi-type heterogeneous networks. Most existing works for community detection is either with undirected links or in homogeneous networks. In this paper, we propose an efficient algorithm OcdRank (Overlapping Community Detection and Ranking), which combines overlapping community detection and community-member ranking together in directed heterogeneous social network. The algorithm has low time complexity and supports incremental update. Experiments show that our method can detect better community structures as compared to other existing community detection methods.

Keywords: Community detection · Directed heterogeneous social network · Ranking

1 Introduction

Community detection[4,5] and ranking[2] in network are two dominating methods in social network analysis. However, both have their own defects[1].

We define X to be the artifact set, x to be one artifact, and Y to be the user set, y to be one user, and G_i to be the i_{th} community. Each artifact belongs to one community (because for example, in microblog, one blog tends to focus only one topic). Each user could belong to multiple communities (user may have many interests).

RankClus[1] combines clustering and ranking together in undirected heterogeneous network. However, its time complexity is high. It distributes users into all communities, which is unreasonable according to the truth.

In our method, we put forward transfer model to pass intermediary variable to connect two seemingly unrelated variables: artifacts and communities.

© Springer International Publishing Switzerland 2015
J. Li and Y. Sun (Eds.): WAIM 2015, LNCS 9098, pp. 490–493, 2015.
DOI: 10.1007/978-3-319-21042-1_47

2   OcdRank

The algorithm is iterative. All the artifacts are divided into k parts (in the first iteration, the distribution is random). Community-based ranking then carry out in every community. With the ranking score of nodes, we use link transfer model to evaluate the correlation coefficient between artifacts and communities. Artifacts and their associated users step into the most similar communities. Every loop the communities will be adjusted and the ranking score of nodes within the community will be changed. When iteration times reach a specified number or clusters changes only by a very small ratio, the iteration will terminate. Then, we get the meaningful communities.

Community-based ranking gives members in the same community a discriminating criterion. If the utilized ranking algorithm is PageRank [2], the score of users in our method can be formulated as

$$r(y'|G_i) = \frac{1-q}{N} + q(\alpha \sum_{x \in Pub(y')} \frac{r(x|G_i)}{Out(x|G_i)} + (1-\alpha) \sum_{y \in Fol(y')} \frac{r(y|G_i)}{Out(y|G_i)}) \qquad (1)$$

Where q is damping factor to make all the nodes in network can be accessed. And N is the number of nodes in community G_i. $Out(x|G_i)$ is the out degree of x in condition of community G_i and $Out(y|G_i)$ is the out degree of y accordingly. $Pub(y')$ is the artifact set y' published, $Fol(y')$ is the user set following y'. $\alpha \in (0,1)$ determines how much weight to put on each factor based on one's belief.

After getting the ranking score of every object in community G_i, we use transfer model to evaluate the correlation coefficient of artifacts and communities.

$$r(x|G_i) = \sum_{y \in (Dir(x) \cap Y')} r(y|G_i) \qquad (2)$$

Where $Dir(x)$ is the users directed to artifact x, Y_i is the user set, $r(y|G_i)$ is the community-based ranking score of user y. Since every artifact only has relation with one community, but the associated users of it could be in different communities.

The correlation coefficient of artifacts and communities can be viewed as a decomposed vector. Every community could be formed as the normalized sum of all artifacts in it. Let $v(x) = (r(x|G_1'), r(x|G_2'), \dots, r(x|G_k'))$ be the vector of artifact x, the vector of a community G_i' $(i = 1, 2, \dots, k)$ is

$$v(G_i') = \sum_{x \in X'} v(x) r(x|G_i') \qquad (3)$$

$v(G_i')$ is the vector of G_i', $r(x|G_i')$ is the community-based ranking score of x. Each artifact and the associated users can be assigned to the nearest community according the cosine similarity scores.

When new artifact needs to be assigned to one community, there will be three considerations: 1) The artifact point to another one. We can directly assign it to the community where the pointed artifact is. 2) The associated users are partly in the user set Y. The correlation coefficient of artifact and communities can be computed directly by equation (2). The iteration goes on as our algorithm described. 3) There's no associated user in the user set Y. A new community should be created.

3   Experiments

We use Twitter (http://arnetminer.org/heterinf) and Weibo datasets. The baselines are BigClam[4], Coda[5] and RankClus[1] respectively. BigClam runs on undirected networks while Coda directed. Both are overlapping community detection methods in homogeneous network. RankClus[1] works in undirected heterogeneous network.

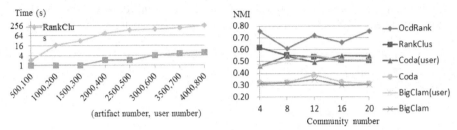

(a)   Running time of OcdRank and RankClus   (b) performance of different algorithms

Fig. 1. Comparison between OcdRank and other algorithms

Normalized Mutual Information (NMI) is an information-theoretic measure of similarity between two partitioning of a set of elements[3].

Fig. 1 (a) is the comparison of running time between RankClus and OcdRank.

We use Weibo dataset with different ground truth of different number of communities. For BigClam and Coda, we treat the heterogeneous networks as homogeneous. We use network of user links to compare with the ground truth of user network (BigClam(user) and Coda(user)). The parameter α of OcdRank is 0.5. We run each algorithm 10 times and get mean value. The result is shown in Fig. 1(b).

In Weibo, the rank result in each community is interesting. The users with higher ranking scores are usually the official accounts in same special field.

4   Discussion

OcdRank works on directed bi-type heterogeneous social networks. The ranking of users based on community can be used for experts finding. The ranking of artifacts can be used for semantic community detection. For very large datasets, running on distributed system is one of our future works. Another future work is semantic community detection, which would utilize the semantic information of artifacts.

Acknowledgments. This research is supported by the Natural Science Foundation of China (Grant No. 61300003), Research Foundation of China Information Technology Security Evaluation Center (No. CNITSEC-KY-2013-018) and Research Foundation Program of Ministry of Education & China Mobile (MCM20130361).

References

1. Sun, Y., Han, J., Zhao, P., Yin, Z., Cheng, H., Wu, T.: Rankclus: integrating clustering with ranking for heterogeneous information network analysis. In: Proceedings of the 12th International Conference on Extending Database Technology: Advances in Database Technology, pp. 565–576. ACM (2009)
2. Brin, S., Page, L.: The anatomy of a large-scale hypertextual Web search engine. Computer Networks and ISDN Systems **30**(1), 107–117 (1998)
3. Lancichinetti, A., Fortunato, S., Kertész, J.: Detecting the overlapping and hierarchical community structure in complex networks. New Journal of Physics **11**(3), 033015 (2009)
4. Yang, J., Leskovec, J.: Overlapping community detection at scale: a nonnegative matrix factorization approach. In: Proceedings of the Sixth ACM International Conference on Web Search and Data Mining, pp. 587–596. ACM (2013)
5. Yang, J., McAuley, J., Leskovec, J.: Detecting cohesive and 2-mode communities indirected and undirected networks. In: Proceedings of the 7th ACM International Conference on Web Search and Data Mining, pp. 323–332. ACM (2014)

Efficient MapReduce-Based Method for Massive Entity Matching

Pingfu Chao[1,2], Zhu Gao[2], Yuming Li[1,2], Junhua Fang[1,2], Rong Zhang[1,2(✉)], and Aoying Zhou[1,2]

[1] Institute for Data Science and Engineering, New York, USA
{rzhang,ayzhou}@sei.ecnu.edu.cn
[2] Shanghai Key Laboratory of Trustworthy Computing, Shanghai, China
{51121500001,10132510331,51141500019,52131500020}@ecnu.cn

Abstract. Most of the state-of-the-art MapReduce-based entity matching methods inherit traditional Entity Resolution techniques on centralized system and focus on data blocking strategies in order to solve the load balancing problem occurred in distributed environment. In this paper, we propose a MapReduce-based entity matching framework for processing semi-structured and unstructured data. We use a Locality Sensitive Hash (LSH) function to generate low dimensional signatures for high dimensional entities; we introduce a series of random algorithms to ensure that similar signatures will be matched in reduce phase with high probability. Moreover, our framework contains a solution for reducing redundant similarity computation. Experiments show that our approach has a huge advantage on processing speed whilst keeps a high accuracy.

1 Introduction

Entity matching aims to identify entities referring to the same real-world object. However, the rapid growth of web data and User Generated Content (UGC) brings new challenges for entity matching. For instance, in the scenario of C2C (Customer to Customer) online markets, as the rarity of descriptions, missing of uniform schema or intended errors generated by users, tradition entity matching methods are not able to get good match performance.

Though MapReduce provides a new platform for solving massive entity matching problem, new challenges occur: load balancing problem and network transmission cost. Blocking-based entity matching algorithms have been presented to deal with the imbalance problem. Some of the most influential works include sorted neighborhood-based and load-balanced entity matching in Dedoop[3], and document-similarity computation[1]. But for processing non-structured data, these kinds of work meet high network cost and computation cost.

This paper sketches out a random-based framework for entity matching based on MapReduce for semi-structured and unstructured data. Inspired by previous studies, our method expects to reduce both the computation cost and network transmission cost whilst promises the processing performance. We convert high

© Springer International Publishing Switzerland 2015
J. Li and Y. Sun (Eds.): WAIM 2015, LNCS 9098, pp. 494–497, 2015.
DOI: 10.1007/978-3-319-21042-1_48

dimensional entity features into low dimensional bit vector by Locality Sensitive Hash (LSH) function in map phase[4], which reduces the network transmission cost dramatically. We do t rounds of random permutations to those bit vectors. It helps to make similar items paired with high probabilities. Our random-based design can also ensure load-balanced during matching process. Finally, we design a new solution for removing redundant computation in reduce phase.

2   MapReduce-Based Entity Matching Framework

Our entity matching framework is shown in Fig. 1. We represent each entity by its high dimensional feature vector generated from the structured, unstructured or semi-structured description data, if any. These vectors are the input of our MapReduce job, as shown in Fig. 2. The first round of MapReduce job implements the Entity Matching job, while the second one realizes the Redundancy Control.

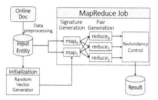

Fig. 1. Framework of random-based entity matching on MapReduce

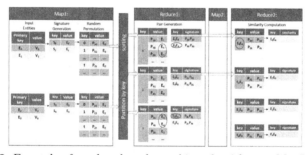

Fig. 2. Example of random-based matching algorithm on MapReduce

Entity Matching. The input is a set of (*key, value*) pairs with the entity ID E_u as its key and its k-dimension vector V_u as its value. In map phase, we generate a signature for each item u using the LSH function h_r defined in Eqn.1. We generate a random k-dimension vector set V_r with $|V_r| = d$. Calculating the hash values between u and every vector in V_r using h_r, we get a d-bits vector S_u as the signature for item u, $d \ll k$. Then we apply t rounds of random permutations to every signature S_u and get t different d-dimension bit vectors $\{P_{u1}, P_{u2}, ..., P_{ut}\}$. We regard this result as our map output. So for each entity u, we have t different map outputs as (i, P_{ui}, E_u), in which i refers the permutation series number $(i \in t)$, P_{ui} refers the i_{th} permutation result, and E_u is the entity ID.

In Reduce phase, each reducer receives permuted signatures of the same series number. It sorts all signatures and generates pairs between each signature and its m nearest neighbors. Then we calculate the hamming distance of every pair. We output the entity pairs with their similarities as $(E_u E_v, similarity)$ with $u < v$.

$$h_r(u) = \begin{cases} 1 & r.u \geq 0 \\ 0 & r.u < 0 \end{cases} \tag{1}$$

We use the LSH function preserving cosine similarity [2] to generate a signature S_u for each entity u. Since the signature carries most of the characteristics of a vector, we can measure the similarity of two vectors by comparing their signatures. We use the hamming distance between two signatures to represent the similarity, which is reasonable and well proved [4]. In reduce phase, we propose a random permutation algorithm inspired by PLEB algorithm[4] to ensure entities with high similarity to be paired with high probabilities.

Redundancy Control. There can be many duplicated pairs in different groups during reduce phrase as marked in Fig. 2. It may cause significant redundant computation cost. We introduce an extra MapReduce job to reduce duplication. In reduce phase of the first MapReduce job, we remove the similarity computation step, and directly send all the pair-wise data to the second MapReduce job. The second map job does nothing. After the shuffle phase, all pairs with the same entity IDs are grouped together. So we pick one pair of permuted signatures in the group and calculate its hamming distance on behalf of the others. At last, we output the similarity $(E_u E_v, similarity)$ as our result.

3   Experiments

We run experiments on a 22-node HP blade cluster. Each node has two Intel Xeon processors (E5335 2.00GHz) with four cores and one thread per core, 16GB of RAM, and two 1TB hard disks. All nodes run CentOS 6.5, Hadoop 1.2.1, and Java 1.7.0. We use CiteSeerX data set, which contains nearly 1.32 Million citations of total size 2.89 GB in XML format. Each citation includes *record ID, author, title, date, page, volume, publisher, etc* and also *abstract*. We compare the performance of our algorithms with Document Similarity Self-Join (DSSJ)[1] and Dedoop[3]. We use *accuracy* and *run-time* metrics to evaluate performance.

In order to measure the accuracy, we manually generate a validation set which contains 200 records. We output the top 10, 20 and 50 similar pairs for each algorithm. Since Dedoop compares all possible pairs and calculates cosine similarity directly, Dedoop is the best as in Fig.1. Ours achieves better accuracy than DSSJ with much less computation cost as in Fig.3. For processing speed, since Dedoop and DSSJ generate enormous size of pairs, they cost much network transmission and bring big burden for in memory processing as in Fig. 3. In our experiment, the transmission data generated by Dedoop or DSSJ is up to several

Table 1. Accuracy Comparison

Name	Top 10	Top 20	Top 50
DSSJ	90%	95%	94%
Ours	90%	100%	94%
Dedoop	100%	100%	100%

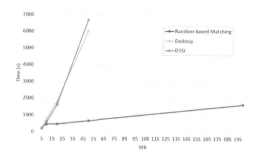

Fig. 3. Run-time Comparison

terabyte for 200MB source data. However, our algorithm is significantly faster than Dedoop, and far more stable even dealing with gigabytes of input data.

4 Conclusion

In this paper, we study the problem of matching the entities with high-dimensional feature vectors based on MapReduce. We take the MapReduce framework as our programming model and point out the two major challenges met on this model, which were load balancing problem and network transmission cost. We propose a random-based matching method to solve the matching problem. We use LSH function to generate signatures for entities and based on random permutations, we can promise similar candidate to be paired with high probabilities. Given the proposed algorithm, we implement it in Hadoop and compare with the other algorithms. We achieve much lower computation cost while still keep high accuracy.

Acknowledgment. This work is partially supported by National Basic Research Program of China (Grant No. 2012CB316200), National Science Foundation of China (Grant No.61232002, 61402180 and 61332006), and Key Program of Natural Science Foundation of Yunnan Province under grant No. 2014FA023.

References

1. Baraglia, R., De Francisci Morales, G., Lucchese, C.: Document similarity self-join with mapreduce. In: Proc. of Data Mining (ICDM), pp. 731–736. IEEE (2010)
2. Charikar, M.S.: Similarity estimation techniques from rounding algorithms. In: Proc. of the Thiry-fourth Annual ACM symposium on Theory of Computing, pp. 380–388. ACM (2002)
3. Kolb, L., Thor, A., Rahm, E.: Dedoop: efficient deduplication with hadoop. Proc. of VLDB **5**(12), 1878–1881 (2012)
4. Ravichandran, D., Pantel, P., Hovy, E.: Randomized algorithms and nlp: using locality sensitive hash function for high speed noun clustering. In: Proc. of ACL, pp. 622–629. Association for Computational Linguistics (2005)

The Role of Physical Location
in Our Online Social Networks

Jia Zhu[1], Pui Cheong Gabriel Fung[2(✉)], Kam-fai Wong[2,3,4], Binyang Li[5],
Zhixu Li[6], and Haoye Dong[1]

[1] School of Computer Science, South China Normal University, Guangzhou, China
[2] Department of Systems Engineering and Engineering Management, The Chinese
University of Hong Kong, Hong Kong, China
pcfung@se.cuhk.edu.hk
[3] MoE Key Laboratory of High Confidence Software Technologies, Beijing, China
[4] Shenzhen Research Institute, The Chinese University of Hong Kong,
Hong Kong, China
[5] The University of International Relations, Beijing, China
[6] School of Computer Science and Technology, Soochow University, Suzhou, China

1   Introduction

One of the most important properties of social networking sites is its reachability – no physical location constraint. In addition, all social networking sites allow us to search people with common interests, so we can find friends anywhere in the world easier than ever. With the help of social media, it seems that expaning our social networks is physical location independent. Motivated by the above observations, we study the role of physical location in social media. If physical location is no longer a barrier and physical interaction can be ignored, then our online social networks should have the following characteristics: (1) A number of our friends are from different places in the world other than the places that we have been; (2) A number of our friends are not from our physical social circles – they are not our colleagues, not our high school friends, etc.

2   Definition and Data Collection

Let $G = (U, E)$ be a social network. $U = \{u_1, u_2, \ldots u_n\}$ is a set of vertices and $E = \{e_1, e_2, \ldots e_m\}$ is a set of edges. A vertex represents an individual and an edge represents two individuals are friends in G. The one-hop network of u_i is called friendship network of u_i, and is denoted by F_i. u_i's social circles is a collection of the names of organizations that are listed in u_i education history and/or employment history. Given u_i and u_j, u_i can reach u_j if and only if u_j's profile is not hidden to u_i. We randomly picked 20,000 user profiles from Facebook and crwal their friends and finally obtained round 2 million profiles.

2.1   Education History and Employment History Analysis

Question: if u_x and u_y are friends, then what is the probability that u_x and u_y are from: (1) the same school? (2) the same employer? (3) both? If the probabilities

© Springer International Publishing Switzerland 2015
J. Li and Y. Sun (Eds.): WAIM 2015, LNCS 9098, pp. 498–501, 2015.
DOI: 10.1007/978-3-319-21042-1_49

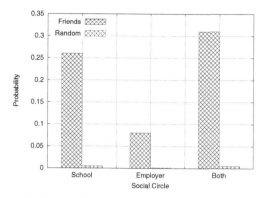

Fig. 1. The probabilities that two persons are from the same social circle

are high, then physical social circle plays an important role in the structure of a social network. Figure 1 shows the result. The probability of "friend bar" is far more than the "random bar".

Limitations: (1) Not many users disclosed their employment histories (15%) and education histories (28%). We may consider the result in this section is a baseline case. (2) two friends who are studying/working in the same school/company does not necessarily mean they *must* know each other via in-person. Yet, this does not affect our conclusion: physical location plays an important role in online social network.

2.2 Home Town and Current City Analysis

Question: if u_x and u_y are friends, then what is the probability that u_x and u_y have: (1) the same home town and/or the same current city? (2) the same home town, current city, school and/or employer? Figure 2 shows the results. From the findings we obtained, we conclude that geographical location does play a significant role in our online social network, even thought social media in theory does not have any physical boundary.

Limitation: Facebook allows users to specific their home towns and current cities but does not provide any information about what other cities a user has been. E.g., u_a and u_b are friends. u_a's home town and current city are both San Jose and u_b's home town and current city are both Phoenix. Suppose u_a and u_b knew each other because they both lived in Rochester for a while. In their Facebook profiles, we only know u_a and u_b are friends but never realize they both stay in the same place. We may underestimate the importance of geographical location.

2.3 Language Analysis

Question: given a user, what is the probability that a friend of this user uses the same language setting? Figure 3 shows the result. Overall, given a random user, 83% of his/her friends will use the same language code. This percentage

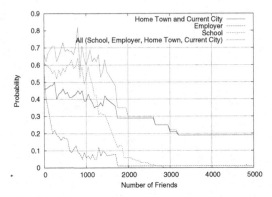

Fig. 2. The relationship between the number of friends and the probability of friends came from the same locality

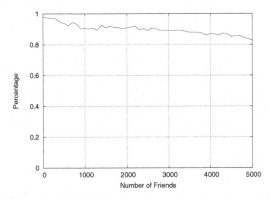

Fig. 3. The probability of friends having the same language setting

is surprisingly high. Almost for all languages, the probability that two random users who are friends will set the same language as their default one will always be > 70%. There are a few exceptions. The most obvious one is Malaysian. There are 2,386 users set Malaysian as their default Facebook language. For a user who set Malaysian as his/her default language, only 20.1% of his/her friends will also set Malaysian as default, where as 70.8% of his/her friends would use English. These finding strongly suggested that language also plays a significant role in determining how our social networks in a social media are shaped. In addition, language is somehow also location dependent.

2.4 Further Discussion

Question: what is the probability that two random people are friends given that they have common interest? If the probability is high, it may implies social media is an important place for people to find and make friends with common interest. It may then suggest that physical interaction may not be too important in some cases. We randomly join 300 groups that are related to some particular interests in Facebook and then compute the probability that two random people from the same group are friends. Figure 4 shows the result. The larger the group, the less likelihood that two random people from the group are friends. Perhaps most people joining some common-interest-groups is for sharing and obtaining information, but not for acquiring new friends.

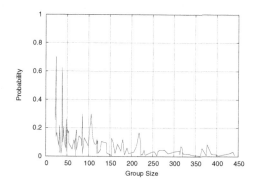

Fig. 4. Probability of two random people from a group are friends

3   Conclusion

There is indeed a very strong relationship between physical locations and our online social networks. In theory there is no physical boundary in social media, few of us will try to utilize this property when we use social media. The reachability of social media is strong but the expandability is limited. This may further suggest that most people use social media sites are not aimming at expanding their networks, but reconnecting some lost networks or maintaining their existing ones. The problem similar to, but in fact quite distinct, to ours is the problem related to homophily [1]: given two people are connected, are they similar to each other? Their primary goal not to answer questions such as: what is the role an attribute (e.g., physical location) plays in the structure of a network?

Acknowledgement. This research is partially supported by General Research Fund of Hong Kong (417112), Shenzhen Fundamental Research Program (JCYJ20130401172046450), RGC Direct Grant (417613), Fundamental Research Funds for the Central Universities (3262014T75).

References

1. McPherson, M., Smith-Lovin, L., Cook, J.: Birds of a feather: Homophily in social networks. Annual Review of Sociology **27**, 415–444 (2001)

Hierarchical Community Evolution Mining from Dynamic Networks

Yonghui Zhang, Chuan Li$^{(\boxtimes)}$, Yanmei Li, Changjie Tang, and Ning Yang

College of Computer Science, Sichuan University, Chengdu 610065, China
zhangyonghui14@163.com, lcharles@scu.edu.cn

Abstract. Research on community evolution contributes to understanding the nature of network evolution. Previous community evolution studies have two defects: (1) the algorithms do not have sufficient stability or cannot handle the radical structure change of communities, and (2) they cannot reveal the evolutionary regularities with multiple levels. To solve these problems, this paper proposes a new method for mining the evolution of communities from dynamic networks. Experiments demonstrate that compared with traditional methods, our work significantly improves the algorithm performances.

Keywords: Dynamic networks · Community evolution

1 Introduction

Currently, the method used to study community evolution in dynamic networks can be roughly divided into two categories. The first type of method is called independent clustering [1-4]. First, a series of network snapshots is taken from the dynamic network, and then the community detection algorithm is applied to get the community structure on each network snapshot. However, most community detection algorithms are unstable and easy to fall into local optimum, i.e., even small changes in network connection will lead to a very different community structure. Especially for agglomerative hierarchical clustering algorithms, when a small amount of the network edges are removed, the community structure found by these algorithms will be of great change [5].

The second type of approach to study community evolution is evolutionary clustering [6-8]. This type of method uses temporal smoothness assumption that community structure does not change dramatically in continuous timestamps. Evolutionary clustering method, by taking into account the temporal information in network evolution, overcomes the problem of randomness caused by the community detection algorithm or the noise of the network. However, because of the use of temporal smoothing hypothesis, these methods cannot capture the dramatic changes of community structure due to external events, which is a common defect to such methods.

This paper is supported by NSFC No. 61103043, and 61173099. And Chuan Li is the associate author.

J. Li and Y. Sun (Eds.): WAIM 2015, LNCS 9098, pp. 502–505, 2015.
DOI: 10.1007/978-3-319-21042-1_50

To solve these problems and explore the issue of community evolution better, this paper proposes a new method for mining the evolution of communities from dynamic networks.

2 Hierarchical Community Mining

Hierarchical clustering process is as follows. First, each edge is assigned to an individual community; the height of the community at this time is 0. Then each time the most two similar communities are selected and merged. In the single-link hierarchical clustering, the similarity between two communities is the similarity of the most similar edges in the two communities. 1 minus this similarity is the height of the combined community. If there are multiple communities with the same similarity, then randomly select two communities to merge. The process is repeated until all the edges are combined into a community. Single-link hierarchical clustering will get a dendrogram shown in Fig. 1. Apply hierarchical clustering to the dynamic network, and we will get a dynamic dendrogram.

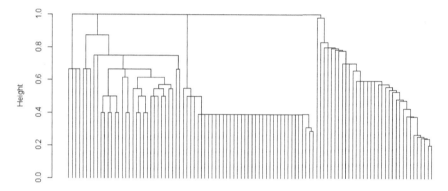

Fig. 1. A Dendrogram Created by Link Communities Method

2.1 Community Screening

The result of hierarchical clustering algorithm is a complete community hierarchy, which inevitably contains a large number of meaningless communities such as trivial communities and redundant communities. These communities are not only pointless for the analysis of community evolution, but also affect the efficiency of our algorithm, and therefore they need to be removed from the collection of communities.

Definition 1: *trivial community*. Given a community $c_{t,i}$ at time t, if $|c_{t,i}| \leq 2$, then this community is a trivial community.

Definition 2: *redundant community*. Let community $c_{t,j}$ and community $c_{t,k}$ have the same parent community $c_{t,i}$, i.e. $c_{t,i} = \Phi_t(c_{t,j}, c_{t,k})$. If both of the following conditions:

(1) $\left|c_{t,j}\right| > 2$, (2) $\left|c_{t,k}\right| \leqslant 2$ or $\left|c_{t,k}\right|/\left|c_{t,j}\right| < \alpha$ (in this paper, we set α to 0.05) are sa-tisfied, then community $c_{t,j}$ is a redundant community.

Redundant communities refer to the communities that are very similar with other communities, and can be substituted by them. As shown in Fig. 2, in the hierarchical clustering process, community B and community C are merged into community A. It can be seen that community B, which is merely one edge and one node fewer than community A, is very similar with community A. The evolution of community B and community A will be very similar as well, so in community evolution analysis com-munity B is redundant and can be replaced with the community A.

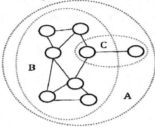

Fig. 2. An Example of Redundant Community

3   Experimental Evaluation

In this paper, we use Enron (the emails that sent among the employees of Enron Cor-poration) and Facebook (the comments that left by a portion of Facebook users on the Facebook wall) data set in the following experiments.

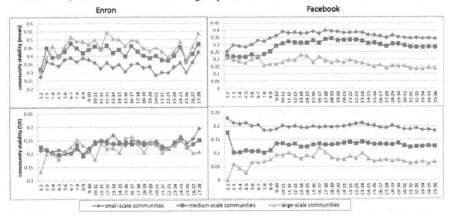

Fig. 3. The Variation of Community Stability of Three Kinds of Scale with Time

In order to study the community size effects on the community evolution, according to the number of nodes in the communities, the communities are divided into small-scale communities(the number of node is less than 10), medium-scale communities(the num-ber of node is greater than 10 and less than 100) and large-scale communities(the number of node is greater than 100). Fig. 3 shows the variation of the community stability of three kinds of scale with time. The horizontal axis is the time axis t-t+1; the vertical axis of the upper three figures is the mean of all the communities' stability $I(c_{t,i}, c_{t+1,i})$ from time t to t+1, and the vertical axis of the lower three figures is the standard deviation of all the communities' stability $I(c_{t,i}, c_{t+1,i})$ from time t to t+1.

As can be seen from Fig. 3, in Enron dataset, small-scale communities are most unstable while large-scale communities are most stable. There is no significant difference in the standard deviation of community stability between communities of three kinds of scale. However, in Facebook dataset, the mean of community stability of small-scale communities is the highest and the standard deviation is also the highest, indicating that small-scale communities as a whole is relatively stable, and there exist many very stable communities and very unstable communities at the same time. The mean of community stability of large-scale communities is the lowest and the standard deviation is also the lowest, which means that the stability of large-scale communities is relatively close.

4   Summary

This paper discusses the issue of new community evolution method with multiple hierarchies and proposes a stable method to enable the analysis of the community evolution. Experiments show that compared with traditional methods, our work significantly improves the algorithm performances.

References

1. Palla, G., Barabási, A.L., Vicsek, T.: Quantifying social group evolution. Nature **446**(7136), 664–667 (2007)
2. Asur, S., Parthasarathy, S., Ucar, D.: An event-based framework for characterizing the evolutionary behavior of interaction graphs. ACM Transactions on Knowledge Discovery from Data (TKDD) **3**(4), 16 (2009)
3. Takaffoli, M., Sangi, F., Fagnan, J., et al.: Community evolution mining in dynamic social networks. Procedia-Social and Behavioral Sciences **22**, 49–58 (2011)
4. Bródka, P., Saganowski, S., Kazienko, P.: GED: the method for group evolution discovery in social networks. Social Network Analysis and Mining **3**(1), 1–14 (2013)
5. Hopcroft, J., Khan, O., Kulis, B., et al.: Natural communities in large linked networks. In: Proceedings of the Ninth ACM SIGKDD International Conference on Knowledge Discovery and Data Mining, pp. 541-546. ACM (2003)
6. Chakrabarti, D., Kumar, R., Tomkins, A.: Evolutionary clustering. In: Proceedings of the 12th ACM SIGKDD International Conference on Knowledge Discovery and Data Mining, pp. 554-560. ACM (2006)
7. Lin, Y.R., Chi, Y., Zhu, S., et al.: Facetnet: a framework for analyzing communities and their evolutions in dynamic networks. In: Proceedings of the 17th International Conference on World Wide Web, pp. 685-694. ACM (2008)
8. Yang, T., Chi, Y., Zhu, S., et al.: Detecting communities and their evolutions in dynamic social networks—a Bayesian approach. Machine learning **82**(2), 157–189 (2011)
9. Ahn, Y.Y., Bagrow, J.P., Lehmann, S.: Link communities reveal multiscale complexity in networks. Nature **466**(7307), 761–764 (2010)

Exploiting Conceptual Relations of Sentences for Multi-document Summarization

Hai-Tao Zheng[✉], Shu-Qin Gong, Ji-Min Guo, and Wen-Zhen Wu

Tsinghua-Southampton Web Science Laboratory, Graduate School at Shenzhen,
Tsinghua University, Shenzhen, China
{zheng.haitao,guojm14,wuwz12}@sz.tsinghua.edu.cn,
gongshuqin90@gmail.com

Abstract. Multi-document Summarization becomes increasingly important in the age of big data. However, existing summarization systems do not or implicitly consider the conceptual relations of sentences. In this paper, we propose a novel method called Multi-document Summarization based on Explicit Semantics of Sentences (MDSES), which explicitly take conceptual relations of sentences into consideration. It is composed of three components: sentence-concept graph construction, concept clustering and summary generation. We first obtain sentence-concept semantic relation to construct a sentence-concept graph. Then we run graph weighting algorithm to get ranked weighted sentences and concepts. Besides, we obtain concept-concept semantic relation for concepts clustering to eliminate redundancy. Finally, we conduct summary generation to get informative summary. Experimental results on DUC dataset using ROUGE metrics demonstrate the good effectiveness of our methods.

Keywords: Multi-document summarization · Sentence-concept graph · Concept clustering · Summary generation

1 Introduction

Multi-document summarization is to produce a summary from a set of documents which describe the same topic, and a variety of document summarization methods have been developed recently [1,2]. As documents and concepts have the same characteristic that focus on an issue, documents are the reflection of concepts to some extent. However, existing methods do not or implicitly reflect the conceptual relation of sentences, while we explicitly construct the relation between sentences and concepts, which better reflect the relations of sentences in the concept degree. In this paper, we propose a novel method called Multi-document Summarization based on Explicit Semantics of Sentences (MDSES). The contributions of our work are summarized as follows: 1) We explicitly consider the conceptual relations of sentences in the task of multi-document summarization. 2) We propose a novel method called MDSES which utilizes explicit semantics of sentences. 3) We exploit sentence-concept semantic relation and concept-concept semantic relation which is based on Wikipedia textual content and hyperlink structure to eliminate redundancy. 4) Experimental results on the DUC dataset verify the effectiveness of MDSES compared with baselines.

J. Li and Y. Sun (Eds.): WAIM 2015, LNCS 9098, pp. 506–510, 2015.
DOI: 10.1007/978-3-319-21042-1_51

2 Multi-document Summarization Based on Explicit Semantics of Sentences

MDSES is composed of three components: sentence-concept graph generation, concept clustering and summary generation. First, we parse $D = \{d_1, d_2, ..., d_l\}$ to sentence set $S = \{s_1, s_2, ..., s_n\}$. Then we map sentences to Wikipedia concepts based on Wikipedia textual content, and we can get concept set $C = \{c_1, c_2, ..., c_m\}$. After the mapping procedure, we can get sentence-concept relation M_{sc}. We exploit M_{sc} to construct sentence-concept graph $G = \{S, C, E\}$. Then we run graph weighting algorithm on G to get ranked weighted sentence set S' and concept set C'. Since similar sentences map to similar concepts, we need cluster concepts to detect redundancy. First, we compute concept-concept relation M_{cc} based on Wikipedia hyperlink structure. Then we clustering concepts based on M_{cc} to concept clusters CC. At last we conduct summary generation to generate summary. Fig.1 presents the overall framework of MDSES.

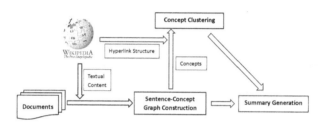

Fig. 1. The framework of MDSES method

Sentence-Concept Graph Construction: We use Explicit Semantic Analysis (ESA) [3] to mapping sentences to Wikipedia concepts. First, we build an inverted index of Wikipedia, which maps each term into a list of concepts in which it appears. We define $s_i = \{t_1, t_2, ..., t_n\}$ be input text, and define $\langle v \rangle = \{v_1, v_2, ..., v_n\}$ be its TFIDF vector, where v_i is the weight of term t_i. let $\langle k \rangle = \{k_1, k_2, ..., k_m\}$ be an inverted index entry for term t_i, where k_j quantifies the strength of association of term t_i with Wikipedia concept c_j, $\{c_j \in c_1, ..., c_m\}$. Then, the weight w_{ij} between s_i and c_j is defined as $\sum_{t_i \in s_i} v_i * k_j$. After the mapping procedure, we can get a sentence-concept relation represented by M_{sc}. Then we use S and C as vertex, M_{sc} as the weighted edge E to get the sentence-concept graph $G = \{S, C, E\}$. Define $weight(s)$ as the weight of s and $weight(c)$ as the weight of c, and we initialize each s in S with the score $\frac{1}{\sqrt{n}}$, then we calculate the weight of c_i and s_j iteratively as follows: $weight^{(k+1)}(c_i) = \sum_{s_j \in S} w_{ji} weight^{(k)}(s_j)$, $weight^{(k+1)}(s_j) = \sum_{c_i \in C} w_{ji} weight^{(k)}(c_i)$. In order to guarantee the convergence of the iterative, the weight of vertex is normalized after each iteration. We iterate the procedure until it reaches convergence, and we can get ranked weighted sentence set S' and ranked weighted concept set C'.

Concept Clustering: In order to get concept-concept relation M_{cc}, we adopt Wikipedia Link Vector based Measure (WLVM) [4]. For each concept, we build a vector space model, which using link counts weighted by the probability of each link occurring. This probability is defined by the total number of links to the target concept over the total number of concepts. Thus if t is the total number of concepts within Wikipedia, then the weighted value w for the link $a \rightarrow b$ (a is the source concept and b is the target concept) is defined as $w(a \rightarrow b) = |a \rightarrow b| * log(\sum_{x=1}^{t} \frac{t}{|x \rightarrow b|})$. We define all n target concepts $\{l_i | i = 1..n\}$ found within the links contained in concepts c_1 and c_2. The vector for each concept c_i is given by $\overrightarrow{c_i} = \{w(c_i \rightarrow l_1), w(c_i \rightarrow l_2), ..., w(c_i \rightarrow l_n)\}$. Our similarity measure for the concepts is then given by the cosine similarity between their vectors. After we compute the similarity between concepts, we can obtain M_{cc}. For concept clustering, we use the conventional Hierarchical Agglomerative Clustering (HAC) algorithm to cluster concepts based on M_{cc} and get concept clusters $CC = \{cc_1, cc_2, ..., cc_k\}$. In HAC algorithm, we finished the cluster merging when the similarity between concept clusters falls below a given threshold (0.3 is an empiric value in our study).

Summary Generation: First, we rank CC based on the ranked weighted C'. For each cluster cc, we compute its weight $weight(cc) = \sum_{c_i \in cc} weight(c_i)$. Then we rank concept clusters in descending order of weight to get ranked concept clusters CC'. For each cluster, we select one concept c with the highest weight to represent the cluster based on the ranked C', and we can get representative ranked concept set $RC = \{c_1, c_2, ..., c_k\}$. At last, we select the first concept c from RC, and get sentence set S_c linked to c based on G, then we select the sentence with the highest weight in S_c based on S' to generate summary. Iterate the procedure until the length of summary is reached.

3   Experiment

In our experiments, we use DUC[1] 2004 dataset and ROUGE [5] as evaluation metric. For baselines, we choose MMR-MD [6] and TextRank [1] for they are two classical methods. With the ubiquity of mobile internet and people's reading habit on mobile devices, different length of summaries are also important. We investigate the performances of MDSES under different summary length, which ranging from 60 words to 100 words. For evaluation, we use ROUGE-1 and ROUGE-SU4, and Fig.2 are comparison at different summary length.

From Fig.2 we have three obvious: (1)All the curves are incremental curves and the reason is also obvious. It is because that with the length of summary increases, the summary will contain more important information. (3)The gradient of MDSES is higher than TextRank and MMR-MD, which means that our

---

[1] http://duc.nist.gov/

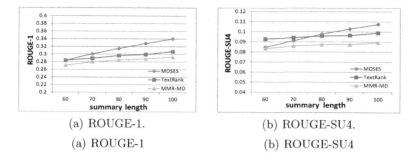

(a) ROUGE-1.   (b) ROUGE-SU4.

(a) ROUGE-1   (b) ROUGE-SU4

Fig. 2. Comparison at different summary lengths

MDSES tends to have more advantages when the length of summary increases. With the summary length increases, TextRank and MMR-MD select redundancy sentences to summary while MDSES selects sentences in concept degree which leading to more coverage of summary. (3)TextRank outperforms MDSES in Fig. 2(b) when the summary length is less than 80 words. The reason may be that the summary length is too small, the importance of conceptual relations of sentences is restricted. But while the summary length is appropriate or big, the good effect of exploiting conceptual relations of sentences for multi-document summarization appears.

4   Conclusion

In this paper we propose a novel method MDSES, it contains three components: sentence-concept graph construction, concept clustering and summary generation. Unlike with other methods, we explicitly construct the relation between sentences and concepts, which better reflect the relations of sentences in the concept degree. Experimental results verify the effectiveness of MDSES.

Acknowledgments. This research is supported by the 863 project of China (2013AA013300), National Natural Science Foundation of China (Grant No. 61375054) and Tsinghua University Initiative Scientific Research Program (Grant No.20131089256).

References

1. Mihalcea, R., Tarau, P.: TextRank: Bringing order into texts. Association for Computational Linguistics (2004)
2. Gong, S., Qu, Y., Tian, S.: Summarization using wikipedia. In: TAC 2010 Proceedings (2009)
3. Gabrilovich, E., Markovitch, S.: Computing semantic relatedness using wikipedia-based explicit semantic analysis. In: IJCAI, vol. 7, pp. 1606–1611 (2007)

4. Milne, D.: Computing semantic relatedness using wikipedia link structure. In: Proceedings of the New Zealand Computer Science Research Student Conference (2007)
5. Lin, C.Y.: Rouge: a package for automatic evaluation of summaries. In: Text Summarization Branches Out: Proceedings of the ACL 2004 Workshop, pp. 74–81 (2004)
6. Goldstein, J., Mittal, V., Carbonell, J., Kantrowitz, M.: Multi-document summarization by sentence extraction. In: Proceedings of the 2000 NAACL-ANLPWorkshop on Automatic Summarization, vol. 4, pp. 40–48 (2000)

Towards Narrative Information Systems

Philipp Wille(✉), Christoph Lofi, and Wolf-Tilo Balke

Institut für Informationssysteme, TU Braunschweig, Braunschweig, Germany
{wille,lofi,balke}@ifis.cs.tu-bs.de

Abstract. Narrative interfaces promise to improve the user experience of interacting with information systems by adapting a powerful communication concept which comes natural in human interaction. In this paper, we outline how such a narrative information system which answers queries using elaborate stories can be realized. We show how to construct query-dependent plot graphs from unstructured data sources. Also, we conduct an extended user study on our prototype implementation in which users adapt and personalize the plots of query-dependent stories. These studies provide further insights into which parts of a story are relevant and can be used as a starting point for either personalizing stories or for crafting better user-independent stories in later works.

Keywords: Narrative information systems · Narrative exploration · Storytelling

1 Introduction

Information systems form the backbone of most Web applications enabling access to a vast amount of digital information. However, interfaces to this information still lack intuitiveness in the sense of how queries and query results are communicated. Especially when faced with complex and multi-facetted queries like "What will diabetes mean for my life?" many state-of-the-art techniques like keyword-based document retrieval are still insufficient as users have to laboriously work through multiple documents. Thus, a core challenge in current information systems research is naturally providing relevant information as opposed to retrieving relevant documents [1].

Stories are particularly well suited to communicate complex and multi-facetted information in an easy-to-understand and easy-to-memorize fashion. They exhibit a plot that covers a series of events in a comprehensible way to help the audience understand their connections and as research in cognitive sciences suggests, this type of packaging information is especially easy for people to digest. The most memorable stories however are interactively tailored for the audience, and a good storyteller takes feedback and directive cues of her audience under consideration.

In this paper, we focus on the core-problem of building a Narrative Information System, namely the construction and interpretation of *query-dependent plot graphs*. To illustrate this concept, figure 1 shows a sample plot graph for a query about *diabetes*. This graph tells an overview story of the topic, briefly touching related topics like *obesity* and *prediabetes* and relating them to the query topic through sub-topics

© Springer International Publishing Switzerland 2015
J. Li and Y. Sun (Eds.): WAIM 2015, LNCS 9098, pp. 511–515, 2015.
DOI: 10.1007/978-3-319-21042-1_52

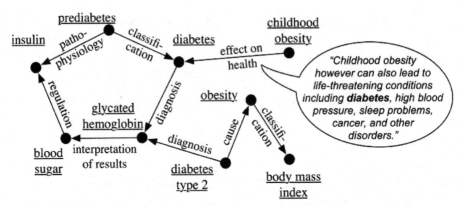

Fig. 1. Simplified example plot graph for a general query about *diabetes*

such as *diagnosis* and *causes*. While the shown plot graph looks similar to knowledge graphs as for example in [2, 3], the core difference is that the shown plot graph can be transformed into a textual representation (story) via *story synthesis* (the problem of finding a suitable linearization of the plot graph) and has text fragments for each node or edge available for this purpose.

2   Modelling Narrative Queries

Plot graphs have been a common representation for stories in previous works [4, 5], however, in contrast to other recent publications on storytelling ([2, 6]), we do not limit plots to just temporally coherent narrations (e.g. news stories) or causal event chains. Instead, we define them as sets of covered topics that may be connected by sub-topics. To help with *story synthesis*, we annotate each node and edge with one or more *text snippets* representing brief descriptions or explanations.

A plot graph G_P is given by $G_P = (S, T, lb_S, snip_S, lb_T, snip_T)$ with S being a finite set of nodes (representing topics), and T being a finite set of edges $T \subseteq S \times S$ representing transitions (via a sub-topic). Furthermore, we have a finite set L_S of node labels (topic names), and a finite set L_T of edge labels. Then $lb_S \subseteq S \times L_S$ is a relation assigning labels to nodes, and $lb_T \subseteq T \times L_T$ is a relation assigning labels to transitions (therefore allowing multiple edges with different labels between two topics). Finally, we have a set of text snippets TS, and the relations $snip_S \subseteq S \times TS$ and $snip_T \subseteq T \times TS$ assigning snippets to nodes (topics) and transitions (sub-topics).

Our system considers only a finite number of different plot graphs which is restricted by the data sources available to it. This *universe of discourse* containing all topics and transitions which could be used in plots constructed by our system is represented by a *universe graph* G_U that is defined analogously to definition 1 with each possible plot graph G_P being an edge-induced subgraph of G_U. Given a universe graph, a user can search its collection by issuing a query featuring a topic T_C. Based on that central topic, a subgraph of the universe graph is selected as a plot graph that contains topics and transitions *relevant with respect to the query topic*.

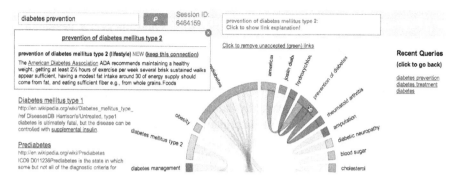

Fig. 2. Screenshot of the user interface of *Narrex* as used in our user study

Since users try to satisfy a fuzzy information need in facetted-search, our system cannot provide one correct answer for a query, but instead has to generate an individual plot graph for each individual user. This is done by interactively expanding a plot graph based on user feedback. We refer to such an interactively generated plot graph as a query-dependent *discourse graph* G_D that contains only nodes and edges that are *relevant* to a query Q and the user. It is also an edge-induced subgraph of G_U.

Based on discourse graphs, we implemented a Narrative Exploration prototype called *Narrex* (see figure 2). As a representative corpus, we used the full English Wikipedia (as of September 4, 2013), chose all Wikipedia articles as valid topics and took all sensible first-level headings of each article as sub-topics (e.g. see figure 1). Topic and sub-topic text snippets were selected using the multi-aspect summarization method proposed in [1] that generates topic- and sub-topic-specific summaries for a given query topic. Each sub-topic specific summary text contains relevant topics related to the query topic via the sub-topic that we find with automated link detection [7]. A discourse graph built this way contains the query topic as seed node and all relevant related topics as additional nodes connected by its sub-topics. All items are annotated with their respective summaries. Using the user interface as shown in figure 2, users of Narrex can interactively and narratively explore the underlying text corpus.

3   User Study and Conclusions

We conducted a user study with 100 native English speakers on the CrowdFlower platform. Since we aim at improving the intuitiveness of processing multi-facetted queries, we asked the participants to educate themselves about *the most important life changes a person has to make in case that he/she was diagnosed with diabetes*, a query that stems from the facetted search task of the TREC 2009 data set and that is quite representative for what we envision our system to be able to do as there are several different stories from different perspectives that could be told to satisfy a user's information need. The participants were split in two groups with the first using Narrex and the second working on a state-of-the-art keyword-based system. We examined the following issues: (1) what subsets of the discourse graph represent

Table 1. Group means (M/SE) for the questions of the third experiment

	Question a)	Question b)	Question c)	Question d)
Narrex	2.91 / 0.27	2.44 / 0.20	2.71 / 0.26	2.91 / 0.23
Keyword-based	2.87 / 0.26	2.61 / 0.28	2.71 / 0.25	3.03 / 0.28

good plot graphs of stories users actually consider interesting and helpful, (2) how informative are the query-dependent story graphs generated in this way for answering a query compared to having verbose information available in form of relevant full documents.

In the first part we studied how our study participants built plot graphs with Narrex to get an understanding which stories are considered good and which topics and transitions are deemed relevant by logging their interactions with Narrex. The generated individual plot graphs differed strongly in the set of contained story items and transitions. However, all these plots share a core set of transitions and topics considered relevant by many participants that comprises a plot giving an overview of the task, whereas the less chosen transitions and topics are part of more specific interest plots.

In experiment 2, both groups were asked to rate the following four questions on a scale from 1 (best) to 6 (worst), so we could measure the user satisfaction with either system: (a) how satisfied were you with the information display, (b) to what extend could you answer the main question, (c) how satisfied are you with the knowledge you got in relation to the main question, (d) can you imagine using this type of information display to answer questions like "how does living on my own will affect my life", "how does my life change if I got pregnant", or "how does my life change if my wife is diagnosed with cancer". The results are shown in table 1. We analyzed the user ratings for each question individually with group means and compared them with an independent sample t-test. Conditions of normal distribution are assumed due to a sample size of at least 30 per group, homogeneity of variances was given (Levene's). All differences between the group means were not significant.

Our results show that even naïve prototypes of Narrative Information Systems can already keep up with keyword search for multi-facetted queries. Also, they dramatically reduce the cognitive burden by directly presenting relevant information in form of relationships between topics and by reducing the texts the user has to read. These findings suggest that further research in Narrative Information Systems is worthwhile.

References

1. Song, W., Yu, Q., Xu, Z., Liu, T., Li, S., Wen, J.-R.: Multi-aspect query summarization by composite query. In: 35th Int. ACM SIGIR, pp. 325-334. ACM, New York (2012)
2. Shahaf, D., Guestrin, C., Horvitz, E.: Trains of thought: generating information maps. In: 21st Int. Conf. on World Wide Web (WWW), Lyon, pp. 899-908 (2012)
3. Shahaf, D., Guestrin, C., Horvitz, E.: Metro maps of science. In: 18th ACM SIGKDD Int. Conf. on Knowledge Discovery and Data Mining (KDD), Beijing, pp. 1122-1130 (2012)
4. Riedl, M.O., Thue, D., Bulitko, V.: Game AI as storytelling. In: Artificial Intelligence for Computer Games, pp. 125-150. Springer, New York (2011)

5. Kasch, N., Oates, T.: Mining script-like structures from the web. In: 1st Int. Workshop on Formalisms and Methodology for Learning by Reading, LA, California, pp. 34-42 (2010)
6. Radinsky, K., Davidovich, S., Markovitch, S.: Learning causality for news events prediction. In: Proc. 21st Int. Conf. on World Wide Web (WWW), Lyon, pp. 909-918 (2012)
7. Milne, D., Witten, I.H.: Learning to link with wikipedia. In: 17th Conf. on Information and Knowledge Management (CIKM), Napa Valley, California, pp. 509-518 (2008)

Maximizing the Spread of Competitive Influence in a Social Network Oriented to Viral Marketing

Hong Wu[1,2], Weiyi Liu[1], Kun Yue[1(✉)], Weipeng Huang[1], and Ke Yang[1,3]

[1] Department of Computer Science and Engineering, School of Information and Engineering,
Yunnan University, Kunming, China
kyue@ynu.edu.cn

[2] College of Computer Science and Engineering, Qujing Normal University, Qujing, China

[3] Head Office, Fudian Bank, Kunming, China

Abstract. In this paper, we discuss the maximization of the spread of competitive influence when multiple companies market competing product using social network. We first extend the linear threshold model by incorporating the competitive influence spread, obtaining the extended linear threshold model (ELTM). We then give the objective function of selecting the optimal seed set of competing products in a social network, and the objective function is monotone and submodular under the ELTM, thus a greedy algorithm could achieve $1-1/e$ approximation ration (where e is the base of natural logarithm). Accordingly, preliminary experimental results verify the feasibility of our method.

Keywords: Social network · Viral marketing · Competitive influence spread · Extended linear threshold model · Approximation algorithm

1   Introduction

With the increasing popularity of online social networks, such as Facebook, Twitter, etc., online social networks have become an important platform to spread opinion, news, information of the products, etc. Many researchers have studied the influence spread of online social networks by focusing on influence maximization. Kempe et al. [1] formulated the influence maximization as a discrete optimization problem and proposed two influence diffusion models: linear threshold model and independent cascade model. The seminal work given by Kempe et al. [1] motivates various studies on influence maximization in social networks [2, 3, 4].

One important application of influence maximization in social network is viral marketing [5, 6]. There are already instances of successful viral marketing campaigns in real life (e.g. the Nike commercial) that use social networking websites such as *orkut.com* and *facebook.com* to market products [7]. In reality, there exist two or more products competing with one another, exactly the problem that we will solve in this paper.

Motivated by the above scenarios, several recent studies have looked into competitive influence diffusion. Bharathi et al. [8] studied the game theory based competitive influence spread under the extended independent cascade model. Budak et al. [9]

© Springer International Publishing Switzerland 2015
J. Li and Y. Sun (Eds.): WAIM 2015, LNCS 9098, pp. 516–519, 2015.
DOI: 10.1007/978-3-319-21042-1_53

discussed the problem of minimizing the devastating effects of misinformation campaigns under an extension of IC model. In these extensions of the IC model, the node only can be activated by one of in-neighbors, and the influence accumulation of activated in-neighbors of a certain node has not been considered.

In this paper, we first construct the competitive influence spread model based on the (LT) model by incorporating the competitive influence spread and considering the nature of competitive influence spread, and the objective function is monotone and submodular under the ELTM. Thus, we can use the greedy algorithm to select the seed set of A, which could approximate the optimal algorithm with $1-1/e$ approximation ration.

2   The Extended Linear Threshold Model

We define the ELTM as an extension of the LT model as follows. In $G=(V, E)$, $I_B \subseteq V$, I_B denotes the B's seed set, and S_A denotes the A's seed set ($S_A \in \mathbb{W}_B$). The weight of edge (u, v) is $w_{uv}^A = w_{uv}^B = 1/d(v)$, where $d(v)$ is the in-degree of node v. Each node has two activated thresholds θ_v^A and θ_v^B uniformly at random from $[0, 1]$. In reality, the products of different companies have different influence degrees for consumers. In this paper, we define the influence degree of A as q^A, and that of B as q^B.

For node v, we describe the above ELTM as the following four active cases.

(a)The weight of A-activated in-neighbors of node v multiplies q^A, which is greater than or equal to the A-activated threshold θ_v^A, but the weight of B-activated in-neighbors of node v multiplies q^B, which is less than the B-activated threshold θ_v^B. That is $\sum w_{uv\,t}^A * q^A \geq \theta_v^A$ and $\sum w_{uv\,t}^B * q^B < \theta_v^B$. Then, the activated probability of node v can be written as $ap_{t+1}^A(v) = 1$ and $ap_{t+1}^B(v) = 0$.

(b) The weight of A-activated in-neighbors of node v multiplies q^A, which is less than the A-activated threshold θ_v^A, but the weight of B-activated in-neighbors of node v multiplies q^B, which is greater than or equal to the B-activated threshold θ_v^B. That is $\sum w_{uv\,t}^A * q^A < \theta_v^A$ and $\sum w_{uv\,t}^B * q^B \geq \theta_v^B$. Then, the activated probability of node v can be written as $ap_{t+1}^A(v) = 0$ and $ap_{t+1}^B(v) = 1$.

(c) If the weight of A-activated (or resp. B-activated) in-neighbors of node v multiplies q^A (or resp. q^B), which is greater than or equal to the A-activated threshold θ_v^A (or resp. the B-activated threshold θ_v^B), i.e, $\sum w_{uv\,t}^A * q^A \geq \theta_v^A$ and $\sum w_{uv\,t}^B * q^B \geq \theta_v^B$, then the bigger weight to v dominates. If they are equal, then the node v will randomly choose A or B. In this paper, we consider that A dominates the influence spread. Then, we obtain $ap_{t+1}^A(v) = 1$ and $ap_{t+1}^B(v) = 0$.

(d) The weight of A-activated (or resp. B-activated) in-neighbors of node v multiplies q^A (or resp. q^B), which is less than the A-activated threshold θ_v^A (or resp. B-activated threshold θ_v^B), then we have $ap_{t+1}^A(v) = ap_{t+1}^B(v) = 0$.

3   Approximation Idea for Maximizing the Competitive Influence Spread

Informally, the competitive influence spread problem is an optimization problem, in which the social network $G=(V, E)$, the initial B seed set, and an integer k have been given. We want to find a seed set of A of size k such that the expected number of activated nodes by A is maximized. More precisely, let I_B denote the seed set of B, and S_A ($S_A \subset W_B$) denote the seed set of A. The problem of selecting the optimal seed set S_A of size k is to maximize the influence of seed set of A, denoted as $\sigma(S_A, I_B)$:

$$S^* = \underset{S_A \subset V \backslash I_B, |S_A = K|}{\arg \max} \ \sigma(S_A, I_B) \tag{1}$$

Selecting the optimal seed set of A is NP hard, while knowing the seed set of B.

The objective function $\sigma(S_A, I_B)$ is monotone and submodular under the ELTM and $\sigma(\phi, I_B)=0$. Based on the result in [10], we can obtain the approximate result by using the greedy algorithm.

4   Experimental Results

We use Cit-HepTh data set, the Arxiv High Energy Physics paper citation network. We chose 1000 nodes from the network, and generate synthetic network, which is called as Cit-HepTh-new. We use the weighted cascade model [1] to generate the influence weight of edges.

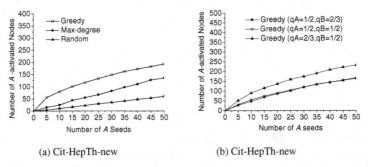

(a) Cit-HepTh-new   (b) Cit-HepTh-new

Fig. 1. Influence spread of different algorithms and influence spread of A with greedy algorithm under different influence degrees

In this experiment, we select 10 seeds with random algorithm as the initial seeds of B and select 50 seeds of A to maximize the spread of A. It can be seen from Fig. 1(a) that the greedy algorithm outperforms the max-degree algorithm and the random algorithm.

We further compare the number of A-activated nodes of greedy algorithm with different influence degrees of A, which includes three cases: (a) $q^A=1/2$, $q^B=2/3$, (b) $q^A=1/2$, $q^B=1/2$ and (c) $q^A=2/3$, $q^B=1/2$, shown in Fig. 1(b). As expected, when the influence degree of A is larger relatively, more nodes become A-activated. This is because in this case the influence of A can traverse long paths, and thus more nodes will become A-activated.

5   Conclusions

Competitive influence spread is important and challenging in the field of social network. In this paper, we extended the LT model to incorporate the competitive influence spread. Selecting the optimal seed set of A is NP hard, when the seed set of B has been known. The objective function of selecting the optimal seed nodes is monotone and submodular under the ELTM, then using the well-known results on submodular functions [10], the greedy algorithm can be used to approximate the optimal result with $1-1/e$.

Acknowledgements. This work was supported by the National Natural Science Foundation of China (61163003, 61472345, 61232002), the Natural Science Foundation of Yunnan Province (2014FA023, 2011FB020), the Program for Innovative Research Team in Yunnan University (XT412011), the Yunnan Provincial Foundation for Leaders of Disciplines in Science and Technology (2012HB004), and the Science Research Foundation of Education Department of Yunnan Province (2014C134Y) .

References

1. Kempe, D., Kleinberg, J.M., Tardos, E.: Maximizing the spread of influence through a social network. In: KDD, pp.137–146 (2003)
2. Jung, K., Heo, W., Chen, W.: Scalable and robust influence maximization in social networks. In: ICDM, pp. 918 – 923 (2012)
3. Zhuang, H., Sun, Y., Tang, J., Zhang, J., Sun, X.: Influence maximization in dynamic social networks. In: ICDM, pp. 1313–1318 (2013)
4. Ma, Q., Ma, J.: An efficient influence maximization algorithm to discover influential users in micro-blog. In: Li, F., Li, G., Hwang, S.-w., Yao, B., Zhang, Z. (eds.) WAIM 2014. LNCS, vol. 8485, pp. 113–124. Springer, Heidelberg (2014)
5. Domingos, P.: Mining social networks for viral marketing. IEEE Intelligent Systems **20**(1), 80–82 (2005)
6. Long, C., Wong, R.C.W.: Minimizing seed set for viral marketing. In: ICDM, pp. 427–436 (2011)
7. Johnson, A.: nike-tops-list-of-most-viral-brands-on-facebook-twitter (2010). http://www.kikabink.com/news/
8. Bharathi, S., Kempe, D., Salek, M.: Competitive influence maximization in social networks. In: Deng, X., Graham, F.C. (eds.) WINE 2007. LNCS, vol. 4858, pp. 306–311. Springer, Heidelberg (2007)
9. Budak, C., Agrawal, D., Abbadi, A.E.: Limiting the spread of misinformation in social networks. In: WWW, pp. 665–674 (2011)
10. Nemhauser, G., Wolsey, L., Fisher, M.: An analysis of the approximations for maximizing submodular set functions – I. Mathematical Programming, 265–294 (1978)

SEMR: Secure and Efficient Multi-dimensional Range Query Processing in Two-tiered Wireless Sensor Networks

Lei Dong[1,2], Jianxiang Zhu[1,2], Xiaoying Zhang[1,2], Hong Chen[1,2],
Cuiping Li[1,2], and Hui Sun[1,2(✉)]

[1] Key Laboratory of Data Engineering and Knowledge Engineering
of Ministry of Education, Beijing, China
sun_h@ruc.edu.cn
[2] Renmin University of China, Beijing 100872, China
donglei1001@163.com, zjxruc@foxmail.com, xiaoyingzhang1987@126.com,
chong@ruc.edu.cn, cuiping_li@263.net

Abstract. Wireless sensor network (WSN) is an important part of Internet of Things. Due to abundant resources and high efficiency, many future large-scale WSNs are expected to follow a two-tiered architecture, where resource-rich storage nodes are placed in the upper tier and resource-limited sensor nodes are placed at the lower tier. However, the architecture threatens network security. Thus it is challenging to process range query while protecting sensitive data from adversaries. Most existing relate work focuses on single-attribute privacy-preserving range query. In practical applications, sensor nodes usually sense multiple attributes. In this paper, we propose a secure and efficient range query protocol for multi-dimensional data in two-tiered wireless sensor networks. A specified pair of matrices is used to compute characteristic values of sensed data and encrypt range queries. Finally, experimental results confirm the efficiency of our protocol.

Keywords: Multi-dimensional range query · Wireless sensor networks · Privacy preservation

1 Introduction

Wireless sensor network (WSN), which is treated as an indispensable part for Internet of Things (IoTs), has been widely used in many applications.

Hui Sun—This work is supported by the National High Technology Research and Development Program of China (863 Program) (No. 2014AA015204), the National Basic Research Program of China (973 Program) (No. 2014CB340402), the National Natural Science Foundation of China (Grant Nos. 61070056, 61272137, 61202114). This work was partially done when the authors visited SA Center for Big Data Research hosted at Renmin University of China. This Center is partially funded by the Chinese National "111" Project "Attracting International Talents in Data Engineering and Knowledge Engineering Research".

© Springer International Publishing Switzerland 2015
J. Li and Y. Sun (Eds.): WAIM 2015, LNCS 9098, pp. 520–524, 2015.
DOI: 10.1007/978-3-319-21042-1_54

Most future large-scale WSNs will follow a two-tiered architecture which consists of a large amount of resource-limited sensor nodes at the lower tier and several resource-rich storage nodes at the upper tier. The sink is reliable and we do not consider compromised sensor nodes. However, since storage nodes are responsible for storing data and processing queries, they are more liable to be compromised. Therefore, it is challenging to preserve data privacy as well as achieving efficient performance and precious results.

Range query is one of important queries. Secure range query for 1-dimensional data have been well addressed [1–3], however, very few works focus on secure range query for multi-dimensional data. In this paper, we propose Secure and Efficient Multi-dimensional Range query processing (SEMR) in two-tiered WSNs. A specified pair of matrices is used to compute characteristic values of sensed data and encrypt range queries. Observations show that our protocol outperforms previous work in efficiency, accuracy and privacy.

2   Secure and Efficient Multi-dimensional Range Query Processing

A z-dimensional data item D can be denoted by $(d^1, \ldots, d^z)(1 \leqslant l \leqslant z)$, where d^l is the value for the l-th dimension. We suppose sensor node S_i collects n z-dimensional data $D_1, D_2, \ldots, D_n$ at epoch t. S_i constructs the *data vector* $v_{i,j}{}^l = ((d_j{}^l)^2, d_j{}^1, 1)$ for each dimension $d_j{}^l$ of D_j, then builds matrix $V_{i,j}$ for D_j by using *data vectors* $v_{i,j}$ as rows of $V_{i,j}$. Then S_i obtains *characteristic value matrix* $CV_{i,j}$ of data D_j through matrix multiplication as follows:

$$CV_{i,j} = V_{i,j} \cdot W \tag{1}$$

Obviously $CV_{i,j}$ is a $z \times 3$ matrix. The message that S_i sends to the storage node ST is

$$S_i \to ST : i, t, \{(D_1)_{k_i}, \ldots, (D_n)_{k_i}\}, \{CV_{i,1}, \ldots, CV_{i,n}\}$$

A z-dimensional range can be denoted as $\{[a^1, b^1], \ldots, [a^z, b^z]\}$. The sink constructs a *query vector* $q^l = (1, -(a^l + b^l), a^l b^l)$ for each dimensional range $[a^l, b^l](1 \leqslant l \leqslant z)$ and builds matrix Q by using *query vectors* q as rows of Q. Then the sink transforms Q to TQ through

$$TQ = M \cdot Q^T \tag{2}$$

It can be seen that TQ is a $3 \times z$ matrix. Then the sink sends TQ to storage node ST. While receiving the range query message, ST computing the product $p_{i,j}$ of data $d_{i,j}$ of sensor S_i by multiplying $CV_{i,j}$ and $TQ[a, b]$ as

$$P_{i,j} = CV_{i,j} \cdot TQ, \tag{3}$$

The product $P_{i,j}$ is a $z \times z$ matrix. Note that the main diagonal element on the l-th row is the range criterion for l-th dimension value.

Theorem 1. *Let $z \times z$ matrix P be defined as Equation (3) for z-dimensional data D. D is in z-dimensional range $\{[a^1, b^1], \ldots, [a^z, b^z]\}$ if and only if the the elements on main diagonal of P are all non-positive.*

Proof. For the l-th element pd^l on main diagonal of P, note that

$$pd^l = cv^l \cdot tq^l = (v^l \cdot W) \cdot (M \cdot (q^l)^T) = v^l \cdot (q^l)^T$$
$$= (d^l)^2 - d^l(a+b) + ab = (d^l - a)(d^l - b)$$

Thus, $pd^l \leqslant 0 \Leftrightarrow d^l$ is in $[a^l, b^l]$, $pd^l > 0 \Leftrightarrow d^l$ is out of $[a^l, b^l]$.

Therefore the storage node ST can filter data according to **Theorem 1**. Then ST sends the query result to the sink. The complexity of range query process based on SEMR is $O(zn)$.

Our basic protocol SEMR can ensure the result accuracy and process in low complexity on the premise of preserving privacy, however, the characteristic value matrix in our basic protocol also has a possible security vulnerability. For instance, if the adversary captures the storage node and possibly derives W, then M and all actual data can be computed. One way to avoid this vulnerability is that each sensor node S_i introduces the randomly generated parameters $\{r_{i,j}^1, \ldots, r_{i,j}^z\}$ ($r \in \mathbb{N}^+$) into data vector $v_{i,j}^1, \ldots, v_{i,j}^z$ respectively to form $V_{i,j}$. This parameter introduction does not affect the range query processing since it is sign-preserving for $P_{i,j}$.

3   Performance Evaluation

To evaluate the performance of SEMR and two prior works, BBP [3] and SafeQ[4], we implement these schemes and perform communication cost comparison based on data set LUCE (Lausanne Urban Canopy Experiment) [5]. Note that BBP works as single-tired sensor network even though it has storage nodes. Thus, to avoid extra communication cost for broadcast, we adapt BBP for two-tired sensor network, which is denoted as CBBP (Changed BBP).

The schemes are implemented on OMNet++4.1, a widely used simulator for WSNs. The area of sensor network is set to 400 $meters \times$ 400 $meters$. Sensor nodes are uniformly deployed in the network. Assume the network is separated

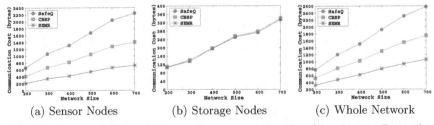

(a) Sensor Nodes   (b) Storage Nodes   (c) Whole Network

Fig. 1. Impact of Network Size on Communication Cost (Epoch=30s, Dim=3)

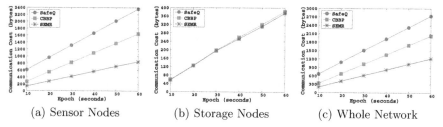

Fig. 2. Impact of Epoch on Communication Cost (400 Sensor Nodes, Dim=3)

(a) Sensor Nodes   (b) Storage Nodes   (c) Whole Network

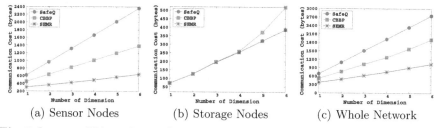

(a) Sensor Nodes   (b) Storage Nodes   (c) Whole Network

Fig. 3. Impact of Dimension on Communication Cost (400 Sensor nodes, epoch = 30s)

into 4 identical regions and one storage node is placed at the center of each region. The transmission radius of each sensor node is 50 meters. We use 128-bit DES as the data encryption for SEMR, SafeQ and CBBP. According to [4] and [3], we also use 128-bit HMAC for SafeQ and 91-bit Bloom filter for CBBP. We generated 200 random range queries for each epoch. The results state that our SEMR reduces more communication cost, achieves higher efficiency than SafeQ and CBBP.

4 Conclusion

In this paper, we present a secure and efficient multi-dimensional range query protocol — SEMR in two-tiered WSNs. SEMR preserves data privacy, enables storage nodes to select the correct result and supports the sink to detect compromised storage nodes. Simulation results confirm the high efficiency and accuracy of SEMR. Our future work will focus on improving SEMR to detect collusion attacks in WSNs.

References

1. Chen, F., Liu, A.X.: Privacy and integrity-preserving range queries in sensor networks. IEEE/ACM Transactions on Networking **20**, 1774–1787 (2012)
2. Zhang, X., Dong, L., Peng, H., Chen, H., Li, D., Li, C.: Achieving efficient and secure range query in two-tiered wireless sensor networks. In: IEEE/ACM International Symposium on Quality of Service (2014)

3. Li, G., Guo, L., Gao, X., Liao, M.: Bloom filter based processing algorithms for the multi-dimensional event query in wireless sensor networks. Journal of Network and Computer Applications **37**, 323–333 (2014)
4. Chen, F., Liu, A.X.: Safeq: secure and efficient query processing in sensor networks. In: IEEE International Conference on Computer Communications, pp. 2642–2650 (2010)
5. Luce deployment. http://lcav.epfl.ch/cms/lang/en/pid/86035

A Sampling-Based Framework for Crowdsourced Select Query with Multiple Predicates

Jianhong Feng[(✉)], Huiqi Hu, Xueping Weng, Jianhua Feng, and Yongwei Wu

Department of Computer Science, Tsinghua University, Beijing 100084, China
{fengjh11,hhq11}@mails.thu.edu.cn, {fengjh,wuyw}@tsinghua.edu.cn,
wxping715@gmail.com

Abstract. In this paper, we consider the crowdsourced select query with multiple predicates. We find that different predicates have different selectivities. An important problem is to determine a good predicate order. However it is rather hard to obtain an optimal order. To address this problem, we propose a sampling-based framework to find a high-quality order. We devise a minimum random selection method by randomly selecting the predicate sequence. Since minimum random selection randomly selects predicate permutations over predicates, which may bring large cost, we propose a filtering based algorithm to further reduce the cost. We evaluate our method using a real-world dataset. Experimental results indicate that our methods significantly reduce the monetary cost.

1 Introduction

crowdsourced select query with multiple predicates aims to find all objects from a data collection by asking crowdsourcing workers to identify whether each object satisfies every query predicate. We find that different predicates have different selectivities and if we first verify a highly selective predicate, we can avoid checking other predicates for those objects that do not satisfy the predicate and thus can significantly reduce the cost. An important problem is to determine a good predicate order. However, it is usually very expensive to obtain the optimal predicate order. To address this problem, we propose a sampling-based framework to determine a good predicate order.

We first determine an predicate order on sampling objects, then the predicate order is deployed as our global order. To reduce the cost, we propose two methods to compute the predicate order. The first method, called minimum random selection, generates predicate order by choosing the predicates sequence in randomly predicates arrangements. The second method, namely filtering based, further reduces the cost by selecting the better predicate order.

2 The Sampling-Based Framework

We propose the sampling-based framework. To answer a given select query over global dataset, a random subset of objects is chosen and published on the crowdsourcing platform to generate the optimal predicate order. Then the rest of

© Springer International Publishing Switzerland 2015
J. Li and Y. Sun (Eds.): WAIM 2015, LNCS 9098, pp. 525–529, 2015.
DOI: 10.1007/978-3-319-21042-1_55

objects are asked by adopting this predicate order. To generate an optimal predicate order is a key component in the sampling-based framework. Therefore, next we formally define the optimal predicate order determination problem.

Definition 1. *Given a set of sampling objects $\mathcal{R}$ and a select query with n predicates, $\{\mathcal{A}_1(o) = a_1, \mathcal{A}_2(o) = a_2, \cdots, \mathcal{A}_n(o) = a_n\}$, we aim to generate a predicate order π_* which has the minimum cost to answer the query.*

3   Minimum Random Selection Algorithm

The minimum random selection (denoted by MRS) includes two steps: (1) it randomly generates m permutations $\pi_1, \cdots, \pi_m$ over attributes $\mathcal{A}_1, \cdots, \mathcal{A}_n$. (2) The algorithm calculates the cost of each permutation and selects the one with the minimum cost as the optimal predicate order. We calculate the cost of a permutation π_i by adding up the cost on each attribute. Suppose $\pi_i^1, \cdots, \pi_i^n$ are the attributes of π_i and we ask questions on crowd with sequence $\pi_i^1, \cdots, \pi_i^n$. The algorithm first asks all the objects whether they satisfy attribute π_i^1 and obtains the candidate set $\mathcal{Q}_1$. Next it asks π_i^2 from the candidate objects in $\mathcal{Q}_1$ and obtains $\mathcal{Q}_2$. Then it continues to ask all the remaining attributes and calculates the total cost $Cost(\pi_i)$. Notice that there exist a large amount of duplicated questions if we ask m permutations separately. To avoid asking duplicated questions, we utilize an auxiliary matrix M to record the results that we have retrieved.

Next we analyze the approximate ratio that π_* compared with the real optimal predicate order, denoted by $\hat{\pi}^*$. Given $k \in \mathbb{Z}$ and $\eta \in [0, 1]$, To ensure π_* is within the k best orders with probability larger than η, then we must select at least m random permutations and m satisfies $1 - \prod_{i=0}^{m-1} \frac{n! - k - i}{n!} \geq \eta$.

4   A Filtering Based Selection Algorithm

A limitation of the minimum random selection algorithm is that it selects permutations randomly over predicates. Therefore, we propose the filtering based selection algorithm. We maintain a filtering profile to record the selectivity of each predicate. The profile is updated with the temporal results obtained by crowd. To generate the next permutation, we filter out the non-optimal permutations by considering the predicate selectivity. At last we select the permutation with the minimum cost.

4.1   Predicate Selectivity

Let $P(\mathcal{A}_j)$ denotes the probability that an object satisfies predicate $\mathcal{A}_j$. $P(\mathcal{A}_j)$ can represent the selectivity of a predicate. $P(\mathcal{A}_j)$ can be estimated by maximizing the likelihood of observed objects by $P(\mathcal{A}_j) = \frac{N(a_j)}{N}$, where N is the number of observed objects over $\mathcal{A}_j$, $N(a_j)$ is the number of objects satisfies $\mathcal{A}_j(o) = a_j$. We make the equation as a validated function only if N is larger than a certain threshold. If we only observer a small proportion of objects over $\mathcal{A}_j$ (let C denote this proportion, $0 < C < 1$), we can not utilize them to calculate the predicate selectivity. The threshold is set to be $C \times |\mathcal{R}|$.

4.2 Permutation Selection with Predicate Selectivity

Given a permutation, if there exist two predicates $\mathcal{A}_i, \mathcal{A}_j$ in the profile with $P(\mathcal{A}_i) < P(\mathcal{A}_j)$ and $\mathcal{A}_j$ is in front of A_i in the permutation, then the permutation is called *conflict*. There are two processing steps to generate permutations without conflicts. First we generate a random permutation. Next we check the permutation with the profile and obtain a proper permutation. In the second step, there are two possible situations. (1) The permutation is conflicting with the profile. Therefore, it is not a proper permutation and we discard it. (2) The permutation does not conflict with the profile, then we ask questions.

We propose two methods to decide the number of permutations:

- $OS - m$. The first method selects a fixed number of orders m, where m is still computed by minimum random selection.
- $OS - k/\eta$. In this method, we dynamically determine m during the process of order selection. To make the best selected order within the k optimal order with probability larger than η, we only need to randomly select an permutation among $\frac{k}{\eta}$ optimal permutations.

5   Experiment

5.1 Experimental Setup

Dataset. People consists of 1000 photos of people from the Attributes-Dataset [1]. We verified five attributes of each photo using CrowdFlower.

Evaluation Metrics. (1) Sampling Cost is the number of tasks to obtain the predicate order. (2) Total Cost is the number of tasks to perform select query in the predicate order.

Comparing methods. We used the Brute-Force method as baseline, which enumerates all the permutations, calculates their cost and selects the best one (denoted by BF). We compared performance between our approaches and BF. To evaluate Total Cost of our approaches, we first obtained optimal predicate order using BF on the entire dataset. Then we compared the Total Cost between optimal predicate order (denoted by *Optimal*) and our selected predicate order.

5.2 Sampling Cost and Total Cost Analysis

The Sampling Cost and Total Cost of our methods on the dataset are shown in Figure 1. We have the primary observations as following: (1) the Sampling Cost

---

[1] http://www.cs.berkeley.edu/~lbourdev/poselets/attributes_dataset.tgz

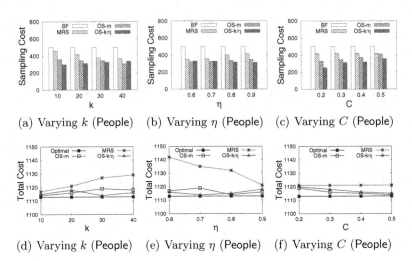

(a) Varying k (People)   (b) Varying η (People)   (c) Varying C (People)

(d) Varying k (People)   (e) Varying η (People)   (f) Varying C (People)

Fig. 1. Sampling Cost and Total Cost Evaluation (Default values: $k = 20$, $\eta = 0.9$ and $C = 0.3$)

of both $OS - m$ and $OS - k/\eta$ was always less than that of both BF and MRS, while the Total Cost of both $OS - m$ and $OS - k/\eta$ was almost the same as that of $Optimal$ and outperformed MRS. That is because both $OS - m$ and $OS - k/\eta$ utilized the selectivity to select "better" permutations which required less questions. (2) Sampling Cost of MRS was less than that of BF. Additionally, with smaller k and larger η, the Total Cost of MRS was close to $Optimal$ because of taking advantage of both k and η to control the quality of the predicate order.

6   Related Work

The most common method of crowdsourced select query is sampling which can significantly save the cost [1–3]. Marcus et al. [1] primarily studied how to make use of human instinct to design the interface of tasks. Trushkowsky et al. [2,3] used species estimation techniques to process select query with a single predicate.

7   Conclusion

In this paper, we studied the problem of query optimization for crowdsourced select query with multiple predicates. We proposed a minimum random selection algorithm to obtain the optimal predicate order. In order to further reduce the cost, we proposed the filtering based selection method. Experimental results indicated that our proposed methods significantly reduce the cost.

Acknowledgement. This Work is supported by National High-Tech R&D (863) Program of China (2012AA012600), the NSFC project (61472198), and the Chinese Special Project of Science and Technology (2013zx01039-002-002).

References

1. Marcus, A., Karger, D.R., Madden, S., Miller, R., Oh, S.: Counting with the crowd. PVLDB **6**(2), 109–120 (2012)
2. Trushkowsky, B., Kraska, T., Franklin, M.J., Sarkar, P.: Getting it all from the crowd. CoRR, abs/1202.2335 (2012)
3. Trushkowsky, B., Kraska, T., Franklin, M.J., Sarkar, P.: Crowdsourced enumeration queries. In: ICDE, pp. 673–684 (2013)

A Customized Schema Design Framework for Multi-tenant Database

Jiacai Ni$^{(\boxtimes)}$, Jianhua Feng, and Yongwei Wu

Department of Computer Science and Technology,
Tsinghua University, Beijing 100084, China
njc10@mails.tsinghua.edu.cn, {wuyw,fengjh}@tsinghua.edu.cn

Abstract. Existing multi-tenant database systems provide little support to consider high performance, good scalability, little space consumption and full customization at the same time. In this paper, we propose a *customized* database schema design framework which supports schema customization without sacrificing performance and scalability. We propose the interactive-based method to help tenants better design their customized schemas and integrate the schemas. We propose the graph partition method to reorganize integrated tables. Experimental results show that our method achieves better performance and higher scalability with schema customization property than state-of-the-art methods.

1 Introduction

Multi-tenant database application has attracted more and more attention from the leading IT companies and academic field. As the software model, it amortizes the cost of hardware, software and professional services to a large number of users and significantly reduces per-user cost. As the service provided by the multi-tenant system is usually topic related, the system predefines some tables called *predefined tables*. In the predefined tables, the system defines some attributes called *predefined attributes*. However, in order to design the schema which fully satisfies tenants' requirements, the tenants require to configure their customized tables or add some new customized attributes in predefined tables. The tables newly configured are called *customized tables*, and the attributes configured by the tenants both in predefined and customized tables are called *customized attributes*. Providing a high-degree customization to satisfy each tenant's needs is one of the important features in the multi-tenant database and the well-known leaders in multi-tenant database systems, e.g., Salesforce, Oracle, SAP, IBM and Microsoft, start to provide such features to boost their competitiveness. However, one big challenge is to devise a high-quality database schema with good performance, low space and high scalability supporting full customization.

To our best knowledge, state-of-the-art approaches on the multi-tenant schema design can be broadly divided into three categories [3,4]. The first one is Independent Tables Shared Instances (ITSI). It maintains a physical schema for each customized schema and this method has the scalability problem when the service faces large numbers of tenants[1]. The second method is Shared Tables

© Springer International Publishing Switzerland 2015
J. Li and Y. Sun (Eds.): WAIM 2015, LNCS 9098, pp. 530–534, 2015.
DOI: 10.1007/978-3-319-21042-1_56

Shared Instances (STSI). In this method different customized tenants share one table and this method has poor performance and consumes more space since it involves large numbers of NULLs. The third method is to simplify the service and do not allow tenants to precisely configure schemas. Thus existing schema designs did little work to consider the performance, scalability and the customization at the same time. To address these limitations, we propose a *customized* framework. We not only allow tenants to fully configure customized schemas but also recommend appropriate customized schemas. We propose the graph partition method to do optimization. Experimental results show that our method with full schema customization achieves high performance and good scalability with low space.

2   Customized Schema Design

Customized schemas in the multi-tenant database pose great challenges. For customized attributes in predefined tables, if we put them into predefined tables, the predefined tables can be extremely wide. And each customized attribute is usually shared by a small number of tenants then the table is sparse containing large numbers of NULLs. It decreases the performance and wastes space. On the contrary, if we maintain one extension table for the customized attributes, the number of NULLs is reduced. However, similar to ITSI, the scalability becomes the bottleneck. While for customized tables, similar to customized attributes in predefined tables, if the system maintains one table for each customized table the scalability is the problem. In this paper we focus on how to design high-quality physical tables for customized schemas.

2.1   Cluster-Based Schema Integration

In the typical multi-tenant applications, the data type is very expressive then we can use the data type information to build a feature vector for each customized table and then cluster them into similar groups. Each bit in the table feature vector represents the number of corresponding data type and we use the cosine function to compute the table feature vector and set it as customized table similarity. Next we use the hierarchical agglomerative clustering method to cluster customized tables. Based on the clustering results we integrate customized schemas into one integrated table for each cluster. In the attribute integration process, the attributes can only be integrated between the one with the same data type. In order to further ensure the accuracy, we take the constraint information the uniqueness, nullable and data length into account and the instance information data cardinality is also considered. We build the Attribute Feature Vector FV_a for each customized attribute. For each customized attribute, the corresponding attribute feature vector has 4 dimensions, the first $FV_{a_{(0)}}$ is to reflect the uniqueness, unique is 1 otherwise 0. The second $FV_{a_{(1)}}$ is to reveal whether is nullable, 1 means it may have NULL value, or else 0. The third $FV_{a_{(2)}}$ represents data length, and the fourth $FV_{a_{(3)}}$ is data cardinality. When we integrate the attributes in each cluster, we sum different dimension information similarity of the attribute feature vector as the final similarity value. For example, when we deicide whether to integrate attributes a_i and a_j, if only one of the two attributes are unique, then the $sim(FV_{a_{i(0)}}, FV_{a_{j(0)}})$ similarity is 0, otherwise it is 1. The value $sim(FV_{a_{i(1)}}, FV_{a_{j(1)}})$ is computed similarly. For the latter two parts, we both set the ratio of the minor value to the max value as the similarity value. For each attribute we must integrate it to the most similar integrated attribute above a threshold with the same data type.

2.2   Interactive-Based Schema Integration

Besides the basic cluster-based schema integration, we can further improve the integration quality and efficiency. When the tenant starts to configure customized schema, the system can recommend some appropriate customized tables or attributes. Meanwhile, the system can record the selection of different tenants to integrate the schema manually online. This interactive integration produces no interruption and enhances the experience of tenants. When a tenant begins to use the system, the system has no information of the tenant to refer. Confronting with the "cold start", we adopt the statistical information from the other tenants to rank customized schemas. For each customized table, the system records the number of tenants who configure to use, and the number is called *voting*. The *voting* information also applies for customized attributes. The system integrates the customized schemas in the same cluster and in each cluster different customized tables are very similar to each other, then based on the clustering result we select one table as the *representative table* for each cluster and the system recommends the *representative tables* from different clusters. Each cluster selects the table with the highest voting and set it as the *representative table*. Meanwhile, in each cluster we calculate the voting of all the customized tables and set it as `Table_Total`. According to `Table_Total`, we rank *representative tables*. A higher voting of `Table_Total` means that customized tables in this cluster are utilized by more tenants, then the representative table has a higher rank.

2.3   Optimizing Integrated Schema

After integrating customized schemas from different tenants, the integrated tables are still sparse and wide and we need to reorganize the integrated tables. Therefore, the optimization for customized schema becomes the integrated tables partition problem. Ideally, we formulate it as a *graph partition problem*. An integrated table can be modeled as a graph, each vertex is an integrated attribute and each edge is the similarity between two integrated attributes. We want to partition the graph into k components. We can put those integrated attributes shared by similar tenants together. We record the configuration of each integrated attribute, build integrated attribute configure vector CV_a and use Jaccard function to measure the similarity of different integrated attributes. In the integrated table graph, given two integrated attributes v_i and v_j, then the similarity is computed $\text{SIM}(v_i, v_j) = \text{JAC}(CV_i, CV_j)$. Since k represents the number

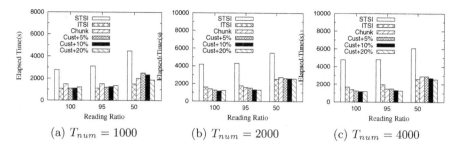

Fig. 1. Performance Comparisons II (Different Reading Ratio)

of tables which can represent scalability, then when we specify k we compute it like this. We define $\mathcal{S}$, where $\mathcal{S} = \frac{k}{T} \times 100\%$, T refers to the number of total customized tables in this cluster. Thus we use $\mathcal{S}$ to compute k. For instance, if $\mathcal{S}=5$ then the scalability improves by 20. Thus we can use different $\mathcal{S}$ to compute the corresponding k, and the scalability improvement is also clear.

3 Experimental Study

Based on the typical tables and data type information in Salesforce we design our fully customized multi-tenant benchmark. When generating customized tables we define following parameters. The number of total tenants is T_{num}. The average number of attributes for each customized table satisfies the normal distribution $N(\mu, \sigma^2)$, where μ represents the average number of attributes and σ is variance and set as 2 and we use MySQL database. Figure 1 compares the performance of our customized framework with STSI, ITSI and Chunk Tables [1]. We use $\mathcal{S}$ to reflect the scalability improvement of the system. Here we utilize Cust+5% to represent our customized framework when $\mathcal{S}=5$, then Cust+10% and Cust+20% are easy to understand. The workload contains 8000 operations. We vary the number of tenants and then compare the performance under different workloads. Here μ is set 15. Based on the analysis in the YCSB benchmark [2], we vary the ratio of reading operations. We observe that STSI performs worst in all the cases. When the number of tenants is small, ITSI has some priority, however, when the number of tenants is larger, scalability becomes the problem. Chunk table method scales well but still performs worse than our method.

4 Conclusion

In this paper, we propose the customized schema design framework for multi-tenant database. We allow tenants to fully configure customized schemas and propose the interactive-based method to help tenants precisely design their schemas. We finally optimize the integrated schemas. Experimental results show that our framework with full customized property achieves high performance and good scalability with low space.

534 J. Ni et al.

Acknowledgement. This Work is supported by National High-Tech R&D (863) Program of China (2012AA012600), the NSFC project (61472198), and the Chinese Special Project of Science and Technology (2013zx01039-002-002).

References

1. Aulbach, S., Grust, T., Jacobs, D., Kemper, A., Rittinger, J.: Multi-tenant databases for software as a service: schema-mapping techniques. In: SIGMOD Conference, pp. 1195–1206 (2008)
2. Cooper, B.F., Silberstein, A., Tam, E., Ramakrishnan, R., Sears, R.: Benchmarking cloud serving systems with ycsb. In: SoCC, pp. 143–154 (2010)
3. Hui, M., Jiang, D., Li, G., Zhou, Y.: Supporting database applications as a service. In: ICDE, pp. 832–843 (2009)
4. Ni, J., Li, G., Zhang, J., Li, L., Feng, J.: Adapt: adaptive database schema design for multi-tenant applications. In: CIKM, pp. 2199–2203 (2012)

A Multi-attribute Probabilistic Matrix Factorization Model for Personalized Recommendation

Feng Tan, Li Li$^{(\boxtimes)}$, Zeyu Zhang, and Yunlong Guo

Department of Computer Science, Southwest University, Chongqing 400715, China
{tf0823,lily,haixzzy,zqlong}@swu.edu.cn

Abstract. Recommender systems can interpret personalized preferences and recommend the most relevant choices to the benefit of countless users in the era of information explosion. Attempts to improve the performance of recommendation systems have hence been the focus of much research effort. Few attempts, however, recommend on the basis of both social relationship and the content of items users have tagged. This paper proposes a new recommending model incorporating social relationship and items' description information with probabilistic matrix factorization called SCT-PMF. Meanwhile, we take full advantage of the scalability of probabilistic matrix factorization, which helps to overcome data sparsity as well. Experiments demonstrate that SCT-PMF is scalable and outperforms several baselines (PMF, LDA, CTR) for recommending.

1 Introduction

With the development of the Web 2.0 and exponential growth of personal information on the network, recommender systems have been widely applied to make people's lives more convenient. What's more, more information can be used to create a superior recommendation system with greater interactions among users and increasing use of tag recommendation systems. To date however, few attempts incorporate interactions or connections among users and effective products' (call items in this paper) information (e.g., tags) into recommendation.

Users' interests can be affected by their behaviors in a social context. They have a tendency to group together with common interests and interact with each other based on their latent interests, which is called interest-based social relationship in this paper. In summary, our main contributions are as follows:

- We obtain interest-based social relationships among users with users' similarities. In order to show users' interests to recommending items, we employ a linear function combining user' own interests and those interests who influenced him in an interest-based social network.
- We propose a probabilistic model to incorporate the interest-based social relationship and the information of items for recommendation. The model is called SCT-PMF. Model parameters are adaptive and the optimal parameters are controlled by a heuristic method which is called 'generate and test'.

© Springer International Publishing Switzerland 2015
J. Li and Y. Sun (Eds.): WAIM 2015, LNCS 9098, pp. 535–539, 2015.
DOI: 10.1007/978-3-319-21042-1_57

- We evaluate our approach SCT-PMF on two real-world data sets. The results confirm that our model outperforms the baselines with the same settings.

2   SCT-PMF

2.1   Problem Definition

Assuming an **interest-based social network** is modelled as a graph $G = (U, V, E_u, E_v)$, and E_u defined as S $=\{s_{ik}\}_{N^r*N^r}$, is called a social matrix, and the edge in E_v, defined by R $= \{r_{ij}\}_{N^r*M}$, is referred as a rating matrix. Meanwhile, we define **a document** $D_j = (\text{w}_{j1}, \text{w}_{j2}, ..., \text{w}_{jN^w})$ that contains the bag-of-words including the title, keywords, abstract and tags of the product as the content information of item j, where w_{jl} ($l \in [1, 2, ..., N^w]$) denotes the lth word of item j.

This paper builds a social network based on items to discuss how much influence the social networks have on recommending. We define Jaccard similarity [1] between users to measure the common interest relationship (to grant the **social matrix** S). In order to define our model more realistically, we consider each user's particular taste in the context of her/his social relationship network in our model. The interest-based social relationship of u_i^* is defined by Equation (1).

$$u_i^* = \alpha u_i + (1 - \alpha) \sum_{t \in T(i)} s_{it} u_t \tag{1}$$

In Equation (1) u_i is a latent vector, denoting the interests of u_i. α and $1 - \alpha$ represent the weight of u_i particular interests and his interests influenced by the friends in an interest social network respectively. s_{it} ($S(u_i, u_t)$) denotes the similarity between u_i and u_t. $T(i)$ is the set of u_i 's friends, where the number of $T(i)$ depends on the similarity between u_i and his friends such as $s_{it} > 0.5$.

2.2   SCT-PMF

This paper exploits the LDA and PMF to incorporate the content information of items and users' social relationships, called SCT-PMF model, to enhance recommendation's predictive accuracy. The new model offers a probabilistic foundation in dealing with different factors, which gives rise to much flexibility for modelling various real recommendation situations. The graphical model of SCT-PMF is shown in Fig.1.

We assume the content information of items is from the topic model in Fig.1. More specifically, we draw topic proportions $\theta_j \sim Dirichlet(\xi)$ and item latent offset $\varepsilon_j \sim N(0, \lambda_v^{-1} I_K)$. Thus item j's latent vector is regarded as $v_j = \varepsilon_j + \theta_j$. Each word w_{jl} draws topic assignment $z_{jl} \sim Mult(\theta)$ and word assignment $w_{jl} \sim Mult(\beta_{zjl})$ in item j (document).

Based on interest-based social relationships' definition of Equation (1) and content information of items, we could obtain the following Equation (2).

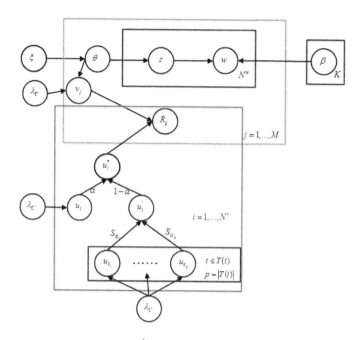

Fig. 1. Model of the SCT-PMF

$$p(R|S,U^*,V,\sigma_R^2)p(U^*|\sigma_U^2)p(V|\sigma_V^2)$$

$$= \prod_{i=1}^{N^r}\prod_{j=1}^{M}\left[N(r_{ij}|u_i^{*T}v_j,\sigma_R^2)\right]^{I_{ij}^R}\prod_{i=1}^{N^r}\prod_{j=1}^{M}N(u_i^*|0,\sigma_U^2 I)N(v_j|\theta_j,\sigma_V^2 I)$$

$$= \prod_{i=1}^{N^r}\prod_{j=1}^{M}\left[N(r_{ij}|(\alpha u_i+(1-\alpha)\sum_{t\in T(i)}s_{it}u_t)^T v_j,\sigma_R^2)\right]^{I_{ij}^R}*$$

$$\prod_{i=1}^{N^r}\prod_{j=1}^{M}N(u_i^*|0,\sigma_U^2 I)N(v_j|\theta_j,\sigma_V^2 I)$$

$$(2)$$

Then we take the method presented in Wang et al. [2] for approximation.

3   Experiments

3.1   Data Sets, Baselines and Evaluation Metrics

BibSonomy[1] [3] and CiteUlike[2] [2,4] are used to test the effectiveness of our approach. And PMF [5], LDA [6] and CTR [2] are used as baselines for comparison. To evaluate our approach, mean average precision at top H (MAP@H) and recall at top H (Recall@H) in users' actually recommendation lists are used in this paper.

---

[1] http://www.kde.cs.uni-kassel.de/bibsonomy/dumps
[2] http://www.datatang.com/data/45466

3.2   Experimental Results and Discussion

In order to make the experiments comparable, the same configuration applies to the proposed model and the baselines as well.

Performance Comparison. In order to show the effectiveness of our method, which displays the superior performance than other three baseline models, we present the performance results with well tuned parameter H and fixed other parameter with $\lambda_U = 0.01$, $\lambda_V = 0.1$, $\alpha = 0.05$ in Bibsonomy data set and $\lambda_U = 0.01$, $\lambda_V = 10$, $\alpha = 0.2$ in CiteUlike data set, respectively. And our approach (SCT-PMF) outperforms other methods (PMF, LDA and CTR) no matter how the tuned parameter H is changing. It illustrates that the proposed model is scalable.

Impact of Social Relationship. To testify the influence of social relationships in the recommendation system, let $\lambda_U = 0.01$, $\lambda_V = 0.1$, $H = 200$ in the Bibsonomy data set and $\lambda_U = 0.01$, $\lambda_V = 10$, $H = 200$ in the CiteUlike data set, respectively. We then observe the effect of varying parameter α. As a result, social relationships have a great influence on recommending items to users from our experiments.

Impact of Content Information. In order to explain how the value of λ_V affects the performance, we hold other variables constant as $\lambda_U = 0.01$, $\alpha = 0.05$, $H = 200$ in Bibsonomy and $\lambda_U = 0.01$, $\alpha = 0.2$, $H = 200$ in CiteUlike, respectively. In short, the content of items strongly influence the performance of recommendation in an aspect of our experiments.

4   Conclusion

In this paper, we proposed an extended probabilistic matrix factorization model incorporating social relationships and elaborated content-based information to enhance the recommendation performance. The experimental results show that our approach has achieved a great improvement on predictive accuracy of book recommendation. In the paper, users' social relationships are considered, but no further actions regarding information diffusion or propagation between users are counted. Furthermore, social relationships may include trust or distrust information, which are worthy of future study.

References

1. Jaccard, P.: Etude comparative de la distribution florale dans une portion des alpes et du jura: Impr (1901)
2. Wang, C., Blei, D.M.: Collaborative topic modeling for recommending scientific articles. In: Proceedings of the 17th ACM SIGKDD International Conference on Knowledge Discovery and Data Mining, pp. 448–456. ACM (2011)
3. Benz, D., Hotho, A., Jäschke, R., Krause, B., Mitzlaff, F., Schmitz, C., Stumme, G.: The social bookmark and publication management system bibsonomy. The VLDB Journal The International Journal on Very Large Data Bases **19**(6), 849–875 (2010)

4. Wang, H., Chen, B.P., Li, W.J.: Collaborative topic regression with social regularization for tag recommendation. In: Proceedings of the 23rd International Joint Conference on Artificial Intelligence. AAAI Press (2008)
5. Mnih, A., Salakhutdinov, R.: Probabilistic matrix factorization. In: Advances in Neural Information Processing Systems, pp. 1257–1264 (2007)
6. Blei, D.M., Ng, A.Y., Jordan, M.I.: Latent dirichlet allocation. The Journal of Machine Learning Research **3**, 993–1022 (2003)

A Novel Knowledge Extraction Framework
for Resumes Based on Text Classifier

Jie Chen[1], Zhendong Niu[1,2]([✉]), and Hongping Fu[1]

[1] School of Computer Science and Technology, Beijing Institute of Technology,
Beijing 100081, China
[2] Information School, University of Pittsburgh, Pittsburgh, PA 15260, USA
`zniu@bit.edu.cn`

Abstract. In the information age, there are plenty of resume data in the internet. Several previous research have been proposed to extract facts from resumes, however, they mainly rely on large amounts of labeled data and the text format information, which made them limited by human efforts and the file format. In this paper, we propose a novel framework, not depending on the file format, to extract knowledge about the person for building a structured resume repository. The proposed framework includes two major processes: the first is to segment text into semi-structured data with some text pretreatment operations. The second is to further extract knowledge from the semi-structured data with text classifier. The experiments on the real dataset demonstrate the improvement when compared to previous researches.

Keywords: Resume fact extraction · Text classifier · Knowledge Extraction

1 Introduction

In the information age, Internet-based recruiting platforms become increasingly crucial as the recruitment channel[1]. Normally, human resources department can receive numerous resumes from job seekers everyday. It is an extra work for them to record these resume data into database. Meanwhile, job seekers may use diverse resume formats and varying typesetting to gain more attention. As a result, most of the resumes are not written in accordance with a standard format or a specific template file. Although unique writing format can bring a better experience for reader, it's really harmful to data mining and candidates searching system. They can decrease the success rate of recommending recruits who meet the employer's requirements perfectly, and on the contrary, may bring some data noise. Thus, it is important to extract the valuable information of each resume.

In the field of web data extraction, most of the proposed research methods have focused on how to extract the main content from news web pages. Compared to news web page, resumes have some other unique properties. First, different

© Springer International Publishing Switzerland 2015
J. Li and Y. Sun (Eds.): WAIM 2015, LNCS 9098, pp. 540–543, 2015.
DOI: 10.1007/978-3-319-21042-1_58

people have different writing style for personal resume, but the content are all the same items including their personal information, including contacts, educations, work experiences etc. Second, in the same module of one resume, the writing style is shared among different items. In other words, resumes share the document-level hierarchical contextual structure[4], because items in one module sharing the same writing style can make the whole resume more comfortable. Above all, resumes can be segmented into several groups, and knowledge can be identified based on the text classifier constructed with the major elements of each resume.

In this paper, we propose an effective framework to extract the detailed knowledge from resumes without the limit of file types and too much human efforts to label data.

2   Our Approach

Different with previous research, we construct the knowledge extraction framework based on text classifier for resumes with two steps, which are semi-structured information extraction step and detailed information identification step. The detail of our approach are described as follows.

2.1   Pipeline of Resume Knowledge Extraction Framework

The pipeline of our framework for resume knowledge extraction without the text format information for building a structured resume repository with an example is presented in Fig.1.

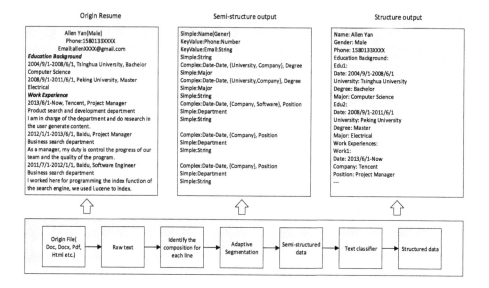

Fig. 1. The pipeline of our framework and an example

First, the input of the framework is raw text, which is parsed from the origin resume files. Tika[1] is used to get the pure text content and styles are removed from the files including table cells, font styles, images etc.

Second, each line of the text is segmented into phrase and a basic text classifier is used to predict the label, such as university, date, number etc. With the help of phrase classifier, the composition information of each line can be determined and the semi-structured information will be the input of the next step. In order to distinguish the difference among lines, the line with only one word is tagged Simple, the line with two words and a common between them is tagged KeyValue and others are tagged Complex.

Last, name entities are matched to the candidates profile based on the information statistics and text classifier.

2.2   Construct the Text Classifier

It's known that each resume consist of someone's education information including job information, personal contact information, certificates etc. Inspired by the components of a resume, we prepared a lot of name entities, such as name of university, name of company, job positions, department etc., which are easy to collect from the internet without human efforts to label. These name entities are used to train a basic multi-class classifier, which covers the important part of a resume. In our experiments, the Naive Bayes algorithm is used to train the classifier.

3   Experiments

For evaluating the proposed framework, we tested it on a real world dataset. Education and work experiences parts are extracted from each resume. Since the extracting algorithms are independent on the test corpus, we compared our approach with PROSPECT[2] and CHM[3], which also use the natural language processing method to extract the detail knowledge from the resume text. However, only two parts' data were provided by these two models, we compared them respectively.

Table 1. Education classification

	PROSPECT	CHM	Our approach
Precision	0.94	0.71	0.912
Recall	0.902	0.77	0.701
F-1	**0.921**	0.73	0.792

---

[1] http://tika.apache.org

Table 2. Work experiences classification

	PROSPECT	Our approach
Precision	0.790	0.873
Recall	0.780	0.720
F-1	0.785	**0.789**

Table 1 shows the results about education classification and Table 2 shows the results about work experiences. The PROSPECT's precision and recall are higher than our framework in the education block classification, the main reason is the difference of test corpus. The application scene of PROSPECT focus on the resumes of software engineers with IT professionals. The advantage of such a limited fields of resumes is that these resumes always cover a limited major, which help to increase the precision and recall for classifier. Face to the work experience block, this advantage is very small and the reason also explain the low precision and recall.

4   Conclusion

In this paper, we propose a novel framework to extract the knowledge from raw resume text. This work aim to improve the accuracy of building resume repository for head-hunters and companies focus on recruiting. The experiments show that this framework encouragingly comparable to other previous research based on sequence labeling model. In future, we are planing to work on the identification of more detailed knowledge from the resume text and investigate the difference of solution on Chinese and English resumes.

Acknowledgments. This work is supported by the National Natural Science Foundation of China (No. 61370137) and the 111 Project of Beijing Institute of Technology.

References

1. Al-Otaibi, S.T., Ykhlef, M.: A survey of job recommender systems. International Journal of the Physical Sciences **7**, 5127–5142 (2012)
2. Singh, A., Rose, C., Visweswariah, K., Chenthamarakshan, V., Kambhatla, N.: PROSPECT: A System for Screening Candidates for Recruitment. In: Proceedings of the 19th ACM International Conference on Information and Knowledge Management, pp. 659–668. ACM (2010)
3. Yu, K., Guan, G., Zhou, M.: Resume Information Extraction with Cascaded Hybrid Model. In: Proceedings of the 43rd Annual Meeting on Association for Computational Linguistics, pp. 499–506. Association for Computational Linguistics (2005)
4. Maheshwari, S., Sainani, A., Reddy, P.K.: An Approach to Extract Special Skills to Improve the Performance of Resume Selection. In: Kikuchi, S., Sachdeva, S., Bhalla, S. (eds.) DNIS 2010. LNCS, vol. 5999, pp. 256–273. Springer, Heidelberg (2010)

SimRank Based Top-k Query Aggregation for Multi-Relational Networks

Jing Xu, Cuiping Li, Hong Chen, and Hui Sun[✉]

Key Lab of Data Engineering and Knowledge Engineering of MOE,
Renmin University of China, Beijing 100872, China
{ruc_xujing,licuiping,chong,sun_h}@ruc.edu.cn

Abstract. SimRank is one measure that compute the similarities between nodes in applications, where the returning of top-k query lists is often required. In this paper, we adopt SimRank as the similarity computation measure and re-write the original inefficient iterative equation into a non-iterative one, we call it Eigen-SimRank. We focus on multi-relational networks, where there may exist different kinds of relationships among nodes and query results may change with different perspectives. In order to compute a top-k query list under any perspective especially compound perspective, we suggest dynamic updating algorithm and rank aggregation methods. We evaluate our algorithms in the experiment section.

Keywords: Top-k list · SimRank · Rank aggregation · Multi-relational networks

1 Introduction

Computing the most similar k nodes w.r.t. a given query node, based on similarity measures such as RWR[12], PPR[14], and SimRank[13], has been studied a lot[3,12,19,20]. Recently, a novel query called **perspective-aware query** has been proposed to provide more insightful information to users[22]. In this paper, we follow perspective-aware query and focus on multi-relational network[1,4–6,17]. Figure 1 shows an example. We extract three relationships denoted by different colors and call them the base perspectives(R, G, B, for short), which help construct the hierarchy showed in figure 1(c). We call higher hierarchy perspectives compound perspectives. We are interested in *how to quickly compute a top-k query list under any compound perspective?*

Rank aggregation was first studied to solve the WEB spam problem[8] to overcome inherent search engine bias, and over the years, it has been studied a lot[11]. We draw an analogy between top-k query lists obtained under some perspectives and biased lists returned by search engines, then top-k query list under compound perspective is compared to the list returned by rank aggregation methods. What's more, based on Eigen-SimRank model, we develop a

© Springer International Publishing Switzerland 2015
J. Li and Y. Sun (Eds.): WAIM 2015, LNCS 9098, pp. 544–548, 2015.
DOI: 10.1007/978-3-319-21042-1_59

(a) A Network G (b) Perspective Graph G' (c) All Perspectives

Fig. 1. Example of Coauthor Relationship

dynamic updating algorithm. In this circumstance, we regard the network structure of compound perspective as the result of network structure of one perspective changes as time goes by.

To be specific, the contributions of this paper are summarized as below:

- To accelerate the computing process, we re-write the SimRank equation into a non-iterative one, and apply the Eigen-SimRank model to perspective-aware top-k query in multi-relational networks, to provide more insightful information to user.
- We adopt dynamic updating algorithm and rank aggregation methods to make the query computation especially under compound perspectives more efficient.

2   Eigen-SimRank Model

Let $\mathbf{W}$ be the column-normalized matrix of the adjacent matrix of a network graph G. As has been proved in[7], that by applying the well-known *Sylvester Equation*[15], and vec operator[2], the original SimRank equation can be written as:

$$vec(\mathbf{S}) = (1 - c)(\mathbf{I} - c(\tilde{\mathbf{W}} \otimes \tilde{\mathbf{W}}))^{-1}vec(\mathbf{I}) \qquad (1)$$

where $\mathbf{I}$ is an identity matrix, $\tilde{\mathbf{W}}$ is the transpose of $\mathbf{W}$. $\otimes$ and $vec()$ are two operators[2]. According to the *Sherman-Morrison Lemma*[21], we have:

$$(\mathbf{I} - c(\tilde{\mathbf{W}} \otimes \tilde{\mathbf{W}}))^{-1} = \mathbf{I} + c\mathbf{P}(\mathbf{Q}^{-1} - c\mathbf{P}^{-1}\mathbf{P})^{-1}\mathbf{P}^{-1} = \mathbf{I} + c\mathbf{P}(\mathbf{Q}^{-1} - c\mathbf{I})^{-1}\mathbf{P}^{-1}$$

By equation (1), we have our non-iterative Eigen-SimRank computation equation:

$$vec(\mathbf{S}) = (1 - c)(\mathbf{I} + c\mathbf{P}(\mathbf{Q}^{-1} - c\mathbf{I})^{-1}\mathbf{P}^{-1})vec(\mathbf{I}) \qquad (2)$$

where $\mathbf{P} = \mathbf{P}_{\tilde{\mathbf{W}}} \otimes \mathbf{P}_{\tilde{\mathbf{W}}}$, $\mathbf{P}^{-1} = \mathbf{P}_{\tilde{\mathbf{W}}}^{-1} \otimes \mathbf{P}_{\tilde{\mathbf{W}}}^{-1}$ and $\mathbf{Q} = \mathbf{Q}_{\tilde{\mathbf{W}}} \otimes \mathbf{Q}_{\tilde{\mathbf{W}}}$. $\mathbf{P}_{\tilde{\mathbf{W}}}$ and $\mathbf{Q}_{\tilde{\mathbf{W}}}$ are eigenvalue matrix and eigenvector matrix[9] of $\tilde{\mathbf{W}}$.

Based on equation (2), our algorithm stores the matrices $\mathbf{P}$, $\mathbf{L}$, and $\mathbf{V}$ in pre-computation phase for query processing phase to use.

3   Dynamic Updating

In order to get the similarity scores between any two nodes at each time step t, we need to update matrices $\mathbf{P}^t$, $\mathbf{L}^t$, $\mathbf{V}^t$ efficiently based on $\mathbf{P}^{t-1}$, $\mathbf{L}^{t-1}$, $\mathbf{V}^{t-1}$

and $\Delta\tilde{\mathbf{W}}$. As $\tilde{\mathbf{W}}^t = \tilde{\mathbf{W}}^{t-1} + \Delta\tilde{\mathbf{W}} = \mathbf{P}_{\tilde{\mathbf{W}}}{}^{t-1}\mathbf{Q}_{\tilde{\mathbf{W}}}{}^{t-1}(\mathbf{P}_{\tilde{\mathbf{W}}}{}^{-1})^{t-1} + \Delta\tilde{\mathbf{W}}$, then $(\mathbf{P}_{\tilde{\mathbf{W}}}{}^{-1})^{t-1}\tilde{\mathbf{W}}^t\mathbf{P}_{\tilde{\mathbf{W}}}{}^{t-1} = \mathbf{Q}_{\tilde{\mathbf{W}}}{}^{t-1} + (\mathbf{P}_{\tilde{\mathbf{W}}}{}^{-1})^{t-1}\Delta\tilde{\mathbf{W}}\mathbf{P}_{\tilde{\mathbf{W}}}{}^{t-1}$, let $\mathbf{C} = \mathbf{Q}_{\tilde{\mathbf{W}}}{}^{t-1} + (\mathbf{P}_{\tilde{\mathbf{W}}}{}^{-1})^{t-1}\Delta\tilde{\mathbf{W}}\mathbf{P}_{\tilde{\mathbf{W}}}{}^{t-1}$. Compute the low-rank approximation for $\mathbf{C}$. Suppose $\mathbf{C} = \mathbf{xyz}$, then we have $\mathbf{P}_{\tilde{\mathbf{W}}}{}^t = \mathbf{P}_{\tilde{\mathbf{W}}}{}^{t-1}\mathbf{x}$, $\mathbf{Q}_{\tilde{\mathbf{W}}}{}^t = \mathbf{y}$, $(\mathbf{P}_{\tilde{\mathbf{W}}}{}^{-1})^t = \mathbf{z}(\mathbf{P}_{\tilde{\mathbf{W}}}{}^{-1})^{t-1}$. At last, $\mathbf{P}^t = \mathbf{P}_{\tilde{\mathbf{W}}}{}^{t-1}\mathbf{x} \otimes \mathbf{P}_{\tilde{\mathbf{W}}}{}^{t-1}\mathbf{x} = \mathbf{P}^{t-1}\mathbf{X}$, where $\mathbf{X} = \mathbf{x} \otimes \mathbf{x}$. It is the same way that we update $\mathbf{L}^t$ and $\mathbf{V}^t$.

4   EXPERIMENT RESULTS

We adopt two rank aggregation methods, namely Borda method and footrule algorithm[8]. We encoded them in C program, and evaluate the effectiveness and efficiency along with Eigen-SimRank model and dynamic updating algorithm. We set the default value of parameter $c = 0.2$. Table 1 and table 2 show the details of our datasets.

Table 1. Synthetic Datasets

Graph	Node Number	Edge Number
G_1	1K	5K
G_2	2K	15K
G_3	3K	25K
G_4	4K	35K
G_5	5K	45K

Table 2. DBLP Dataset

Perspective	Edge Number
DB	37225
(DB DM)	54107
(DB DM IR)	76503
(DB DM IR AI)	110486

Effectiveness Evaluation. We adopt the widely-used NDCG measure[7]. Figure 2(a) and 2(b) shows the NDCG scores, with dynamic updating algorithm achieves the highest accuracy around 90%, and on average 85%. Although Borda method and Footrule algorithm perform not as good as dynamic updating algorithm, they still hold the average accuracy up to 80% and 77%, respectively. As for DBLP dataset, for NDCG@25, Borda method and Footrule method achieve 85% and 84%, respectively.

Efficiency Evaluation. Figure 2(c) - 2(e) display the runtime on synthetic datasets, and figure 2(f), 2(g) on DBLP dataset correspondingly. "E-S", "D U", "B", "F" are short for Eigen-SimRank and dynamic updating, Borda and Footrule respectively. On the whole, Borda method and Footrule algorithm perform better than both Eigen-SimRank model and dynamic updating algorithm. However, Eigen-SimRank is not so bad considering the better performance when compared to NI_Sim algorithm[7] when k is larger than 15 (k is a parameter) as Figure 2(f) showed.

5   Conclusion

We choose SimRank as the proximity measure, and re-write the original iterative equation into a non-iterative one, named Eigen-SimRank, we apply it to

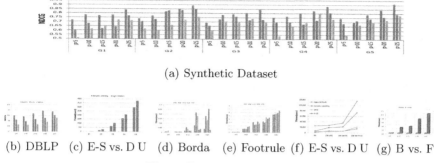

(a) Synthetic Dataset

(b) DBLP   (c) E-S vs. D U   (d) Borda   (e) Footrule   (f) E-S vs. D U   (g) B vs. F

Fig. 2. Experiments Results

perspective-aware query, in order to compute top-k query under any compound perspective, we suggest four solutions, they are directly computing using Eigen-SimRank, dynamic updating algorithm, Borda method, and Footrule algorithm. The experiment results show the more efficiency one method is, the less accuracy it achieves, however, the average accuracy is still acceptable considering the acceleration in computation.

Acknowledgments. This work was supported by National Basic Research Program of China (973Program) (No.2014CB340402, No. 2012CB316205), National High Technology Research and Development Program of China (863 Program) (No.2014AA015204), NSFC under the grant No.61272137, 61033010, 61202114, 61165004 and NSSFC (No: 12&ZD220). It was partially done when the authors worked in SA Center for Big Data Research in RUC. This Center is funded by a Chinese National 111 Project Attracting.

References

1. Ahn, Y.Y., Bagrow, J.P., Lehman, S.: Link communities reveal multiscale complexity in networks. Nature, 761–764 (2010)
2. Laub, A.J.: Matrix Analysis for Scientists and Engineers. Society for Industrial and Applied Mathmatics (2004)
3. Avrachenkov, K., Litvak, N., Nemirovsky, D., Smirnova, E., Sokol, M.: Quick detection of top-k personalized pagerank lists. In: Frieze, A., Horn, P., Pralat, P. (eds.) WAW 2011. LNCS, vol. 6732, pp. 50–61. Springer, Heidelberg (2011)
4. Berlingerio, M., Coscia, M., Giannotti, F.: Finding and characterizing communities in multidimensional networks. IEEE, 490–494 (2011)
5. Berlingerio, M., Coscia, M., Giannotti, F., Monreale, A., Pedreschi, D.: Foundations of multidimensional network analysis. IEEE, 485–489 (2011)
6. Cai, D., Shao, Z., He, X., Yan, X., Han, J.: Community mining from multi-relational networks. In: Jorge, A.M., Torgo, L., Brazdil, P.B., Camacho, R., Gama, J. (eds.) PKDD 2005. LNCS (LNAI), vol. 3721, pp. 445–452. Springer, Heidelberg (2005)
7. Li, C., Han, J., He, G., Jin, X., Sun, Y., Yu, Y., Wu, T.: Fast Compoutation of SimRank for Static and Dynamic Information Networks. EDBT (2010)
8. Dwork, C., Kumar, R., Naor, M., Sivakumar, D.: Rank Aggregation Methods for the Web. WWW (2001)

9. Harville, D.A.: Matrix Algebra from a Statistician's Perspective
10. Lizorkin, D., Velikhov, P., Grinev, M., Turdakov, D.: Accuracy estimate and optimization techniques for simrank computation. VLDB (2008)
11. Schalekamp, F., van Zuylen, A.: Rank Aggregation : Together We're Strong
12. Fujiwara, Y., Nakatsuji, M., Onizuka, M., Kitsuregawa, M.: Fast abd exact top-k search for random walk with restart. PVLDB, 442–453 (2012)
13. Jeh, G., Widom, J.: Simrank: a measure of structural-context similarity. KDD (2002)
14. Jeh, G., Widom, J.: Scaling personalized web search. WWW, 271–279 (2003)
15. Benner, P.: Factorized solution of sylvester equations with applications in control. MTNS (2004)
16. Diaconis, P.: Group Representation in Probability and Statistics. IMS Lecture Series 11 (1988)
17. Rodriguez, M.A., Shinavier, J.: Exposing multi-relational networks to single-relational network analysis algorithms. J. informetrics, 29–41 (2010)
18. Sarkar, P., Moore, A.W., Prakash, A.: Fast incremental proximity search in large graphs. ICML, 896–903 (2008)
19. Sarkar, P., Moore, A.W.: Fast nearest-neighbor search in disk-resident graphs. KDD, 513–522 (2010)
20. Sun, Y., Han, J., Yan, X., Yu, P.S., Wu, T.: Pathsim: Meta pathbased top-k similarity search in heterogeneous information networks. PVLDB, 992–1003 (2011)
21. Piegorsch, W., Casella, G.: Invertint a sum of matrices. SIAM Rev., 32 (1990)
22. Zhang, Y., Li, C., Chen, H., Sheng, L.: On Perspective-Aware Top-k Similarity Search in Multi-Relational Networks. DASFAA, 171–187 (2014)
23. Zhao, P., Li, X., Xin, D., Han, J.: Graph cube: on warehousing and olap multidimensional networks. SIGMOD, 853–864 (2011)

On Dynamic Top-k Influence Maximization

Hao Wang[1], Nana Pan[2], U Leong Hou[2(✉)], Bohan Zhan[2], and Zhiguo Gong[2]

[1] State Key Laboratory for Novel Software Technology,
Nanjing University, Nanjing, China
hwang2@cs.hku.hk

[2] Department of Computer and Information Science,
University of Macau, Macau, China
{mb15572,ryanlhu,mb15460,fstzgg}@umac.mo

Abstract. This paper studies the top-k influence maximization problem together with network dynamics. We propose an incremental update framework that takes advantage of smoothness of network changes. Experiments show that the proposed method outperforms the straightforward solution by a wide margin.

1 Introduction

Finding those people who are the most influential in a social network is important in viral marketing as it helps companies to identify proper users for promoting their products and services. Motivated by this, the problem of *top-k influence maximization* (IM) has been studied and developed extensively (see, e.g., [1]). Kempe et al. [4] prove this problem NP-hard. Leskovec et al. [5] propose an algorithm, CELF, to solve this problem via Monte Carlo simulations with approximation ratio arbitrarily close to $1 - e^{-1}$. Recently, Chen et al. [1] improve the performance of CELF; their method, MIXEDGREEDY, is reported to be one order of magnitude faster.

Prior methods focus on *static* networks. In this paper, we aim at monitoring the evolution of the top-k influential users in *dynamic* networks. Applying existing algorithms at every time step is a straightforward solution, however it is inefficient as it ignores the smoothness of network changes. Our finding is that the top-k set of influential users is relatively stable even if the network changes frequently over time. Based on such observations, our objective in this paper is to study an incremental processing framework which maintains the top-k result subject to the network changes.

This work is supported in part by grant 61432008 from the National Natural Science Foundation of China, and in part by grant MYRG109(Y1-L3)-FST12-ULH and MYRG2014-00106-FST from UMAC RC.

H. Wang—Part of the work was done when this author was a PhD student at the University of Hong Kong.

J. Li and Y. Sun (Eds.): WAIM 2015, LNCS 9098, pp. 549–553, 2015.
DOI: 10.1007/978-3-319-21042-1_60

2   Background

A *social network* is a weighted directed graph $G = (V, E)$ where each vertex u is a *user* and each edge $e_{u,v}$ is the *social relation* weighted by an *activation probability* $p_{u,v} \in [0, 1]$. With a *seed* of one or several vertices, the *independent cascade (IC)* *model* simulates information diffusion in discrete steps. At each step, each newly-activated user attempts to activate each of her inactive outgoing neighbors, and such attempts succeed with corresponding activation probabilities; the diffusion process terminates if no more activation is possible. Eventually, we get a *diffusion subgraph* $G_S = (V_S, E_S)$ of G, consisting of all active users and edges of successful activation. The *influence* of S in G is defined as $I(S|G) = \mathbb{E}[|V_S|]$, which is usually estimated by Monte Carlo simulations.

1: **for** $j = 1, 2, \cdots, M$ **do**
2:     **for each** $e_{u,v} \in E$ **do** Add $e_{u,v}$ into E_S with probability $p_{u,v}$
3:         $\boldsymbol{r}_j \leftarrow$ REACHABILITY(G_S) $\triangleright$ $G_S = (V, E_S)$; *Cohen's estimation [3]*
4: Set $c_v \leftarrow (r_{1v} + r_{2v} + \cdots + r_{Mv})/M$ for all $v \in V$; Feed max-heap H with (c_v, v) pairs
5: $(I_1, v_1) \leftarrow H.\text{pop\_max}()$; $\Phi_1 \leftarrow \{v_1\}$
6: **for** $i = 2, 3, \cdots, k$ **do**
7:     **while** $v \leftarrow H.\text{max}().\text{ID}$ has not yet been evaluated with Φ_{i-1} **do**
8:         $c_v \leftarrow$ MONTECARLO$(\Phi_{i-1} \cup \{v\}, G, M) - I_{i-1}$; update v in H with new key c_v
9: $(c_{\max}, v_{\max}) \leftarrow H.\text{pop\_max}()$; $\Phi_i \leftarrow \Phi_{i-1} \cup \{v_{\max}\}$; $I_i \leftarrow c_{\max} + I_{i-1}$

Let Φ_i be the top-i set of influential users. Algorithm MIXEDGREEDY [1] takes k iterations to construct a chain of $\Phi_1 \subset \Phi_2 \subset \cdots \subset \Phi_k$; during the i-th iteration, it adds into Φ_{i-1} the vertex v_i^*, the one that maximizes $C_i(\cdot) \stackrel{\text{def}}{=} I(\Phi_{i-1} \cup \cdot) - I(\Phi_{i-1})$. Specifically, to compute Φ_1 MIXEDGREEDY generates *random subgraphs* G_S (Line 2), on which calculating the influence becomes counting the reachability (Line 3). Then, Lines 6-9 are based on the following lemma on $C_i(\cdot)$, which directly follows the *submodularity* of $I(\cdot)$.

Lemma 1. *For any $j < i$ and $v \notin \Phi_{i-1}$, $C_i(v) \leq C_j(v)$.*

Dynamic top-k Influence Maximization. In this paper we target top-k IM with network dynamics. Formally, assume that over a certain time period a social network $G = (V, E)$ evolves to $G' = (V, E')$; then given a chain of top-k sets of G, $\Phi_1 \subset \Phi_2 \subset \cdots \subset \Phi_k$, we seek an *efficient* solution to finding Φ_i', the top-k sets of G'.

3   Our Approach

We build our solution upon the stability of the top-k sets. Specifically, we progressively verify whether $\Phi_i' = \Phi_i$. Once $\Phi_i' \neq \Phi_i$, then $\Phi_i', \Phi_{i+1}', \cdots, \Phi_k'$ are recomputed. For verification, we distinguish the *strengthened edges*, $\Delta E^+ =$

$\{e_{u,v} | \Delta p_{u,v} > 0\}$, and the *weakened edges*, $\Delta E^- = \{e_{u,v} | \Delta p_{u,v} < 0\}$. On verifying each Φ_i, we perform two consecutive phases: (i) Apply ΔE^+ to G, resulting in $G^+ = (V, E^+)$; verify Φ_i in G^+; (ii) Apply ΔE^- to G^+, resulting in $G' = (V, E')$; verify Φ_i in G'.

Phase 1 of Verifying $\Phi_1 = \{v_1^*\}$. We estimate how ΔE^+ affects the influence of $v \in V$, i.e., we estimate $\Delta I(v) \stackrel{\text{def}}{=} I(v|G^+) - I(v|G)$ for all $v \in V$. A strengthened edge $e_{u,v}$ provides influence increase to a vertex w if and only if v is reachable from w in G^+ but not in G. Our approach is to consider the effects of ΔE^+ on those random subgraphs G_S generated by MixedGreedy, identify those *affected* vertices, and update their reachabilities. The following procedure serves this purpose by generating *delta random subgraphs* $G_\delta = (V_S, E_S; E_\delta)$, where E_δ is a random sample of ΔE^+. Starting from the endpoints of E_δ, we perform two probabilistic traversals in G^+: a *backward* one finding $V_b \subseteq V$ (Lines 3-4), and a *forward* one finding $V_f \subseteq V$ (Line 5). Let G_S be any supergraph of G_δ. We say that G_S is *consistent* with G_δ when, for any edge e visited in Lines 3-5, $e \in G_S$ if and only if $e \in G_\delta$. Let $\Delta r(v|\cdot) \stackrel{\text{def}}{=} r(v|\cdot) - r(v|\cdot \setminus E_\delta)$. We have the following lemma.

1: **for each** $e_{u,v} \in \Delta E^+$ **do** Add $e_{u,v}$ into E_δ with probability $\Delta p_{u,v}$
2: initialize $E_S \leftarrow E_\delta$; $V_b \leftarrow \{u | \exists v \text{ s.t. } e_{u,v} \in E_\delta\}$; $V_f \leftarrow \{u | \exists v \text{ s.t. } e_{v,u} \in E_\delta\}$
3: **while** $\exists u \in V_b$ for which Line 5 not done **do**   ▷ *probabilistic backward traversal*
4:     **for each** edge $e_{v,u} \notin E_S$ **do** add $e_{v,u}$ to E_S and v to V_b with probability $p_{v,u}$
5: Perform a *probabilistic forward traversal* (symmetric to Lines 3-4), expanding E_S and V_f **return** $G_\delta = (V_b \cup V_f, E_S; E_\delta)$

Lemma 2. *If $v \notin V_b$ then $\Delta r(v|G_S) = 0$; if $v \notin V_f$ then v does not contribute to $\Delta r(w|G_S)$ of any vertex $w \in V$. Therefore, $\forall v \in V$, $\Delta r(v|G_\delta) = \Delta r(v|G_S)$.*

Lemma 2 implies that the impact of ΔE^+ is completely limited within the delta random subgraph G_δ. Given the interrelationship between influence and reachability, to get $\Delta I(\cdot)$ values it is sufficient to track reachability changes within G_δ. Eventually we check whether $I(v_1^*|G^+)$ is the largest.

Phase 2 of Verifying $\Phi_1 = \{v_1^*\}$. Recall Lines 6-9 of MixedGreedy. We essentially rely on upper-bounding the contribution $C_i(\cdot)$ to prune less influential vertices. In this phase, since from G^+ to G' there are only weakened edges, the results of Phase 1 (i.e., $I(v|G^+)$ values) are natural upper bounds of $C_1(v|G') = I(v|G')$. Using these upper bounds, we employ a procedure similar to Lines 6-9 of MixedGreedy to verify whether $I(v_1^*|G') \geq I(v|G^+) \geq C_1(v|G') = I(v|G')$ for any $v \neq v_1^*$. If yes, then we safely conclude that $\Phi_1' = \Phi_1 = \{v_1^*\}$; otherwise we recompute $\Phi_1', \Phi_2', \cdots, \Phi_k'$.

Verifying $\Phi_i = \Phi_{i-1} \cup \{v_i^*\}$. As before, the verification of Φ_i is done in two phases. We check the validity of Φ_i by comparing the *exact value* of $C_i(v_i^*|\cdot)$ with some *upper bound* of $C_i(v|\cdot)$ for any other v, in G^+ and G' respectively.

In addition to the *submodular* upper bounds $C_i(v|\cdot) \leq C_{i-1}(v|\cdot)$ (Lemma 1), we develop the following *incremental* upper bounds.

Lemma 3. $\forall v \notin \Phi_{i-1}$, $C_i(v|G^+) \leq C_i(v|G) + \Delta I(v)$, and $C_i(v|G') \leq C_i(v|G^+) + \Delta I'(\Phi_{i-1})$, where $\Delta I(\cdot) \overset{def}{=} I(\cdot|G^+) - I(\cdot|G)$ and $\Delta I'(\cdot) \overset{def}{=} I(\cdot|G^+) - I(\cdot|G')$.

Incremental upper bounds are closely related to network changes. If the network changes are minor, for example if there are few strengthened edges ($G^+ \approx G$) or weakened edges ($G' \approx G^+$), the incremental upper bounds could be much more tighter than the submodular ones, thus a pruning procedure similar to Lines 6-9 of MIXEDGREEDY could be more efficient. In practice, we use whichever the better of submodular and incremental upper bounds.

4   Experiments

In this section we experimentally evaluate our method, DYNAG, in comparison of MIXEDGREEDY [1], where for the dynamic influence maximization problem MIXEDGREEDY is executed at every time step. We implemented DYNAG in C++ and reused the source of MIXEDGREEDY (in C). All experiments were run on a Linux server equipped with an Intel Core i7 3.2GHz CPU.

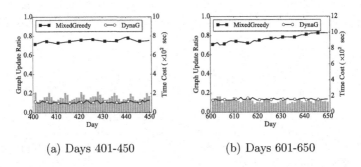

(a) Days 401-450   (b) Days 601-650

Fig. 1. Computational cost of MIXEDGREEDY and DYNAG on Brightkite

Dataset. We use the Brightkite dataset [2] for the experiments. The dataset consists of (i) a social network of users and their friendships, and (ii) user check-in records of the form (u, ℓ, t), meaning that user u visited location ℓ on the t-th day. Brightkite involves 58,228 users, 214,078 friendship links, and 2,627,870 check-in records in a period of 942 days. We take 2 temporal samples from Brightkite, each consisting of 50 consecutive days, and use a *recommendation model* (RM) to transform each sample into an evolving network. RM assumes that, when a user u visits a location ℓ for the first time, she will recommend ℓ to all of her friends who haven't visited ℓ yet. If later on some friend of u, say v, does visit

ℓ, then we assume that v is influenced by u. Hence, if up to the t-th day u has recommended $N_{u,v;t}$ locations to v in total, and among these recommendations $n_{u,v;t}$ have actually been followed by v, then the activation probability can be defined as the ratio $p_{u,v;t} = n_{u,v;t}/N_{u,v;t}$.

For each of the two networks we run MIXEDGREEDY and DYNAG independently to find the top-10 influential sets on each day and compare their running time. Figure 1 shows the results. The horizontal axes show the temporal interval of the sample. The left vertical axes represent daily update ratio (gray bars), i.e., the percentage of edges of which the weights changed on the day. The lines represent running time of MIXEDGREEDY and DYNAG, with values shown on the right vertical axes. As we can see, in all cases DYNAG outperforms MIXED-GREEDY.

References

1. Chen, Y.-C., Peng, W.-C., Lee, S.-Y.: Efficient algorithms for influence maximization in social networks. KAIS **33**(3), 577–601 (2012)
2. Cho, E., Myers, S.A., Leskovec, J.: Friendship and mobility: user movement in location-based social networks. In: KDD (2011)
3. Cohen, E.: Size-estimation framework with applications to transitive closure and reachability. JCSS **55**(3), 441–453 (1997)
4. Kempe, D., Kleinberg, J.M., Tardos, É.: Maximizing the spread of influence through a social network. In: KDD (2003)
5. Leskovec, J., Krause, A., Guestrin, C., Faloutsos, C., VanBriesen, J.M., Glance, N.S.: Cost-effective outbreak detection in networks. In: KDD (2007)

Community Based Spammer Detection
in Social Networks

Dehai Liu[1], Benjin Mei[1], Jinchuan Chen[2]($\boxtimes$), Zhiwu Lu[1], and Xiaoyong Du[1]

[1] School of Information, Renmin University of China, Beijing, China
{liudehai,meibenjin,luzhiwu,duyong}@ruc.edu.cn
[2] DEKE(MOE), Renmin University of China, Beijing, China
jcchen@ruc.edu.cn

Abstract. Social networks (SNS in short) such as Sina Weibo have now become one of the most popular Internet applications. A major challenge to social networks is the huge amount of spammers, or fake accounts generated by computer programmes. Traditional approaches in combating spammers mostly lie on user features, such as the completeness of user profiles, the number of users' activities, and the content of posted tweets etc. These methods may achieve good results when they are designed. However, spammers will evolve in order to survive. Hence the feature-based approaches will gradually lose their power. After careful analysis on a real SNS, we find that people will rebuild in cyber social networks their communities in the physical world. Interestingly, spammers also have to construct their communities in order to hide themselves, because separated users will be easily detected by anti-spam tools. Moreover, it is very hard for spammers to sneak into normal users' communities since a community member needs to have links with most other members. Based on this observation, we propose a novel approach to judge a user by the features of both himself/herself and the accounts within the same communities as him/her.

1 Introduction

In China, Sina Weibo[1] (weibo.com) is one of the largest social networks, which has more than 500 million registered user accounts. More and more people are rebuilding their social circles in this cyber world. The success of social networks brings a huge commercial opportunity. Many companies are trying to promote their products over social network.

An important threat to SNS is the existence of spammers. In this paper, the term "spammers" means fake (computer-generated) users in SNS, who want to make profits by exploiting the implicit trust relationships between users. Millions of spammers are pouring into social networks. For the celebrities who have thousands of followers, the average ratio of spammers is higher than 50 percent [2].

The existing works for combating spammers in SNS mainly depend on users' features [3]. These features may be identified by experts, e.g. the number of

---

[1] "Weibo" is the Chinese pronunciation of micro-blog.

© Springer International Publishing Switzerland 2015
J. Li and Y. Sun (Eds.): WAIM 2015, LNCS 9098, pp. 554–558, 2015.
DOI: 10.1007/978-3-319-21042-1_61

followers, the information shown in user profiles, and the content of micro-blogs etc. The feature-based methods may perform quite well when they are designed. However, spammers are also generated by clever people. They will evolve in order to survive in SNS. In fact, according to our experience, on average we need about 30 seconds to decide whether an account is spammer or not. Feature-based methods cannot evolve quickly and keep up with the changes of spammers, and they will gradually lose their power.

Another kind of works focus on the links [1]. This method chooses a set of seeds including both spammers and normal users, assigns trust and un-trust scores to the neighbors of the seeds, and further spreads these scores over the whole network. Finally, it ranks all users according to these scores. The link-based methods do not depend the features, and can still work when spammers evolve. However, since the number of seeds is very limited compared to the huge size of SNS, the initial scores of the original seeds will be quickly diluted. Therefore many users in the network can only receive an imprecise score which is not accurate enough to label these users as spammer or normal. The link-based approach is more suitable to rank users, as [1] does, but not good at classifying users.

Although spammers are constantly evolving, we find that they cannot change one thing. Note that the relationship of Sina Weibo (also Twitter) is not symmetric. A user follows another one but may not get the link back. In reality, few people would like to follow fake accounts. By analyzing the data crawled from Sina Weibo, we find that on average, more than 90 percent followers of a spammer are also spammers. More importantly, people usually live in several communities, e.g. relatives, classmates, and colleagues etc. In the cyber social network, a user, whether fake or not, is still involved in some communities. Normal users tend to rebuild their relationship in physical world. Spammers need to cluster in order to avoid being detected [1]. Our observations is, although spamemrs may get links with normal users, it is hard for them to join *normal communities*.

Fig. 1 illustrates the communities detected by the algorithm in [5] from a sub-graph of Sina Weibo. Each point stands for a user, with its color specifying the chance of this user to be spammer. Dark color represents spammer, and vice versa. We can see that users are clustered into communities, and the member of each community are almost homogeneous. Hence the communities will be split into two kinds, i.e. spammer communities (boxed in rectangles) and normal communities (boxed in ellipses).

Our general idea is exactly based on this observation. Since spammers usually live in their own communities. We should take into account the features of a user's community friends, i.e. those within the same communities as him/her, when judging the label of this user. Our approach contains two major steps. We first detect communities from a given SNS. Next, we compute the label of each user based on the features of both himself/herself and his/her community friends.

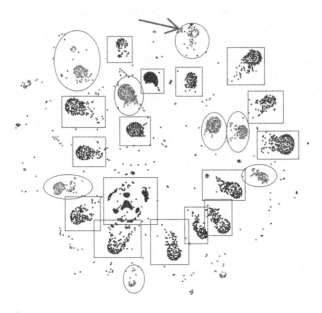

Fig. 1. Communities inside a sub-graph of Sina Weibo

2   Our Proposed Approach

As aforementioned, whether a user is a spammer depends on not only his/her features, but also the features of his/her community friends. Our approach contains two steps. Firstly, we recognize communities in a given social network by a state-of-the-art community detection algorithm[5]. We then assign an initial label for each user by a standard SVM. Secondly, we iteratively refine a user's label by combining his/her label and the labels of his/her community friends until meeting a stable status. We name this approach C^2IR since it utilizes $\underline{c}$community and $\underline{c}$lassifier, and $\underline{i}$teratively $\underline{r}$efines user labels.

Detecting Communities. We use the method of overlapping community detection [5] to get the communities from a given social network. The rationale of this method lies in the fact that the larger chance two nodes belong to a same community, the higher probability there exists an edge between them. Hence the best solution of constructing communities in a network should maximize the likelihood of the existing network structure.

Calculating Initial User Labels. We construct a feature-based SVM classifier to judge each user and give him/her an initial label. We can manually choose some important features like the number of followers, the frequency of posting tweets, and whether a user is VIP and so on. Also, we can adopt the matrix factorization method to build latent features like the work in [7] does. The labels predicted by SVM will be further refined by utilizing community information.

Refining User Labels. Suppose after community detection, we recognize K overlapped communities from the network. According to the theory of *hypergraph*[6], each recognized community composes a *hyperedge* connecting all of its members. Therefore, the original social network could be transformed into the following hypergraph.

$$\mathcal{G} = (\mathcal{U}, \mathcal{C}) \qquad (1)$$

Here $\mathcal{U}$ is the set of users and $\mathcal{C}$ is the set of recognized communities.

The introduction of hypergraph is to model two users' similarity in terms of communities, the foundation of our proposed approach. In our problem, a desired *community similarity* measure should satisfy a basic requirement. *The more communities both u_i and u_j belong to, the larger value of their similarity is.*

In this work, we measure the community similarities by utilizing the symmetric Laplacian matrix L of the hypergraph $\mathcal{G}$. The Laplacian matrix L is a $n \times n$ matrix representing the similarity between users through the recognized communities. For any pair of accounts u_i and u_j, the more communities they both belong to, the larger value $l_{i,j}$ will be.

Thus, we can use L to compute a new set of labels E_{t+1} based on E_t and E_0 ($t = 0, 1, \cdots$). A user's new label is obtained by combining his initial label and the influence of his community friends.

3   Conclusion and Future Work

In this work, we propose to detect spammers in Sina Weibo by utilizing the influence of communities. We propose a method to measure user similarities in terms of communities and, based on which, design two simple but effective algorithms for discovering spammers and spammer communities. According to the results over a real dataset, our method can outperform the state-of-the-art ones. In the future, we will study how to extract the features of a community and detect spammers in the community level.

This work is supported by the Fundamental Research Funds for the Central Universities, the Research Funds of Renmin University of China (No. 14XNLQ06). The authors would also like to thank Sa Shi-Xuan Research Center of Big Data Management and Analytics for its supports.

References

1. Ghosh, S., Viswanath, B., Kooti, F., Sharma, N.K., Korlam, G., Benevenuto, F., Ganguly, N., Gummadi, K.P.: Understanding and combating link farming in the twitter social network. In: WWW 2012, pp. 61–70. ACM, New York (2012)

2. guancha.cn. Detecting spammers by an automatic tool (2013). http://www.guancha.cn/society/2013_11_14_185757.shtml (in Chinese)
3. Hu, X., Tang, J., Zhang, Y., Liu, H.: Social spammer detection in microblogging. In: IJCAI (2013)
4. Page, L., Brin, S., Motwani, R., Winograd, T.: The pagerank citation ranking: Bringing order to the web. Technical Report 1999-66, Stanford InfoLab (November 1999)
5. Yang, J., Leskovec, J.: Overlapping community detection at scale: a nonnegative matrix factorization approach. In: WSDM 2013, pp. 587–596 (2013)
6. Zhou, D., Huang, J., Schölkopf, B.: Learning with hypergraphs: clustering, classification, and embedding. In: NIPS, pp. 1601–1608 (2006)
7. Zhu, Y., Wang, X., Zhong, E., Liu, N., Li, H., Yang, Q.: Discovering spammers in social networks. In: AAAI Conference on Artificial Intelligence (2012)

LDM: A DTD Schema Mapping Language Based on Logic Patterns

Xuhui Li[1,2]([✉]), Yijun Guan[2], Mengchi Liu[2], Rui Cai[2], Ming Zhong[2], and Tieyun Qian[2]

[1] State Key Lab. of Software Engineering, Computer School, Wuhan University, Wuhan, China
lixuhui@whu.edu.cn
[2] School of Information Management, Wuhan University, Wuhan, China
mengchi@sklse.org, guanyijun111@163.com, {cairui,clock,qty}@whu.edu.cn

Abstract. Existing XML schema mapping languages often follow the approach for the relational schema mappings to conjunctively combine the correspondences between tuples of elements. However, this approach is inconvenient to present the complex mappings which often involve the hierarchical and heterogeneous data. In this paper, we propose a new pattern-based DTD schema mapping language named LDM. The views in LDM adopt the logical as well as the structural operators of compose the patterns for matching the data elements, so as to flexibly specify the structural relationships among the elements and uniformly organize them in a logical structure. Under the support of a deductive restructuring mechanism of the logical structure, the data elements can be coherently reorganized and thus the mappings can be specified declaratively. We describe the design issues of LDM and demonstrate its features with certain typical mappings on integrating DTD schemas.

1 Introduction

Existing studies on DTD schema mappings often follow the conventional approaches in relational schema mapping, specifying the mapping as the correspondence between conjunctive tuples of the source and the target schemas. These studies adopt structural patterns, either path-like ones [1–4] or tree-like ones [5–7] to parse the document structure and combine the matching elements conjunctively. However, this tuple-oriented schema mapping is insufficient to specify certain complex XML mapping requests. The difficulties often lie in the following aspects. Firstly, the subordination indicated in the nested lists can hardly be specified using simple structural patterns. Secondly, to differentiate the elements by their values or by their references is often a dilemma. Thirdly,

---

This research is supported by the NSF of China under the contracts No.61272110, No.61272275, No.61202100 and No.61202036, the Open Fund. of Shanghai Key Lab. of Intelligent Info. Processing under contract No.IIPL-2011-002, and the China Postdoctoral Science Foundation under contract No. 2014M562070.

© Springer International Publishing Switzerland 2015
J. Li and Y. Sun (Eds.): WAIM 2015, LNCS 9098, pp. 559–562, 2015.
DOI: 10.1007/978-3-319-21042-1_62

it often needs redundant mappings to deal with the heterogeneous elements for similar integration purposes.

In this paper, we propose a new DTD mapping language named LDM (Logic DTD Mapping) to overcome the insufficiencies. LDM is a pattern-based DTD schema mapping language whose patterns employ both the structural and the logical operators to flexibly present the relationships among the elements to be mapped. In comparison with existing studies, our work on LDM makes the following contributions. a) We propose the structural pattern specific to the DTD schema and seamlessly integrate the structural patterns and the logical patterns into the expressive pattern-based views named LDV (standing for Logical DTD View) to present the mappings on DTDs. b) We adopt a new structure of the variables named the matching term to coherently organize the data elements through the patterns, and deploy a simple but expressive restructuring mechanism to facilitate presenting the data transformation in the mapping.

2   LDM: Logic DTD Mapping Language

The LDM language is a mapping language using the Logic DTD Views (LDV), a pattern-based view for specifying the logical and the structural relationships among the DTD document elements, to denote the correspondence between the data elements in the source and target DTD documents. The syntax of LDV and LDM is specified in Fig.1.

$$
\begin{aligned}
e &:= /l \mid //l \mid /l(x) \mid //l(x) \mid \#x \\
v_e &:= e \mid (e, c) \mid e[v] \mid \textbf{\textit{null}} \\
v_s &:= v_e \mid v_s;v_s \mid <v_s> \mid v_s|v_s \mid (v_s{:}v) \\
v &:= v_s \mid v{*}v \mid (v{*}v, c) \mid \{v\} \mid (\{v\},c)
\end{aligned}
$$

$$
\begin{aligned}
m &:= \textbf{from } sv \textbf{ to } (d) \; v \\
sv &:= (d)v \mid sv \mid sv \mid (sv * sv, c)
\end{aligned}
$$

Fig. 1. The Syntax of Logic DTD Views and Logic DTD Mapping Language

We use l, x, c, d to denote the labels, the variables, the constraints and the DTD schemas in a LDV. A LDV v can be an element view v_e, a structure view v_s or a composite view combining the sub views with the operators "*" and "{}" and an optional constraint. An element view v_e is to match an XML element with an element pattern e and indicate the element content of interest for the mapping. Here e contains either the location prefix "/" to match a child element with the label l or the prefix "//" to match a descendant element with the label l, and it uses the optional variable x to represent the element content for further use in the mappings or the constraints. A structure view introduces three structural composite operators, that is, the adjacent operator ";", the sequence set operator "<>" and the exclusive option operator "|", for combining the element views. The three kinds of views essentially specify the structure of common DTD schemas, which enables the subtle differentiation of data elements. A structure view can also be extended wtih another view in the form of $(v_s:v)$, indicating that the fragment matching the view v_s would be further matched with the view v. A composite view can be a conjunctive view $v_1 * v_2$ or a set view $\{v\}$. The conjunctive view $v_1 * v_2$ indicates that the data respectively matching v_1 and v_2 be associated conjunctively. The set view $\{v\}$ indicates that the data matching the view v be gathered as a set of matching results of the view v. The elements in the set are required to be different from each other and the set should cover all the valid elements.

A LDM mapping is specified as a **from-to** statement. The **from** and the **to** clause contains the DTD schemas attached with the LDVs. Especially, in the **from** clause the DTD schemas can be combined conjunctively or disjunctively with the operators "*" and "|", which works like the composite views.

LDM introduces the **matching term**, an expression composed of the variables and the composite connectives "*", "|" and "{}", to indicate the logical structure of the composite binding of a LDV. For a LDV v, the matching term $m(v)$ can be derived from v by maintaining the variables and the conjunctive, disjunctive and set operators and reducing the structural operators to the conjunctive, the disjunctive and the set operators, e.g., , $mt(v_1;v_2) = mt(v_1)*mt(v_2)$.

LDM adopts a rewriting system for the matching terms, as shown in Fig.2, to enable the elements in the source schemas be the deductively restructured to the ones in the target schemas, which provides a of the source a more flexible way than the conventional variable tuples to present the correspondences between the elements in the mappings. This idea comes from our previous study on XML query and for the further theoretical properties please refer to [9] .

$p * /\!\!| \ p' \to p' * /\!\!| \ p$ *(conj/option-commutation)*
$p \to p^* p$ *(conj-duplication)* $p * p' \to p \backslash p'$ *(conj-reduction)*
$p * \epsilon \leftrightarrow p$ *(empty-red-ext)* $p \leftrightarrow p \mid p'$ *(option-red-ext)*[†]
$p * (p_1 \mid p_2) \to p^* p_1 \mid p^* p_2$ *(option-distribution)*
$p * \{p'\} \to \{p * p'\}$ *(set-distribution)*
$\{ \{ p \} \} \to \{ p \}$ *(set-flattening)*
$\{p^* p'\} \to \{f(p)\%^*\{p^* p'\}\}$ *(set-folding)*[‡]
[†]p and p' are disjunctive in original term.
[‡]*Grouping the set into a collection by the distinct values of f(p).*

Fig. 2. Restructuring rules for matching terms

For example, the branches of a company need to be gathered from the schema *s.dtd* to the schema *t.dtd*. *s.dtd* contains the *branches* element consisting of a flat list of the company branches which contains the sub-elements *bno*, *bname*, and *subbno* and actually forms a recursive hierarchy indicated by *bno* and *subbno*. In *t.dtd*, the *branch* elements are explicitly hierarchically organized with the *sub-branches* sub-elements. The mapping is listed below.

```
from (s.dtd) /branches[</bno(bn);/bname(bm);(</subbno(sn)>|null)>]
to (t.dtd) {//branch[(/bno(bn1);/bname(bm1) *
                     (/sub-branches[</branch[/bno(sn1)]>]|null),
                     bn1=(bn\bm)% and bm1=(bm\bn)% and
                     %({sn\(bm*bn)})={sn1})]}
```

3   Conclusion

In this paper, we introduced a pattern-based mapping language named LDM to present the DTD schema mapping as the correspondence between different but compatible structures of the variables composed of the logical operators. LDM adopts the structural as well as the logical operators to compose the sub-patterns and the associating semantic constraints, and is convenient to present complex mapping requests on the hierarchical and heterogeneous data. Due to the space restriction, we omitted the details of the design issues, the semantics and typical examples of the language in this article, and readers can refer to the related reports[9,10].

The core of the mapping language is being implemented. We are testing various constraints and designing more expressive pragmatic paradigms to present typical mapping requests which can be processed efficiently, and we would further study the theoretical issues of constraints so as to find a formal definition of the efficient constraints and its common processing methods. Additionally, we would study query rewriting of LDM programs for processing the mappings, especially for GAV-style XML integration.

References

1. Aguilera, V., Cluet, S., Milo, T., Veltri, P., et al.: Views in a large-scale XML repository. VLDB J. **11**(3), 238–255 (2002)
2. Fan, W., Garofalakis, M.N., Xiong, M., Jia, X.: Composable XML integration grammars. In: CIKM 2004, pp. 2–11 (2004)
3. Poggi, A., Abiteboul, S.: XML data integration with identification. In: Bierman, G., Koch, C. (eds.) DBPL 2005. LNCS, vol. 3774, pp. 106–121. Springer, Heidelberg (2005)
4. Fan, W., Bohannon, P.: Information preserving XML schema embedding. ACM TODS **33**(1) (2008)
5. Arenas, M., Libkin, L.: XML data exchange: consistency and query answering. Journal of ACM **55**(2) (2008)
6. Amano, S., Libikin, L., Murlak, F.: XML Schema Mappings. In: PODS 2009, pp. 33–42 (2009)
7. Bonifati, A., Chang, E., Ho, T., Lakshmanan, L., et al.: Schema mapping and query translation in heterogeneous P2P XML databases. VLDB J. **19**, 231–256 (2010)
8. Li, X., Zhu, S., Liu, M., Zhong, M.: Presenting XML schema mapping with conjunctive-disjunctive views. In: Wang, J., Xiong, H., Ishikawa, Y., Xu, J., Zhou, J. (eds.) WAIM 2013. LNCS, vol. 7923, pp. 105–110. Springer, Heidelberg (2013)
9. Li, X., Liu, M., et-al.: XTQ: A Declarative Functional XML Query Language. CoRR abs/1406.1224 (2014)
10. Li, X., Liu, M., et-al.: LDM: A DTD Schema Mapping Language Based on Logic Patterns. Tech. Report (2015). http://www.sklse.org:8080

Effective Sampling of Points of Interests on Maps Based on Road Networks

Ziting Zhou[1], Pengpeng Zhao[1（⊠）], Victor S. Sheng[2], Jiajie Xu[1], Zhixu Li[1], Jian Wu[1], and Zhiming Cui[1]

[1] School of Computer Science and Technology, Soochow University, Suzhou 215006, People's Republic of China
carol_zzt@163.com, {ppzhao,xujj,zhixuli,jianwu,szzmcui}@suda.edu.cn
[2] Computer Science Department, University of Central Arkansas, Conway, USA
ssheng@uca.edu

Abstract. With the rapid development of location-based services, it is particularly important for start-up marketing research to explore and characterize points of interests (PoIs) such as restaurants and hotels on maps. However, due to the lack of a direct access to PoI databases, we have to rely on existing APIs to query PoIs within the region and calculate the PoI statistics. Unfortunately, public APIs generally impose a limit on the maximum number of queries. Therefore, we propose effective and efficient sampling methods based on a road network to sample PoIs on maps and give unbiased estimators to calculate the PoI statistics. Experimental results show that compared with the state-of-the-art methods, our sampling methods improve the efficiency of aggregate statistical estimation.

Keywords: Sampling · Aggregate Estimation · Road Networks

1 Introduction

On map services such as Google Maps, we can obtain the aggregate statistics (e.g., sum, average and distribution) of PoIs (e.g., restaurants) to get more benefit. For this work, we need to acquire all PoIs within a region. However, the fact is most map service suppliers wouldn't like to provide the entire PoI databases. In addition, the vast majority of public APIs restrict the number of queries per day for each user, and the largest number of results returned. Therefore, we have to estimate aggregate statistics by sampling methods. The existing sampling methods have been proved to sample PoIs with biases which require a large number of queries to eliminate the biases. Besides, Wang et al. [5] proposed new methods to remove sampling bias which randomly samples some fully accessible regions from an area of interest and collects PoIs within sampled regions.

On the whole, previous researches only stay in the level of PoI information. Since it is to sample PoIs on the map, we can use information about the related road networks to further improve the efficiency of aggregate statistical

J. Li and Y. Sun (Eds.): WAIM 2015, LNCS 9098, pp. 563–566, 2015.
DOI: 10.1007/978-3-319-21042-1_63

estimation. According to observations, in general, the areas with intensive road networks such as urban districts, have dense distributions of PoIs. So we can conduct preprocessing by utilizing the information about road networks. On the one hand, the road network information can be easily obtained, and don't occupy the online query time. On the other hand, the road network information is relatively static, rather than obviously dynamic. In this paper, we propose sampling methods that make use of the information of road networks to preprocess a large area of interest to estimate aggregate statistics.

Related Work. Several papers have dealt with the problem of crawling and downloading information present in search engine [1], unstructured Hidden Web [2] and structured Hidden Web [3]. Some of these methods depend on knowing the exact number of results for a query, which is not available in our setting. For estimating structured Hidden Web, the sampling methods proposed in [4] have a limitation inputs which must be specified as categorical input and need a large number of queries to reduce biases. So these methods cannot be directly applied into the application for sampling PoIs on maps.

2   Problem Statement

If A is the area of interest, P is the set of PoIs within the region A. Sum aggregate is $f_s(P) = \sum_{p \in P} f(p)$, where $f(p)$ is the target function of a PoI p . Average aggregate is $f_a(P) = \frac{1}{|P|} \sum_{p \in P} f(p)$, where $|P|$ is the number of all PoIs in P. In addition, let $\theta = (\theta_1, \ldots, \theta_J)$ be the distribution of set of PoIs which has several labels as $\{l_1, \ldots, l_J\}$, where θ_j is the fraction of PoIs with label l_j . If $L(p)$ is the label of a PoI p to specify a certain property, so $\theta_j = \frac{1}{|P|} \sum_{p \in P} I(L(p) = l_j), 1 \leq j \leq J$, where $I(L(p) = l_m)$ is the indicator function that equals one when predicate $L(p) = l_m$ is true, and zero otherwise.

3   Sampling Methods to Estimate Aggregate Statistics

RRZI_URSP. From the initial region A, the random region zoom-in (RRZI) uniformly divides the current region into two non-overlapping sub-regions, and then randomly selects a non-empty sub-region as the next region with same probability to zoom in and query. Repeat this procedure recursively until a fully accessible region which includes PoIs less than k is observed. To correct sampling bias, we add a counter Γ to record the probability of sampling a sub-region.

Because the information of road networks is local,we can divide a region further by utilizing its road network to decrease the number of queries. However, public map APIs might impose a limit on the size of the input region.To solve this problem, we can select a small sub-region from a large region as the input of RRZI. Wang et al. [5] introduces a uniform region sampling (URS). It divides a large region into 2^L small sub-regions, where L represents the number of iterative divisions. Let B_L denote the 2^L regions set, and URS repeats to sample regions from B_L at random until a non-empty region is observed.

We extend URS to utilize the intersection points of road networks, called URSP. We further integrate URSP into RRZI to obtain a new sampling method, called RRZI_URSP. URSP randomly selects a non-empty region b from B_L, and then explores the intersection points within b. The region b whose number of intersection points is greater than the threshold h needs to further divide into two non-overlapping sub-regions. URSP repeats this procedure recursively until a region t whose number of intersection points is not more than h. Then RRZI_URSP utilizes RRZI to sample a fully accessible region from the region t.

RRZIC_URSP. Different from RRZI, RRZIC requires map services can return a query including the number of PoIs within the input search region. Compared to RRZI_URSP, RRZIC_URSP has same preprocessing URSP and different RRZIC. Initially we set the URSP output region Q as the RRZIC input region A. If the number of PoIs within region Q is greater than the threshold k, the region Q is divided into two non-overlapping sub-regions Q_0 and Q_1 uniformly. z, z_0 and z_1 present the number of PoIs within the region Q, Q_0 and Q_1 respectively. Then RRZIC selects Q_0 with a probability z_0/z to further explore. RRZIC repeats this procedure until a fully accessible region is observed.

RRZI_URSR/RRZIC_URSR. Analogously, we extend URS to utilize the edges of road networks, called URSR. Then, we further integrate URSR into RRZI/RRZIC to obtain a new sampling method, called RRZI_URSR/RRZIC_URSR.

4   Experiments

In our experiments, we use the real-world road networks and PoIs of Beijing as datasets which includes three types of information (vertices, edges and geos).

Estimating $n(A)$. We evaluate the performance of the RRZI group methods for estimating $n(A)$, the total number of PoIs within the area A. $NRMSE(\hat{n}(A)) = \sqrt{E[(\hat{n}(A) - n(A))^2]}/n(A)$ is a metric to measure the relative error of an estimate $\hat{n}(A)$ with respect to its true value $n(A)$. In Fig.1, we can find the NRMSEs of both RRZI_URSP and RRZI_URSR are much less than that of RRZI_URS, about decreasing 20% when $L < 20$. When L is very large, the effect of preprocessing is not obvious because the numbers of the intersection points or edges of roads within sub-regions by URS are relative small.

Estimating Average and Distribution Statistics. Fig.2 shows the results for estimating the average cost and distribution aggregate of restaurant-type PoIs for different methods. We can clearly see that RRZI_URSP and RRZI_URSR are more accurate than RRZI_URS, and RRZIC_URSP and RRZIC_URSR are more accurate than RRZIC_URS. From this figure, we can also see that the RRZIC group methods perform better than their corresponding methods in the RRZI group.

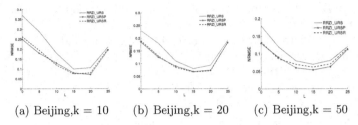

(a) Beijing,k = 10   (b) Beijing,k = 20   (c) Beijing,k = 50

Fig. 1. NRMSEs of estimates of $n(A)$

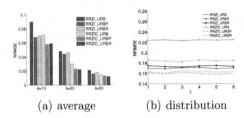

(a) average   (b) distribution

Fig. 2. NRMSEs of estimates of the average cost and distribution aggregate

5   Conclusion

In this paper, we proposed four effective sampling methods, which utilize the information of road networks to obtain unbiased estimations of PoI aggregate statistics, and conducted several experiments to investigate the effectiveness of our proposed methods. Thus, it is visible that offline road network information has predictable positive effect on sampling PoIs on maps.

Acknowledgment. This work was partially supported by Chinese NSFC project (61170020, 61402311, 61440053), and the US National Science Foundation (IIS-1115417).

References

1. Bar-Yossef, Z., Gurevich, M.: Mining search engine query logs via suggestion sampling. Proceedings of the VLDB Endowment **1**(1), 54–65 (2008)
2. Jin, X., Zhang, N., Das, G.: Attribute domain discovery for hidden web databases. In: Proceedings of the 2011 ACM SIGMOD International Conference on Management of data, pp. 553–564. ACM (2011)
3. Sheng, C., Zhang, N., Tao, Y., Jin, X.: Optimal algorithms for crawling a hidden database in the web. Proceedings of the VLDB Endowment **5**(11), 1112–1123 (2012)
4. Wang, F., Agrawal, G.: Effective and efficient sampling methods for deep web aggregation queries. In: Proceedings of the 14th International Conference on Extending Database Technology, pp. 425–436. ACM (2011)
5. Wang, P., He, W., Liu, X.: An efficient sampling method for characterizing points of interests on maps. In: 2014 IEEE 30th International Conference on Data Engineering (ICDE), pp. 1012–1023. IEEE (2014)

Associated Index for Big Structured and Unstructured Data

Chunying Zhu[1], Qingzhong Li[1(✉)], Lanju Kong[1], Xiangwei Wang[2],
and Xiaoguang Hong[1]

[1] School of Computer Science and Technology, Shandong University, Jinan, China
cyzhu1990@gmail.com, {lqz,klj,hxg}@sdu.edu.cn
[2] State Grid SHANDONG Electric Power Company, Jinan, China
shandongwangxw@163.com

Abstract. In big data epoch, one of the major challenges is the large volume of mixed structured and unstructured data. Because of different form, structured and unstructured data are often considered apart from each other. However, they may speak about the same entities of the world. If a query involves both structured data and its unstructured counterpart, it is inefficient to execute it separately. The paper presents a novel index structure tailored towards associations between structured and unstructured data, based on entity co-occurrences. It is also a semantic index represented as RDF graphs which describes the semantic relationships among entities. Experiments show that the associated index can not only provide apposite information but also execute queries efficiently.

1   Introduction

Accessing both structured and unstructured data in an integrated fashion has become an important research topic that also gained commercial interest. As stated by Mudunuri[1], in order to achieve useful results, researchers require methods that consolidate, store and query combinations of structured and unstructured data sets efficiently and effectively. The key to speeding up data access is to create a complete index mechanism. There are many sophisticated index structure, such as B+ tree, Hash and their distributed variants[2] towards structured data, inverted index and semantics index[3] used to large scale text. However, index for the combinations of structured and unstructured data is still lacking.

In this paper, we propose an associated index model which can be used to efficient conduct hybrid semantic query leveraging information from structured and unstructured sources as described in [4]. Associated index, is built based on entities co-occurrence (e.g., a person may be occurred in both a document and a database record) between structured and unstructured data. To support semantic query, we also establish the relationships among entities with knowledge of ontology, and define two custom relationships between an entity and its structured and unstructured data source. The index is stored as RDF graphs. Moreover, the index itself can be queried potentially for knowledge discovery and decision making.

© Springer International Publishing Switzerland 2015
J. Li and Y. Sun (Eds.): WAIM 2015, LNCS 9098, pp. 567–570, 2015.
DOI: 10.1007/978-3-319-21042-1_64

2   Associated Index Model

The associated index model is a hybrid index schema over database and documents stored in DFS. As shown in Fig.1. , the model has two index layers: associated index layer and secondary index layer. The associated index layer is represented as RDF graphs, which describe the semantic relationships among entities that extracted from documents, and two custom relationships "InDoc" and "InRow" between an entity and its corresponding structured and unstructured resource identifiers respectively. The resource identifiers namely rowID and termID point to secondary index. The secondary index layer is consists of two separate indexes. One is inverted index for document; the other is index for database, such as B+ tree, hash .etc. When searching for information about a specific entity, the associated index will first be applied to get the resource identifiers through the relationships InDoc and InRow with the entity. Then, use the resource identifiers to search secondary indexes separately to get documents and table records that contain information about this entity.

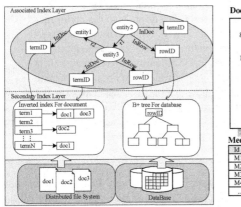

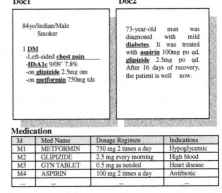

Fig. 1. Associated Index Model   **Fig. 2.** Medical Structured and Unstructured Data

2.1   Associated Index Structure

The associated index, as we call it, is to associate unstructured and structured data based on entity co-occurrences, e.g., entity "Aspirin" is not only contained in document Doc2, but also described by a record "M4" in table Medication. To express the complex relationships explicitly, we store the associated index as RDF graphs. The index graph consists of two types of nodes, namely entity node and resource node. The entity nodes are represented as flat circles, and constructed from the entities mentioned in documents. The resource nodes are represented as rectangles, and derived from structured and unstructured resource identifiers. The edges in the graph represent the semantic relationships specified in knowledge bases between the connecting entity nodes (e.g., disease Diabetes Mellitus has the "treat" relationship to medicine Aspirin). In addition, our custom relationships "InRow" and "InDoc" also represented as edges to denote that

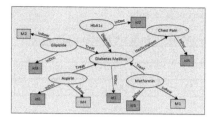

Fig. 3. Index Graph for Data in Fig. 2

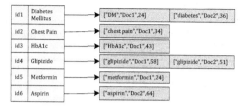

Fig. 4. Entity Inverted Index for Doc1, Doc2

an entity has a related resource in some document or database record. Fig.3. is an associated index graph for documents and structured table in Fig.2.

2.2 Secondary Index Structure

Document indexing is conducted with entity inverted index, which takes into account the extract named entities (called mentions) from documents instead of terms. The inverted list for an entity entry contains one index item per occurrence of mentions that refer to the entity. The index item is represented as *<mention, docID, position>*. As an example, the entity inverted index for the documents Doc1and Doc2 written in natural language in Fig.2 is shown in Fig.4. As for structured data, every record in database can be regard as an entity description. We just use the database's built-in indexing mechanism such as B+ tree to create index on primary key or pseudo ROWID for these records.

3   Experiments

Data Collection. The structured dataset is collected from an open access medicine database (www.accessmedicine.mhmedical.com), which contains about 8,000 drug records. For each drug name, we executed a Web search on PubMed[12], and collected the top five medical literatures as unstructured dataset.

Experimental Design. (1)Index storage costs. We tested against three differently sized subsets of dataset collection: 350MB (PM1), 700MB (PM2) and 1G(PM3). The results of index creation experiments are shown in Table 1. Both the index size and creation time grow linearly with size of dataset. Though it is time-consuming to create the index, it can be an offline job. (2) Query performance. We manually constructed five queries according the complexity (Q1<Q2<Q3<Q4<Q5), the queries' execution time against dataset PM1, PM2, PM3 is shown in Fig.5. The result shows that all queries can be executed efficiently in seconds as dataset size grows. (3) Accuracy. In order to prove the advantage of the associated index, we also compare it with separate index mechanism. The separate index mechanism employs two separate indexes: inverted index for documents and B+ tree index on the drug database. Then we construct 10 queries and execute them against the two different index mechanisms. The comparison results in Fig.6 demonstrate that the association index can help query find more comprehensive information than separate indexes.

Table 1. Index Creation Experiment Results

Dataset	Number of RDF Triples	Index Size (MB)	Index Creation Time (sec)
PM1	849,593	175	423
PM2	1,400,873	312	875
PM3	2,699,380	604	1540

Fig. 5. Query Execution Times for Dataset   **Fig. 6.** Accuracy Comparison on 10 Queries

4   Conclusions and Future Work

The paper presented an associated index model that tends to index combinations of structured and unstructured data based on entity co-occurrences. Experimental results demonstrated that our proposed index achieved good performance and high accuracy of query processing. However, we are still far from creating big associated index on large-scale structured and unstructured data. In the future, we will study how to construct the index under distributed platform and how to maintain the index.

Acknowledgments. This work is funded by National Natural Science Foundation of China under Grant No.61303085; Natural Science Foundation of Shandong Province of China under Grant No.ZR2013FQ014; Science and Technology Development Plan Project of Shandong Province No. 2014GGX101047, No. 2014GGX101019; The Technology Project of State Grid No.2012GWK515;

References

1. Mudunuri, U.S., Khouja, M., Repetski, S.: Knowledge and Theme Discovery across Very Large Biological Data Sets Using Distributed Queries: A Prototype Combining Unstructured and Structured Data. PLoS One 8(12), e80503 (2013)
2. Wu, S., Jiang, D.W., Ooi, B.C., Wu, K.: Efficient B-tree based indexing for cloud data processing. Proc. of the VLDB Endowment 3(1), 1207–1218 (2010)
3. George, T., Iraklis, V., Kjetil, N.: SemaFor: semantic document indexing using semantic forests. In: CIKM, pp. 1692–1696 (2012)
4. Markus, G., Andreas, R., Helmut, B.: Bridging structured and unstructured data via hybrid semantic search and interactive ontology-enhanced query formulation. Knowl. Inf. Syst. 41(3), 761–792 (2014)
5. PubMed. http://www.ncbi.nlm.nih.gov/pubmed

Demo Papers

Web Knowledge Base Improved OCR Correction for Chinese Business Cards

Xiaoping Wang[✉], Yanghua Xiao, and Wei Wang

Fudan University, Shanghai 201203, China
xiaopingwang@fudan.edu.cn

Abstract. In the field of Optical Character Recognition(OCR), improving the recognition accuracy has been extensively studied in the past decades. In this paper, different from previously published model-based correction methods, Knowledge Base was applied to OCR correcting system from the perspective of linked knowledge. A pipelined method integrating selectivity-aware pre-filtering, text-level and image-level comparison was explored to identify the best candidate with better efficiency and accuracy. For more reliable comparison of company, the weighted coefficients derived from Wikipedia were applied to distinguish the different importance. Moreover, traditional Levenshtein distance was generalized to Image-based Levenshtein measure to better distinguish strings with similar text similarity. The experimental results demonstrated that the proposed system could perform more effectively than the baseline case.

1 Introduction

The performance of Optical Character Recognition (OCR) is often impacted by weak illumination, noise, and skew. Most of the error correction methods address on Natural Language Processing (NLP) or Machine Learning techniques. However, they might not perform well for business card. For instance, it's difficult to determine the correct address among various candidates valid for models without resorting to additional information. Nowadays, there exist diverse Knowledge Bases (KBs) such as web encyclopedias (e.g. Wikipedia[1]), and Point of Interest (POI) database. It is plausible that the "linked" knowledge within the KBs be explored to improve the OCR accuracy further.

The main contributions of this work were: Based on our knowledge, it's the first work for the OCR correction of Chinese business cards from the point of view of KB. A pipelined framework of correction method including selectivity-aware pre-filtering, text-level correction and image-level correction was proposed. Selectivity-aware pre-filtering was tested to exclude irrelevant records quickly while retain the possible candidates. To be flexible for address integrity, a robust similarity measuring method integrating Dynamic Time Warping (DTW) with Jaccard/Levenshtein measure was developed. To make the company comparison more reliable, a strategy of weighting importance based on Wikipedia KB was conducted. To distinguish the candidates with similar text similarity, traditional Levenshtein distance was generalized to Image-based Levenshtein measure.

---

[1] Http://zh.wikipedia.org

© Springer International Publishing Switzerland 2015
J. Li and Y. Sun (Eds.): WAIM 2015, LNCS 9098, pp. 573–576, 2015.
DOI: 10.1007/978-3-319-21042-1_65

2   Methods

We focused on the correction of company-address pair. The biggest challenge was similarity computation to handle the format diversity or importance imbalance. Related key techniques were emphasized below.

2.1   Address/Company Similarity

To compute the similarity reasonably for addresses with different integrity, we decomposed address by its hierarchy at first. Then, DTW and Jaccard/Levenshtein measure were performed. Since different part within company name played different roles in comparison, Part of Speech Tagging was applied for segmentation and feasible weights were then assigned using Inverse Document Frequency(IDF) derived from Wikipedia KB. The details of weight computing is illustrated in Formula.1. An example was shown in Table.1.

$$w_{i,j} = \frac{idf_i \cdot idf_j}{\sum_{x,y=1}^{N} idf_x \cdot idf_y} \tag{1}$$

$S_1 = Similarity(\text{Company1,Company2}), S_2 = Similarity(\text{Company1,Company3})$

Company1: 上海万得信息技术有限公司
Company2: 上海万得技术
Company3: 上海优创信息技术有限公司

Table 1. Comparison example

Method	S_1	S_2
1 DTW(Levenshtein)	0.600	0.800
2 Weighted DTW (Levenshtein)	0.710	0.597

2.2   Text-level/Image-level Correction and Pipelined Framework

Although the incorrectly recognized characters are different from the correct ones in text, they might be similar in image. Hence, a pipelined correction method using both text similarity and image similarity was applied. For text-level comparison, DTW combined with Jaccard/Levenshtein measure was applied. For image-level comparison, 2D Discrete Cosine Transform features and intersecting features were applied. To integrate the image similarity with text comparison, we proposed Image-based Levenshtein measure. The proposed distance between strings a, b is generalized in Formula.2. An example was shown in Table.2.

Our pipelined correction method was illustrated in Fig.1. To leverage speed and accuracy, more latter the step was, more complicated the method would be.

Company1: 方得信息技术有限公司
Company2: 万得信息技术有限公司
Company3: 汉得信息技术有限公司

Table 2. Comparison example

Method	S_1	S_2	Method	S_1	S_2
Text(DTW+Jaccard)	0.833	0.833	Text(DTW+Levenshtein)	0.875	0.875
Image(DTW+Jaccard)	0.917	0.833	Image(DTW+Levenshtein)	1.000	0.875

$$LSH_{a,b}(i,j)= \begin{cases} max(i,j) & if \quad min(i,j) = 0 \\ min \begin{cases} LSH_{a,b}(i-1,j)+1 \\ LSH_{a,b}(i,j-1)+1 \\ LSH_{a,b}(i-1,j-1)+C[Sim(I(a_i),I(b_j)) < T] \end{cases} & otherwise \end{cases}$$

$$(2)$$

Selectivity-aware Pre-filtering	Address Similarity Computing	Company Similarity Computing	Joint Similarity Computing
Step 1: Key part matching	Step 2: DTW + Jaccard (Text-level)	Step 3: DTW + Levenshtein (Text-level)	Step 4: DTW + Levenshtein (Image-level)

Fig. 1. Framework of the pipelined correction method

3 Experiments and Discussions

Baseline: A novel method based on Online Spelling Suggestion[1] which harnesses an internal database containing a huge collection of terms and n-gram words statistics was used as the baseline. To be more effective on Chinese text, we improved the baseline to a strengthened "Did you mean" correction method by using Baidu's engine as the complement of Google's engine.

Schemas: Two schemas were conducted: 1) Schema.1: correction on original OCR outputs. 2) Schema.2: correction on customized OCR outputs (to be comparable with current OCR system, we retained the error rate less than 10% for characters while kept the statistics of right/wrong OCR cases invariant).

Dataset: 204 Shanghai business cards were collected for test. The POI KB derived from Baidu map covers 330 thousands records of Shanghai related points.

Discussions: The OCR accuracy is relatively lower on schema 1. Besides our rigid definition of "Right/Wrong", insufficient trained data was another factor.

Table 3. Statistics in Address-Company joint pair

Status			Wrong Joint Cases Number	Percentage (%)	
				Absolute	Relative
Before correction			167	81.86	—
After correction	Schema.1	Baseline	158	77.45	94.61
		Our method	78	38.24	46.71
	Schema.2	Baseline	137	67.16	82.04
		Our method	11	5.39	6.59

On schema 2, even under less noisy condition comparable with current OCRs, the system could further improve the correction accuracy.

Different from most of previous publications, this work performed data correction from the perspective of linked knowledge and demonstrated its effectiveness in reducing errors further in both noisy and less-noisy condition.

References

1. Bassil, Y., Alwani, M.: Ocr post-processing error correction algorithm using google online spelling suggestion. arXiv preprint arXiv:1204.0191 (2012)

Shortest Path and Word Vector Based Relation Representation and Clustering

Xiaoping Wang$^{(\boxtimes)}$, Yanghua Xiao, and Wei Wang

Fudan University, Shanghai 201203, China
xiaopingwang@fudan.edu.cn

Abstract. Relation representation plays an important role in text understanding. In this paper, different from previously published supervised methods or semi-supervised methods, an new method of relation representation and clustering based on shortest path and word vector was proposed. By accumulating the word vector along the shortest path within dependency tree, we can not only obtain the essential representation of the relation, but also can map the relation into semantic space simultaneously. Therefore, reliable distance between any two relations could be measured. Moreover, further applications such as relations clustering can be performed conveniently by direct analysis on the collection of vectors.

1 Introduction

Upon the era of big data, it's urgent to extract useful information automatically from vast amount of data. As for the text data, relation extraction crucial for natural language understanding have been paid much attention. Primarily, the approaches of relation extraction can fall into two categories: supervised methods and semi-supervised methods. The formers require the relations comparison to be exact string matching. Therefore, it's sensitive to diverse word forms such as polysemy or synonyms. While for the supervised methods, relations are extracted by manually defined features which need to be trained and classified. Therefore, it's difficult to measure their distance directly.

Aim to handle the defects discussed above, this paper proposed a new method for relation representation by integrating dependency trees and Word2vec method[1]. It primarily advantaged in that relation was directly represented by accumulating words vector along the path so that further applications such as relations clustering could be performed conveniently and flexibly by any data mining techniques.

2 Methods

2.1 Semantic Relation Representation

To characterize the essential relation between the queried entities, we use Stanford typed dependencies[1] for grammatical analysis. Then, by viewing the dependencies tree as a graph, the relation between any two entities can be obtained by applying shortest path searching such as Dijkstra algorithm[2].

---

[1] Http://nlp.stanford.edu

© Springer International Publishing Switzerland 2015
J. Li and Y. Sun (Eds.): WAIM 2015, LNCS 9098, pp. 577–580, 2015.
DOI: 10.1007/978-3-319-21042-1_66

In Natural Language Processing related applications, a fundamental requirement is to reliably compute the semantic distance between given words or phrases. Typical previously publications apply the tree distance or Information Content by the aid of hierarchy concepts such as WordNet, or measuring their similarity in semantic space such as TF-IDF, Explicit Semantic Analysis (ESA)[3]. While the former method is usually impacted by corpus insufficiency and diversity in words forms, ESA describes the word based on huge-dimension Wikipedia[2] concepts and brings unbearable computation. Compared with ESA, Word2vec provides an efficient implementation of learning high-quality vector representations of words with Skip-gram model. Due to its simple model architectures and evidently lower computational complexity compared with other methods, it can compute accurate word vectors from huge dataset with billions of words. More attractively, the word offset technique enables Word2vec to exhibit linear structure in word representations. Therefore, it can identify linguistic regularities and perform simple algebraic operations on the word vectors which inspires our relation representation by accumulating the vectors along the shortest path.

2.2   Relation Clustering

For deeper text understanding and knowledge acquisition, further techniques especially relations clustering should be addressed. On the basis of our relation representation, any data mining techniques can be applied on the collection of relations to implement further applications such as clustering. Here, the method of K-means clustering was employed.

3   Experiments

3.1   System Implementation and Data Preparation

The system was implemented by using Java7.0 and developed in the environment of Eclipse 4.2.1. And Prefuse[3] was employed for graphical presentation.

To deal with the problem of data sparsity, We used Wikipedia corpus and extracted 43335 thousand sentences in total. Also, we obtain a training corpus derived from 3.06 million Wikipedia normal pages for the Word2vec training. The dimension of trained vector was set to 500 in our experiments.

3.2   Relation Clustering and Discussion

We manually selected 10 relationships in Freebase as follows:

- /location/country/capital
- /people/person/nationality
- /film/director/film

---

[2] Http://zh.wikipedia.org
[3] Http://prefuse.org

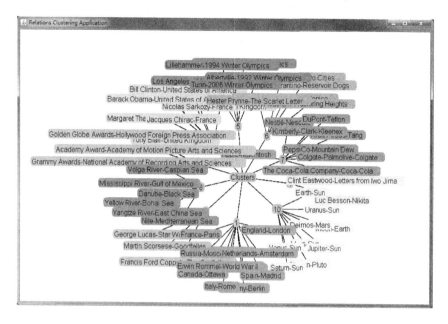

Fig. 1. Relations Clustering

- /olympics/olympic_host_city/olympics_hosted
- /book/book_character/appears_in_book
- /business/company_brand_relationship
- /geography/river/mouth
- /award/award/presented_by
- /military/military_person/participated_in_conflicts
- /astronomy/orbital_relationship/orbits

For each relation, we randomly chose 1~10 entity pairs and obtained 59 pairs in total. 13.4 thousand qualified sentences were hit within the sentences corpus. Subsequently, relations clustering was performed by using K-means method based on Cosine measure. Experimental results were shown in Fig.1(the class number was set to 10 and pairs in same color belong to identical cluster). Compared with the benchmark(Freebase relations), our accuracy reaches 93.2%.

4   Conclusions

In this paper, an new method based on shortest path and word vector was put forward for better reliability in relation representation and clustering. It primarily advantaged in: 1)Convenient and flexible processing on vectors. 2)Reliable relations comparison based on semantic representation. 3)Evidently lightweight compared with millions of concepts-based methods like ESA.

References

1. Mikolov, T., Sutskever, I., Chen, K., Corrado, G.S., Dean, J.: Distributed representations of words and phrases and their compositionality. In: Advances in Neural
   Information Processing Systems 26: 27th Annual Conference on Neural Information
   Processing Systems 2013. Proceedings of a meeting held December 5-8, Lake Tahoe,
   Nevada, United States, pp. 3111–3119 (2013)
2. Cormen, T.H., Leiserson, C.E., Rivest, R.L., Stein, C.: Introduction to Algorithms,
   3rd. edn. MIT Press (2009)
3. Gabrilovich, E., Markovitch, S.: Computing semantic relatedness using wikipedia-
   based explicit semantic analysis. IJCAI **7**, 1606–1611 (2007)

CrowdSR: A Crowd Enabled System for Semantic Recovering of Web Tables

Huaxi Liu, Ning Wang$^{(\boxtimes)}$, and Xiangran Ren

School of Computer and Information Technology, Beijing Jiaotong University, Beijing, China
{13120407,nwang,13125181}@bjtu.edu.cn

Abstract. Without knowing any semantic of tables on web, it's very difficult for web search to take advantage of those high quality sources of relational information.We present CrowdSR, a system that enables semantic recovering of web tables by crowdsourcing. To minimize the number of tuples posed to the crowd, CrowdSR selects a small number of representative tuples by clustering based on novel integrative distance. An evaluation mechanism is also implemented on Answer Credibility in order to recommend related tasks for workers and decide the final answers for each task more accurately.

1 Introduction

Structured information of tables including the table schema and entity column could be used to find related tables and learn binary relationships between multiple columns[1]. But most of the web tables are lack of header rows[2]. State-of-the-art technology used to recover semantics of web table is not yet able to provide satisfactory accuracy[3][4]. We develop a hybrid machine-crowdsourcing framework that leverages human intelligence to settle the problem. CrowdSR has the following unique features. (1) CrowdSR presents the worker a small number of representative tuples in the table by clustering based on novel integrative distance. (2) CrowdSR prompts workers with candidate lists of concepts by machine-based algorithms. (3) CrowdSR implements an evaluation mechanism on Answer Credibility to recommend the most related tasks for workers and decide the final answers more accurately.

2 CrowdSR Design and Implementation

2.1 CrowdSR Architecture

CrowdSR is implemented in JSP with SQL Server database as back-end. Fig.1 depicts the system architecture.

User Interface is used to interact with users. Basically, DB stores answers collected from the crowd, targeting information and details about each user and task. Task builder receives requests from requesters and build tasks. Crowd Manager constantly receives an updated list of online workers from targeting information in DB. Once a worker logs in, the Task Recommender recommends a list of tasks for him based on

© Springer International Publishing Switzerland 2015
J. Li and Y. Sun (Eds.): WAIM 2015, LNCS 9098, pp. 581–583, 2015.
DOI: 10.1007/978-3-319-21042-1_67

his answer credibility. When he is going to complete a task, the Tuple Cluster clusters similar tuples and the task is enriched with a set of candidate answers from the Semantic Recovery Component. While a task's deadline arrives, received answers are passed to Answers Voter which decides the final answer based on our voting mechanism.

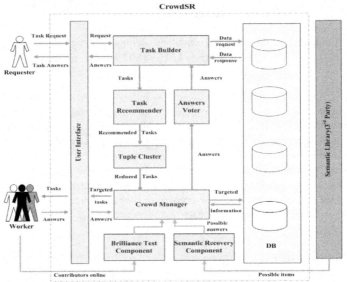

Fig. 1. CrowdSR architecture

2.2   Semantic Recovery Based on Probase

To help workers to annotate column labels and identify subject keys, CrowdSR prompts them with candidate lists of concepts by machine-based algorithms, which leverage Probase, a knowledge base consisting of a huge number of concepts, instances, attributes and relationships extracted from the web[4].

2.3   CAID Algorithm

In CrowdSR we use a clustering algorithm named CAID(Clustering Algorithm based on Integrative Distance), which is an improved K-means algorithm to cluster similar tuples in a web table and prompt workers with representative tuples that are nearest to each of k cluster centers.

A web table is composed of columns with various data types. In order to make the calculation of distance between two tuples effective, integrative distance is proposed to combine Euclidean distance with Jaccard similarity on different attributes. To improve the clustering precision, the distance function is biased towards significant attributes which are typical to the concept of table by putting more weight on them.

2.4   Evaluation Mechanism Based on Answer Credibility

Answer Credibility is modeled to evaluate the probability of which a worker's intuitive answer comes to be the final answer for a task. It's influenced by three factors which are the setting of expertise, brilliance test and practical performance.

When a new worker registers in CrowdSR, he's required to set his expertise and join the brilliance test to evaluate his acquaintance with fields. Then his answer credibility is dynamically changed with the practical performance of doing tasks. We establish a novel evaluation mechanism based on answer credibility, which is used to recommend related tasks for workers and decide the final answers for each task.

3   Demonstration Scenario

We plan to demonstrate the CrowdSR system with the following scenario:

Publish Tasks: Requesters could select tables for publishing tasks and they're also required to assign several fields for each task.

Semantic Annotation: Workers complete tasks via a question-choice game where they are shown representative values of each column, along with a set of candidate table headers and subject column, and are asked to select ones most matched. They could check the alteration of his answer credibility based on his performance.

Brilliance Test: When a new worker registers in CrowdSR, he's invited to participate in the brilliance test to evaluate his acquaintance with fields and could see his field credibility on each field by a histogram.

Acknowledgment. This work is supported by National Natural Science Foundation of China (Grant No. 61370060, 61300071).

References

1. Limaye, G., Sarawagi, S., Chakrabarti, S.: Annotating and searching web tables using entities, types and relationships. In: VLDB (2010)
2. Cafarella, M.J., Halevy, A., Wang, Z.D., Wu, E., Zhang, Y.: Web tables: exploring the power of tables on the web. In: VLDB (2008)
3. Deng, D., Jiang, Y., Li, G., Li, J., Yu, C.: Scalable column concept determination for web tables using large konwledge bases. In: VLDB (2013)
4. Wang, J., Wang, H., Wang, Z., Zhu, K.Q.: understanding tables on the web. In: ER (2012)

A Personalized News Recommendation System Based on Tag Dependency Graph

Pengqiang Ai[1], Yingyuan Xiao[1,2(✉)], Ke Zhu[1], Hongya Wang[3], and Ching-Hsien Hsu[1,4]

[1] Tianjin University of Technology, Tianjin 300384, China
yyxiao@tjut.edu.cn
[2] Tianjin Key Lab of Intelligence Computing and Novel Software Technology, Tianjin 300384, China
[3] Donghua University, Shanghai 201620, China
[4] Chung Hua University, Hsinchu 30012, Taiwan

Abstract. The tags of news articles give readers the most important and relevant information regarding the news articles, which are more useful than a simple bag of keywords extracted from news articles. Moreover, latent dependency among tags can be used to assign tags with different weight. Traditional content-based recommendation engines have largely ignored the latent dependency among tags. To solve this problem, we implemented a prototype system called PRST, which is presented in this paper. PRST builds a tag dependency graph to capture the latent dependency among tags. The demonstration shows that PRST makes news recommendation more effectively.

1 Introduction

With the advance of World Wide Web, more and more people are accustomed to reading news through news websites, like Google News and Yahoo News. However, faced with the floods of news, people may realize that it is difficult to find news articles they are interested in from the news websites. Personalized recommendation technology is the most effective way to solve the problem of information overload [1, 2] so far. Among the news recommendation algorithms, content-based algorithms are widely used [1, 2]. Content-based techniques often involve extracting keywords from news articles and modeling user interest model [3, 4].

However, in quite a lot of scenario, simply representing the news articles by a bag of words is insufficient [3]. The tags of a news article give readers the most important and relevant information regarding this news article, which is more useful than a bag of keywords extracted from news article. Besides, each news article often has a key tag which shows the topic of this news. In most case, whether a reader is interested in a news article depends on the key tag. Therefore, the key tag plays an important role in news recommendation. Assume that a news article is labeled by 3 tags: *Russia, fire disaster, forest*. If a reader doesn't care about *fire disaster*, he/she would not read this news article. Therefore, *fire disaster* is the key tag of this news article for the reader. This also means that different tags should be assigned with different weights in the

© Springer International Publishing Switzerland 2015
J. Li and Y. Sun (Eds.): WAIM 2015, LNCS 9098, pp. 584–586, 2015.
DOI: 10.1007/978-3-319-21042-1_68

same news article. In this paper, we implemented a personalized news recommendation prototype system called PRST, which uses a vector of tags to represent the news content and user profile. Furthermore, we evaluate the importance of each tag by building a weight dependency graph among different tags.

2   Architecture and Key Technology

Figure 1 illustrates the architecture of our system.

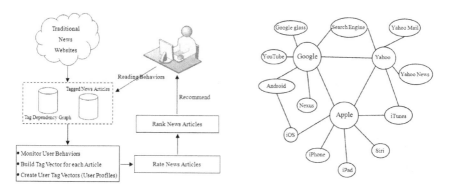

Fig. 1. The architecture of PRST   **Fig. 2.** The dependency among some tags

Some tags always emerge together. For example, *iPhone* and *Apple* often appear together in some news articles, as usually means the latent dependency between the two tags. Figure 2 shows the dependency among some tags.

A tag dependency graph is defined as a triple $G = (V, E, W)$, where V is a set of the tags crawled from news websites, E denotes the set of edges connecting two tags in V, and W represents the set of the weights of these tags in V. Only when two tags come together in the same news article is there an edge in E connecting the two tags.

For any $tag_i \in V$ in $G = (E, V, W)$, it's easy to get $degree(tag_i)$ and $idf(tag_i)$, where $degree(tag_i)$ denotes the degree of tag_i and $idf(tag_i)$ represents the inverse document frequency of tag_i, which is calculated by Equation 1.

$$idf(tag_i) = log \frac{|D|}{|D(tag_i)|} \tag{1}$$

Here, $|D|$ is the total number of news articles and $|D(tag_i)|$ denotes the amount of those news articles containing tag_i. The weight of tag_i is calculated as:

$$\omega(tag_i) = \frac{1}{degree(tag_i)} \cdot idf(tag_i) \tag{2}$$

For each $tag_i \in V$, $degree(tag_i)$ and $idf(tag_i)$ are calculated and stored. Therefore, a news article n_j can be denoted as a tag vector: $Vector(n_j) = \{\omega(tag_1), \omega(tag_2), \cdots\}$. For a given reader u_k, its user profile $P(u_k)$ is also a tag vector calculated by Equation 3.

$$P(u_k) = \frac{1}{|N(u_k)|} \sum_{n_j \in N(u_k)} Vector(n_j) \qquad (3)$$

Here, $N(u_k)$ denotes the collection of those news articles the reader u_k has read in the past. In PRST, we use the cosine similarity to measure the similarity between n_j and $P(u_k)$, which is defined as:

$$sim(P(u_k), n_j) = \frac{P(u_k) \cdot Vector(n_j)}{\|P(u_k)\| \times \|Vector(n_j)\|} \qquad (4)$$

For a given reader u_k, these news articles with the higher $sim(P(u_k), n_j)$ are recommended to u_k.

3   Demonstration

Fig. 3. The user interface of PRST

The interface of PRST is shown in Figure 3. When a user registers as a reader in PRST and logins onto PRST for the first time, PRST will record reading behaviors of the user. Assume that the user only clicked news about *Google* and *Facebook*. When the user logins onto PRST again, other news articles regarding *Google* and *Facebook* are recommended according to the cosine similarity between these news articles and the user.

Acknowledgment. This work is supported by the NSF of China (No. 61170174, 61370205) and Tianjin Training plan of University Innovation Team (No.TD12-5016) .

References

1. Liu, J., Dolan, P., Pedersen, E.R.: Personalized news recommendation based on click behavior. In: Proc. of IUI, Hong Kong, China (2010)
2. Kompan, M., Bieliková, M.: Content-based news recommendation. In: Buccafurri, F., Semeraro, G. (eds.) EC-Web 2010. LNBIP, vol. 61, pp. 61–72. Springer, Heidelberg (2010)
3. Li, L., Wang, D., Li, T., et al.: Scene: a scalable two-stage personalized news recommendation system. In: Proc. of SIGIR, New York, USA (2011)
4. Wei, X., Croft, W.B.: LDA-based document models for ad-hoc retrieval. In: Proc. of SIGIR, New York, USA (2006)

SmartInt: A Demonstration System for the Interaction Between Schema Mapping and Record Matching

Jun Jiang, Zhixu Li[(⊠)], Qiang Yang, Pengpeng Zhao, Guanfeng Liu, and Lei Zhao

School of Computer Science and Technology, Soochow University, Suzhou, China
{20134527003,20144227003}@stu.suda.edu.cn
{zhixuli,ppzhao,gfliu,zhaol}@suda.edu.cn

Abstract. Schema Mapping and Record Matching are two necessary steps in merging multiple data sources with different schemas. While schema mapping unifies the schemas of different data sets, record matching finds out all pairs of linked instances between the data sets. So far, the two processes are well-studied separately, but no effort has been made in the interaction between them. In this demonstration, we present the SmartInt system which performs schema mapping and record matching interactively in merging data sets, and we show that schema mapping and record matching can benefit each other in reaching a better integration performance. We will demonstrate, step by step, how the interaction works to improve the integration quality.

Keywords: Data integration · Scheme mapping · Record matching

1   Introduction

Data Integration combines data from multiple sources to provide users a coherent data store [4]. With intensified diversity of data in various data sources, the integration of data is becoming increasingly difficult. In particular, the diversity of data across different relational databases are mainly in two dimensions: the inconsistency of the schemas and the inconsistency of the attribute values. Thus, in order to merge multiple data sources, one should deal with two problems: (1) Schema Mapping (SM for short): finding out the attributes referring to the same one across different data sources. (2) Record Matching (RM for short): matching the records referring to the same entity across different data sources.

So far, plenty of work has been done on SM [5] and RM [2]. The SM approaches include string similarity based matching [3] and attribute value set based mapping [6], while RM approaches are mainly based on string similarities [1]. However, we noticed that all existing efforts took the two tasks as independent steps in data integration, and no study has been conducted on the interaction between them in integrating data from multiple sources.

In this demonstration, we present the SmartInt system which performs SM and RM interactively in integrating multiple data sources, and we show that SM

© Springer International Publishing Switzerland 2015
J. Li and Y. Sun (Eds.): WAIM 2015, LNCS 9098, pp. 587–589, 2015.
DOI: 10.1007/978-3-319-21042-1_69

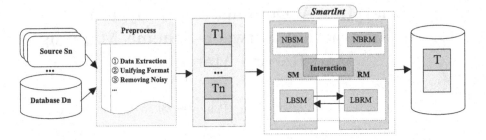

Fig. 1. The System Overview of SmartInt

and RM can benefit each other in reaching a better integration performance, i.e., higher precision and recall for both SM and RM. We will demonstrate how the two tasks be performed interactively to improve the integration quality.

2   System Overview

The system overview of SmartInt is depicted in Fig. 1. There are three modules in SmartInt: the SM module, the RM module, and the interaction module.

(1) SM Module: Initially, we do Name Based SM (NBSM for short), which relies on Edit Distance to detect pairs of attributes in high similarities. In the interaction process, we propose a Link Based SM (LBSM for short) method: if an attribute value v_a of an instance I_a in one table equals to another attribute value v_b of an instance I_b in another table, and assume I_a and I_b refer to the same entity, it is highly possible that v_a and v_b are under the same attributes. If this situation is observed with several pairs of linked entities between two tables, we can determine that the two attributes referring to the same one.

(2) RM Module: Initially, we do Name Based RM (NBRM for short) with the key attribute values. In the interaction process, we propose a Link Based RM (LBRM for short) method: briefly, given two instances I_a in one table and I_b in the other, the more attribute values they share under linked attributes, the more likely they refer to the same entity.

(3) Interaction Module: We do NBSM and NBRM as initial steps, and then perform LBSM and LBRM in turn iteratively until no more links can be detected. While RM gets more linked entities to help find attribute pairs, in return SM gets more linked attribute pairs which also benefits RM in finding entity pairs.

3   Demonstration Scenarios

Our demonstration shows not only the integration result at each step, but also the changes on the integration quality (precision and recall). The data sets used for the demonstration are cellphones collected from several e-commerce websites, such as Tmall, Yesky, PConline and Zol. After some initial cleaning, there are about 15-30 different attributes, and 2000-8000 instances in each database.

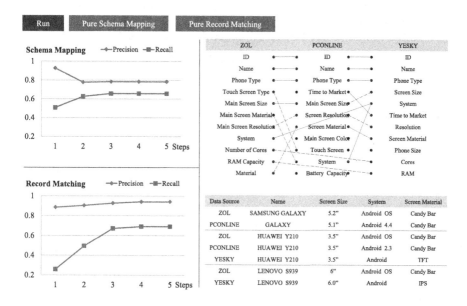

Fig. 2. Demonstration Snapshot of SmartInt

1) Interactive Integration Result. At each step, we show the linkages generated for both schema level & instance level and present the precision & recall of both SM and RM at each step in Fig. 2 .

2) Performance Study. We also compare the precision and recall of our method with state-of-the-art SM and RM methods to show the advantage of SmartInt over the existing methods.

Acknowledgements. This research is partially supported by Natural Science Foundation of China (Grant No. 61472263, 61402313, 61303019).

References

1. Bilenko, M., Mooney, R.J.: Adaptive duplicate detection using learnable string similarity measures. In: Proceedings of the Ninth ACM SIGKDD International Conference on Knowledge Discovery and Data Mining, pp. 39–48. ACM (2003)
2. Dorneles, C.F., Gonçalves, R., dos Santos Mello, R.: Approximate data instance matching: a survey. Knowledge and Information Systems **27**(1), 1–21 (2011)
3. Giunchiglia, F., Yatskevich, M., Shvaiko, P.: Semantic matching: algorithms and implementation. In: Spaccapietra, S., et al. (eds.) Journal on Data Semantics IX. LNCS, vol. 4601, pp. 1–38. Springer, Heidelberg (2007)
4. Han, J., Kamber, M.: Data Mining, Southeast Asia Edition: Concepts and Techniques. Morgan kaufmann (2006)
5. Rahm, E., Bernstein, P.A.: A survey of approaches to automatic schema matching. The VLDB Journal **10**(4), 334–350 (2001)
6. Shvaiko, P., Euzenat, J.: A survey of schema-based matching approaches. In: Spaccapietra, S. (ed.) Journal on Data Semantics IV. LNCS, vol. 3730, pp. 146–171. Springer, Heidelberg (2005)

CDSG: A Community Detection System Based on the Game Theory

Peizhong Yang, Lihua Zhou$^{(\boxtimes)}$, Lizhen Wang, Xuguang Bao, and Zidong Zhang

Department of Computer Science and Engineering,
Yunnan University, Kunming 650091, China
{285342456,272270289}@qq.com, {lhzhou,lzwang}@ynu.edu.cn,
bbaaooxx@163.com

Abstract. Community detection is an important task in social network analysis. A game theory-based community detection system (CDSG) is developed in this demonstration. CDSG uses cooperative and non-cooperative game theory to detect communities. The combination of cooperative and non-cooperative game makes utilities of groups and individuals can be taken into account simultaneously and decreases the computational cost, thus CDSG can detect overlapping communities with high accuracy and efficiency, such that it can effectively help users in analyzing and exploring complex networks.

Keywords: Social network · Community detection · Cooperative game · Non- cooperative game

1 Introduction

Communities are groups (or clusters) of nodes that are densely interconnected, but only sparsely connected with the rest of the network [1]. Detecting communities is a challenging task because there are a great number of nodes and edges in a network and communities are usually overlapped (i.e. nodes simultaneously belong to more than one group). Nowadays, community detection has not been settled satisfactory, although there are many algorithms have been proposed [2].

As a very useful mathematical theory for studying the complex conflict and cooperation amongst rational agents, methods that based on the cooperative and non-cooperative game theory have been used separately to solve community detection problems [3,4] in recent years. However, cooperative game theory-based methods have high efficiency but low accuracy, while non-cooperative game theory-based methods have high accuracy but low efficiency.

In this paper, we develop a game theory-based community detection system (CDSG) that combines cooperative and non-cooperative game theory to detect communities in large networks. Unlike previous works, CDSG plays cooperative and non-cooperative game in two consecutive phases rather than play them separately. CDSG can detect overlapping and non-overlapping communities with high accuracy, it also has high efficiency such that it is applicable for large-scale networks, and it is easy to

© Springer International Publishing Switzerland 2015
J. Li and Y. Sun (Eds.): WAIM 2015, LNCS 9098, pp. 590–592, 2015.
DOI: 10.1007/978-3-319-21042-1_70

use because it does not require a priori knowledge on the number and size of communities that are usually unknown beforehand, thus it is helpful to users analyzing complex networks and improving services.

2   System Overview

The game theory-based community detection system (CDSG) contains five modules that are shown in Figure 1 (a).

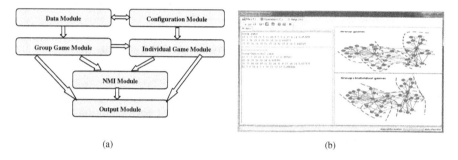

(a)   (b)

Fig. 1. (a) CDSG Overview; (b) CDSG Demonstration

— **Data Module**

The *data module* is used to read real networks, benchmark networks, or purpose-built synthetic networks from files or databases located by users. Each file consists of the number of nodes and a list of edges (source target).

— **Group Game Module**

The *group game module* carries out the cooperative game in which agents are modeled as rational agents trying to achieve and improve group's utilities by cooperating with other agents to form coalitions. Coalitions with fewer agents can merge into larger coalitions as long as the merge operation can contribute to improve the utilities of coalitions merged. The game starts from the nodes as separate coalitions (singletons), coalitions that can result the highest utility increment are iteratively merged into larger coalitions. The game ends when no coalition has an interest in performing a merge operation any further.

— **Individual Game Module**

The *individual game module* carries out the non-cooperative game in which each agent is modeled as a selfish agent, who selects independently coalitions from the result of the group game module to join or leave based on its own utility measurement. Each agent is allowed to select multiple coalitions to join or leave, thus overlapping coalitions can be identified. The game ends when no agent has an interest in changing its coalition memberships any further. Then the collection of coalitions at this time is regarded as the last community structure of a network.

Because the non-cooperative game is played on the basis of the result of cooperative game rather than the initial configuration in which every agent has one community of its own, the number of agents that will change their memberships to improve their utilities will decrease, thus efficiency of non-cooperative game will be improved. The combination of cooperative and non-cooperative game makes utilities of groups and individuals can be taken into account simultaneously, thus the accuracy has been improved.

— **Configuration Module**

The *configuration module* allows users to set parameters needed for carrying out the individual game, i.e. the lower bound of the utility value that an agent can join a new coalition and the upper bound of the utility value that an agent can leave the current coalition that it is in.

— **NMI Module**

The *NMI module* computes the *Normalized mutual information* (*NMI*) between the detected community structure and the underlying ground truth thus the community structures detected by different stages can be compared quantificationally.

— **Output Module**

The *output module* displays the community structures detected by the group game and the individual game. Meanwhile, the utility value of each coalition is presented following with the members of each coalition.

3   Demonstration Scenarios

Here we use the community detection of *Zachary's network of Karate* [5] as an example scenario. Figure 1 (b) shows the results of CDSG: the left of the window shows the member of each community and its utility, while the right of the window visualizes the network and their community structures.

Acknowledgments. This work is supported by the National Natural Science Foundation of China (Grant No. 61472346, 61272126 and 61262069) , Program for Young and Middle-aged Skeleton Teachers, Yunnan University, and Program for Innovation Research Team in Yunnan University (Grant No. XT412011).

References

1. Newman, M.E.J., Girvan, M.: Finding and evaluating community structure in networks. Physical Review E **69**, 026113 (2004)
2. Fortunato, S.: Community detection in graphs. Physics Reports **486**, 75–174 (2010)
3. Chen, W., Liu, Z., Sun, X., Wang, Y.: A Game-theoretic framework to identify overlapping communities in social networks. Data Mining and Knowledge Discovery **21**(2), 224–240 (2010)
4. Zhou, L., Cheng, C., Lü, K., Chen, H.: Using coalitional games to detect communities in social networks. In: Wang, J., Xiong, H., Ishikawa, Y., Xu, J., Zhou, J. (eds.) WAIM 2013. LNCS, vol. 7923, pp. 326–331. Springer, Heidelberg (2013)
5. Zachary, W.W.: An information flow model for conflict and fission in small groups. Journal of Anthropological Research **33**, 452–473 (1977)

Author Index

Printed in the United States
By Bookmasters